The Enzymes

VOLUME XX

MECHANISMS OF CATALYSIS

Third Edition

THE ENZYMES

Edited by

David S. Sigman

Department of Biological Chemistry
and Molecular Biology Institute
University of California
School of Medicine
Los Angeles, California

Volume XX

MECHANISMS OF CATALYSIS

THIRD EDITION

ACADEMIC PRESS, INC.

Harcourt Brace Jovanovich, Publishers
San Diego   New York   Boston
London   Sydney   Tokyo   Toronto

Copyright © 1992, 1990, 1963 by ACADEMIC PRESS, INC.
Copyright renewed 1991 by Paul D. Boyer, Henry Lardy and Karl Myrback
All Rights Reserved.
No part of this publication may be reproduced or transmitted in any form or by any
means, electronic or mechanical, including photocopy, recording, or any information
storage and retrieval system, without permission in writing from the publisher.

Academic Press, Inc.
1250 Sixth Avenue, San Diego, California 92101-4311

United Kingdom Edition published by
Academic Press Limited
24–28 Oval Road, London NW1 7DX

Library of Congress Cataloging-in-Publication Data
(Revised for vol. 20)

The Enzymes.

Vol. 17–   edited by Paul D. Boyer and Edwin G.
Krebs.
   Vol. 19–   edited by David S. Sigman, Paul D.
Boyer.
   Vol. 18–   Published in: Orlando.
   Vol. 19–   published in: San Diego.
   Includes bibliographical references and indexes.
   1. Enzymes–Collected works.   2. Enzymes.   I. Boyer,
Paul D., ed.   II. Krebs, Edwin G., ed.   III. Sigman, D. S.
QP601.E523   574.1'925   75-117107
ISBN   0-12-122720-0

PRINTED IN THE UNITED STATES OF AMERICA
92 93 94 95 96 97   BC   9 8 7 6 5 4 3 2 1

Contents

4. Nucleotidyltransferases and Phosphotransferases: Stereochemistry and Covalent Intermediates

PERRY A. FREY

5. Glycosidases and Glycosyltransferases

GREGORY MOOSER

6. Catalytic Strategies in Enzymic Carboxylation and Decarboxylation

MARION H. O'LEARY

7. Mechanisms of Enzymic Carbon–Carbon Bond Formation and Cleavage

RONALD KLUGER

8. Enzymic Free Radical Mechanisms

EDWARD J. BRUSH AND JOHN W. KOZARICH

9. Molecular Mechanism of Oxygen Activation by *P*-450

YOSHIHITO WATANABE AND JOHN T. GROVES

10. Mechanism of NAD-Dependent Enzymes

NORMAN J. OPPENHEIMER AND ANTHONY L. HANDLON

Preface

This volume completes our two-part miniseries on the "Mechanisms of Catalysis." Paul Boyer and I felt that these two volumes would provide a useful summary of the maturity and complexity of enzymology in this last decade of the twentieth century. They will be of interest to biochemists, structural biologists, organic chemists, and perhaps even a molecular biologist or two.

Volume XIX of "The Enzymes" focused on general principles and approaches. This one focuses on specific reaction types and cofactor catalysis. Although the coverage of topics is necessarily incomplete, the outstanding authors contributing to this volume have brought their subjects clearly into focus. The contributions should provide a useful perspective to graduate students initiating their studies and to established investigators assessing the status of a field. They underscore the importance of interaction between disciplines for understanding the chemistry of biology.

David S. Sigman

1

Transient-State Kinetic Analysis of Enzyme Reaction Pathways

KENNETH A. JOHNSON

Department of Molecular and Cell Biology
The Pennsylvania State University
University Park, Pennsylvania 16802

1

THE ENZYMES, Vol. XX

I. Introduction

Two fundamental questions underscore studies in enzymology and can be addressed by proper kinetic analysis: (1) elucidation of the enzymic reaction pathway to identify reaction intermediates and to specify the steps that limit the rate of turnover and (2) quantitative evaluation of the use of binding energy for enzyme specificity and catalysis. In this chapter, I will describe transient-state kinetic methods and analyses that allow the direct measurement of rate and equilibrium constants governing individual steps of an enzymic reaction pathway. Emphasis will be placed on the application of two rapid mixing methods, stopped-flow and chemical-quench-flow. In the past, two factors have limited the application of transient kinetic methods to enzymology: the amounts of enzyme required to perform such studies and the complexity of the equations describing the time dependence of transient reactions. Substantial improvements in the instrumentation (*1*) and the availability of cloned, overexpressed enzymes have largely eliminated the concerns over quantities of enzyme. The complexities of data analysis have been overcome by advances in computational methods to analyze reaction time courses by numerical integration (*2, 3*).

The last time this topic was reviewed in "The Enzymes" (*4*), the chapter dealt largely with the solution of differential equations necessary to interpret transient kinetic data, and only a small section described computational methods used to fit data by numerical integration. To a great extent, work was dependent on the simplification of reaction pathways to allow explicit solution of the rate equations describing the time dependence of each species in the reaction sequence. Moreover, the need to fit data to an explicit solution greatly restricted the design of experiments. Equilibrium perturbation methods were preferred, in large part, because of the ability to linearize differential equations describing systems near equilibrium. All of that has changed due to the recent advances in computational methods. One is no longer restricted to simplified reaction schemes or tied to tedious analysis of differential equations in seeking solutions of rate equations. Experiments can now be performed without restrictions for the starting conditions of the experiment dictated by the easy solution of the rate equations, and the results can be analyzed by the fitting of the data directly to a kinetic scheme. For these reasons, we are now free to focus in this chapter on the logical design of experiments and attempt to develop an intuitive understanding of the kinetics of enzyme reactions. I will refer to the equations describing the time dependence of reactions only to help in the development of an understanding of multiple-step reaction sequences and in the quantitative analysis of reaction rates in simplified cases. Reference to the literature will be selective rather than exhaustive, drawing on a few examples wherein transient kinetic methods have afforded a quantitatively complete analysis of each step in the reaction sequence to illustrate the conceptual basis for these studies.

II. Rationale for Transient Kinetic Analysis

A. LIMITATIONS OF STEADY-STATE KINETICS

The kinetic analysis of an enzyme mechanism often begins by analysis in the steady state; therefore, we first consider the conclusions that can be derived by steady-state analysis and examine how this information is used to design experiments to explore the enzyme reaction kinetics in the transient phase. It has often been stated that steady-state kinetic analysis cannot prove a reaction pathway, it can only eliminate alternate models from consideration (5). This is true because the data obtained in the steady state provide only indirect information to define the pathway. Because the steady-state parameters, k_{cat} and K_m, are complex functions of all of the reactions occurring at the enzyme surface, individual reaction steps are buried within these terms and cannot be resolved. These limitations are overcome by examination of the reaction pathway by transient-state kinetic methods, wherein the enzyme is examined as a stoichiometric reactant, allowing individual steps in a pathway to be established by direct measurement. This is not to say that steady-state kinetic analysis is without merit; rather, steady-state and transient-state kinetic studies complement one another and analysis in the steady state should be a prelude to the proper design and interpretation of experiments using transient-state kinetic methods. Two excellent chapters on steady-state methods have appeared in this series (6, 7) and they are highly recommended.

For practitioners of steady-state kinetics, the "kinetic mechanism"of an enzyme is defined by specifying the order of addition of substrates and the order of release of products. This analysis defines which reactants are present at the active site of the enzyme during the chemical reaction, for example, by distinguishing Ping-Pong versus sequential mechanisms. However, no information is available to define the steps in the reaction involving enzyme-bound species or to identify the rate-limiting step. Enzyme mutations, pH variation, or alternate substrates can slow the rate of a chemical reaction so that it becomes rate limiting in the steady state, and the examination of isotope effects can probe the extent to which chemistry is rate limiting (8, 8a, 8b). Direct measurements of reactions at the enzyme active site are required in order to establish the kinetic and thermodynamic bases for enzymic specificity and efficiency. We must redefine what constitutes proof of an enzyme reaction mechanism to include direct measurement of individual reactions occurring at the enzyme active site at physiological pH and with the natural substrates.

B. MEANING OF K_m AND k_{cat}

The two kinetic constants, K_m and k_{cat}, are most often misinterpreted as the substrate dissociation constant and the rate of the chemical reaction, respectively. However, this is not always the case, and K_m can be greater than, less than, or equal to the true substrate dissociation constant, K_d. The steady-state kinetic parameters only provide information sufficient to describe a minimal kinetic scheme. In terms of measurable steady-state parameters, a reaction sequence must be reduced to a minimal mechanism (Scheme I),

$$E + S \underset{k_{-1}}{\overset{k_1}{\rightleftharpoons}} E{\cdot}S \overset{k_2}{\rightarrow} E + P$$

SCHEME I

where the steady-state parameters are defined by

$$k_{cat} = k_2; \qquad K_m = (k_2 + k_{-1})/k_1 \tag{1}$$

This simplified model can be understood as resulting from the kinetic collapse of a more realistic mechanism (Scheme II).

$$E + S \underset{k_{-1}}{\overset{k_1}{\rightleftharpoons}} E{\cdot}S \underset{k_{-2}}{\overset{k_2}{\rightleftharpoons}} E{\cdot}P \overset{k_3}{\rightarrow} E + P$$

SCHEME II

The minimal complete model shown in Scheme II leads to the following steady-state parameters:

$$k_{cat} = \frac{k_2 k_3}{k_2 + k_{-2} + k_3}$$

$$K_m = \frac{k_2 k_3 + k_{-1} k_{-2} + k_{-1} k_3}{k_1(k_2 + k_{-2} + k_3)} \tag{2}$$

A reaction pathway containing a single intermediate as shown later in Scheme III (p. 6) leads to even more complex expressions for k_{cat} and K_m:

$$k_{cat} = \frac{k_2 k_3 k_4}{(k_2 + k_{-2})(k_{-3} + k_4) + k_2 k_3 + k_3 k_4}$$

$$K_m = \frac{k_{-1}(k_{-2}(k_{-3} + k_4) + k_3 k_4) + k_2 k_3 k_4}{k_1((k_2 + k_{-2})(k_{-3} + k_4) + k_2 k_3 + k_3 k_4)} \tag{3}$$

Thus k_{cat} and K_m are a function of all the rate constants in the pathway and any simplifying assumptions concerning individual rate constants are likely to be inaccurate. Moreover, the three reaction pathways shown in Schemes I and II, and III are indistinguishable by steady-state methods. Although product inhibition patterns provide evidence for the E·P state, individual kinetic constants cannot be resolved. Schemes II and III reduce to Scheme I under the conditions where k_3, $k_4 \gg k_2$. Steady-state kinetics cannot resolve the three reaction mechanisms because the form of the equation for steady-state kinetics is identical for each mechanism (v = rate):

$$v/[E_0] = \frac{k_{cat}[S]}{K_m + [S]} \tag{4}$$

Only the complexity of the expressions for k_{cat} and K_m differ.

The two fundamental-steady state kinetic constants are the terms k_{cat}, the maximum rate of product formation at saturating substrate, and k_{cat}/K_m, the *apparent* second-order rate constant for *productive* substrate binding. The value of K_m should only be considered as a ratio of the maximum rate of decomposition of the enzyme–substrate complex divided by the *apparent* rate of substrate binding (6). This understanding more accurately reflects the magnitude of K_m as representing the concentration of substrate at which the rate of substrate binding during the steady-state conversion of substrate to product equals the sum of the rates of substrate and product release. The two steady-state kinetic constants provide lower limits for the intrinsic rate constants that govern an enzymic reaction. The value of k_{cat}/K_m represents a lower limit for the rate of substrate binding, and k_{cat} sets a lower limit on the magnitude of any first-order rate constant following the binding of substrate and proceeding to the release of products.

With this perspective, one can summarize the three things learned by steady-state kinetic analysis as follows:

1. The order of binding of substrates and release of products serves to define the reactants present at the active site during catalysis; it does not establish the *kinetically preferred* order of substrate addition and product release or allow conclusions pertaining to the events occurring between substrate binding and product release.

2. The value of k_{cat} sets a lower limit on each of the first-order rate constants governing the conversion of substrate to product following the initial collision of substrate with enzyme. These include conformational changes in the enzyme–substrate complex, chemical reactions (including the formation and breakdown of intermediates), and conformational changes that limit the rate of product release.

3. The value of k_{cat}/K_m defines the apparent second-order rate constant for substrate binding and sets a lower limit on the true second-order rate constant for substrate binding. The term k_{cat}/K_m is less than the true rate constant by a

factor defined by the kinetic partitioning of the E·S complex to dissociate or go forward in the reaction. The magnitude of k_{cat}/K_m is used to quantitate enzyme specificity, but it cannot be used to establish the kinetic and thermodynamic basis for enzyme specificity, a goal that can only be achieved by analysis of individual reaction steps.

C. Goals of Complete Kinetic Analysis

To begin, we will base our analysis on a theoretical enzyme reaction pathway occurring by the sequence in Scheme III,

$$E + S \underset{k_{-1}}{\overset{k_1}{\rightleftharpoons}} E \cdot S \underset{k_{-2}}{\overset{k_2}{\rightleftharpoons}} E \cdot X \underset{k_{-3}}{\overset{k_3}{\rightleftharpoons}} E \cdot P \underset{k_{-4}}{\overset{k_4}{\rightleftharpoons}} E + P$$

Scheme III

where E·X represents an enzyme intermediate that can be defined as a distinct chemical species or a unique kinetically significant conformational state such as one formed by a change in structure of the E·S complex preceding the chemical reaction. Throughout this review, I will use a nomenclature in which k_n and k_{-n} represent the rate constants in the forward and reverse directions, respectively, for the nth step in the reaction sequence and K_n represents the equilibrium constant written in the forward direction.

Enzyme reaction mechanisms involving multiple substrates can be understood intuitively and addressed experimentally in terms of the one-substrate case by examining partial reactions along the sequence, or by considering the kinetics at saturating concentrations of the cosubstrates. Even with this simplification, as we will describe below, all data can and should be fit to the complete reaction pathway involving all substrates, products, and intermediates. Computer simulation based on the complete reaction sequence ensures that the data can be quantitatively understood in terms of the reaction sequence, with no simplifying assumptions. The extent to which all of the rate constants in the pathway are constrained during the fitting process by the experimental evidence is dependent on deriving the minimal mechanism.

According to the reaction sequence defined in Scheme III, the goal of a complete kinetic analysis is to provide estimates for each of the eight rate constants and four equilibrium constants. Measurement of the equilibrium constants can often provide a means to estimate individual rate constants that cannot be measured directly; moreover, redundancy in measurements of rate and equilibrium constants provide a check for internal consistency. In the process of measuring each rate constant, the identities and kinetic competence of potential intermediates will be established, and the resulting free energy profile for the reaction pathway will help to define the relationship between binding energy and catalytic efficiency. This information then serves as a background for analysis of proteins altered by site-directed mutagenesis to examine the roles of individual amino acids or groups of amino acids in substrate binding and transition-state stabilization (8). Alternatively, measurement of the changes in intrinsic rate constants occurring with altered substrates, mutated by chemical synthesis, establishes the kinetic and thermodynamic bases for enzyme specificity. Thus, whether one alters the substrates or the enzyme, it is necessary to assess the effects of such alterations on catalysis by quantitation of the changes in individual steps in the reaction sequence. These thermodynamic terms can then be related to the structural alterations to define the ways in which binding energy is used to carry out selectively catalysis of desired reactions.

The effect of mutations on steady-state parameters is rather boring in most cases: k_{cat} decreases and K_m increases. However, transient-state kinetic analysis has revealed many interesting properties of mutated enzymes (8, 9). For example, in studies on tyrosyl-tRNA synthetase (EC 6.1.1.1, tyrosine-tRNA ligase), Fersht and co-workers have quantitated changes in ground-state and transition-state binding energy by measurement of the binding constants and rate of the chemical reaction in single-turnover experiments for mutations in residues contacting the substrates (9). In recent work on DNA polymerase, we have established the kinetic and thermodynamic bases for fidelity by single-turnover experiments to measure changes in binding energy, rates of substrate binding, and changes in rate of chemical reaction occurring on substitution of incorrect nucleoside triphosphates (10–12). Neither of these goals could have been achieved without direct measurement of events occurring at the enzyme active site by transient kinetic methods.

D. RATIONALE FOR DESIGNING TRANSIENT KINETIC EXPERIMENTS

Although steady-state kinetic methods cannot establish the complete enzyme reaction mechanism, they do provide the basis for designing the more direct experiments to establish the reaction sequence. The magnitude of k_{cat} will establish the time over which a single enzyme turnover must be examined; for example, a reaction occurring at 60 sec^{-1} will complete a single turnover in approximately 70 msec (six half-lives). The term k_{cat}/K_m allows calculation of the concentration of substrate (or enzyme if in excess over substrate) that is required to saturate the rate of substrate binding relative to the rate of the chemical reaction or product release. In addition, the steady-state kinetic parameters define the properties of the enzyme under multiple turnovers, and one must make sure that the kinetic properties measured in the first turnover mimic the steady-state kinetic parameters. Thus, steady-state and transient-state kinetic methods complement one another and both need to be applied to solve an enzyme reaction pathway.

Analysis of steady-state kinetics can also reveal those cases in which rapid mixing methods are likely to fail or at least be more difficult to perform. For example, because one must examine a single enzyme turnover, enzymes with exceedingly high k_{cat} values (greater than 1000 sec^{-1}) are not amenable to rapid mixing methods unless the rates are slowed by working at reduced temperatures. Faster rates of reaction can be measured by temperature jump or pressure jump methods (*13, 14*), but, as we will describe later, mechanistic information is obtained by examining the concentration dependence of the reaction and this is often difficult to obtain by equilibrium perturbation methods. For these reasons, we will restrict our attention to two rapid mixing methods, stopped-flow and chemical-quench-flow, for measurement of the rates of enzyme-catalyzed reactions.

A second limitation of the single-turnover or presteady-state burst experiment arises with those enzymes that exhibit a high K_m relative to the accessible concentrations of enzyme. For example, in studies of ATPases, the background of inorganic phosphate is difficult to get much below 1% of the total ATP. Thus, if the enzyme requires 1mM ATP to saturate the enzyme–substrate complex, it is necessary to use at least 10 μM enzyme to have a signal as large as the background in attempting to measure the rate of hydrolysis at the active site. When the separation of products from the substrates is easier to perform, such as in studies on DNA polymerase, one can accurately quantitate the formation of 100 nM product in the presence of 100 μM substrate (*10*). In ideal cases, experiments can be performed with enzyme in excess over substrate and thereby examine 100% conversion of substrate to product in a single enzyme turnover with a high degree of sensitivity in looking for enzyme intermediates (*3, 15*).

Perhaps the most difficult aspect of learning transient-state kinetic methods is that it is not possible to lay down a prescribed set of experiments to be performed in a given sequence to solve any mechanism. Rather, the sequence of experiments will be dictated by the details of the enzyme pathway, the relative rates of sequential steps, and the availability of signals for measurement of rates of reaction. The latter constraint applies mainly to stopped-flow methods, and less so for chemical-quench-flow methods provided that radiolabeled substrates can be synthesized. Therefore, I will describe the kinetic methods used to establish an enzyme reaction mechanism with emphasis on the direct measurement of the chemical reactions by rapid quenching methods. Stopped-flow methods are useful in instances in which optical signals provide an easy means to measure the rates of individual steps of the reaction.

A reasonable sequence of experiments toward establishing the mechanism of an enzyme-catalyzed reaction can be put forth. In suggesting the order in which experiments should be carried out, the emphasis is in providing a logical progression of information wherein each experiment can be designed based on the quantitative information provided by the previous results. In general, the analysis begins at the ends of the pathway and works toward the center. The following protocol provides an outline for the experiments presented in the order that is most logical, although not all measurements will be possible in all cases and the prescribed order is not absolute.

1. Use steady-state kinetic methods to determine k_{cat} and K_m for the reaction, in each direction if it is reversible, and to determine the sequence of addition of substrates and release of products. In some cases the orders of substrate binding or product release may not be easy to determine unambiguously by steady-state methods, and so further tests by transient kinetic methods may be necessary.

2. Perform equilibrium measurements to establish the overall free energy of the reaction in solution, and, if possible, the internal equilibrium constants for reactions occurring at the active site. In addition, it is often possible to measure the equilibrium dissociation constants for the binding of any substrates that, by themselves, will bind and not react, such as the first substrate to bind in a multiple-substrate mechanism.

3. Measure the rates of substrate binding and dissociation by stopped-flow methods or by substrate trapping methods. This is an optional step in that although information on binding rates can be useful, it is not essential for the design of subsequent experiments.

4. Look for a presteady-state burst of product formation by chemical-quench-flow methods to determine if a step after chemistry is at least partially rate limiting and to then measure the rate and equilibrium constant for the chemistry step.

5. Perform single-turnover experiments with enzyme in excess of substrate to examine more closely the conversion of substrates to products at the active site of the enzyme and to look for intermediates.

6. Put all of the kinetic and equilibrium data together into a single, complete model for the reaction sequence, quantitatively accounting for every experimental result by the complete reaction pathway. Computer simulation can be used to establish that all experimental results can be understood quantitatively in terms of a single mechanism.

It should be noted that a complete, quantitative kinetic analysis can provide the framework with which an investigation proceeds to completion, with other direct, structural methods applied to the problem as they are suggested by the kinetic data. For example, observation and kinetic characterization of an intermediate will define the conditions required to isolate the intermediate and prove its structure by other methods (15, 16). The structural and kinetic methods complement one another and neither can be interpreted rigorously without the other (16, 17).

It should become apparent that when one begins a study, the information required to properly design the experiments is not entirely in hand. The approach must be an iterative one in which initial estimates provide the means to make some initial measurements, and as the information on the kinetics of the reaction is refined, the measurements can be repeated to approach a final convergence of the solution of the mechanism. Such an approach was successful in defining all 12 rate constants in the reaction catalyzed by 5-enolpyruvoylshikimate-3-

phosphate (EPSP) synthase, including identification, isolation, and kinetic characterization of an intermediate formed at the active site (15, 16).

This outline describing the logical progression of experiments will be followed for the remainder of this chapter. We begin with equilibrium measurements.

III. Experimental Methods for Complete Kinetic Analysis

A.   MEASUREMENT OF EQUILIBRIUM CONSTANTS GOVERNING CATALYSIS

For a simple enzyme-catalyzed reaction (Scheme IV) it is often possible to measure each of the equilibrium constants, K_1, K_2, K_3, and $K_{net} = K_1 K_2 K_3$.

$$E + A \underset{k_{-1}}{\overset{k_1}{\rightleftharpoons}} E{\cdot}A \underset{k_{-2}}{\overset{k_2}{\rightleftharpoons}} E{\cdot}P \underset{k_{-3}}{\overset{k_3}{\rightleftharpoons}} E + P$$

SCHEME IV

Because of the conversion of substrate to product and the reverse, it is obviously not possible to measure directly by equilibrium methods the binding constants for substrate and product. However, the binding constants can be estimated from the ratio of the binding and dissociation rates. The methods, interpretation, and limits of such measurements will be described below.

1. Overall Equilibrium

The net overall equilibrium constant for the reaction in solution (K_{net}) and the internal equilibrium constant for reaction at the active site (K_{int}) can be measured most easily using radiolabeled substrates. Thus the problem becomes one of separating substrate and product chromatographically and then quantitating the ratio of $K_{net} = [P]/[S]$ following incubation of the substrate with a trace of enzyme for a time sufficient for the reaction to come to equilibrium. The equilibration time can be estimated from the magnitude of k_{cat}/K_m in the forward and reverse reactions. If the substrate and product concentrations are below their K_m values, then the rate of approach to equilibrium can be approximated by

$$k_{obs} \simeq (k_{cat}/K_m)^{for}[E] + (k_{cat}/K_m)^{rev}[E] \tag{5}$$

The reaction will be 98% complete after six half-lives: $t_{1/2} = \ln 2/k_{obs}$. Alternatively, if the starting concentration of substrate, $[S]_0$, is greater than the K_m, then the time required to approach equilibrium can be approximated from $\Delta t = [S]_0/k_{cat}[E]$. Measurement of the product/substrate ratio at the time calculated and at twice the time will ensure the attainment of equilibrium.

The concentration of enzyme added should be dictated on the one hand by the desire to minimize the time required to reach equilibrium relative to the stability of the reactants and products, and on the other hand by the requirement that the concentration of enzyme be negligible relative to the concentration of the least abundant reactant at equilibrium. Here the need for caution is to ensure that the concentration of substrate or product measured is due to that in solution and not that bound to the enzyme.

2. Internal Equilibrium

The internal equilibrium constant can be measured after finding conditions under which *all* of the substrate and product will be bound to the enzyme. This is done by working at concentrations of enzyme in 5- or 10-fold excess of the dissociation constants for each substrate. Accordingly, the ratio of $[P]/[S]$ measured will reflect the ratio of $[E \cdot P]/[E \cdot S] = K_{int}$. The time required for the reaction to come to equilibrium can be approximated from the relationship $k_{obs} \geq k_{cat}^{for} + k_{cat}^{rev}$ to provide a minimum estimate of the rate of reaction at the active site. Usually the time calculated will be in the millisecond domain, but incubation for 5 sec is more convenient for manual mixing and usually no side products are formed on this time scale. Although in some cases it may be difficult to obtain concentrations of enzyme in excess of the dissociation constants for the substrates and products, the quantitation of the product/substrate ratio can be done quite accurately thanks to the fundamental property of enzyme catalysis that leads to an internal equilibrium constant close to unity for most enzymes (*18–20*).

Similar methods can be applied to enzymes catalyzing the reactions of two or more substrates, although there are actually some advantages afforded by this seemingly greater complexity. For example, consider the Bi Bi ordered reaction sequence (Scheme V).

$$E + A + B \underset{k_{-1}}{\overset{k_1}{\rightleftharpoons}} E \cdot A + B \underset{k_{-2}}{\overset{k_2}{\rightleftharpoons}} E \cdot A \cdot B \underset{k_{-3}}{\overset{k_3}{\rightleftharpoons}} E \cdot P \cdot Q \underset{k_{-4}}{\overset{k_4}{\rightleftharpoons}} E \cdot Q \underset{k_{-5}}{\overset{k_5}{\rightleftharpoons}} E + Q + P$$

SCHEME V

First it should be noted that the equilibrium constants for binding A and Q can be measured by standard equilibrium methods, thus the problem reduces to the central reactions that parallel the simple one-substrate reaction described above. Moreover, one can measure the overall and internal equilibria by placing radio-label in the substrate A and quantitating its conversion to product Q. For the external equilibrium measurement, the ratio of unlabeled $[P]/[B]$ can be altered to bring the ratio of labeled $[Q]/[A]$ close to unity to allow more accurate measurement (*3*), allowing the overall equilibrium constant to be calculated from $K_{net} = [P][Q]/[A][B]$.

The internal equilibrium is also more easily measured by taking advantage of the mass action afforded by working at high concentrations of unlabeled B and P to force the reaction toward the central complexes. Thus by working with enzyme in excess of the limiting radiolabeled substrate, A, and using a high concentration of B to drive the reaction forward and a high concentration of P to drive the reaction toward the reverse, the internal equilibrium constant can be measured directly from the ratio of radiolabeled A and Q: $[Q^*]/[A^*]$ = $[E \cdot P \cdot Q]/[E \cdot A \cdot B]$. In the case of EPSP synthase, the internal equilibrium measurements afforded equilibrium constants both for the formation of an intermediate from substrates and for its breakdown to products (3, 16).

It should also be noted that a multisubstrate enzyme also allows the application of substrate trapping methods to estimate the rates of dissociation of the first substrate to bind to the enzyme. These methods will be described in Section IV,F,2.

The importance of equilibrium measurements cannot be overly stressed. They provide true thermodynamic constants to evaluate the role of substrate binding in catalysis, they provide the background with which kinetic experiments can be properly designed and interpreted to establish the pathway of catalysis, and they provide additional constraints to be used in the fitting of kinetic data.

B.   Rapid Mixing Transient Kinetic Methods

I will describe only briefly the instrumental methods used to collect data and I will limit this discussion to rapid mixing methods (stopped-flow and chemical-quench-flow) because these methods allow measurement of the concentration dependence of reaction rates, providing important mechanistic information (1). Methods of data fitting will be presented at the end of the chapter.

1.   Stopped-Flow Methods

Stopped-flow methods are useful whenever there is an optical signal for the reaction of interest. Absorbance changes have been useful in several reactions, including enzymes containing a reactive heme (21, 22) or pyridoxal phosphate (23–26), or reactions involving NADH (27). Absorbance changes are relatively easy to measure instrumentally, but they depend on changes in extinction coefficient between substrate and product, and the absorbance by the substrate free in solution can limit the sensitivity of the method in attempting to detect changes occurring at the enzyme active site when substrate is in large excess over enzyme.

A stopped-flow experiment is quite simple in principle. The apparatus allows the rapid mixing of two or more solutions, which then flow into an observation cell while the previous contents are flushed and replaced with freshly mixed reactants (Fig. 1). A stop syringe is used to limit the volume of solution expended with each measurement and also serves to abruptly stop the flow and to trigger simultaneously a computer to start data collection. Thus, if one were to watch from the point of view of the photodetector, one would first see the solution flow into the observation cell and then abruptly stop. The reaction is followed as the solution ages after the flow stops. The time resolution of the method

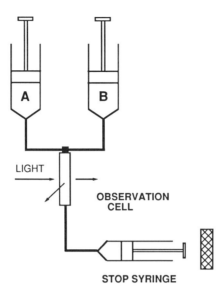

FIG. 1. Schematic of stopped-flow apparatus.

is limited by the time required for the reactants to flow from the point of mixing to the point of observation; this "dead time" is typically on the order of 1–4 msec, depending on the apparatus. Thus, half of the signal is lost at rates of 175 sec^{-1} (4 msec dead time) to 700 sec^{-1} (1 msec dead time). The ability to measure a reaction is limited by the signal-to-noise ratio and so rates in excess of 1000 sec^{-1} have been measured (27) in cases in which the signal was particularly good, even though the measurement was based upon less than half of the original amplitude.

Fluorescence methods provide several advantages over absorbance due to the greater signal-to-noise ratio resulting from the lower background. For example, there are numerous cases in which protein fluorescence methods have allowed measurements of the rates of changes in conformation of the enzyme during substrate binding and catalysis (28, 29). In other cases the use of a fluorescent substrate or product (natural or synthetic) has provided accurate measurement of reactions occurring at the active site (30, 30a). For example, the use of fluorescence energy transfer from the protein to bound NADH has afforded a signal measuring the conversion of NADH to NAD$^+$ at the enzyme active site (27). The major limitation to fluorescence methods is that relatively intense light sources are required (in comparison to absorbance methods) and particular attention must be made to achieve efficient recovery of the light emitted from the sample.

The measurement of signal amplitude is often overlooked in studies of reaction rates; however, important information is contained in the amplitude. For example, the absolute amplitude of formation of a transient intermediate is essential to resolve the kinetics of its formation and decay. In other cases, evidence for a fast phase of a reaction can be obtained by loss of signal amplitude beyond that expected from the observable rate of reaction and the known dead time of the instrument. For example, in studies on the myosin ATPase, evidence for a fast binding reaction was obtained by noting the loss of 40% of the protein fluorescence signal amplitude at a rate of 125 sec^{-1} (28). From the dead time of the instrument, equal to 1.5 msec, one would expect the loss of only 17% of the signal, calculated from $A/A_0 = e^{-kt_d}$, where t_d is the dead time. The greater loss in signal was correlated with the ATP binding rate measured by dissociation of the actomyosin complex, approaching a maximum rate of ~2000 sec^{-1}. The fast phase was confirmed by direct observation at lower temperature.

In order to observe a reaction in a stopped-flow experiment, data must be collected over the proper time window. Because the method spans some five orders of magnitude in time, it is easy to miss a reaction because it was either too fast or too slow to be observed on the time scale selected. The rule is to observe the reaction over six half-lives. An iterative method should be applied. Initial estimates of the rate are used to select the time over which to examine the reaction; the measured rate is then used to adjust the time of data collection, eventually converging on the optimal time according to the rate of the reaction.

2. Chemical-Quench-Flow Methods

The most serious limitation of stopped-flow methods is that one does not always have an optical signal for the reaction of interest and the optical signals cannot be interpreted rigorously if the extinction coefficients of intermediates or products are not known. For example, an enzyme intermediate may have an unknown extinction coefficient, and without an absolute measurement of concentration of the intermediate, one cannot obtain a unique solution to its rate of formation and decay (see below). For these reasons, a direct measurement of the conversion of substrate to product is required. Chemical-quench-flow methods allow such direct measurement of the chemistry of enzyme-catalyzed reactions and can be performed for nearly any reaction. One must recognize that these experiments are based on examining the enzyme as a stoichiometric reactant such that the concentration of enzyme required will depend upon the kinetics of the reaction and the sensitivity of the methods for detection of intermediates or products. Nonetheless, quench-flow experiments can be performed using as little as 20 μl of solution and a complete enzyme pathway can be solved using only 5–10 mg of enzyme.

The chemical-quench-flow apparatus allows the mixing of two reactants, followed, after a specified time interval, by mixing with a quenching agent, usually acid or base, to denature the enzyme and liberate any enzyme-bound intermedi-

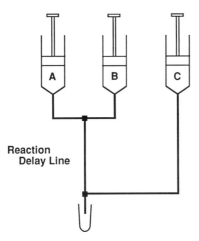

FIG. 2. Standard quench-flow apparatus.

ates or products. A simple quench-flow apparatus is illustrated schematically in Fig. 2.

A precisely controlled motor is used to drive syringes containing enzyme (syringe A), substrate (syringe B), and quench solution (syringe Q). Enzyme and substrate are first mixed and the reaction proceeds as the reaction mixture flows through a reaction loop of defined length. Reaction is terminated when the enzyme–substrate solution is mixed with quenching solution. The time of reaction is determined by the volume of the reaction loop between the two points of mixing and the rate of flow. In practice, the reaction time is varied by changing the length of tubing in the reaction loop and, to a lesser extent, by changing the rate of flow. The quenched sample is then collected and analyzed to quantitate the conversion of substrate to product (or intermediate), usually by chromatographic methods using radiolabeled substrates. The apparatus is then flushed and a new reaction loop is selected to obtain a different reaction time. By selecting various reaction loops, times from 2 to 100 msec can be obtained.

This simple design limits the maximum attainable reaction time by the maximum loop size. Longer reaction times cannot be obtained by using slower flow rates because it is necessary to use rapid rates of flow in order to maintain turbulence and efficient mixing. A 40-cm-long reaction loop, containing 200 μl of solution, provides a 100-msec reaction time. This design also wastes the often precious biological samples, by using enzyme and substrate to push the reactants through the reaction loop: at the end of each push, the reaction loop will contain 200 μl of reactants that must be discarded before the next run. In order to minimize sample volumes, to recover 200% of the reactants, and to achieve reaction

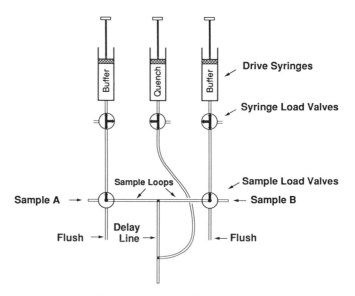

FIG. 3. Pulsed quench-flow apparatus.

times longer than 100 msec, we have redesigned the quench-flow according to the schematic shown in Fig. 3.

In this apparatus, enzyme and substrate are loaded into small loops of tubing containing 40 μl of solution. A three-way valve is then used to connect the loaded sample loop to a drive syringe containing buffer. The drive syringes are then used to force the reactants together and through the reaction loop to the point of mixing with the quenching solution and out into the collection tube. This apparatus allows collection of the entire sample and allows longer reaction times by operating in a push–pause–push mode. The first push is used to mix the reactants into the reaction loop, the pause allows the reaction to continue, and the second, timed push forces the reactants out of the reaction loop to be quenched and expelled into the collection tube. A three-way valve is used to allow loading and flushing of the contents of the reaction and sample loops between runs. By these methods, reaction times from 100 msec to 100 sec or longer can be obtained to complement the reaction times of 2–100 msec obtained in the direct push mode. A computer-controlled stepping motor provides the precise start–stop sequence. The use of an eight-way valve to select the appropriate reaction loop allows any reaction time to be selected conveniently. An entire reaction time course consisting of 25 time points can be collected from 1 ml of reactants in less than 1 hr. This apparatus made possible experiments to establish the reaction pathways of DNA polymerase (10–12), EPSP synthase (3, 31), and tryptophan synthase (32).

IV. Kinetics of Substrate Binding and Catalysis

In this section, I will first describe experiments to measure each step in a simple reaction sequence (Scheme VI),

$$E + S \underset{k_{-1}}{\overset{k_1}{\rightleftharpoons}} E{\cdot}S \underset{k_{-2}}{\overset{k_2}{\rightleftharpoons}} E{\cdot}X$$

SCHEME VI

where $E{\cdot}X$ can represent an enzyme-bound product, intermediate, or a distinct conformation of the enzyme–substrate complex. I will then consider more complex reaction sequences, including the release of products from the enzyme. Multiple-substrate enzyme reaction sequences can be reduced to the simple scheme above by measuring the rate and equilibrium constants for the first substrates to bind and then considering the reaction of the second (or last) substrate according to Scheme VI.

A.   ONE-STEP BINDING

If one has a signal for the binding of S to the enzyme and follows the progress of the reaction as it goes to completion, with substrate in excess over enzyme, the time dependence of the reaction can be fit to a single exponential. If we first consider an irreversible binding reaction

$$E + S \overset{k_1}{\rightarrow} E{\cdot}S \tag{6}$$

then the time dependence of the disappearance of E and appearance of $E{\cdot}S$ can be written:

$$\partial[E]/\partial t = -\partial[E{\cdot}S]/\partial t = -k_1[E][S] \tag{7}$$

If the concentration of substrate is in sufficient excess over enzyme, then [S] can be assumed to be a constant term giving a pseudo-first-order rate constant defined by the product, $k_1[S]$. Under these "pseudo-first-order" conditions the rate equation can be integrated to yield the time dependence of the reaction:

$$
\begin{aligned}
[E] &= [E_0]e^{-k_1[S]t} \\
[E{\cdot}S] &= [E_0](1 - e^{-k_1[S]t})
\end{aligned}
\tag{8}
$$

Thus, the free enzyme disappears and the enzyme–substrate complex appears with a time constant defined by the pseudo-first-order rate constant, $k_{obs} = k_1[S]$, obtained by fitting the reaction time course to a single exponential. The half-time for the reaction can be solved according to the time required to get $[E] = \frac{1}{2}[E_0]$, which gives $t_{1/2} = \ln 2/k_{obs}$.

When the concentrations of enzyme and substrate are comparable, the differential equations cannot be solved explicitly because both the enzyme and sub-strate concentrations vary with time. Under these and other conditions, computer simulation by numerical integration has been essential (see Section V,B).

We can now consider the fully reversible binding reaction. Under pseudo-first-order conditions with substrate in excess over enzyme, the rate equations can be integrated, yielding a form similar to Eq. (8):

$$[E]/[E]_0 = 1 - \frac{K_1[S]}{K_1[S] + 1}(1 - e^{-k_{obs}t})$$

$$[ES]/[E]_0 = \frac{K_1[S]}{K_1[S] + 1}(1 - e^{-k_{obs}t}) \tag{9}$$

$$k_{obs} = k_1[S] + k_{-1}$$

where the amplitude term, $K_1[S]/(K_1[S] + 1) = [S]/(K_d + [S])$, is defined by the equilibrium concentration of ES. The first surprise is that the observed rate of reaction, as an exponential approach to equilibrium, is equal to the sum of the forward and reverse rates. This is a general rule that should be committed to memory because it applies to all reaction kinetics involving the approach to equilibrium. The basis for this phenomenon becomes apparent when one realizes that the amplitude of the reaction is reduced by the reverse reaction. When the reverse rate is negligible, the reaction goes to 100% completion and the observed rate is simply equal to $k_1[S]$. However, the back reaction reduces the amplitude of the reaction and thereby shortens the half-time for reaching equilibrium.

There are two important results from this analysis. First, the rate constants for binding and dissociation can be obtained from the slope and intercept, respectively, of a plot of the observed rate versus concentration. In practice this is possible when the rate of dissociation is comparable to $k_1[S]$ under conditions that allow measurement of the reaction. At the lower end, resolution of k_{-1} is limited by the concentration of substrate required to maintain pseudo-first-order kinetics with substrate in excess of enzyme and by the sensitivity of the method, which dictates the concentration of enzyme necessary to observe a signal. Under most circumstances, it may be difficult to resolve a dissociation rate less than 1 sec^{-1} by extrapolation of the measured rate to zero concentration. Of course, the actual error must be determined by proper regression analysis in fitting the data, and these estimates serve only to illustrate the magnitude of the problem. In the upper extreme, dissociation rates in excess of 200 sec^{-1} make it difficult to observe any reaction. At a substrate concentration required to observe half of the full amplitude, where $k_1[S] = k_{-1}$, the reaction would proceed toward equilibrium at a rate of 400 sec^{-1}. Thus, depending upon the dead time of the apparatus, much of the reaction may be over before it can be observed at the concentrations required to saturate the enzyme with substrate.

Within these limits, estimates of both the *on* and *off* rates can be obtained from a single experiment. However, in most instances other direct methods of estimating dissociation rates must be considered. Two methods are available and will be described below. First, substrate competition experiments can be performed by displacing the substrate with a tighter binding inhibitor or alternate substrate under conditions in which the rate-limiting step is the release of the bound substrate (Sections III,D,5 and III,D,6). Second, with multiple-substrate enzymes, the rate of dissociation of the first substrate can be estimated from substrate trapping experiments, measuring the kinetic partitioning between the forward reaction after binding of the second substrate and the reverse reaction by dissociation of the bound substrate (Section IV,F,2).

B.   Two-Step Reaction Kinetics

The key to obtaining mechanistic information from transient kinetic methods is to examine the concentration dependence of the rates and amplitudes of the reaction. The concentration dependence reflects changes in the kinetics of binding and can reveal the presence of steps subsequent to binding that limit the rate of the observed reaction. The rate of binding is expected to increase linearly with increasing concentration of substrate with no signs of curvature. Curvature in the concentration dependence of the rate of a reaction is indicative of a two-step mechanism approaching a maximum rate that is limited by a first-order isomerization of the enzyme–substrate complex. For example, the binding of a substrate to an enzyme often occurs in two steps: the formation of a collision complex is followed by a change in state of the enzyme–substrate complex according to the following sequence:

$$E + S \underset{k_{-1}}{\overset{k_1}{\rightleftharpoons}} E{\cdot}S \underset{k_{-2}}{\overset{k_2}{\rightleftharpoons}} E{\cdot}X$$

where $E{\cdot}X$ may represent a distinct conformational state of the enzyme-bound substrate, an intermediate, or the enzyme-bound product of the chemical reaction. The observed kinetics of reaction will depend on the relative magnitudes of the four rate constants. We will first consider the rapid-equilibrium case and will then consider the complete solution to the two-step mechanism with all four rate constants.

1.   *Rapid-Equilibrium Binding*

In many cases the rate of dissociation from the collision complex ($E{\cdot}S$), defined by k_{-1}, is much faster than k_2, the rate-limiting chemical reaction or isomerization leading to tighter binding. Under these conditions the binding in step 1 comes to equilibrium on a time scale much faster than the first-order isomerization. Accordingly, the first step can be simplified as a rapid-equilibrium reaction mechanism (Scheme VII).

$$E + S \overset{K_1}{\rightleftharpoons} E{\cdot}S \underset{k_{-2}}{\overset{k_2}{\rightleftharpoons}} E{\cdot}X$$

SCHEME VII

The solution of Scheme VII provides the time dependence of each of the three species:

$$[E]/[E]_0 = \{1 - (K_1[S] + K_1 K_2[S])/\alpha\}(1 - e^{-\lambda t})$$
$$[ES]/[E]_0 = (K_1[S]/\alpha)(1 - e^{-\lambda t}) \tag{10}$$
$$[EX]/[E]_0 = (K_1 K_2[S]/\alpha)(1 - e^{-\lambda t})$$

where the denominator of the amplitude term is defined by

$$\alpha = 1 + K_1[S] + K_1 K_2[S]$$

The preexponential amplitude terms represent the concentration of enzyme in each form at equilibrium; for example, $[EX]/[E]_0 = K_1 K_2[S]/(1 + K_1[S] + K_1 K_2[S])$. The concentration dependence of the rate of reaction to form $E{\cdot}X$ follows a hyperbola, which is a function of the saturation of the initial collision complex:

$$\lambda = \frac{K_1 k_2[S]}{K_1[S] + 1} + k_{-2} \tag{11}$$

The form of the equation exactly parallels that for Michaelis–Menten kinetics and for a similar reason. The increase in rate as a function of concentration reflects the saturation of the $E{\cdot}S$ collision complex; at the upper limit, the rate approaches the maximum rate of reaction. There are three important differences that distinguish these kinetic measurements from steady-state parameters: (1) the hyperbola is a function of the true dissociation constant, $K_d = 1/K_1$, because only a single turnover is measured; (2) the maximum rate provides a direct measure of the sum of the rate constants, $k_2 + k_{-2}$; and (3) the intercept on the y axis is equal to the rate constant, k_{-2}, defining the dissociation rate.

One should note that according to this mechanism the apparent second-order rate constant for substrate binding is given by the product $K_1 k_2$ and the rate of dissociation of substrate will be limited by the rate constant k_{-2}. Thus the ratio of the *off* rate divided by the *on* rate is equal to the true dissociation constant even for this two-step reaction.

2. *Generalization: Kinetic Consequences of Rapid-Equilibrium Steps*

Whenever there is a step in a reaction sequence that can be considered to be in a rapid equilibrium, the two species in equilibrium become kinetically linked such that they are maintained in a constant proportion to each other. The rate of reaction for species to leave the equilibrium is defined by the fraction of species in the reactive form times the intrinsic rate constant. The two (or more) enzyme species involved in the rapid equilibrium maintain a constant ratio as the reaction proceeds to completion.

3. *Examples: EPSP Synthase, Myosin, Tyrosyl-tRNA Synthetase, and DNA Polymerase*

The binding of shikimate 3-phosphate (S3P) and glyphosate (a potent, commercially important herbicide) to EPSP synthase provides a good example of two-step reaction kinetics with the first step in rapid equilibrium (*33*). S3P binds to the enzyme in a rapid equilibrium with rates of $k_{on} = 6.5 \times 10^8 \ M^{-1} \ sec^{-1}$ and $k_{off} = 4500 \ sec^{-1}$ (*3*). Glyphosate binds only to the enzyme–S3P complex and induces a change in protein fluorescence. Thus the kinetics follow a two-step reaction sequence (Scheme VIII),

$$E + S \underset{k_{-2}}{\overset{K_1}{\rightleftharpoons}} E \cdot S \overset{k_2[G]}{\to} E \cdot S \cdot G$$

SCHEME VIII

where the observable signal is a function of the formation of E·S·G. If S3P (S) and enzyme (E) are preincubated to allow saturation of E·S, and then mixed with glyphosate (G), the reaction follows a single exponential (Fig. 4A). As shown in Fig. 4B, the rate of the fluorescence change increases linearly with increasing concentration of glyphosate following the equation $k_{obs} = k_2[G] + k_{-2}$. The slope defines the second-order rate constant, $k_2 = 7.8 \times 10^5 \ M^{-1} \ sec^{-1}$, and the intercept defines the dissociation rate of $k_{-2} = 0.12 \ sec^{-1}$. Of course, the errors in estimating k_{-2} are quite large and the best estimate is based on a calculation of the *off* rate from the K_d and the *on* rate.

If the reaction is initiated by mixing various subsaturating concentrations of S3P with the enzyme and a fixed concentration of glyphosate, then the observed rate for the two-step reaction is defined by

$$k_{obs} = \frac{K_1 k_2[S][G]}{K_1[S] + 1} + k_{-2} \tag{12}$$

A plot of rate as a function of S3P concentration follows a hyperbola (Fig. 4C) with an apparent $K_d = 1/K_1$, and a maximum rate defined by $k_2[G] + k_{-2}$. Thus, Eq. (12) reduces to the simple $k_{obs} = k_2[G] + k_{-2}$ at saturating concentrations of S3P.

In the example involving the binding of S3P and glyphosate, it was relatively easy to resolve the two-step reaction kinetics because of the separate concentration dependence for each step; the first step depended on S3P and the second on glyphosate. Moreover, substrate trapping studies established that S3P dissociates at a very fast rate (*3*), justifying the rapid equilibrium assumption. With the

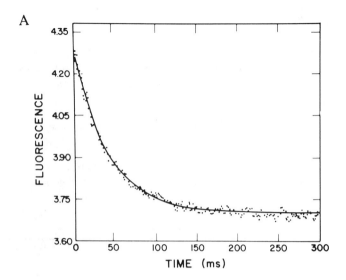

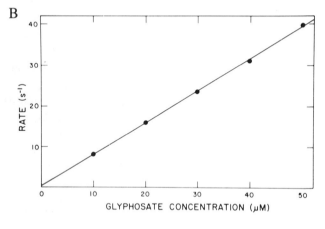

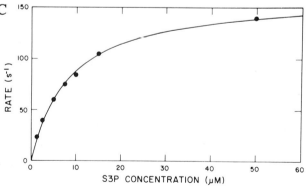

binding of a single substrate, it is reasonable to suggest that there is always some rearrangement of the E·S complex following the initial collision and leading to two-step binding kinetics. The important question becomes one of the rate of the rearrangement relative to the time resolution of the methods and the rate of catalytic turnover. On the one extreme, if the rearrangement following the initial collision simply involves the rotation of several bonds in the substrate, it may occur at rates in excess of 10^9 sec^{-1}, and would not be observed. In other cases, a conformational rearrangement after substrate binding may limit the rate of the chemical reaction. In practice, rapid mixing methods are limited to observation of rates slower than 1000 sec^{-1}, and so many fast rearrangements may go undetected and be of little physiological relevance.

A two-step binding mechanism is often invoked to explain an apparent second-order rate constant slower than diffusion. Estimates of the rate of diffusion-limited binding of a substrate to an enzyme active site vary considerably, but fall in the range of approximately 10^9 M^{-1} sec^{-1} (5). The apparent second-order rate constant for ATP binding to myosin has been measured to be 2×10^6 M^{-1} sec^{-1} and is attributed to an initial weak collision complex with $K_1 = 10^3$ M^{-1} followed by an isomerization at a rate of $k_2 = 2000$ sec^{-1}, providing an apparent second-order rate constant by the product $K_1 k_2$. Thus, if ATP binds at a rate of 10^8 M^{-1} sec^{-1}, it will dissociate from this weak collision complex at a rate of 10^5 sec^{-1}, effectively coming to equilibrium at a rate much faster than the first-order isomerization, leading to tighter binding.

In the myosin ATPase pathway, a second conformational change occurs at a rate of 100 sec^{-1}, which is followed by a faster chemical reaction (30). In the case of myosin, the conformational changes are responsible for the coupling of ATP binding energy to force production. For other enzymes, similar conformational rearrangements may be more common than is often recognized and may limit the observed rate of chemical reaction. For example, in extensive studies on tyrosyl-tRNA synthetase, the reaction of ATP with tyrosine to form the tyrosyl-AMP plus pyrophosphate was originally measured in single-turnover experiments and was described as the rate of the chemical reaction (34, 35). Analysis of the crystal structure and quantitation of the kinetic effects of mutations in residues near the active site have provided evidence for a large change in struc-

FIG. 4. Kinetics of glyphosate binding to EPSP synthase. (A) The change in protein fluorescence at 340 nm was measured by stopped-flow methods after mixing enzyme (4.5 μM) with glyphosate (30 μM) and S3P (250 μM). The smooth line shows a fit to a single exponential with a rate of 4.5 sec^{-1}. (B) The glyphosate concentration dependence of the observed rate was measured in the presence of a saturating concentration of S3P (250 μM). (C) The S3P concentration dependence of the rate was measured in the presence of a fixed concentration of glyphosate (200 μM). Reproduced with permission from (33).

ture of a mobile loop that appears to limit the rate of the chemical reaction (8, 36). Single-turnover kinetic analysis of DNA polymerase has also shown that the maximum rate of incorporation of dNTP is limited by a conformational change in the E·DNA·dNTP complex (10–12). This two-step binding mechanism largely accounts for the extraordinary specificity in recognition of the correctly base-paired dNTP during DNA replication.

C. EXPLICIT SOLUTION FOR TWO-STEP REACTIONS

We can first consider a simple two-step, irreversible mechanism:

$$E + S \xrightarrow{k_1} E\cdot S \xrightarrow{k_2} E\cdot X$$

For this simplified pathway, the time dependence of reaction is a double exponential. That is, for each species, E, E·S, and E·X, the time dependence follows an equation of the form

$$[E_i] = \alpha_1 e^{-\lambda_1 t} + \alpha_2 e^{-\lambda_2 t} + \gamma \tag{13}$$

where the observable rates are $\lambda_1 = k_1$ and $\lambda_2 = k_2$. This pathway is perhaps easy to relate to because each observable reaction rate corresponds to an intrinsic rate constant. However, it is rare that enzyme-catalyzed reactions are irreversible, and so this limiting case is not often useful.

For the more realistic, two-step reversible sequence

$$E + S \underset{k_{-1}}{\overset{k_1}{\rightleftharpoons}} E\cdot S \underset{k_{-2}}{\overset{k_2}{\rightleftharpoons}} E\cdot X$$

the solution to the time dependence of each enzyme species still follows a double exponential according to Eq. (15), but the amplitudes and the rates, λ_1 and λ_2, are dependent on all four rate constants. The full solution, including the amplitude terms for each species, is too complex to be useful except for the calculation of the predicted time course from a given set of kinetic constants, a task that can be done more easily by computer simulation as described below. The complete solution has been published elsewhere (1) and will not be repeated here. However, there are several important lessons to be learned from examination of the form of the equations for the rate of reaction, which can be applied in principle to any two-step reaction sequence.

The values of the rates λ_1 and λ_2 are defined by the roots of the quadratic equation:

$$2\lambda_{1,2} = (k_1 S + k_{-1} + k_2 + k_{-2})$$
$$\pm \{(k_1 S + k_{-1} + k_2 + k_{-2})^2 \tag{14}$$
$$- 4[k_1 S(k_2 + k_{-2}) + k_{-1}k_2]\}^{1/2}$$

$B^2 - 4\varphi$ use $\lambda_1 + \lambda_2 = k_1 s + k_{-1} + k_2 + k_{-2}$

$\lambda_1 \lambda_2 = k_1 s (k_2 + k_2) + k_{-1} k_2$

(see Fersht)

Equation (14) can be simplified by the square root approximation (1) to yield

$$\lambda_1 \simeq k_1[S] + k_{-1} + k_2 + k_{-2} \tag{15}$$

$$\lambda_2 \simeq \frac{k_1[S](k_{-2} + k_2) + k_{-1}k_{-2}}{k_1[S] + k_{-1} + k_2 + k_{-2}}$$

The fast phase of the reaction, λ_1, is equal to the sum of all four intrinsic first-order rate constants, defining the rate of formation of E·S and the decay of E. The substrate concentration dependence of the rate is a straight line with a slope equal to k_1 and an intercept equal to $k_{-1} + k_2 + k_{-2}$. The slow phase defines the rate of decay of E·S and the rate of formation of E·X, and the rate of slow decay of E if it is noticeably biphasic. The substrate concentration dependence of the slow reaction approximates a hyperbola with an apparent K_d defined by

$$K_d^{app} \simeq \frac{k_{-1}k_{-2}}{k_1(k_2 + k_{-2})} = \frac{K_d k_2}{k_2 + k_{-2}} \tag{16}$$

and a maximum rate equal to $k_{max} = k_2 + k_{-2}$. The limiting slope at low substrate concentration defines the apparent second-order rate constant for binding,

$$k_{on} \simeq \frac{k_1[S](k_2 + k_{-2})}{k_{-1} + k_2 + k_{-2}} \tag{17}$$

and the intercept on the y axis is equal to the dissociation rate,

$$k_{off} \simeq \frac{k_{-1}k_{-2}}{k_{-1} + k_2 + k_{-2}} \tag{18}$$

Thus estimates of the apparent dissociation constant by calculation of the ratio of k_{off} divided by k_{on} will be equal to the K_d for step 1 if k_{-2} is large (relative to k_2 and k_{-1}), and will define the net dissociation constant for the two-step process, $K_d = 1/(K_1 K_2)$, if k_{-2} is small relative to k_2. In any case, the estimated K_d will be within the limits

$$1/K_1 \geq K_d^{app} \geq 1/(K_1 K_2) \tag{19}$$

and the calculated value will be an accurate estimate of the apparent K_d involving the actual distribution of species E·S $\rightleftharpoons$ E·X at equilibrium.

1. *Example: Tryptophan Synthase*

At first glance, Eq. (15) appears too complex to allow measurement of individual reaction rate constants. However, as we illustrate with this example, it is possible to extract estimates of all four rate constants from an analysis of the concentration dependence of the observed rates. The time dependence of reaction of serine with pyridoxal phosphate at the β-site of tryptophan synthase provides a good example of two-step reaction kinetics because of the unique optical signals on reaction with pyridoxal phosphate (PLP). The serine first reacts with PLP to form an external aldimine (E·X), which then reacts further to form an aminoacrylate (E·A) (Scheme IX). The formation and decay of the aminoacrylate can be monitored by absorbance or fluorescence changes (23–26, 32). Figure 5A shows the time dependence of a change in fluorescence (exciting at 405 nm and observing emission at 500 nm). The data can be fit to a double exponential to extract the rates of each phase of the reaction (32). The concentration dependence of the fast and slow rates is shown in Fig. 5B. The fast phase follows a linear concentration dependence, extrapolating to a nonzero intercept in the limit of zero substrate concentration and showing no signs of reaching a maximum rate at high serine concentration. The slow phase approximates a hyperbolic function, reaching a maximum rate of 55 sec^{-1}.

These data fit two-step reaction kinetics according to Scheme IX.

$$E + S \underset{k_{-1}}{\overset{k_1}{\rightleftharpoons}} E{\cdot}X \underset{k_{-2}}{\overset{k_2}{\rightleftharpoons}} E{\cdot}A$$

SCHEME IX

The fast phase of the reaction occurs at a rate equal to the sum of all four rates in the pathway according to Eq. (15) (λ_1). The slope of the line defines $k_1 = 0.135 \ \mu M^{-1} \ sec^{-1}$ and the intercept is equal to the sum $k_{-1} + k_2 + k_{-2} = 75 \ sec^{-1}$. The slow phase of the reaction can be approximated by Eq. (15) (λ_2), which reaches a maximum rate equal to the sum $k_2 + k_{-2} = 55 \ sec^{-1}$. Comparison of the intercept of the plot of the fast phase with the maximum rate of the slow phase yields an estimate of $k_{-1} = 20 \ sec^{-1}$ from the difference. This analysis leads to definition of k_1, k_{-1}, and the sum $k_2 + k_{-2}$. Resolution of the contributions of the forward and reverse rates to the net rate, $k_2 + k_{-2}$, is based on an estimate of the absolute amplitude of the slow phase of the reaction and a fit to the concentration dependence of the slow reaction. Complete solution of the two-step reaction kinetics yielded all four rate constants: $k_1 = 0.135 \ \mu M^{-1} \ sec^{-1}$, $k_{-1} = 20 \ sec^{-1}$, $k_2 = 45 \ sec^{-1}$, and $k_{-2} = 10 \ sec^{-1}$.

In principle, the reaction of serine with the enzyme should have been written to include the formation of the collision complex between serine and the enzyme (E·S) (Scheme X).

$$E + S \underset{k_{-1}}{\overset{k_1}{\rightleftharpoons}} E{\cdot}S \underset{k_{-2}}{\overset{k_2}{\rightleftharpoons}} E{\cdot}X \underset{k_{-3}}{\overset{k_3}{\rightleftharpoons}} E{\cdot}A$$

SCHEME X

However, because the rate of the fast·phase increased linearly without signs of curvature at high concentrations of serine, there is little information to define the binding of serine in the initial collision complex. The data suggest that the initial binding of serine is weak ($K_d > 20 \ mM$) and the rate of its reaction to form E·X is fast, compared to the attainable binding rates. For example, if the

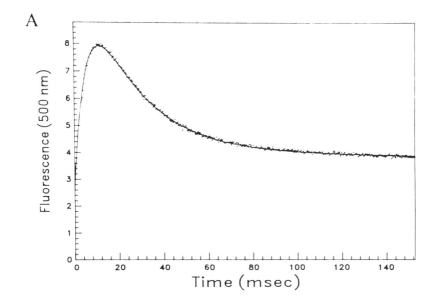

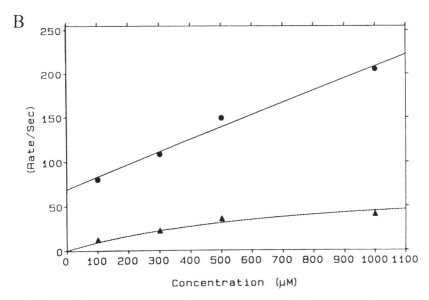

FIG. 5. Kinetics of serine reaction with tryptophan synthase. (A) The change in fluorescence at 500 nm was measured by stopped-flow after mixing enzyme (2 μM) with serine (500 μM). The smooth line represents a fit to a double exponential with rates of 150 and 40 sec^{-1}. (B) The serine concentration dependence of the fast and slow phases of the reaction are fit to the equations as described in the text. Reproduced with permission from (*32*).

K_d = 10 mM for the binding of serine, then k_2 = 1350 sec^{-1} is consistent with the observed rate of binding as the ratio k_2/K_d, and no curvature would have been apparent in the concentration dependence of the observed rate of the fast phase of the reaction at the highest concentrations of serine explored.

There have been significant advances toward the use of rapid-scanning stopped-flow, wherein the instrument can measure the absorption spectrum of the sample every 5 msec (23, 24). This is of obvious value in helping to identify the species by their characteristic absorption spectra. However, it should also be apparent from the above analysis that the rates of reaction that are extracted by measurement at different wavelengths should be identical, provided that each reaction rate is represented. For example, in studies of the pyridoxal phosphate-catalyzed reaction of tryptophan synthase (23), measurements at 460 nm gave a single exponential at a rate of 136 sec^{-1}, while measurements at 430 nm gave a single exponential at a rate of 14 sec^{-1}. At 454 nm, a double exponential (rise–fall) was observed, which is a function of both rates. Thus the three experiments served to measure the same kinetically linked processes, with the different wavelengths providing a signal weighted toward one or both of the phases.

2.   *Simulations of Two-Step Reactions*

In Fig. 6, we show simulations of the time dependence of E, E·S, and E·X according to the simple mechanism

$$E + S \underset{k_{-1}}{\overset{k_1}{\rightleftharpoons}} E \cdot S \underset{k_{-2}}{\overset{k_2}{\rightleftharpoons}} E \cdot X$$

with the rate constants defined in the figure legend. These simulations illustrate how the time dependence of the concentrations of each of the enzyme species is linked to other species in the pathway. Simulations were performed at four different concentrations of S. In Figs. 6B and 6C, the rise and fall of the intermediate species, E·S, can be clearly seen. At lower concentrations of S (Fig. 6A), the formation of E·S is rate limiting and so only a small amount of E·S ever accumulates (note longer time scale); E decays and E·X is formed according to a single exponential. At higher concentrations of S, the rate of formation of E·S is much greater than the reaction to form E·X ($k_1[S] \gg k_2, k_{-1}$), and so E·S rises very rapidly and then decays in a single exponential as E·X is formed at a rate equal to $k_2 + k_{-2}$. Thus at both low and high concentrations of S, the intermediate species, E·S, is kinetically transparent if one measures the rate of formation of E·X which follows a single exponential.

At moderate concentrations of S, where the rate constants $k_1[S]$ and the sum $k_2 + k_{-2}$ are *comparable*, one sees the rise and fall of the E·S species with the rate and amplitude of formation of E·S dependent on the magnitude of $k_1[S]$ relative to k_2 (Figs. 6B and 6C). The formation of E·X proceeds with a lag phase

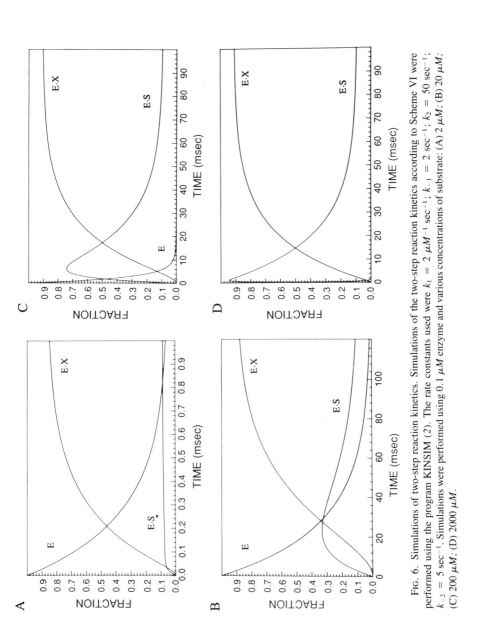

Fig. 6. Simulations of two-step reaction kinetics. Simulations of the two-step reaction kinetics according to Scheme VI were performed using the program KINSIM (2). The rate constants used were $k_1 = 2\ \mu M^{-1}\ \text{sec}^{-1}$; $k_{-1} = 2\ \text{sec}^{-1}$; $k_2 = 50\ \text{sec}^{-1}$; $k_{-2} = 5\ \text{sec}^{-1}$. Simulations were performed using 0.1 μM enzyme and various concentrations of substrate: (A) 2 μM; (B) 20 μM; (C) 200 μM; (D) 2000 μM.

that is a function of the rate of formation of E·S. The maximum rate of formation of E·X occurs when the peak in [E·S] is reached and the rate of reaching the peak in [E·S] is equal to the rate of approach to the maximum rate of formation of E·X. In mathematical terms, the time dependence of decay of E, formation of E·S, and formation of E·X each follow double exponentials [Eqs. (13) and (15)] and the rate constants governing each species are the same. The rate constant governing the rate of rise of E·S is the same as the rate constant defining the lag phase in the formation of E·X, and the rate of decay of E·S equals the rate of formation of E·X. Similarly, the fall of E is governed by the same two rate constants. At each concentration, the differences in the three curves for E, E·S, and E·X are only in the amplitude terms.

3. Generalizations: Time Dependence of Two-Step Reactions

There are two general conclusions from this analysis that can be applied to any multistep reaction sequence. First, the amplitude terms for each species, E, E·S, and E·X, are a function of all four rate constants. The rise and fall of the intermediate species, E·S, will be described by two exponential terms, one with a negative amplitude defining the rise and one with a positive amplitude defining the fall. At high substrate concentration (Fig. 6D), the amplitude governing the formation of E·S is large, but the rate of reaction is too fast to be observed, leading to collapse of the kinetics to resemble a one-step reaction. Under these conditions, the lag in formation of E·X is still there, but it is extremely short. Thus at low and high concentrations of S, the reaction to form E·X follows a single exponential; it is only at intermediate concentrations, where k_1 [S] is comparable to $k_2 + k_{-2}$, that a double exponential is observed with a lag phase in the formation of E·X. Stated in other terms, the observation of a lag in the kinetics implies that there are at least two steps in a reaction sequence that are comparable (within a factor of 10 in rate).

The second general conclusion from this analysis is that the time dependence for each of the species is governed by the same two rates, λ_1 and λ_2. Thus the fast phase of the fall of E and the rise of E·S and the lag prior to formation of E·X are all governed by one rate, λ_1. The fall of E·S and rise of E·X are governed by the other rate, λ_2. The time dependence of formation and decay of each species are kinetically linked to other species in the reaction sequence, such that they are all governed by the same set of rate constants. This is true for any mechanism.

4. Resolving Rapid Equilibrium Steps

It is often difficult to distinguish whether a two-step reaction can be fit assuming a rapid-equilibrium first step (Scheme VII) or whether it must be considered in terms of two steps of comparable rate (Scheme VI). The rapid-equilibrium case is just one extreme, where $k_{-1} \gg k_2$, but in reality one must consider the case where the two rates are comparable. If the only observable signal is a function of the formation of E·X, then the rate of reaction, measured as a fit to a single exponential, will have a hyperbolic concentration dependence for either mechanism [Eq. (11) versus Eq. (15)]. In the rapid-equilibrium case, the fit to the hyperbola will provide an estimate of the K_d for the first binding step. However, if the reaction is not sufficiently rapid, the apparent K_d will be a function of all four rate constants [Eq. (15)]. There is one diagnostic feature of the reaction kinetics that distinguishes the two mechanisms. In the rapid-equilibrium case, the time dependence follows a single exponential at all substrate concentrations. In the general case, the time dependence of reaction will show a distinct lag when the rates of the initial binding and dissociation ($k_1[S]$ and k_{-1}) are comparable to the rate of relaxation of E·S to E·X ($k_2 + k_{-2}$). Careful examination of the reaction time course may reveal a lag in the reaction kinetics at low substrate concentrations. The lag is most prominent when the first reaction is 3- to 10-fold slower than the second reaction and will disappear when $k_1[S] > k_2 + k_{-2}$ as the substrate concentration increases. When two reactions in series occur at exactly the same rate, the kinetics are difficult to distinguish from a single exponential, and the observed rate of reaction is approximately half of the rate of either reaction.

5. Kinetics of Ligand Exchange

We can now consider the kinetics of ligand exchange with competition of two ligands for a single enzyme site according to the complete two-step reaction sequence (Scheme XI).

$$E \cdot A \underset{k_{-1}}{\overset{k_1}{\rightleftharpoons}} E + A; \qquad E + B \underset{k_{-2}}{\overset{k_2}{\rightleftharpoons}} E \cdot B$$

SCHEME XI

In general, the appearance of E·B will follow a double exponential with rates of

$$\lambda_1 \simeq k_1 + k_{-1}[A] + k_2[B] + k_{-2}$$

$$\lambda_2 \simeq \frac{k_1(k_{-2} + k_2[B]) + k_{-1}[A]k_{-2}}{k_1 + k_{-1}[A] + k_2[B] + k_{-2}}$$

(20)

As the concentration of B increases, the first relaxation, λ_1, will become too fast to measure and λ_2 will approach a maximum rate equal to k_1. In practice, extrapolation of the rate of exchange to an infinite concentration of B provides an estimate of the rate of dissociation of the E·A complex, k_1. This can often be accomplished by stopped-flow methods if the binding of A or B induces a distinct change in protein fluorescence.

6. *Example: Dihydrofolate Reductase*

The binding of substrates to dihydrofolate reductase follows the simple, one-step association reaction, which can be easily measured by the quenching of protein fluorescence that occurs on binding (27). Tetrahydrofolate (THF), for example, binds with a second-order rate constant of 25 μM^{-1} sec^{-1} and dissociates at a rate of 1.4 sec^{-1}. The slow rate of dissociation was accurately measured by mixing E·THF with methotrexate (MTX), a tight-binding inhibitor that quenches the protein fluorescence to a greater extent than THF, thus providing a signal in proceeding from E'·THF → E''·MTX. Under the appropriate conditions, the rate of the observable reaction is limited by the rate of dissociation of THF from the enzyme according to Scheme XII.

$$\text{E}'\cdot\text{THF} \underset{k_1}{\overset{k_{-1}}{\rightleftharpoons}} \text{E} + \text{THF}; \qquad \text{E} + \text{MTX} \underset{k_{-2}}{\overset{k_2}{\rightleftharpoons}} \text{E}''\cdot\text{MTX}$$

SCHEME XII

The experiment serves to measure k_{-1} because the binding of MTX, at the concentration used, is faster than the rebinding of THF ($k_2[\text{MTX}] \gg k_1[\text{THF}]$), and the binding of MTX is largely irreversible ($k_2[\text{MTX}] \gg k_{-2}$).

7. *Conformation Change Prior to Substrate Binding*

As an alternative to the substrate-induced conformational change that occurs after substrate binding, the mechanism according to Scheme XIII has been considered, with a conformational change preceding binding.

$$\text{E} + \text{S} \underset{k_{-1}}{\overset{k_1}{\rightleftharpoons}} \text{E}' + \text{S} \underset{k_{-2}}{\overset{k_2}{\rightleftharpoons}} \text{E}'\text{S}$$

SCHEME XIII

This mechanism can account for the saturation in the rate of binding with increasing substrate concentration, where the maximum rate is defined by k_1. It can also provide a rationale for an apparent second-order rate constant lower than the diffusion limit by a factor equal to the fraction of enzyme in the E' state. How then can we distinguish this reaction pathway from the induced-fit model considered earlier (Scheme VI)?

First, if we consider that step 1 is a rapid equilibrium such that the E $\rightleftharpoons$ E' equilibrium is maintained throughout the progress of the reaction, then the rate of the forward reaction will be defined by the rate constant, k_2, times the fraction of enzyme in the E' state. The reaction will approach equilibrium with a single exponential defined by $k^{\text{obs}} = K_1 k_2[\text{S}]/(K_1 + 1) + k_{-2}$. By this model, the apparent second-order rate constant for substrate binding is given by $K_1 k_2/(K_1 + 1)$. However, the rate will continue to increase with increasing substrate concentration and not reach a maximum according to this mechanism, with the rate constants k_1 and k_{-1} much larger than $k_2[\text{S}]$, the assumption for a rapid-equilibrium mechanism. Thus, this model can account for any linear concentration dependence where the rate of binding is substantially below the diffusion limit. This mechanism may account for the remarkably low rate of phosphate binding seen for a number of enzymes, with values in the range of $10^4 M^{-1} sec^{-1}$ (3, 37). The rebinding of phosphate, the product of the reaction, may indeed be constrained by the small fraction of enzyme in a proper conformation required for the enzyme–product state.

If the rate of the conformational change is slow enough to be observed, then one must consider the complete solution to the two-step reaction sequence, with the substrate concentration dependence in rate constant k_2 rather than k_1. Thus the time dependence of the reaction will follow a double exponential according to Eq. (13) but with the appropriate substitutions of the rate equations [Eq. (15)] to yield:

$$\lambda_1 \simeq k_1 + k_{-1} + k_2[S] + k_{-2}$$

$$\lambda_2 \simeq \frac{k_1(k_{-2} + k_2[S]) + k_{-1}k_{-2}}{k_1 + k_{-1} + k_2[S] + k_{-2}}$$

(21)

To a first approximation, the fast phase will be due to the binding of S to E', and the slow phase will then proceed by a rate that is limited by the rate constant k_1 at high substrate concentration. If the signal, such as a change in protein fluorescence, measures the sum of E' + E'S, then the fast phase will not be observable and the data will follow a single exponential. If the signal measures the formation of E'S, then the reaction will show two distinct phases and the amplitude of the fast phase will be a function of the fraction of enzyme in the E' state. The form of the equations is identical to that observed for the model invoking a conformational change subsequent to S binding. The two models cannot be distinguished on the basis of rate data alone, although the observation of different reaction rates with different substrates can be used to argue against this mechanism. In addition, the concentration dependence of the amplitude terms of the two phases could be used to distinguish the two models in some cases.

This example illustrates one of the limitations to the interpretation of stopped-flow kinetic data that must be kept in mind. The problem arises largely due to the fact that the extinction coefficients for kinetic intermediates may not always be known. Rigorous interpretation of kinetic data requires a knowledge of the absolute concentrations of species. For these reasons, chemical-quench-flow data can be more rigorously interpreted. On the whole, unless extinction coefficients of intermediates and products are known, stopped-flow experiments are much more difficult to interpret rigorously than quench-flow results. Stopped-flow experiments produce data of higher precision and can measure faster rates of reaction, but the quench-flow methods have distinct advantages by providing the absolute amplitudes of known chemical reactions.

D.   Rules Governing Number of Exponential Terms

The integrated rate equations describing enzyme reaction pathways under first-order (or pseudo-first-order) conditions will always be a sum of exponential terms:

$$[E_i] = \sum \alpha_n e^{-\lambda_n t} + \gamma \qquad (22)$$

Therefore, it is important to ask how many exponential terms can be expected for a given mechanism. The maximum number of exponentials defining the time dependence of a reaction can be estimated from the following rules:

1. There will be one exponential term for each step in a sequential reaction, with exceptions as defined below.
2. If substrate is in excess, product release from E·P to regenerate E does not lead to another exponential phase. Under these conditions, product release leads to the linear steady-state phase of the reaction and the rate of release contributes to the net rate of approach to steady state.
3. Any reaction that is much faster than the step preceding it will not be observable as a distinct step, but will occur at the rate of the preceding step. In mathematical terms, the fast step still leads to an exponential function in the integrated equation, but the term drops out of the expression at short times due to a fast rate.
4. Rapid-equilibrium steps do not contribute a kinetic step; the species in rapid equilibrium can be grouped together as a single, kinetic unit.

Thus according to these rules, the formation of E·P will follow a single exponential for the reaction sequence according to Scheme XIV, even though there are three steps.

$$E + S \overset{K_1}{\rightleftharpoons} E{\cdot}S \underset{k_{-2}}{\overset{k_2}{\rightleftharpoons}} E{\cdot}P \overset{k_3}{\to} E + P$$

Scheme XIV

At saturating [S] ([S] $\gg 1/K_1$), the maximum rate of approach of E·P to its steady-state concentration is $k_2 + k_{-2} + k_3$ for this model.

Scheme XV can be understood in terms of the simplified mechanism,

$$E + S \underset{k_{-2}}{\overset{\lambda_1}{\rightleftharpoons}} E{\cdot}P \overset{k_3}{\to} E + P$$

Scheme XV

where the reaction of S to form E·P is approximated by the hyperbolic relationship:

$$\lambda_1 = \frac{K_1 k_2}{K_1[S] + k_2} \qquad (23)$$

1. *Kinetics of Enzyme Inactivation*

   The usefulness of knowing that the reactions of individual species are kinetically linked is illustrated by studies on the time dependence of enzyme inactivation in the presence of a slow-binding inhibitor or inactivator, for example. The reaction can be considered as

$$E_1 \xrightarrow{k_1} E_2$$

where E_1 and E_2 represent two enzyme forms catalyzing a reaction at two different rates:

$$S \xrightarrow{k_2[E_1]} P; \quad S \xrightarrow{k_3[E_2]} P$$

By measuring the time dependence of product formation, the change in activity of the enzyme can be examined. The common method of data analysis is based on estimating the enzyme activity from the tangent to the curve at each time of reaction in a plot of product concentration versus time. However, this tedious and error-prone method of analysis can be avoided by recognizing that the time dependence of product formation will occur by an exponential followed by a linear phase:

$$[P] = A_0(1 - e^{-k_1 t}) + k_{ss}t \qquad (24)$$

where k_{ss} represents the final steady-state rate. Thus, the rate of inactivation can be derived directly by analysis of the kinetics of product formation, without the intermediate step of attempting to estimate the rate of product formation at each time point along the progress of the inactivation. The rate of approach to the final linear phase, obtained as a fit to the exponential, will define the kinetics of inactivation directly, provided that the requirements of steady-state initial-velocity kinetics are maintained throughout the reaction time course. The amplitude of the exponential phase will depend on the rate of inactivation relative to the rate of catalysis by the fully active enzyme. Similar analysis can be applied to the case of a slow activation of an enzyme.

2. *Kinetics of Competing Reactions*

   The kinetics of the partitioning of an enzyme-bound substrate to two products illustrates how the linkage between the two reactions leads to a single exponential. Consider the kinetics of a reaction pathway:

$$ES \xrightarrow{k_2} EP; \quad ES \xrightarrow{k_3} EQ$$

According to this pathway, the rate of disappearance of ES and the rates of formation of EP and of EQ are each governed by a rate equal to the sum $k^{obs} = k_2 + k_3$. The kinetic partitioning between EP and EQ will only be revealed by the relative amplitudes for formation of each product. For example, if $k_2 = 10$ sec^{-1} and $k_3 = 20$ sec^{-1}, then the reactions will follow an exponential with $k^{obs} = 30$ sec^{-1}, but one-third of the product will be P and two-thirds will be Q.

V. Measurement of Reaction Rates at Enzyme Active Sites

A. PRESTEADY-STATE BURST KINETICS

When enzyme is first mixed with substrates (in excess over the enzyme), one can often observe a burst of product formation at a rate faster than steady-state turnover. This presteady-state burst is due to the accumulation of product at the active site of the enzyme. On quenching the reaction mixture with a denaturant to stop further reaction, the enzyme-bound product is liberated and the quantitation of product includes the sum of that bound to the enzyme and free in solution at the time of the quench.

Burst kinetics are often described in terms of a two-step irreversible mechanism (5, 38).

$$E{\cdot}S \xrightarrow{k_2} E{\cdot}P \xrightarrow{k_3} E + P$$

The equation describing the time dependence of formation of product will be given by an exponential followed by a linear phase

$$\frac{[P]_{obs}}{[E_0]} = \frac{[EP] + [P]}{[E_0]} = A_0(1 - e^{-\lambda t}) + k_{cat}t \qquad (25)$$

where the rates and amplitudes are defined by

$$\lambda = k_2 + k_3$$

$$A_0 = [k_2/(k_2 + k_3)]^2 \qquad (26)$$

$$k_{cat} = k_2 k_3/(k_2 + k_3)$$

The presteady-state burst will be followed by steady-state turnover at a rate given by k_{cat}. The presteady-state burst of product formation will occur at a rate defined by the sum of the rates of the chemical reaction and product release. The amplitude is also a function of both rate constants, k_2 and k_3. Thus, the amplitude of the burst can be predicted from the rate of the burst and the rate of steady-state turnover. Although this model can account for burst kinetics, it is often inadequate due to the assumed irreversibility of the chemical reaction. The internal equilibrium arising from the reverse of the chemical reaction (k_{-2}) reduces the amplitude of the burst to less than predicted by Eq. (26).

For a pathway including a reversible chemical reaction (Scheme XVI),

$$E \cdot S \underset{k_{-2}}{\overset{k_2}{\rightleftharpoons}} E \cdot P \overset{k_3}{\rightarrow} E + P$$

SCHEME XVI

the time dependence of product formation still follows the general equation for burst kinetics

$$[P]_{obs}/[E_0] = A_0(1 - e^{-\lambda t}) + k_{cat}t \tag{27}$$

where the rates and amplitudes are defined by

$$\lambda = k_2 + k_{-2} + k_3$$

$$A_0 = k_2(k_2 + k_{-2})/(k_2 + k_{-2} + k_3)^2 \tag{28}$$

$$k_{cat} = k_2k_3/(k_2 + k_{-2} + k_3)$$

The rate of the burst is given by the sum of all three rate constants, and the amplitude is defined by the term, A_0. Thus, compared to the irreversible pathway, the rate of the burst will be faster and the amplitude will be lower, both due to the reversal of the chemical reaction. Estimates of all three rate constants can be obtained by proper fitting of the rate and amplitude of the presteady-state burst.

As the rate of product release increases, the amplitude of the burst decreases due to a lowering of the concentration of E·P present during steady-state turnover. It is important to remember that the rate of steady-state turnover is a simple function of $[E \cdot P]_{ss}$; namely, $k_{cat} = k_3[E \cdot P]_{ss}$. Thus, to a first approximation, the value of k_3 can be estimated from the slope of the time course divided by the intercept obtained by extrapolation of the linear phase to the y axis. This is not quite true because the intercept on the y axis, defining the amplitude of the burst, is less than the actual concentration of E·P present during the steady state: $[E \cdot P]_{ss} = k_2/(k_2 + k_{-2} + k_3)$. The burst amplitude is lower than the actual steady-state concentration of E·P by a factor equal to $(k_2 + k_{-2})/(k_2 + k_{-2} + k_3)$. Thus, a faster rate of product release, k_3, relative to the rate of the chemical reaction $k_2 + k_{-2}$, leads to a lower burst amplitude in the extrapolation.

The amplitude of the burst is governed by the rate of formation of E·P relative to the rate of release of P and the rate of back reaction to form E·S. The following rules can be applied to evaluate the amplitude of the burst:

1. The amplitude will be 1 per enzyme site provided that the formation of product is much faster than the rate of product release ($k_2 \gg k_3$) and the internal equilibrium favors E·P ($k_2 > k_{-2}$).

2. The amplitude will be reduced to be less than $[E \cdot P]_{ss}$ due to the competi-

tion between the rates of product formation and product release. As the rate of product release increases, the rate of approach toward steady state increases as the amplitude of the burst decreases. The amplitude will always fall between 0 and 1.

3. An amplitude less than that expected due to the observed rate of steady-state turnover and rate of the burst can be attributed to the internal equilibrium. Thus, as k_{-2} increases, the amplitude of the burst decreases and the rate of the burst increases.

4. If there is no burst, then either the chemical reaction or a step preceding the chemistry is rate limiting. Alternatively, the internal equilibrium may favor substrates ($K_1 \ll 1$); however, this is less likely due to the principles governing catalysis that tend to bring the equilibrium for a reaction closer to unity at the active site of the enzyme than in solution (*18*). In addition, a complete kinetic analysis would include examination of the burst in each direction, which could reveal an unfavorable equilibrium in one direction and a full burst amplitude in the opposite direction.

Because the amplitude of the burst is less than or equal to the concentration of enzyme sites, these experiments must be performed using enzyme concentrations that will produce measurable amounts of product in the first turnover. In order to saturate the rate of substrate binding so that chemistry, not binding, limits the rate of the burst, high concentrations of substrate must be used. The major experimental limitation of the method is due to the problems associated with measurement of less than one product per enzyme site with a background of excess substrate. Depending on the chromatographic resolution of the product from substrate, the ability to observe and measure a burst of product formation may be limited by accessible concentrations of enzyme.

Two assumptions underlie the derivations of the above equations. The first is that the rate of substrate binding is fast and is sufficiently favorable to saturate the enzyme sites on a time scale faster than catalysis. The second is that the release of products is presumed to be irreversible so that the concentration of product in solution can be neglected. In practice, neither of these assumptions limit the design or interpretation of experiments provided that the data are analyzed by numerical integration rather than by relying upon the explicit solutions provided here. By numerical integration, estimates of the rate constants governing the binding of substrate and product can be entered into the computer and included as part of the fitting process. The equations shown here should be used only to provide an initial fit to the data by nonlinear regression. These initial estimates of the constants can then be refined by more complete analysis based on computer simulation (*3*).

Even with computer simulation, it is still a concern that the rate of substrate binding must be faster than catalysis if the intent of the experiment is to measure the rate of the chemical step. If substrate binding is much faster than catalysis ($k_1[S] \gg k_2$), the reaction proceeds toward steady state with a single exponential decay of E·S to E·P. However, if the concentration of substrate is lower, the rate of reaction to form E·P will be reduced according to the equations described for the two-step binding, and this will lead to a corresponding reduction in the amplitude of the burst. In the lower limit, the rate of the burst can provide a direct measurement of the rate of substrate binding, provided that product release is sufficiently slow. At moderate substrate concentrations, where $k_1[S]$ is only marginally faster than k_2 (less than 10-fold difference), there will be a lag in the production of E·P. In this case, it is still possible to extract reliable estimates of the rate of catalysis by including the rate of substrate binding in the data analysis by computer simulation (2, 3, 39).

B. PULSE-CHASE EXPERIMENTS

It is often possible to measure the kinetics of substrate binding by performing a kind of millisecond pulse-chase experiment in the quench-flow apparatus. The enzyme is first mixed with radiolabeled substrate, and then after a period of milliseconds is mixed with an excess of unlabeled substrate. After incubation for a brief period sufficient to convert all of the enzyme-bound substrate to products (six to eight turnovers), the reaction is stopped by the addition of acid, base, or other suitable quenching reagent. During the chase period, any tightly bound radiolabeled substrate is given sufficient time to be converted to product, while any loosely bound substrate or unbound substrate is diluted out by the excess of unlabeled material. According to this experimental protocol, one can measure the time dependence of formation of an enzyme–substrate complex. Of course, the method only permits detection of the fraction of enzyme–substrate complex that proceeds in the forward reaction, defined by the ratio $k_2/(k_2 + k_{-1})$ according to Scheme VI. The kinetics of reaction define the rate of formation of tightly bound substrate. Comparison of the reaction kinetics obtained in the pulse-chase experiment with that of a conventional acid-quench experiment can provide direct evidence for the formation of a tight enzyme–substrate complex preceding the chemistry step. For example, a pulse-chase experiment provided evidence for a tightly bound deoxynucleoside triphosphate involved in base pair recognition during DNA polymerization (10) and during ATP hydrolysis by myosin (28) or dynein (37, 39), as described below (Section IV,B,1).

It is important to stress that I have used the term "tight" binding of the E·S complex in this context to refer only to the rate at which the substrate dissociates from the enzyme relative to the rate of reaction to form products. It is often the case that the binding of substrate in an initial, weak collision complex induces a change in enzyme conformation, resulting in a slower rate of substrate release and tighter binding (29, 40). Nonetheless, it is important to keep in mind the distinction between thermodynamic and kinetic stability. The experiment provides direct evidence for the kinetic partitioning of the E·S complex. It is only when this information is combined with measurements of the forward rate constants that estimates of thermodynamic stability can be obtained.

Examples: Mechanochemical ATPases

Studies on the dynein ATPase presented a particular challenge to the measurement of a presteady-state burst (*39*). During the course of studies on dynein, it was shown that molecular weight per ATPase site was approximately 750,000 and the maximum concentration of enzyme sites that could be obtained in solution was 1 μM. The steady-state turnover rate was 8 sec^{-1} and the second-order rate constant for binding ATP was 4×10^6 M^{-1} sec^{-1}. Thus a concentration of 30–50 μM ATP was required to obtain a sufficiently fast rate of ATP binding to observe a burst of product formation. Higher concentrations of ATP could not be employed because of the background due to the contaminating products of hydrolysis in every ATP preparation. A burst was observed, as shown in Fig. 7. In these experiments, the binding of ATP was still partially rate limiting and so the formation of product was biphasic with a slight lag phase. Data were fit using a minimal mechanism including all three steps—binding, chemistry, and product release (*39*).

Studies on kinesin established that the rate of ADP release limits steady-state turnover to an exceedingly slow rate in the absence of microtubules (0.009 sec^{-1}) (*41*). Therefore, as long as the rate of substrate binding ($k_1[S]$) is larger than the rate of ADP release, a burst can be observed. Hackney *et al.* took advantage of this fact to estimate the rate of ATP binding by measuring the rate of the burst at very low substrate concentrations, in the nanomolar range (*41*).

C.   SINGLE-TURNOVER EXPERIMENTS

Because of the factors that reduce the amplitude in a presteady-state burst experiment and the difficulty in resolution of the product (or intermediates) from excess substrate, it is often desirable to use single-turnover methods. These experiments are performed with enzyme in excess over substrate to allow the direct observation of the conversion of substrates to intermediates and products in a single pass of the reactants through the enzymatic pathway. Unlike the presteady-state burst experiments, the kinetics are free of complications resulting from the steady-state formation of products, which limits the resolution of the burst kinetics and the detection of any intermediates above the background of excess substrates and products.

In a single-turnover experiment with enzyme in excess, the kinetics of the reaction are different than with substrate in excess. The rate of substrate binding

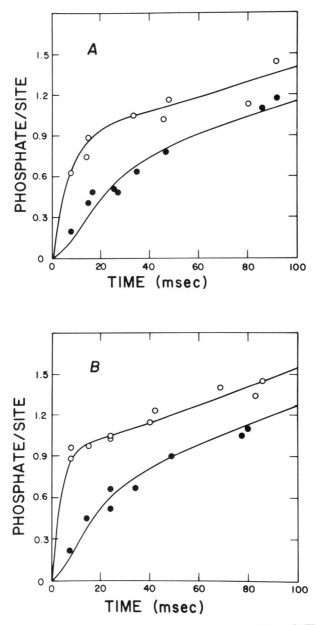

FIG. 7. Dynein ATPase burst kinetics. The kinetics of a presteady-state burst of ATP binding (○) and hydrolysis (●) were determined at two ATP concentrations: (A) 30 μM and (B) 50 μM. The data fit rate constants of $k_1 = 4.7 \mu M^{-1} sec^{-1}$, $k_2 = 55 sec^{-1}$, $k_{-2} = 10 sec^{-1}$, and $k_3 = 8 sec^{-1}$ according to Scheme IV. Reproduced with permission from (39).

is governed by a pseudo-first-order rate constant defined by the product, $k_1[E]$. Thus if substrate binding is rate limiting, the rate of disappearance of substrate, obtained as an exponential fit to the time dependence of the reaction, will be independent of substrate concentration, but directly proportional to enzyme concentration. This is an important point that can be used to advantage in the design of experiments, although it also provides the major limitation of the method. The accessible concentrations of enzyme may limit the magnitude of $k_1[E]$ necessary to saturate the rate of substrate binding to measure the rate of the chemical reaction.

D.   OBSERVATION OF ENZYME INTERMEDIATES

The two-step reaction sequence discussed above can be translated directly to understand the reactions occurring at the enzyme site describing the formation and decay of an enzyme intermediate. For example, the enzyme sequence in Scheme XVII

$$E + S \underset{k_{-1}}{\overset{k_1}{\rightleftharpoons}} E{\cdot}S \underset{k_{-2}}{\overset{k_2}{\rightleftharpoons}} E{\cdot}I \underset{k_{-3}}{\overset{k_3}{\rightleftharpoons}} E{\cdot}P$$

SCHEME XVII

will be reduced kinetically to Scheme XVIII

$$E{\cdot}S \underset{k_{-2}}{\overset{k_2}{\rightleftharpoons}} E{\cdot}I \underset{k_{-3}}{\overset{k_3}{\rightleftharpoons}} E{\cdot}P$$

SCHEME XVIII

at concentrations of enzyme sufficient to saturate the rate and equilibrium for substrate binding in an experiment with enzyme in excess. Any enzyme concentration dependence observed for the rate of formation of the enzyme intermediate will reflect a rate-limiting substrate binding. At concentrations of enzyme sufficient to bind substrate at a rate faster than catalysis ($k_1[E] > k_2$), the kinetics of the reactions at the active site will then follow the questions described for a two-step reaction sequence [Eq. (15)].

Any fast reaction that follows a slow reaction will occur at the rate of the slow reaction. Consider the formation of an enzyme intermediate according to Scheme XVIII. This enzyme-bound intermediate (E·I) will be invisible kinetically and thermodynamically if $k_3 \gg k_2$ and $k_{-2} \gg k_{-3}$. Thus, one can only define the kinetically significant intermediates or conformational states. Transition states or extremely reactive intermediates cannot be directly observed; their presence can only be inferred by knowledge of the chemistry of the reaction. Alternatively, the use of alternate substrates or analogs has provided evidence for an intermediate by slowing the rate of the second step. For example, in studies on chymotrypsin, evidence in favor of an acyl intermediate was derived largely using reactive substrate analogs (*42, 43*). In the normal physiological reaction, the formation of the acyl–enzyme intermediate is slow and is followed by a faster hydrolysis reaction. Therefore the acyl–enzyme cannot be observed. However, synthetic substrates were prepared that reacted rapidly to form the acyl–enzyme intermediate but were hydrolyzed more slowly, leading to the accumulation and identification of the intermediate.

Unique solution to the time dependence of formation and decay of the intermediate depends on an absolute measurement of its concentration. This is illustrated in Fig. 8, comparing the results of a simulation for a two-step irreversible mechanism,

$$E{\cdot}S \xrightarrow{k_2} E{\cdot}I \xrightarrow{k_2} E{\cdot}P$$

with the rate constants $k_2 = 100$ sec^{-1} and $k_3 = 20$ sec^{-1} (Fig. 8A) or with $k_2 = 20$ sec^{-1} and $k_3 = 100$ sec^{-1} (Fig. 8B). Note the lower amplitude for formation of the intermediate in the simulation in Fig. 8B. The same two simulations are shown superimposed in Fig. 8C, where a fivefold higher scaling factor (extinction coefficient) was used for the intermediate in the simulation involving the slower rate for k_2. This example illustrates clearly that the two pathways are indistinguishable unless the absolute concentration of the intermediate is known. The difficulty arises when the intermediate is unstable and its absolute concentration cannot be determined. However, even in this case, rapid-quench methods can measure the time dependence of formation and decay of the intermediate provided that it decomposes in the quench to yield products that can be uniquely identified (*3, 44*).

Example: EPSP Synthase

The kinetics of formation and decay of a tetrahedral intermediate at the active site of EPSP synthase has been examined in both the forward and reverse directions, as shown in Fig. 9. The reaction catalyzed by EPSP synthase is an ordered, Bi Bi mechanism (Scheme XIX).

$$E + S3P + PEP \underset{k_{-1}}{\overset{k_1}{\rightleftharpoons}} E{\cdot}S3P + PEP \underset{k_{-2}}{\overset{k_2}{\rightleftharpoons}} E{\cdot}S3P{\cdot}PEP$$

$$E{\cdot}S3P{\cdot}PEP \underset{k_{-3}}{\overset{k_3}{\rightleftharpoons}} E{\cdot}I \underset{k_{-4}}{\overset{k_4}{\rightleftharpoons}} E{\cdot}EPSP{\cdot}P_i$$

$$E{\cdot}EPSP{\cdot}P_i \underset{k_{-5}}{\overset{k_5}{\rightleftharpoons}} E{\cdot}EPSP + P_i \underset{k_{-6}}{\overset{k_6}{\rightleftharpoons}} E + EPSP + P_i$$

SCHEME XIX

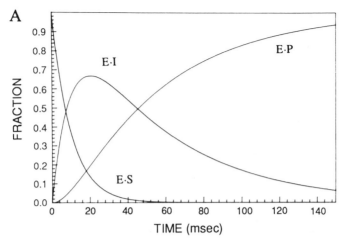

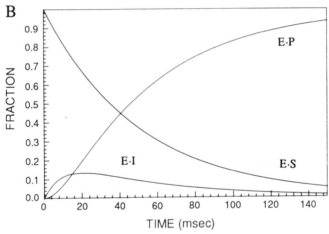

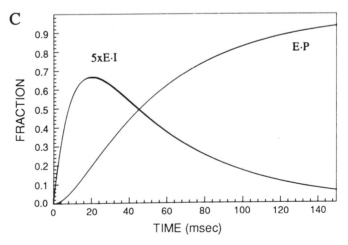

Figure 9A shows the reaction in the forward direction in a single-turnover experiment tracing the conversion of radiolabeled phosphoenol pyruvate (PEP) to intermediate (I) and then EPSP with enzyme in excess over the limiting substrate. The enzyme was first saturated with S3P and the reaction was initiated by the addition of labeled PEP. The intermediate was formed within 3 msec and decayed to EPSP over the next 30–40 msec. Figure 9B shows the reverse reaction, which was initiated by mixing radiolabeled EPSP with enzyme and an excess of phosphate. The kinetics show the transient appearance of the intermediate in equilibrium with EPSP at the active site; the intermediate and EPSP decay in constant proportion as the products, PEP and S3P, are formed. The fit to the data led to a unique set of rate constants accounting for the time dependence of appearance and disappearance of the intermediate in both the forward and reverse reactions, with $k_3 = 1200$ sec^{-1}, $k_{-3} = 100$ sec^{-1}, $k_4 = 320$ sec^{-1}, and $k_{-4} = 240$ sec^{-1} Initial fitting of the rapid-quench data was based on use of an acid quench, which caused the intermediate to break down to yield pyruvate. Subsequent work, using a quench with neat triethylamine, allowed isolation and identification of the intermediate and measurement of the internal equilibrium constants (K_3, K_4) for its formation at the active site (3, 16).

The reactions in each direction illustrate the need for computer simulation to interpret quantitatively the reaction kinetics. In the forward direction, the binding of PEP was largely rate limiting and constrained by other experiments to be no greater than $k_2 = 15$ μM^{-1} sec^{-1}. Thus, in the fitting of the data, the rate of reaction to form the intermediate was limited to a rate of 300 sec^{-1} for PEP binding ($k_2[E]$) followed by 1200 sec^{-1} for the chemical reaction. Combined with the measurement of the internal equilibrium constant (K_3) and the rate of the single turnover in the reverse direction, confidence in the estimate of $k_3 = 1200$ sec^{-1} is beyond what could have been achieved by a single experiment. It is the combination of all experiments together that allows the unique fit to the reaction kinetics and defines the pathway.

In the reverse reaction, driving the synthesis of S3P and PEP from EPSP and phosphate, the amplitude of the product formation was altered because of the formation of a dead-end complex with S3P and phosphate bound to the active site. The time dependence of this inhibition during the single turnover to form E·S3P would lead to a model too complex to solve explicitly. The reaction time

---

FIG. 8. Importance of amplitude information. The kinetics of a two-step irreversible reaction sequence were simulated using KINSIM (2). (A) $k_1 = 100$ μM^{-1} sec^{-1}; $k_2 = 20$ sec^{-1}. (B) $k_1 = 20$ μM^{-1} sec^{-1}; $k_2 = 100$ sec^{-1}. (C) The results of the two simulations from **A** and **B** are superimposed with a scaling factor of 5× multiplied by the concentration of E·I for the second simulation (**B**). These simulations show that the two kinetic pathways in **A** and **B** are indistinguishable in the absence of absolute amplitude information.

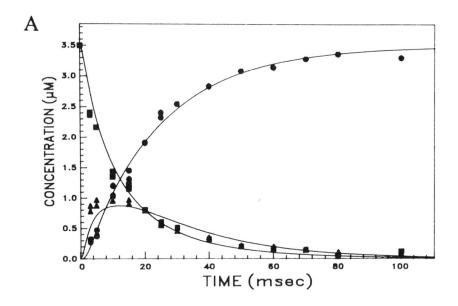

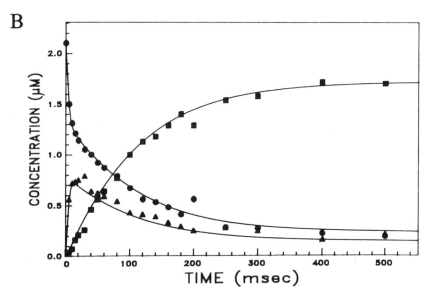

FIG. 9. EPSP synthase single-turnover kinetics and EPSP synthase reaction pathway. (A) The disappearance and formation of PEP (●), EPSP (■), and intermediate (▲) were monitored in the reverse direction. The reaction was initiated by mixing enzyme (10 μM) and S3P (100 μM) with radiolabeled PEP (3.5 μM). (B) The disappearance and formation of EPSP (●), PEP (■), and intermediate (▲) were monitored in the reverse direction. The reaction was initiated by mixing enzyme (10 μM) with phosphate (7.5 μM) and radiolabeled EPSP (2.1 μM). The curves were calculated by computer simulation using the fill kinetic pathway shown in Scheme XIX and the 12 individual rate constants (3). Reproduced with permission from (3).

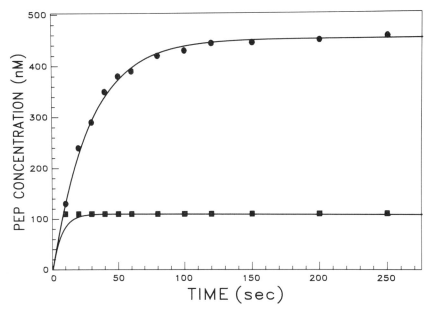

FIG. 10. Kinetics of phosphate binding to EPSP synthase. The reaction was run in the absence (●) and presence (■) of unlabeled PEP. The reaction was initiated by mixing enzyme (5 μM) with radiolabeled phosphate (0.5 μM) and excess EPSP (1 mM). The curves were calculated by computer simulation using the full kinetic pathway shown in Scheme XIX and the 12 individual rate constants (3). Reproduced with permission from (3).

course is easy to analyze by including in the computer simulation the reaction steps leading to formation of the nonproductive complex.

The rate of phosphate binding to EPSP synthase was measured by a single-turnover experiment in the reverse reaction shown in Fig. 10. The experiment was initiated by mixing an excess of enzyme and EPSP with a trace of labeled phosphate (<1 μM). Under these conditions, the rate of formation of PEP was limited by the rate of phosphate binding to the enzyme–EPSP complex. However, in the absence of unlabeled PEP, the reaction did not go to completion. Successful execution of this experiment required the addition of the unlabeled reaction product (PEP) in order to ensure that the release of radiolabeled PEP was irreversible. In the absence of unlabeled PEP, the reaction came to equilibrium short of complete conversion of radiolabeled phosphate to PEP. The addition of unlabeled PEP "pulled" the reaction to completion by dilution of the radiolabeled PEP. Computer simulation was required to analyze quantitatively the reaction time course. Conventional data fitting to the time dependence of the reaction gives a rate of approximately 0.035 sec^{-1}. The simple interpretation would then lead to calculation of a second-order rate constant for phosphate binding of 0.007 μM^{-1} sec^{-1}. However, computer simulation leads to an estimate of 0.07 μM^{-1} sec^{-1}. The difference is due to the fact that the computer simulation accounts for all of the steps of the reaction. Although this experiment provides information that is largely a function of the rate of phosphate binding, the reverse reaction and the other steps in the pathway, which are measured more accurately by other experiments, have a secondary effect on the time dependence of the reaction. Inclusion of the complete reaction sequence in the fitting of these data provides the most accurate estimate of the binding rate that is consistent with all of the experimental evidence.

E. KINETICS OF SUBSTRATE CHANNELING

The channeling of metabolic intermediates between pairs of sequential enzymes in a reaction pathway is a controversial topic (45–47). Kinetic evidence for the direct passage of a metabolite from one enzyme to the next has been based largely on indirect tests in the steady state (48–50). Recently the problem has been addressed by direct measurement of the rate of exchange (51–53). A proper comparison with rates of substrate binding and dissociation establishes whether the rates of exchange are consistent with the dissociation–rebinding pathway, or whether one is forced to suggest a direct transfer between the two enzyme active sites.

The kinetics of exchange of a single ligand between two enzymes can be considered according to the pathway shown in Scheme XX.

$$E_1 \cdot A \underset{k_{-1}}{\overset{k_1}{\rightleftharpoons}} E_1 + A; \qquad E_2 + A \underset{k_{-2}}{\overset{k_2}{\rightleftharpoons}} E_2 \cdot A$$

SCHEME XX

The complete solution will again follow a double exponential, which can be reduced to a single exponential depending on the rates of reaction.

$$\lambda_1 \simeq k_1 + k_{-1}[E_1] + k_2[E_2] + k_{-2} \tag{29}$$

$$\lambda_2 \simeq \frac{k_1(k_{-2} + k_2[E_2]) + k_{-1}[E_1]k_{-2}}{k_1 + k_{-1}[E_1] + k_2[E_2] + k_{-2}}$$

Because it is unlikely that the rates of reaction will differ greatly in comparing the two enzymes and because there are practical limits to the concentration of enzyme E_2 that can be used, the complete solution to the two-step reaction must be considered in analyzing data.

Application of these methods to the channeling of NADH between glycerol phosphate dehydrogenase (GDH) and lactate dehydrogenase (LDH) has produced conflicting results (51, 54). The initial results showed that the rate of exchange was faster than the rate of dissociation of NADH from GDH, and therefore it was concluded that NADH must have been transferred directly from the active site of GDH to LDH (54). However, it is important to note that the rate of ligand exchange, according to the equations described above, will be greater than the dissociation rate. At the concentrations of LDH accessible in solution, the displacement NADH from GDH is incomplete and the reaction reaches an equilibrium. The observed rate of approach to equilibrium is a function of the sum of the forward and reverse rates. By overlooking this fundamental property of transient kinetics, the authors were led to the wrong conclusion (54). A more rigorous analysis has shown that the rates of transfer are entirely consistent with a free diffusion model, involving dissociation of NADH from GDH and rebinding to LDH, rather than requiring a direct transfer (51).

Example: Tryptophan Synthase

The kinetics of metabolite channeling have been examined most thoroughly by application of single-turnover kinetic methods in the case of tryptophan synthase. Tryptophan synthase catalyzes the last two reactions in the pathway for tryptophan biosynthesis as shown in Fig. 11 (55). Indole, the product of the cleavage of indoleglycerol phosphate (IGP) at the α-site, is thought to pass through a channel in the enzyme to react with serine at the β-site in a pyridoxal phosphate-dependent reaction. Solution of the crystal structure revealed the presence of a 25-Å-long tunnel through the enzyme connecting the α- and β-sites (55). Evidence for the passage of indole through the tunnel was obtained by stopped-flow studies with analogs (56, 57) and analysis of the kinetics of tryptophan formation from indoleglycerol phosphate in a single turnover with enzyme in excess over the radiolabeled IGP (32). As shown in Fig. 12, the formation of tryptophan occurs without significant appearance of indole from IGP. This result implies that indole must diffuse to the β-site and react quite rapidly, with an overall rate of least 1000 sec^{-1}. Measurement of the rate of reaction of indole added from solution established a rate of only 40 sec^{-1} under conditions identical to the reaction with IGP. Computer simulation established that if the indole had to dissociate from the α-site, diffuse through solution, and rebind to the β-site, then a large fraction of indole should have accumulated in the single-turnover experiment, as shown by the dashed line in Fig. 12. Thus, the kinetic measurements establish that the indole must have passed through the tunnel to account for the exceedingly fast reaction rate and the negligible accumulation of indole (32).

Further analysis of the reaction kinetics showed that there is a communication from the β-site to the α-site (32). In the absence of serine, the α-site is relatively inactive in catalyzing the cleavage of IGP (0.16 sec^{-1}). The reaction of serine with pyridoxal phosphate at the β-site to form the highly reactive aminoacrylate

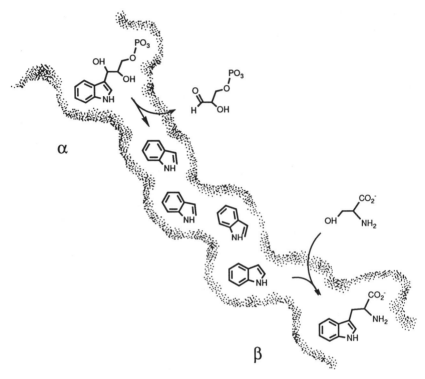

Fɪɢ. 11. Channeling in the tryptophan synthase reaction. Indoleglycerol 3-phosphate is cleaved to indole and glyceraldehyde 3-phosphate at the α-site. Indole is then passed through the hydrophobic channel to the β-subunit where it reacts with serine to form tryptophan. This schematic was drawn from data presented in (55).

leads to a 200-fold activation of the rate of IGP cleavage at a distance 25 Å away at the α-site. When serine and IGP are added simultaneously to the enzyme, there is a lag in the kinetics of IGP cleavage and tryptophan formation, which parallels the kinetics of formation of the aminoacrylate. When serine is pre-incubated with the enzyme, the lag is no longer seen when the reaction is initiated by adding IGP. Thus, direct analysis of the reaction kinetics revealed novel features of substrate channeling. The channeling of indole from the α-site to the β-site is efficient for two reasons. First, the intersubunit communication precludes the formation of indole in the absence of serine and serves to keep the reactions at distant sites in phase. Second, the passage of the indole through the tunnel and its reaction with the aminoacrylate to form tryptophan are much faster than the rate of release of indole from the α-site. The application of single-

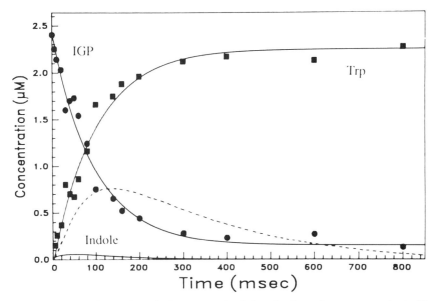

FIG. 12. Tryptophan synthase single turnover. A solution of serine and enzyme was mixed with indoleglyerol phosphate (IGP) to initiate the reaction. The disappearance and formation of IGP (●) and tryptophan (Trp; ■) were monitored. The reaction was initiated by mixing enzyme (20 μM) and serine (10 mM) and radiolabeled IGP (2.4 μM). The curves were calculated by computer simulation using the full kinetic pathway (32). For the dashed line, the curves were simulated using a rate of indole binding equal to 20 μM^{-1} sec^{-1} (40 sec^{-1} at 20 μM enzyme), predicting a substantial accumulation of indole. Reproduced with permission from (32).

turnover kinetic methods to other enzyme systems thought to exhibit channeling should reveal whether these properties are of general importance in governing channeling of metabolic intermediates.

F. KINETIC PARTITIONING

The kinetic partitioning of enzyme intermediates is an important principle, and the rules governing kinetic partitioning are quite simple. The fractional yield of a given reaction is given simply as the rate of the desired reaction divided by the sum of the rates of all reactions involving the intermediate. For example, consider the forked reaction pathway in Scheme XXI.

$$\text{E·S} \underset{k_{-1}}{\overset{k_1}{\rightleftharpoons}} \text{E·I} \overset{k_2}{\underset{k_3}{\substack{\nearrow \text{E + P} \\ \searrow \text{E + Q}}}}$$

SCHEME XXI

According to this mechanism, the fractional yield of the intermediate to form P is given by the ratio $k_2/(k_2 + k_{-1} + k_3)$. The importance of kinetic partitioning will be illustrated by two examples.

1. Substrate Trapping Experiments

For a sequential bisubstrate enzyme, the rate of dissociation of the first substrate can be estimated by substrate trapping methods. The rationale for this experimental approach is shown in Scheme XXII. The enzyme is first preincubated with radiolabeled substrate A* and is then mixed with an excess of unlabeled substrate A and substrate B to initiate the reaction. The recovery of radiolabeled product is a function of the kinetic partitioning of the enzyme-bound substrate between dissociation to yield free S and forward reaction with substrate B to yield product P.

$$
\begin{array}{c}
\text{E·A*} \underset{k_{-2}}{\overset{k_2[\text{B}]}{\rightleftharpoons}} \text{E·A*·B} \rightarrow \text{E·Q*·P} \rightarrow \text{E} + \text{Q*} + \text{P} \\
k_{-1} \downarrow \\
\text{E} + \text{A*}
\end{array}
$$

SCHEME XXII

Quantitative analysis of the reaction is based on examining the effect of increasing concentrations of B on the recovery of radiolabeled product. In the limit, extrapolating to infinite concentration of B, one expects 100% conversion of the enzyme-bound radiolabeled substrate to product. Recoveries less than 100% have been attributed to dissociation of A* from the ternary E·A*·B complex, nonproductive binding of A* in the E·A* complex, or an appreciable fraction of dead enzyme. In addition, the analysis is dependent on an accurate knowledge of the equilibrium constant for the binding of the substrate A to the enzyme and of the concentration of *active* enzyme sites. Independent of these concerns, one can estimate the rate of dissociation of A from the enzyme by measurement of the concentration of B required to trap half of the maximal amount of radiolabeled A*. At this concentration, the rate dissociation of A* from the E·A* complex is equal to the rate of binding of B. Such analysis was used to estimate the rates of dissociation of S3P and EPSP from the enzyme EPSP synthase (3).

2. DNA Polymerase Error Correction

DNA polymerases contain two active sites, a polymerase site that catalyzes the reversible, template-directed elongation of DNA, and the exonuclease site, which catalyzes the hydrolysis of the 3′-terminal base. During processive synthesis, DNA replication occurs by the consecutive addition of bases without the intervening dissociation of the DNA, and the DNA can migrate between the polymerase site and the exonuclease site without dissociating from the enzyme (12). After the addition of each base pair, the enzyme–DNA complex partitions

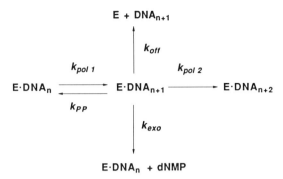

FIG. 13. DNA polymerase kinetic partitioning. The pathway shows the potential reactions during DNA replication: polymerization (k_{pol}), dissociation of the E–DNA complex (k_{off}), pyrophosphorolysis (k_{pp}), and exonuclease digestion of the 3'-terminal base (k_{exo}). Reproduced with permission from (11).

according to the kinetics of each of the optional pathways involving (1) continued polymerization, (2) reversal of polymerization by pyrophosphorolysis, (3) migration of the DNA to the exonuclease site, or (4) dissociation of the DNA from the enzyme, as illustrated in Fig. 13. This kinetic partitioning governs the fidelity and efficiency of DNA replication. After the incorporation of a correct base, the probability of continued incorporation is defined by the rate of polymerization divided by the sum of all other rates. According to the kinetic constants established for T7 DNA polymerase summarized in Table I (10–12), the probability of extension is quite high, with dissociation only once out of 1500 bases, and the exonuclease removes less than 0.1% of the correct bases. After an incorporation of the wrong base, the kinetic partitioning changes dramatically so that

TABLE I

DNA POLYMERASE KINETIC CONSTANTS[a]

Rate constant	Correct base (sec^{-1})	Incorrect base (sec^{-1})
$k_{pol\,1}$	300	0.002
$k_{pol\,2}$	300	0.013
k_{PP}	1	<0.0001
k_{off}	0.2	0.4
k_{exo}	0.2	2.8

[a] The kinetic constants governing the partitioning of the enzyme–DNA complex during normal polymerization (Correct) and with a mismatched base pair (Incorrect) are summarized (Ref). The rate constants are defined according to the pathway given in Fig. 13.

most of the mismatches are removed rather than extended, because the polymerase rate becomes very slow and the exonuclease rate is accelerated. This kinetic partitioning accounts for the high fidelity of replication catalyzed by T7 DNA polymerase and establishes the basis of the selectivity of the exonuclease in correcting errors.

G.   ISOTOPE AND pH EFFECTS

The effect of pH variation and isotope (or elemental) substitution on reaction kinetics has been used in the steady state to explore the roles of active site acid/base catalysts and to attempt to define the nature of the transition state (*8a, 8b, 58*). Each of these methods also depends on the extent to which the rate of the chemical reaction is rate limiting in the steady state. If some other step limits the rate of steady-state turnover, then changes in the rate of the chemical reaction will be obscured. Use of pH variation or isotope effects in transient kinetic experiments has been useful in a number of cases (*27*), especially where it has been possible to examine directly the rate of the chemical reaction at the enzyme active site. In these cases, the effect of pH or isotope substitution can be interpreted directly in terms of the effect on a single reaction.

The use of pH variation and isotope effects in transient kinetics can be illustrated with a recent study on dihydrofolate reductase. Analysis by steady-state methods had indicated an apparent pK_a of 8.5 that was assigned to an active site aspartate residue required to stabilize the protonated state of the substrate (*59*). In addition, it was shown that there was an isotope effect on substitution of NADPD (the deuterated analog) for NADPH at high pH but not at low pH, below the apparent pK_a. This somewhat puzzling finding was explained by transient-state kinetic analysis. Hydride transfer, the chemical reaction converting enzyme-bound NADPH and dihydrofolate to NAD^+ and tetrahydrofolate, was shown to occur at a rate of approximately 1000 sec^{-1} at low pH. The rate of reaction decreased with increasing pH with a pK_a of 6.5, a value more in line with expectations for an active site aspartate residue. As shown in Fig. 14, there was a threefold reduction in the rate of the chemical reaction with NADPD relative to NADPH. Thus direct measurement of the chemical reaction revealed the full isotope effect.

In the steady state, the release of products is rate limiting (14 sec^{-1}) and largely independent of pH. Figure 15 combines the pH dependence of the chemical reaction, product release, and steady-state turnover. At low pH, the chemical reaction is fast and product release limits the steady-state rate so there is no elemental effect observed in the steady state. As the pH is increased, there is a crossover in the identity of the rate-limiting step, so that at high pH the chemical reaction becomes rate limiting for steady-state turnover, leading to an observable isotope effect. The apparent pK_a of 8.5 in the steady state represents

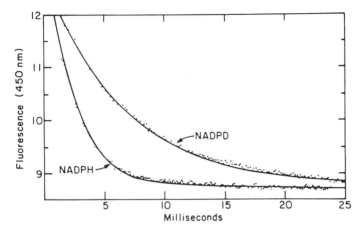

FIG. 14. Isotope effect on the rate of hydride transfer. The rate of hydride transfer to dihydrofo-late catalyzed by dihydrofolate reductase (15 μM) was measured by fluorescence energy transfer, exciting the protein at 280 nm and observing emission by NADPH at 450 nm. The reaction with NADPH occurred at a rate of 450 sec^{-1}, followed by a linear phase at 12 sec^{-1}, as shown by the smooth line. The rate of the burst observed with NADPD, the deuterium analog, occurred at 150 sec^{-1}. Reproduced with permission from (27).

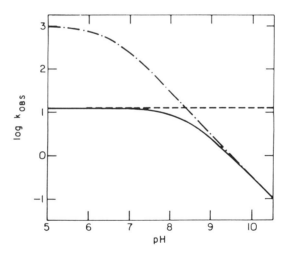

FIG. 15. The pH dependence of a reaction catalyzed by dihydrofolate reductase. The observed rate of hydride transfer (–·–) is compared with the rate of product release (---) and k_{cat}(—) on a log scale as a function of pH. The break in the rate of steady-state turnover at pH 8.5 is due to a change in the rate-limiting step from product release to hydride transfer. Reproduced with permission from (27).

the crossover point in switching from rate-limiting product release to rate-limiting chemistry.

This study has important lessons for enzyme kinetic analysis. The use of pH variation and examination of isotope effects can be a powerful combination to explore the chemistry of enzyme-catalyzed reactions and to dissect the contributions of individual reaction steps to the net steady-state turnover (27). Examination of the effects of pH on each step of the reaction pathway could resolve the contributions of ionizable groups toward ground-state binding energy and transition-state stabilization. The use of isotope effects by transient-state kinetic methods is more limited than in the steady state due to the errors involved in comparing two rate measurements. In the steady state, the ratio method has allowed isotope effects of less than 1% to be measured accurately (8a, 58). By transient-state kinetics, one would require at least a 10–20% change in rate to demonstrate a convincing difference between two rate measurements in most instances.

The use of a rate effect on the chemical reaction due to an elemental or isotope substitution has been used to test the extent to which an observed rate is a function of the chemical reaction. For example, in recent studies on DNA polymerase, use of dTTP(αS) in place of dTTP in a single-turnover experiment has provided evidence for a rate-limiting conformational change that precedes a much faster chemical reaction. Substitution of dTTP(αS) for dTTP should have resulted in a 100-fold reduction in the rate of reaction if chemistry were rate limiting (10, 11). The small (threefold) effect provided evidence to suggest that chemistry was not rate limiting, arguing indirectly for a protein conformational change involved in recognition of the correct base pair. Subsequent analysis has questioned the expectation of a 100-fold reduction in rate with the thio analog. The original estimation of a sulfur elemental effect by Benkovic and Shray (60) was based on the hydrolysis of phosphate triesters, while a recent study by Hershlag et al. (61) has indicated only a 4- to 10-fold sulfur effect for phosphate diesters. A more complete kinetic analysis of DNA polymerase mechanisms has substantiated the evidence for a conformational change following substrate binding, based on pulse-chase experiments (10, 11) and the observation of a protein fluorescence change by stopped-flow (62). Thus, the thio analog work may have given the right answer for the wrong reason.

VI. Methods of Data Fitting

The point of a transient kinetic experiment is to establish the reaction pathway by examination of the concentration dependence of the rate of reaction. It is the goal of proper data analysis to ferret out the best fit to the data to establish the reaction mechanism containing the minimal number of steps. I will describe two approaches to data analysis: first, the conventional fitting of data to equations derived by integration of simple rate equations, and second, by computer simulation based on numerical integration of the equations describing the chemical reaction mechanism. In practice, both methods should be applied, with initial analysis by conventional data fitting, followed by refinement and convergence to a complete kinetic description of a reaction pathway by computer simulation. The standards set for this analysis now require the fitting of all data directly to a complete mechanism with no simplifying assumptions (*2, 3, 10–12*).

A. CONVENTIONAL DATA FITTING

The conventional data analysis involves the fitting of data to an equation describing the time dependence of the reaction, leading to the best estimates for the constants defining the equations. Analytical solutions to most simple reaction sequences can be obtained (*1, 5, 63*). Solutions of differential equations describing the series of first-order (or pseudo-first-order) reactions will always be a sum of exponential terms [Eq. (22)]. Thus for a single exponential, the fitting process provides the amplitude (A), the rate of reaction (λ), and the end point (C)

$$Y = Ae^{-\lambda t} + C$$

A proper fitting must include the minimization of the sum square error between the data and the calculated curve, allowing all three constants to float in converging to the optimal fit by nonlinear regression methods involving an iterative approach to the best fit. There are numerous reasons why the older methods involving linearization of the equations in a logarithmic form should not be used: (1) Logarithmic plots compress the most accurate data collected during the early stages of the reaction; (2) a value for the reaction endpoint must be assumed and errors in estimating the endpoint are propagated to alter the calculated rate; and (3) experimental values less than the endpoint cannot be included on a logarithmic plot, thus eliminating data collected late in the reaction. Modern computational methods based on nonlinear regression solve all of these problems by estimating the endpoint as a fitted parameter and by allowing the inclusion of data spanning six half-lives of the reaction (98% completion) even with a signal-to-noise ratio of 1.

For more complex kinetics, the data can be fit to an equation including two or more exponential terms. A double-exponential equation is defined by

$$Y = Ae^{-\lambda_1 t} + Be^{-\lambda_2 t} + C$$

The distinction between a single- and double-exponential fit is sometimes obvious, but in other cases, in which the difference in rates of reaction between the two phases is small (less than fourfold), it can be difficult to resolve the two rates. In these cases, the evaluation of the goodness of fit and the justification for including a second exponential term cannot be based solely on a reduction in the sum square error. Visual evaluation of the fit is required, looking for systematic deviation of the fit from the data. The final fitting of the data should include computer simulation to ensure that the model predicts the observed concentration dependence of the rates and amplitudes of each phase and to justify the data fitting in terms of the reaction mechanism.

Presteady-state burst kinetics can be fit to an equation of the form

$$Y = A(1 - e^{-\lambda t}) + k_2 t$$

Fitting to this equation provides the rate and amplitude of the burst followed by the steady-state turnover rate during the linear phase. These rates and amplitudes can then be quantitatively evaluated in terms of the pathway and kinetic constants for the reaction as described above.

B.   COMPUTER SIMULATION OF COMPLEX KINETICS

Computer simulation has become increasingly important because it allows for analysis without any simplifying assumptions and includes amplitude information in the fitting process. Although any method can lead to errors in interpretation, computer simulation has been viewed with suspicion because the errors are made at the beginning of the analysis in writing the mechanism and choosing the best rate constants and it is difficult to document all of the logic that led to the convergence of the best fit to define the minimal mechanism. Nonetheless, the stringency with which a given model is fit to kinetic data is considerably higher by use of computer simulation because of the need to account for the concentration dependence of both the rate and amplitude of a reaction. Moreover, while there may be arguments as to whether a given mechanism can quantitatively account for a given set of kinetic data, there can be no such arguments with the proper use of computer simulation. The potential errors in conventional analysis of kinetic data arise from the simplifying assumptions that must be made to derive the appropriate equations; however, with computer simulation, there are no assumptions. The problem then becomes one of establishing a minimal kinetic scheme and quantitatively eliminating other possible mechanisms.

At first, the fitting of data to a reaction sequence may seem impossible, since not enough information is in hand to model the complete reaction. However, the fitting process should be based on identification of those rate constants that are defined by each experiment (*3, 39*). An initial fit to a given experiment to extract the critical rate constants can be based on a model with crude estimates for rate constants that do not affect greatly the fitting of that experiment. The results of fitting each experiment then provide the basis for fitting other experiments to define other rate constants. One can then return to the fitting of the first experiment based upon refined estimates of rate constants for each step in the pathway. With a bit of luck, enough experiments can be performed to constrain the fitting of the data to converge to a single solution. The computer program KINSIM has now been extended to allow the fitting of reaction kinetics by nonlinear regression analysis based upon simulation (64). This analysis provides an estimate of the confidence limits on each kinetic parameter which allows the investigator to recognize when a given model is not sufficiently supported and constrained by the data.

The goal of a complete kinetic analysis is to define the rate and free energy change of each step in the reaction. Because the rates of each reaction in an enzymic pathway are comparable, the measurable events are kinetically linked and sometimes difficult to separate. Therefore, solution of an enzyme mechanism must include a fitting of all experiments to the complete model, including all steps in the pathway. Ideally one should measure each reaction in a sequence and then provide one additional measurement as a check for internal consistency. The two important checks on an enzyme reaction sequence are (1) measurement of the overall free energy change for the reaction in solution and (2) comparison of the predicted and measured steady-state kinetic constants.

VII. Résumé

In this chapter, I have attempted to provide a rationale for the design and interpretation of transient kinetic experiments to establish enzyme pathways by direct analysis of individual reaction steps. In many ways, transient kinetics are straightforward and experimental results can be understood in terms of simple principles derived from first-order kinetics. The application of these methods to enzyme kinetics was illustrated by a number of examples, and interested readers are encouraged to examine the original research publications to obtain a deeper understanding of the reaction kinetics. This is especially useful for those cases in which the kinetic data could be fit to a single, complete model, the most notable being EPSP synthase (3), DNA polymerase (10–12), and tryptophan synthase (32). The examples also served to strengthen the arguments in favor of computer simulation to analyze kinetic data. In many cases, the experiments could not have been interpreted quantitatively by any other means, but even in seemingly simple cases, the refinements resulting from computer simulation are significant. The most important danger in the interpretation of kinetic data stems from the tendency to include an additional reaction step to explain an unusual kinetic result, when often the new result could have been accounted for by a more simple model based on a deeper understanding of the subtleties of the kinetics. Thus, the overriding rule remains: Do not include a step in the mechanism unless there is direct evidence for it or the fitting of the data absolutely requires it. Although computer simulation has been viewed with some caution because it is perhaps too easy to include an unnecessary step, it should also be recognized that computer simulation also permits the quantitative evaluation of the most simple models in fitting kinetic data.

The application of transient kinetic methods to the solution of enzyme mechanisms has increased dramatically due to recent advances in instrumentation and in the overexpression and purification of new enzymes. Transient kinetics are becoming the method of choice for evaluation of site-directed enzyme mutants and for detailed questions regarding the relationships between protein structure and observable function. In conjunction with advances in methods of structural and genetic analyses, transient-state kinetic analysis forms the basis for what might be called the "new enzymology."

ACKNOWLEDGMENTS

I would like to thank Karen S. Anderson (Yale University, New Haven, CT) for her careful reading of this manuscript.

REFERENCES

1. Johnson, K. A. (1986). *In* "Methods in Enzymology" (R. B. Vallee, ed.), Vol. 134, pp. 677–705. Academic Press, New York.
2. Barshop, B. A., Wrenn, R. F., and Frieden, C. (1983). *Anal. Biol.* **130**, 134.
3. Anderson, K. S., Sikorski, J. A., and Johnson, K. A. (1988). *Biochemistry* **27**, 7395.
4. Hammes, G. (1970). "The Enzymes," 3rd Ed., Vol. 2, pp. 65–114.
5. Fersht, A. R. (1985). "Enzyme Structure and Mechanism," 2nd Ed., Freeman, San Francisco, California.
6. Cleland, W. W. (1970). "The Enzymes," 3rd Ed., Vol. 2, pp. 1–64.
7. Cleland, W. W. (1991). "The Enzymes," 3rd Ed., Vol. 19, pp. 99–158.
8. Johnson, K. A., and Benkovic, S. J. (1991). "The Enzymes," 3rd Ed., Vol. 19, pp. 159–211.
8a. Cleland, W. W. (1982). *In* "Methods in Enzymology" (D. L. Purich, ed.), Vol. 87, pp. 625–626. Academic Press, New York.
8b. Cleland, W. W. (1982). *In* "Methods in Enzymology" (D. L. Purich, ed.), Vol. 87, pp. 390–405. Academic Press, New York.
9. Fersht, A. R. (1987). *Biochemistry* **26**, 8031.
10. Patel, S. S., Wong, I., and Johnson, K. A. (1991). *Biochemistry* **30**, 511.
11. Wong, I., Patel, S. S., and Johnson, K. A. (1991). *Biochemistry* **30**, 526.
12. Donlin, M. J., Patel, S. S., and Johnson, K. A. (1991). *Biochemistry* **30**, 538.
13. Del Rosario, E. J., and Hammes, G. G. (1971). *Biochemistry* **10**, 716.
14. Hurst, J. K., and Hammes, G. G. (1969). *Biochemistry* **8**, 1083.
15. Anderson, K. S., Sikorski, J. A., Benesi, A. J., and Johnson, K. A. (1988). *J. Am. Chem. Soc.* **110**, 6577.
16. Anderson, K. S., Sammons, R. D., Sikorski, J. A., Leo, G. E., Benesi, A. J., and Johnson, K. A. (1990). *Biochemistry* **29**, 1460.
17. Barlow, P. N., Appleyard, R. J., Wilson, B. J. O., and Evans, J. N. S. (1989). *Biochemistry* **28**, 7985.
18. Albery, W. J., and Knowles, J. R. (1976). *Biochemistry* **15**, 5631.
19. Burbaum, J. J., Raines, R. T., Albery, W. J., and Knowles, J. R. (1989). *Biochemistry* **28**, 9293.
20. Burbaum, J. J., and Knowles, J. R. (1989). *Biochemistry* **28**, 9306.

21. Andrawis, A., Johnson, K. A., and Tien, M. (1987). *J. Biol. Chem.* **263**, 1195–1198.
22. Cai, D., and Tien, M. (1990 *Biochemistry* **29**, 2085.
23. Drewe, W. F., and Dunn, M. F. (1985). *Biochemistry* **24**, 3977.
24. Drewe, W. F., and Dunn, M. F. (1986). *Biochemistry* **25**, 2494.
25. Lane, A., and Kirschner, K. (1983a). *Eur. J. Biochem.* **129**, 561.
26. Lane, A., and Kirschner, K. (1983b). *J. Biochem. (Tokyo)* **129**, 571.
27. Fierke, C. A., Johnson, K. A., and Benkovic, S. J. (1987). *Biochemistry* **26**, 4085.
28. Johnson, K. A., and Taylor, E. W. (1978). *Biochemistry* **17**, 3432.
29. Holbrook, J. J., Liljas, A., Steindel, S. J., and Rossmann, M. G. (1975). "The Enzymes," 3rd Ed., Vol. 2, pp. 191–268.
30. Rosenfeld, S. S., and Taylor, E. W. (1984). *J. Biol. Chem.* **259**, 11920.
30a. Woodward, S. K. A., Eccleston, J. F., and Geeves, M. A. (1991). *Biochemistry* **30**, 422.
31. Anderson, K. S., and Johnson, K. A. (1990). *Chem. Rev.* **90**, 1131.
32. Anderson, K. S., Miles, E. W., and Johnson, K. A. (1991). *J. Biol. Chem.* **266**, 8020.
33. Anderson, K. S., Sikorski, J. A., and Johnson, K. A. (1988). *Biochemistry* **27**, 1604.
34. Wells, T. N. C., and Fersht, A. R. (1986). *Biochemistry* **25**, 1881.
35. Fersht, A. R., Leatherbarrow, R. J., and Wells, T. N. C. (1987). *Biochemistry* **26**, 6030.
36. Fersht, A. R., Knill-Jones, J. W., Bedouelle, H., and Winter, G. (1988). *Biochemistry* **27**, 1581.
37. Johnson, K. A. (1985). *Annu. Rev. Biophys. Biophys. Chem.* **14**, 161.
38. Gutfreund, H. (1972). "Enzymes: Physical Principles." Wiley (Interscience), New York.
39. Johnson, K. A. (1983). *J. Biol. Chem.* **258**, 13825.
40. Remington, S., Wiegand, G., and Huber, R. (1982). *J. Mol. Biol.* **158**, 111.
41. Hackney, D. D., Malik, A. S., and Wright, K. W. (1989). *J. Biol. Chem.* **264**, 15943.
42. Hartley, B. S., and Kilby, B. A. (1954). *Biochem. J.* **56**, 288.
43. Gutfreund, H., and Hammond, B. R. (1959). *Biochem. J.* **59**, 526.
44. Anderson, K. S., and Johnson, K. A. (1990). *J. Biol. Chem.* **265**, 5567.
45. Srere, P. A. (1987). *Annu. Rev. Biochem.* **56**, 89.
46. Welch, G. R. (1977). *Prog. Biophys. Mol. Biol.* **32**, 103.
47. Keleti, T., and Ovadi, J. (1988). *Curr. Top. Cell. Regul.* p. 1.
48. Weber, J. P., and Bernhard, S. A. (1982). *Biochemistry* **21**, 4189.
49. Srivastava, D. K., and Bernhard, S. A. (1985). *Biochemistry* **24**, 623.
50. Srivastava, D., and Bernhard, S. A. (1986). *Science* **234**, 1081.
51. Chock, B. P., and Gutfreund, H. (1988). *Proc. Natl. Acad. Sci. U.S.A.* **85**, 8870.
52. Kvassman, J., and Petterson, G. (1989). *Eur. J. Biochem.* **186**, 261.
53. Kvassman, J., and Petterson, G. (1989). *Eur. J. Biochem.* **186**, 265.
54. Srivastava, D. K., and Bernhard, S. A. (1987). *Biochemistry* **26**, 1240.
55. Hyde, C. C., Ahmed, S. A., Padlan, E. A., Miles, E. W., and Davies, D. R. (1988). *J. Biol. Chem.* **263**, 17857.
56. Dunn, M. F., Aguilar, V., Brzovic, P., Drewe, W. F., Houben, K. F., Leja, C. A., and Roy, M. (1990). *Biochemistry* **29**, 8598.
57. Houben, K. F., and Dunn, M. F. (1990). *Biochemistry* **29**, 2421.
58. Cook, P. F., and Cleland, W. W. (1981). *Biochemistry* **20**, 1797.
59. Stone, S., and Morrison, J. F. (1984). *Biochemistry* **23**, 2757.
60. Benkovic, S. J., and Schray, K. (1973). "The Enzymes," 3rd Ed., Vol. 8, pp. 201–236.
61. Hershlag, D., Piccirilli, J. A., and Cech, T. R. (1991). *Biochemistry* **30**, 4844.
62. Patel, S., Wong, I., and Johnson, K. A. (1991). Unpublished results.
63. Capellos, C., Bielski, B. H. J. (1972). "Kinetic Systems: Mathematical Description of Chemical Kinetics in Solution" (Wiley-Interscience).
64. Zimmerle, C. J., and Frieden, C. (1989). *Biochemistry J.* **258**, 381–387.

2

Metal Ions at Enzyme Active Sites

JOSEPH J. VILLAFRANCA* • THOMAS NOWAK†

*Department of Chemistry
Pennsylvania State University
University Park, Pennsylvania 16802

†Department of Chemistry and Biochemistry
University of Notre Dame
Notre Dame, Indiana 46556

I. General Considerations

It is estimated that approximately one-third of all enzyme-catalyzed reactions require a metal ion or ions for catalytic activity. The functions of these metal ions can be placed in three broad categories: (1) structural integrity (no specific catalytic function), (2) electron transfer reactions, and (3) electrophilic catalysis. The latter of these broad classes of function/structure will be the focus of this chapter, in which we will explore current data and theories of metalloenzyme catalysis.

THE ENZYMES, Vol. XX

An exhaustive review of all of the types of reactions that are catalyzed by metal-requiring enzymes and the specific functions of these metals, as currently understood, is beyond the scope of this chapter. To complicate this general area of investigation, even within a single group of enzymes, the metal ions may play different roles in the catalytic processes for reaction-related enzymes. Not all enzymes of a specific class necessarily require a cation for activity. In some cases, the roles of the cation may be substituted by specific amino acid residues in the protein. A classic example of such a case is the muscle and yeast fructose-bisphosphate aldolases. The muscle enzyme catalyzes the aldol condensation using a Schiff base intermediate to activate the substrate, whereas the yeast enzyme is a Zn^{2+}-metalloenzyme (1). The cation appears to serve as the electrophile in the activation of the substrate for the same reaction.

Another aspect of catalysis at metal ion sites on enzymes is the selectivity exhibited by some enzymes for the type of cation (monovalent versus divalent) and/or the oxidation state ($+2$ versus $+3$) of the cation. Model studies of metal ion-assisted reactions and of metal–ligand structures will be reviewed to establish trends for these types of reactions and features that direct metal ion preference for certain ligand environments.

II. Properties of Metal Ion–Ligand Complexes

To begin to understand the electrophilic nature of metal ion-assisted catalysis, the properties of groups of metal ions in the periodic table must first be appreciated. The relationship between size (and charge when appropriate) of metal ions, in part, depends on the atomic number of the element, on the most stable electronic configuration of the ion, and finally on the nature of electron–ligand interactions when the outer shell contains partially filled orbitals (transition ion series) of defined geometry. For Group IA, IIA, and IIIA metal ions, the most stable electronic configuration (Nobel gas configuration) dictates that these ions are monovalent, divalent, and trivalent, respectively (Table I). All ions have filled outer shells and possess spherical symmetry. This is also true for the trivalent lanthanide ion series in which the outlying orbitals are filled $5s$ and $5p$ orbitals, and the partially filled $4f$ orbitals are "contracted" and shielded from interaction with ligand atoms. With ions that have spherical symmetry, ligation number and geometry depend on size of ligands, whereas in the transition metal ion series, geometry and number are imposed by metal ion–ligand orbital overlap. The latter effect produces covalent interactions and introduces "classes" of ligands (axial versus equatorial) that dictate the chemistry available at each ligand position. The examples chosen for discussion below describe structural consequences of metal ion selectivity in peptides and proteins for two important

TABLE I

EFFECTIVE IONIC RADII OF SELECTED METAL IONS[a]

Group IA Radius (Å)	Group IIA Radius (Å)	Group IIIA Radius (Å)	Lanthanides Radius (Å)
Li$^+$ 0.76	Mg^{2+} 0.72	Sc^{3+} 0.81	Nd^{3+} 1.05
Na$^+$ 1.02	Ca^{2+} 1.00	Y^{3+} 0.96	Eu^{3+} 1.01
K$^+$ 1.38	Sr^{2+} 1.18	La^{3+} 1.10	Gd^{3+} 1.00
Rb$^+$ 1.52	Ba^{2+} 1.35		Tb^{3+} 0.98

[a] The effective ionic radii for Groups IA and IIA are for six-coordinate complexes, whereas the radii for Group IIIA and the selected lanthanide ions are for seven-coordinate complexes. Data are taken from Shannon (6).

ions, Ca^{2+} and Mg^{2+}, and the results can directly be translated to enzymes for which cation selectivity directly affects catalytic function.

Several proteins show a high selectivity for Ca^{2+} over Mg^{2+} that is especially important because intracellular concentrations of Mg^{2+} are often much higher than Ca^{2+}. A peptide model has been studied and experimental and calculated binding constants and stoichiometries have been compared (2). The peptide is cyclo(-L-Pro-Gly-)$_3$ (PG$_3$) and the crystal structures show that Mg^{2+} is octahedrally coordinated to PG$_3$ in a 1:1 complex, whereas Ca^{2+} is sandwiched between two PG$_3$ molecules in a 1:2 complex. The crystal ionic radii of the two ions provide a starting point for analysis. The smaller radius of Mg^{2+} (0.75 Å) compared to Ca^{2+} (1.15 Å) provides some steric constraints for packing two PG$_3$ molecules around Mg^{2+}. The largest driving force for complex formation is the greater preference for Mg^{2+} over Ca^{2+} to be in water (by 68 kcal/mol) than in the peptide (or protein) complex. This varies whether it is the 1:1 or 1:2 complex (see Ref. 2). The free energy difference for "selectivity" favors Ca^{2+} in the 1:2 complex by -36 kcal/mol over Mg^{2+}, whereas in the 1:1 complex Mg^{2+} is favored by 8 kcal/mol over Ca^{2+}. In a theoretical treatment, Rashin and Hönig (3) have recalculated solvation energies of cations and anions in water and favor the use of covalent radii of ions in estimating the cavity size of binding pockets in complexes. This may be of critical importance in evaluating metal ion binding and thus function in metalloproteins.

An excellent study of the effect of size and charge on metal ion binding to proteins was reported by Falke and co-workers (4) regarding a protein that has one Ca^{2+} binding site and for which the crystal structure is known (5). All seven ligands are protein oxygen atoms that are provided by side chains and backbone groups (Fig. 1). Three side-chain carboxyl groups provide a net -3 charge arising from two monodentate (Asp-134 and Asp-138) and one bidentate (Glu-205)

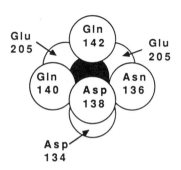

FIG. 1. Representation of the heptagonal array of oxygen ligands around the Ca^{2+} binding site of the galactose-binding protein. The data are from the crystal structure reported in Ref. 5. The dark sphere is the Ca^{2+} and the ligands are monodentate Asp-134 and Asp-138, bidentate Glu-205, the backbone carbonyl of Gln-140, and side-chain carbonyls of Asn-136 and Gln-142.

coordination schemes, while the other three oxygens are from one backbone carbonyl and two side-chain carbonyls (Asn-136 and Gln-142). Spherical ions from Groups IA, IIA, and IIIA and lanthanides were studied (Table II). Monovalent cations (Group IA) uniformly showed weak binding while divalent ions from Group IIA exhibited affinities related to size, with optimal binding between Mg^{2+} (0.81 Å) and Ca^{2+} (1.06 Å). Note that the effective ionic radii reported by different groups differ slightly depending on the source of information quoted. The trends are the same as long as one uses only a single source book for each set of comparisons (6). For trivalent ions (Group IIIA and lanthanides), the optimal size for best binding affinity was between Sc^{3+} (0.81 Å) and Yb^{3+} (0.925 Å). Thus, the metal ion site on the protein selects for highly charged ions of a definite size and demonstrates how ligand arrangement and ligand charge select for the physiologically less abundant ion (Ca^{2+}) over more abundant ions (Mg^{2+}, Na^+, and K^+).

This study was extended to investigate the subtleties of cavity size and charge by mutagenesis of the axial Gln-142 side chain, which provides carbonyl oxygen coordination. When the cavity size was enlarged by shortening the side chain (to Asn or Asp), larger cations bound more tightly by up to 50-fold (Table II). With a change in charge of the side chain (to Glu or Asp), the protein showed a preference for trivalent over divalent ions up to 1900-fold higher than native protein. Thus, selectivity can be altered in a predictable manner and demonstrates that nature has solved this biological specificity problem by engineering ion cavities in proteins to match exactly the ion of preference for the function intended and to exclude more abundant metal ions that may be in the cellular environment.

For a metal ion at a catalytic site of an enzyme, the electrophilic behavior of the ion will depend on the nature of the ligands and to some extent on the ge-

TABLE II

METAL ION DISSOCIATION CONSTANTS OF SITES WT, Q142N, Q142E, AND Q142D[a]

Metal	Effective ionic radius (Å)	K_D (M)			
		WT	Q142N	Q142E	Q142D
Group IA					
Li+	0.84	≥5.9	≥1.6	≥90	≥26
Na+	1.12	≥2.7	≥13	≥26	≥4
K+	1.46	≥3.9	≥1.8	≥4.3	≥13
Rb+	1.56	≥3.3	≥16	≥90	≥2
Group IIA					
Mg^{2+}	0.81	$570 \pm 120 \times 10^{-3}$	≥7	≥1.1	≥1.6
Ca^{2+}	1.06	$13 \pm 2 \times 10^{-6}$	$15 \pm 2 \times 10^{-6}$	$0.18 \pm .2$	$2.8 \pm 0.1 \times 10^{-3}$
Sr^{2+}	1.21	$6.1 \pm 1.2 \times 10^{-3}$	$2.5 \pm 6 \times 10^{-3}$	≥0.33	$28. \pm 2. \times 10^{-3}$
Ba^{2+}	1.38	$\geq 740 \times 10^{-3}$	≥0.33	≥0.76	≥0.35
Group IIIA					
Sc^{3+}	0.81	$170 \pm 30 \times 10^{-6}$	$170 \pm 40 \times 10^{-6}$	$250 \pm 25 \times 10^{-6}$	$150 \pm 10 \times 10^{-6}$
Y^{3+}	0.96	$74 \pm 12 \times 10^{-6}$	$75 \pm 7 \times 10^{-6}$	$390 \pm 80 \times 10^{-6}$	$22 \pm 3 \times 10^{-6}$
La^{3+}	1.10	$1.9 \pm 0.2 \times 10^{-3}$	$400 \pm 20 \times 10^{-6}$	$1.1 \pm 50 \times 10^{-3}$	$380 \pm 10 \times 10^{-6}$
Lanthanides					
Ce^{3+}	1.07	$1.0 \pm 0.1 \times 10^{-3}$	$140 \pm 11 \times 10^{-6}$	$490 \pm 70 \times 10^{-6}$	$122 \pm 10 \times 10^{-6}$
Pr^{3+}	1.06	$730 \pm 80 \times 10^{-6}$	$100 \pm 8 \times 10^{-6}$	$230 \pm 40 \times 10^{-6}$	$54 \pm 8 \times 10^{-6}$
Nd^{3+}	1.05	$510 \pm 50 \times 10^{-6}$	$61 \pm 11 \times 10^{-6}$	$94 \pm 10 \times 10^{-6}$	$10 \pm 1 \times 10^{-6}$
Sm^{3+}	1.02	$400 \pm 30 \times 10^{-6}$	$77 \pm 5 \times 10^{-6}$	$110 \pm 14 \times 10^{-6}$	$18 \pm 3 \times 10^{-6}$
Eu^{3+}	1.01	$230 \pm 20 \times 10^{-6}$	$81 \pm 5 \times 10^{-6}$	$170 \pm 50 \times 10^{-6}$	$36 \pm 11 \times 10^{-6}$
Gd^{3+}	1.00	$270 \pm 15 \times 10^{-6}$	$85 \pm 7 \times 10^{-6}$	$270 \pm 40 \times 10^{-6}$	$34 \pm 8 \times 10^{-6}$
Tb^{3+}	0.98	$150 \pm 50 \times 10^{-6}$	$51 \pm 15 \times 10^{-6}$	$280 \pm 90 \times 10^{-6}$	$49 \pm 10 \times 10^{-6}$
Dy^{3+}	0.97	$20 \pm 1 \times 10^{-6}$	$17 \pm 2 \times 10^{-6}$	$280 \pm 60 \times 10^{-6}$	$14 \pm 4 \times 10^{-6}$
Ho^{3+}	0.96	$18 \pm 1 \times 10^{-6}$	$17 \pm 2 \times 10^{-6}$	$220 \pm 50 \times 10^{-6}$	$9 \pm 2 \times 10^{-6}$
Er^{3+}	0.945	$18 \pm 1 \times 10^{-6}$	$25 \pm 4 \times 10^{-6}$	$280 \pm 50 \times 10^{-6}$	$15 \pm 5 \times 10^{-6}$
Tm^{3+}	0.94	$45 \pm 10 \times 10^{-6}$	$18 \pm 4 \times 10^{-6}$	$410 \pm 80 \times 10^{-6}$	$9 \pm 3 \times 10^{-6}$
Yb^{3+}	0.925	$27 \pm 5 \times 10^{-6}$	$6 \pm 1 \times 10^{-6}$	$420 \pm 90 \times 10^{-6}$	$8 \pm 2 \times 10^{-6}$
Lu^{3+}	0.92	$12 \pm 2 \times 10^{-6}$	$9 \pm 1 \times 10^{-6}$	$500 \pm 10 \times 10^{-6}$	$10 \pm 3 \times 10^{-6}$

[a] Data from Falke et al. (4).

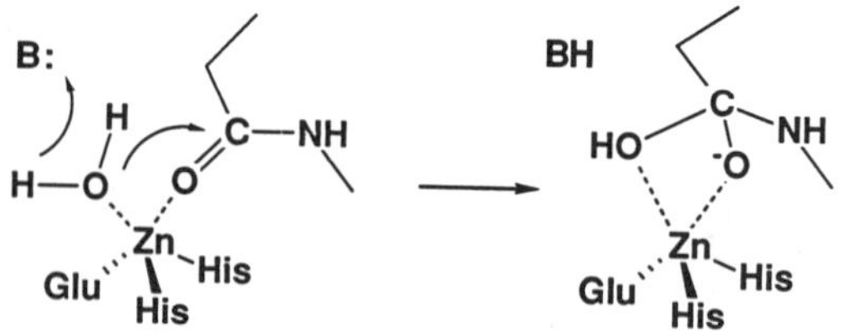

FIG. 2. Mechanism for a Zn^{2+}-peptidase. The water molecule is activated by having its pK_a lowered by coordinatic, ʼ to the Zn^{2+} (assisted by an enzymic base) and the negative charge on the tetrahedral intermediate is stabilized by coordination to the metal ion.

ometry of the site. In many enzymes, the metal ion has coordination sites that are occupied by protein ligands and water molecules in the absence of substrate. The water molecules are displaced or altered by the binding of substrate to produce the catalytically competent complex. As an example, a generic reaction that involves substrate carbonyl binding to the metal ion and subsequent polarization of the carbonyl (the catalytic event) will be considered as a general reaction. If the metal ion is Zn^{2+}, the preferred ligand arrangement will be tetrahedral, and in the case of thermolysin, the protein ligands are two histidyl side chains and one glutamyl residue. Since the imidazole nitrogens are electronegative and the glutamyl is negative, the Zn^{2+} will have some fraction of its net $+2$ charge and would be an effective site for binding the oxygen of a carbonyl group. Attack by a nucleophile at the carbon of the carbonyl group would result in development of a negative charge on the carbonyl oxygen and the Zn^{2+} would assist in the catalysis of this reaction by electrostatic "neutralization" of the charge (Fig. 2). If one of the imidazole ligands were replaced by a carboxyl group (aspartate would fit nicely and should not produce a large change in geometry), the "effective" $+2$ charge on the Zn would be further reduced and the expectation is that the metal would not be as effective a catalyst for the reaction. In addition, the negative charge on the ligand would also be expected to provide charge repulsion for the developing negative charge on the carbonyl oxygen, resulting in additional retardation of the chemical event. These predictions can be tested by mutagenesis experiments and several are underway in other laboratories. Additional complications can arise because other atoms of the ligands [the other oxygen of a carboxyl group or the other nitrogen (—NH) of an imidazole] that are not involved in direct coordination can be bound to the rest of the protein matrix, and disruption of these structural features could also result in alteration of catalysis.

A series of kinetic experiments reported by Herschlag and Jencks (7) addressed the influence of metal ions on phosphoryl transfer reactions for nonenzymic phosphoryl transfer between pyridine and carboxylate ions (and other nucleophiles). The reaction of actetate ion with phosphorylated γ-picoline monoanion and the reverse reaction of γ-picoline with acetyl phosphate dianion are

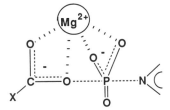

FIG. 3. Structure of the metal ion chelate for the reaction of nucleophiles with carboxy phosphates. The model is taken from the data of Herschlag and Jencks (7).

equally catalyzed by ~20-fold by Mg^{2+}. The catalysis was considered to arise from overcoming unfavorable electrostatic, solvation, and perhaps steric interactions. Comparison among acetate, formate, and bicarbonate anions in this reaction, however, is ascribed to chelation of Mg^{2+} to the transition state, with $K_a^{\ddagger} = 97\ M^{-1}$ and weaker binding to the ground states with $K_a = 5$ and 4.4 M^{-1} for γ-picolinate phosphate and acetyl phosphate, respectively. This type of catalysis can be considered to arise from a large entropic contribution as a result of bringing the reactants together, making the "effective" molarity of the Mg^{2+} equal to ~10 M; the "true" catalytic effect is measured as a comparison of ground- and transition-state structures, with the Mg^{2+} acting as a template to stabilize the assembled transition state, which is the common feature for both the forward and reverse reactions. In the reverse reaction, Mg^{2+} would not chelate to picoline in a ground-state complex, so catalysis in this direction could only arise from transition-state stabilization; the "chelate structure" is shown in Fig. 3 (7). The "effective" molarity of ~10 M can be considered to be a large component of phosphoryl transfer reactions that involve metal ion complexes of nucleotides, e.g., ATP, with a large loss in entropy, even though the transition state for phosphoryl transfer is dissociative. All the auxiliary amino acid side chains that comprise the active site of enzymes that utilize MgATP will contribute to maximizing the exact location for all atoms in the transition state, thus producing a larger loss of entropy and the ~10^9-fold catalysis found in many enzymes.

III. Methods to Study Metal Ion Environments in Proteins

The most precise method to determine the nature of metal ion ligands, the geometry of the metal ion site, and the mode of substrate binding is X-ray crystallography. Lacking such a three-dimensional structure, several methods, including optical spectroscopy, electron paramagnetic resonance (EPR), spectrometry, X-ray absorption, and/or extended X-ray absorption fine structure (EXAFS), can give some information on ligand type provided that a metal ion–ligand interaction is present, with a unique spectroscopic signal due to the choice of metal ion and ligand. Since the ligands are fixed by the protein under investigation, metal ion substitution can be used in favorable cases for this purpose. For optical spectroscopy, the metal ion has to have intense absorption properties in the UV–visible range, which is a property of some transition metals, e.g., Fe^{3+}, Cu^{2+}, and Co^{2+}, but not others, e.g., Zn^{2+} and Mn^{2+} (8). Some information about geometry (octahedral versus tetrahedral) and ligand type (O or N) can be obtained by this method, but in general this is a low-resolution method. All Group IA, IIA, and IIIA metal ions, representing a significant number of biologically important ions (e.g., Na^+, K^+, Mg^{2+}, and Ca^{2+}), are devoid of optical properties. EPR can be used to study paramagnetic metal ions (e.g., Fe^{3+}, Cu^{2+}, and Mn^{2+}) and information can be obtained about ligand types when the ligands contain atoms that possess the property of nuclear spin (e.g., ^{14}N, $I = 1$) (9). More precise information about the number and type of nitrogen-containing ligands can be gathered from spin-echo EPR studies. This has proved to be a very powerful technique to explore Cu^{2+}- and Mn^{2+}-containing enzymes (10–12). When the ligand atoms are oxygen, no information can be directly obtained by EPR studies unless ^{17}O is substituted for natural abundance ^{16}O. Such substitutions can be difficult. The involvement of solvent water molecules at enzyme active sites can be explored by EPR and nuclear magnetic resonance (NMR) methods, and several measurements of the number of water molecules and substrate atoms involved in enzyme–metal–substrate complexes have been reported (13, 14). This subject will be expanded below.

X-Ray absorption methods have been extensively used to study atoms in the first-coordination sphere of metals. The method easily distinguishes N/O from S atoms but unfortunately does not distinguish N from O (15, 16). The limitations of the method extend to the study of metal atoms with a large number of electrons in their outer shell (e.g., Fe, Cu, and Mo) and make it difficult to study "lighter" atoms (e.g., Na and Mg). The advantage of this technique lies in the precise determination of the number of ligands and the distances between the metal ion and the ligand atoms.

NMR can be used to determine paramagnetic metal ion-to-nuclear distances in favorable cases (17). The advantage of this approach is that stable enzyme–metal–substrate (or substrate analog) complexes can be studied for multisubstrate enzymes without the complications of turnover (18), but the limitations are that the substrate or analog must contain nuclei with magnetic spin (e.g., ^1H, ^{31}P, and ^{15}N). The characteristic of the nuclei mentioned is that they have spin $\frac{1}{2}$ and thus have narrow line widths. The entire basis for this method is the induced "broadening" of the line widths of nuclei when they are bound near paramagnetic metal ions; specifically, the paramagnetic contribution to the relaxation parameters ($1/T_1$ and $1/T_2$) is used to calculate metal ion-to-nucleus distances and quite accurate data can be gathered using this method (subject to the conditions and limitations given in Ref. 17). Thus, active site "structures" can be obtained under solution conditions.

The measurements of water proton relaxation rates (PRR) as an NMR application, to study the effects of paramagnetic metal binding to proteins, has been extensively reviewed (*19–21*). This method involves the measurements of interactions of solvent water protons with the metal ion. The information that can be garnered is the electron symmetry of the cation, the alteration of solvation number, and water (proton) exchange rates on cation binding to the enzyme and on ligand binding to form an enzyme–cation–ligand complex. Both NMR methods given above can supply active site distance relationships in the absence of a crystal structure.

IV. Metal Ion-Assisted Phosphoryl Transfer Reactions

A. MECHANISMS OF PHOSPHORYL TRANSFER

Mechanistically related enzymes that are ubiquitous in their requirement for a metal ion are those that catalyze phosphoryl group transfer reactions. These enzymes include the (phospho)kinases and phosphatases, which catalyze the transfer of a phosphoryl group (PO_3) from the γ-position of a nucleotide to an acceptor molecule (the phosphatases are a special case wherein the acceptor is the solvent water); nucleotidyltransferases and nucleases, which catalyze the transfer of a nucleoside phosphodiester to an acceptor molecule (the nucleases and phosphodiesterases are special cases wherein the acceptor is the solvent water); and phosphomutases. In all of these cases, the mechanism of phosphoryl group transfer can occur through one of several postulated pathways. The possible mechanistic extremes are those described as dissociative, analogous to an S_N1 reaction,

$$\text{RO}-\text{P}(\text{O})(\text{O})\text{O} \longrightarrow \text{RO}^- + \left[\text{P}(\text{O})(\text{O})\text{O} \right] \tag{1}$$

and associative, analogous to an S_N2 reaction,

$$\text{RO}-\text{P}(\text{O})(\text{O})\text{OR} \longrightarrow \left[\text{RO}\cdots\text{P}(\text{O})(\text{RO})(\text{O})\cdots\text{A} \right] \tag{2}$$

In Eq. (1), metaphosphate (PO_3) as the leaving group is formed in the transition state. Metaphosphate is formed with a net change in hybridization and planar geometry. This reactive species can then react with the acceptor. In Eq. (2), the nucleophilic acceptor molecule A interacts with the phosphate center, forming a pentacoordinate phosphate transition state. Although studies of model reactions show that the associative mechanism can take place either by direct displacement via an "in-line" process, indicated in Eq. (2), or by an adjacent addition–elimination process, resulting in the alteration of the stereochemistry of the phosphoryl transfer process (22), no evidence exists to date to support an enzyme-catalyzed phosphoryl transfer reaction via an adjacent associative mechanism. In addressing the role of the cations that are required in the catalytic processes for these enzymes, the mechanism of the phosphoryl transfer must first be addressed to understand the function of the cations in the chemical processes.

The enzymes that utilize a nucleotide triphosphate/diphosphate as a substrate appear to universally require a cation. The cation used is most commonly Mg^{2+} or in some cases Ca^{2+} and it forms the cation–nucleotide complex that serves as the substrate. A subgroup of these enzymes requires an additional cation for catalytic activity. These requirements indicate that there are several roles that the cations may play in the catalytic proceses. Within the group of nucleotide-utilizing enzymes, including kinases, ATPases, and synthetases, ATP is the most common of the nucleotides that serve as substrates. Recent methodologies have been developed to determine the nature of the metal–nucleotide complex and investigations of the various enzymes suggest that there is not a universal structure of the substrate that the enzymes in these subclasses prefer. Thus if the functions of the cations are intimately related to the structure of the metal–substrate complex preferred by the enzyme, then there may not be a universal function for the metal ion with a singular group of enzymes. It has been apparent for quite some time that one of the important marvels of biological phenomena is the recognition of molecular size and shape that elicits biological responses. This is not only true in phenomena such as antibody–antigen and ligand–receptor responses but also in enzyme–substrate interactions. With such a consideration in mind, one of the roles of the cation may be to assist in the proper conformation or orientation of the substrate at the catalytic site of the enzyme. Since the energy barriers for polyphosphate reorientation are expected to be low, one of the important functions of the metal ion appears to be structural. Investigations of the structural data for metal–nucleotide complexes for various enzymes for which such analyses have been performed indicate that subtle differences in structure of the substrate are required for different enzymes.

B. MODEL REACTIONS

A recent kinetic study of the catalysis of hydrolysis of phosphorylated pyridines by Mg^{2+} in aqueous solution as a model for enzymic phosphoryl transfer reactions has been reported by Herschlag and Jencks (23). They have provided evidence that a 10^4–10^6 rate enhancement can be obtained in the presence of Mg^{2+}. From the analysis, approximately a 10^4-fold rate enhancement can be attributed to the greater nucleophilicity of the species $Mg(OH)^+$ compared to solvent H_2O. This role of the cation is proposed as a model for increasing the nucleophilicity of the enolate oxygen of enolpyruvate as the substrate in pyruvate kinase and the serine alkoxide in alkaline phosphatase. With both of these enzymes, stabilization of the more nucleophilic anionic species could be elicited at neutral pH as the cation complex of the nucleophile. The pK_a values of each of the free species, the enol of pyruvate and the hydroxymethyl of serine, are such that they are expected to give a vanishingly small concentration of the anionic species at neutral pH, at which the enzymes have catalytic activity. The kinetic results are also modeled by formation of the bimetallic complex consisting of $Mg(OH)^+$ and the Mg^{2+} complex of phosphorylated γ-picoline. In this model, approximately a 10^2 rate enhancement is due to the induction of an intramolecular-like complex, wherein a role of the second cation is to serve as a bridge to link the two substrates, the nucleophile $[Mg(OH)^+]$ and the donor (γ-phosphopicoline). This latter function facilitates the phenomenon of propinquity and has the effect of increasing the effective concentration of the reactants in a second-order reaction. Thus, these elegant model studies have shown that in phosphoryl transfer reactions, the metal ions can play key roles in the activation of the nucleophile, analogous to the role that may be played by a specific base, and can assist in proximity effects when the reacting species are brought together. These concepts are important considerations as these groups of enzymes are investigated.

C. KINETIC ANALYSIS

In the analysis of phosphoryl transfer, the distinction between an "associative" and "dissociative" mechanism is quite clear; however, experimental evidence to distinguish among the two processes has been rather indirect and inconclusive. In either mechanism considered, the phosphate being transferred has the geometry of a metaphosphate, i.e., the phosphate center is planar and the phosphate atom either has trivalent or pentavalent character. The bonding character depends on the reaction being dissociative, i.e., the transition state is metaphosphate, or associative, i.e., the transition state is trigonal bipyrimidal. The electrophilic cation(s) could activate such processes by several methods. Several mechanistic approaches have been applied to begin to address these questions.

Steady-state kinetics have been used to determine the kinetic mechanisms of many of these enzymes. The questions that have been primarily addressed are the sequence of steps that occur in substrate binding prior and subsequent to the catalytic reaction and the potential formation of covalent enzyme intermediates. Classical interpretation of kinetic analyses has been the determination of the relevant reactions occurring via a random or an ordered sequential reaction, or if the reaction is a double-displacement or Ping-Pong reaction. In the former case, phosphoryl transfer occurs in the ternary complex that contains enzyme, phosphoryl donor, and phosphoryl acceptor. In the latter case, enzyme reacts with the phosphoryl donor to yield modified enzyme, usually phosphorylated enzyme. The subsequent reaction of phosphoenzyme with acceptor yields the final product. Although most of the kinases catalyze reactions via a sequential kinetic mechanism and details of the substrate binding vary among enzymes, there are also well-documented cases of these enzymes participating in catalysis via a phosphoenzyme intermediate. The diagnosis of a double-displacement reaction usually comes from obtaining parallel lines from the kinetic data, where substrate is varied as a function of a fixed, variable concentration of the second substrate. These studies are usually performed in the presence of saturating metal ions. Confirmatory kinetic data for these mechanisms have been obtained using equilibrium isotope exchange studies as well as isotope exchange and partial exchange studies. The formation of a transient covalent intermediate in a sequential reaction is difficult to determine. Physical isolation of a phosphoenzyme has been used in a number of cases to indicate a reaction sequence by double displacement. This criterion must be supported by evidence that the species identified and/or characterized is kinetically competent. That is, the phosphoenzyme must be able to react with a substrate acceptor to yield product at a rate at least as fast as the overall turnover number.

D.   STEREOCHEMISTRY OF PHOSPHORYL TRANSFER

A sensitive probe applied to understand the nature of the reaction mechanism of group transfer is the stereochemistry of the overall reaction. The reaction at a phosphoryl center normally is a degenerate question, since a monosubstituted phosphate ester or anhydride is proprochiral at the phosphate center. Phosphate centers at a diester or disubstituted anhydride are prochiral. Two related methods to analyze the stereochemistry at a phosphate center have been developed by the generation of chirality at the phosphorus center. The first approach was developed by Usher et al. (24) and gave rise to the formation of isotopically chiral [$^{16}O,^{18}O$]thiophosphate esters and anhydrides (I). Isotopically chiral [$^{16}O,^{17}O,^{18}O$]phosphates (II) have also been synthesized and the absolute configurations determined. Two primary problems must first be addressed with respect to both of the methods that have been developed: the synthesis of the isotopically pure chiral thiophosphates and phosphates and the analysis of the isotopic chirality of the products. An example of the chiral starting substrates, as developed for ATP, is schematically demonstrated. Ad = adenosine.

(I)   (II)

The shaded oxygen atoms indicate the different isotopes of oxygen (^{16}O, ^{17}O, and ^{18}O). The syntheses and the chiral analyses of these compounds and the related products of the phosphoryl transfer reaction have been developed using chemical and mass spectral or ^{31}P NMR analyses. The results of analyses of a large number of related enzymes indicate that phosphoryl transfer occurs via one of only two routes—retention of chirality or inversion of chirality. Retention of chirality occurs from an even number of stereospecific phosphoryl transfer steps. This series of reactions is what is expected from a double-displacement reaction wherein a nucleophile on the enzyme is activated and enzyme–phosphate is formed as an intermediate. The phosphoryl group is subsequently transferred to the acceptor. Inversion of chirality is expected from an odd number of group transfer steps. Most simply, this would occur from a single phosphoryl transfer step from donor to acceptor via an associative reaction. There is no substantive evidence for an enzyme reaction where a triple-displacement reaction occurs. A summary of the stereochemical course of phosphoryl group transfer reactions for a number of enzymes is presented in Table III (25–56).

The results summarized in Table III suggest that the enzymes that catalyze phosphoryl transfer via an inversion of configuration do so with an "in-line" transfer in a sequential mechanism. The mechanistic pathway is prevalent in the phosphokinases. Although this information does not provide direct evidence for an associative or S_N2 mechanism in contrast to a dissociative mechanism, if the latter process does occur then there is insufficient room at the catalytic site for the metaphosphate to rotate or dissociate and to cause racemization. The observation of a secondary ^{18}O isotope effect less than 1.00 indicates that a dissociative transition state occurs with yeast hexokinase (57). The enzymes that demonstrate retention of configuration do so via double-displacement reactions. Mutases exclusively use this mechanistic pathway.

E.   STRUCTURE OF METAL CHELATES

The analysis of Mg^{2+}–nucleotide complexes to determine the structure of these ligands has been difficult because of the dynamic nature of the cation–ligand complex. A variety of possible complexes at the catalytic site can be formed. These can be monodentate, bidentate, or tridentate. Monodentate complexes can be formed with either the α-, β-, or γ-phosphoryl groups as ligands. Bidentate complexes can be α–β, β–γ, or α–γ. Differences in structural requirements for the substrate complexes for various enzymes imply different roles of the metal ion in catalysis among ATP-utilizing enzymes. Indirect probes for the structures of these complexes have been developed by the synthesis of Cr^{3+} and Co^{3+} complexes of the nucleotides. The usefulness of these complexes is that they are kinetically stable. Instead of cation–ligand exchange rates of about 10^6 sec^{-1}, as found for complexes such as MgATP or CaATP, the exchange rates are about 10^{-6} sec^{-1}, a turnover time of about 2 weeks. Thus it is possible to

TABLE III

STEREOCHEMICAL COURSE OF ENZYME-CATALYZED PHOSPHORYL TRANSFER REACTIONS

Enzyme	Retention of chirality		Inversion of chirality		Ref.
	$[^{16}O, {}^{18}O, S]$	$[^{16}O, {}^{17}O, {}^{18}O]$	$[^{16}O, {}^{18}O, S]$	$[^{16}O, {}^{17}O, {}^{18}O]$	
Phosphokinases					
Nucleoside diphosphate kinase	+				25
Nucleoside phosphotransferase	+				26
Acetate kinase				+	27
Adenosine kinase			+		28
Adenylate kinase			+		29
Creatine kinase				+	30
Glucokinase				+	31
Glycerol kinase			+	+	27, 32, 33
Hexokinase			+	+	27, 32, 34
Phosphoenolpyruvate carboxykinase			+		35, 36
Phosphofructokinase				+	37
Phosphoglycerate kinase			+		38, 39
Polynucleotide kinase			+	+	33, 40, 41
Pyruvate kinase			+	+	27, 32, 42
Ribulose-phosphate kinase			+		43
Thymidine kinase				+	44
Phosphatases					
Acid phosphatase		+			45
Alkaline phosphatase		+			46
ATPase (sarcoplasmic reticulum)	+				47
Glucose-6-phosphatase		+			48
5′-Nucleotidase (snake venom)	+				49
ATPase			+		50
Myosin ATPase			+		51
T GTPase			+		52
PS3 ATPase			+		53
Pyrophosphatase			+		54
Mutases					
Phosphoglucomutase		+			55
Phosphoglycerate mutase (cofactor independent)		+			56
Phosphoglycerate mutase (cofactor dependent)		+			56

form stable diastereomers of the metal–nucleotide complexes and to investigate the stereoselectivity of various enzymes for the particular isomer. Several enzymes demonstrate single-turnover activity with an appropriate diastereomer that serves as a substrate analog. Most studies have been concerned with the determination of the tightest binding or most stereoselective ligand by steady-state kinetics. A detailed review of this approach has been published (58). Diastereomers of ATP and related nucleotide triphosphates occur when bidentate complexes are formed, since the α- and β-phosphates are prochiral centers. The two bidentate diastereomers of the β,γ-Co(NH$_3$)$_4$ATP complexes were prepared, separated, and identified (59). Ultimately, the absolute configurations of the isomers were determined by X-ray analysis of the Co(NH$_3$)$_4$PPP$_i$ complex (60). The Co(NH$_3$)$_4$PPP$_i$ complex was formed by chemical degradation of a specific nucleotide complex. The two diastereomers, indicated as **III** and **IV**, are designated as the Δ and Λ isomers, respectively.

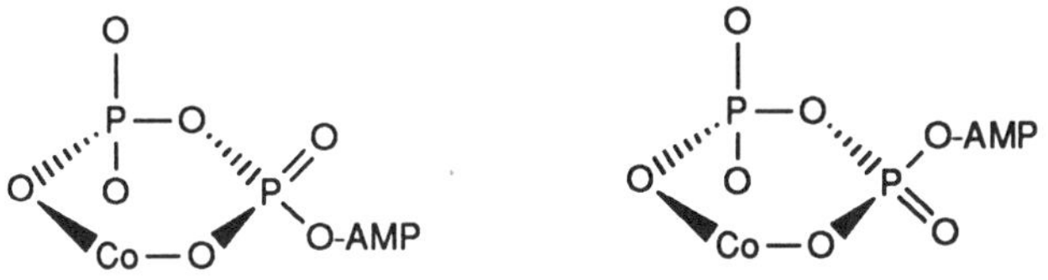

Δ isomer   Λ isomer

III   **IV**

There are two additional conformational isomers of each of the complexes that are formed. Redrawing the Λ isomer such that it is easier to envision the left-handed screw sense of this geometrical isomer, the adenosine group can be found either in a pseudoaxial position, **V** or in a pseudoequatorial position **VI**.

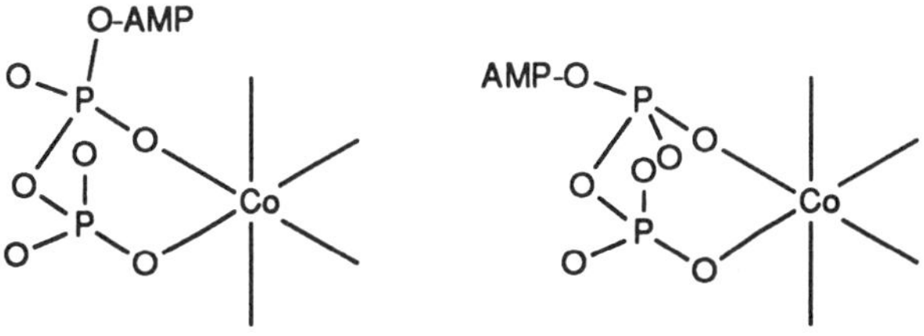

(AMP pseudoaxial)   (AMP pseudoequatorial)

V   **VI**

TABLE IV

STEREOSELECTIVITY OF ENZYMES FOR M^{2+}–NUCLEOTIDE COMPLEXES

	CoATP or CrATP		CoADP or CrADP		
Enzyme	Λ	Δ	Λ	Δ	Ref.
Creatine kinase				+	61, 62
Glycerol kinase	+				61
Hexokinase	+				61
Protein kinase		+			63
Phosphoribosyl- pyrophosphate synthetase		+			64
Pyruvate kinase		+			62
Phosphoenolpyruvate carboxykinase		+			65

Following the preparation of the β–γ complex of $Cr^{3+}ATP$, careful chromatography yields four isomers of the complex, two L and two D. Initial studies were performed with each of the purified diastereomers on hexokinase (61). Analysis demonstrated that only one isomer, Λ, served as a substrate for this enzyme and that the other diastereomer was unreactive. Subsequent studies on other kinases demonstrated varying selectivity for different kinases for the Δ or Λ isomers. These studies were performed with either complexes of ADP or ATP. Selectivity was based on potency of inhibition of the various isomers in many cases and the observation of single-turnover reaction in several other cases. If analogy of these structures with those of MgADP and MgATP complexes holds, individual enzymes have different stereoselectivities for the M–nucleotide structures. A summary of the selectivity of varying M–nucleotide complexes for some enzymes is presented in Table IV (61–65).

F.   THIOPHOSPHATE ANALOGS OF NUCLEOTIDES

An additional method to determine the nature of the metal–nucleotide complex that is preferred by the enzyme is in the study of the effects of cation substitution on the stereoselectivity of chiral-specific thiophosphate nucleotides. Thiophosphate derivatives of cAMP, AMP, ADP, and ATP at the α-, β-, and/or γ-positions were initially introduced by Eckstein and co-workers (66, 67) and expanded to several other nucleotides. Substitution of a phosphate oxygen by sulfur at the α-position of cAMP, ADP, or ATP or at the β-position of ATP gives rise to a new chiral center. Substitution in the α-position of AMP, β-position of ADP, and the γ-position of ATP changes an achiral center to a prochiral center. This substitution gives rise to a small decrease in pK_a for the phosphate, and with many enzymes these thiophosphate analogs have a decreased substrate ac-

TABLE V

STEREOSELECTIVITY OF ENZYMES FOR CHIRAL THIOPHOSPHATES AS SUBSTRATES[a]

Enzyme	NTPαS		NTPβS		Ref.
	S	R	S	R	
Pyruvate kinase	+	−	+	−	66
Acetate kinase	+	(+)	−	+	66
Hexokinase	+	(+)	−	+	68
Creatine kinase	(+)	+	+ (70%)	+ (30%)	66,68
Myokinase	+	(+)	+	+	66
Arginine kinase	−	+			66
Myosin	+	+	+	−	66
Phosphoenolpyruvate carboxykinase	+ (67%)	+ (33%)	+	−	69

[a] The symbols + and − indicate the diastereomers used and not used, respectively, by the enzyme; (+) indicates activity but at a much lower level than the other diastereomer.

tivity. Some enzymes demonstrate rather strict stereoselectivity for the specific diastereomers of the thiophosphates. Other enzymes do not show strict specificity for a diastereomer but do show strong selectivity. This selectivity appears to be a less extreme restriction of the enzyme than for the nucleotide binding site on the enzyme. A summary of some of the enzymes that have been investigated using either ATPS or GTPS and their selectivity for the diastereomers synthesized at the α- and β-positions are indicated in Table V (66–69).

The results from these studies also indicate that some kinases have strict selectivity for the nucleotides as substrates, others are less restrictive, and others are even less restrictive. To further investigate this selectivity of conformation by the enzyme, Jaffe and Cohn (68) have shown that with hexokinase, changing the divalent cation from an oxygen-preferring metal such as Mg^{2+} to a sulfur-preferring metal such as Cd^{2+}, the selectivity of the enzyme for the diastereomer of ATPβS changes from R to S. The selectivity of the enzyme for ATPαS is independent of cation. These results indicate that the enzyme prefers the β–γ bidentate complex of ATP with the R configuration. No coordination at the α-position is indicated. Comparison of the conformation of this isomer with the single selective isomer of $Co(NH_3)_4$ATP that hexokinase utilizes as a substrate indicates that they are the same conformation. The R configuration is shown below:

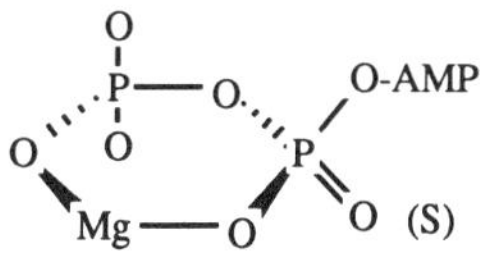

The experimental tools that have been described indicate that within the group of nucleotide-utilizing enzymes (primarily kinases, but synthetases and phosphohydrolases as well), each enzyme appears to have a different selectivity for the metal–nucleotide complex. Some enzymes appear to require a tridentate complex, others require β–γ as the more frequent conformer, and still others require a monodentate complex. The results with some enzymes appear to be not as selective. In these cases, the nucleotide and the metal ion may change structure and hence coordination geometry during the course of the catalytic steps. Thus, since the catalytic process is a dynamic one, the studies described above yield extremely valuable information regarding the stereospecificity in a static case. Insight into the dynamic processes of these cations therefore depends on the ability of the investigators to interpret the nature of these results with respect to function.

The applications of these and additional kinetic, chemical, and physical studies have been made toward a large variety of different enzymes. The depth of investigation and the rigor of these studies vary with the enzymes investigated. In order to describe rigorously the reaction mechanism of any enzyme, we must know the following information: the kinetic scheme of the reaction, that is, the number of steps in the chemical process; the affinity and specificity for substrates; the geometric structure of the substrates and products at the catalytic site along with the structure of the active site of the enzyme; the chemical mechanism for each step in the reaction, including intermediates and transition states; and the rate constants and physical descriptions for these rate constants for each step in the mechanistic process and their relationship to structure. Clearly, such details are lacking for most enzymes investigated. With the recent abundance of X-ray data and results of site-directed mutagenesis, information regarding the structure of the proteins has become sufficiently abundant that our knowledge of structure–function relationships is becoming more sophisticated. As this information is related to metal ion effects, the understanding of the functions of cations in the catalytic processes of selected enzymes is summarized for several groups of enzymes.

G. SPECIFIC ENZYMES

Kinases represent the largest group of phosphoryl transfer enzymes that have been investigated. Several of these enzymes have been investigated in great detail, and the roles of the metal ions in the catalytic processes appear to be well described. Although the mechanisms of these enzymes and thus the roles of the metal ions appear to be very similar on paper, they do not generalize as easily as they are mechanistically classified. Some of the kinases, classically creatine kinase, require a single divalent cation per active site for activity. Others, classically muscle pyruvate kinase, require two divalent cations per active site for activity. Thus we have single- and multiple-cation effects to decipher. Also, creatine kinase forms a "substrate bridge" complex for the formation of the active enzyme–ATP–M^{2+}. This nomenclature indicates the binding of the cation to the enzyme only after the formation of the E–S complex. Pyruvate kinase forms a "metal bridge" complex where the active complex is enzyme–M^{2+}–phospho-enolpyruvate or enzyme–M^{2+}–ATP. This nomenclature indicates that the metal can bind to the enzyme in the absence of substrate and the substrate can interact with the E–M^{2+} complex. The difficulty in making such simplified generalizations is that even in the complex formed with pyruvate kinase, the ATP chelates the second cation analogous to the substrate bridge complexes. Examples of some of the kinases investigated are described in the following sections.

1. *Adenylate Kinase*

Adenylate kinase, or myokinase, can be viewed as the "incestuous" kinase, where MgATP is used to phosphorylate AMP to yield the products ADP and MgADP. The nature of this readily reversible reaction and of the binding sites for the apparently symmetrical catalytic site has been the subject of considerable study and controversy (*70–72*). In this reaction, two unsymmetrical substrate binding sites for ADP are present: one binds ADP and the other binds MgADP. Studies with the exchange-inert metal–nucleotide complexes demonstrate that ATP binds as the bidentate β,γ-complex and the product apparently becomes the monodentate β-complex (*61*). This conclusion is also consistent with the cation-independent stereoselectivity of the ATP(γS) analog, but the selectivity for the R_P diastereomer of ATP(βS) (*66*) and the ^{31}P NMR studies indicate direct chelation of the cation at the β or β,γ of ATP but no direct chelation to P_α (*72*). A likely mode of ATP binding to adenylate kinase is presented in Fig. 4 (*73*). No

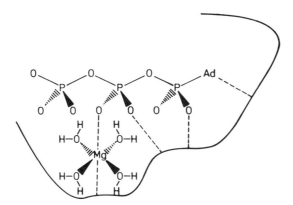

FIG. 4. Postulated mode of binding of MgATP to the active site of adenylate kinase. The metal ion is coordinated to the *pro-R* oxygen atom of the β-phosphate and to 3, 4, or 5 water molecules. Since Mg^{2+} is normally six-coordinated, any unused coordination sites are presumably made up from amino acid side chains of the protein. Ad, Adenosine (*73*).

interaction of the cation with apoenzyme occurs and the cation only interacts with enzyme-bound ATP or one of the two ADPs. Recent studies of selective mutants of adenylate kinase and ^{25}Mg NMR (74) indicate that in the formation of the enzyme–AMP–MgATP complex, the cation also interacts with Asp-93. The proper interaction of this cation and the ATP assists in the structural alignment of the tripolyphosphate chain of the ATP. The cation interaction with the protein assists in this alignment. The absence of the Asp-93 diminishes the ability of the cation to align in an optimal conformation and the result is a significant decrease in k_{cat} with little effect on K_m. There is probably an electronic effect as well, with the cation interaction at the γ-phosphate facilitating its electrophilicity and assisting in charge neutralization and facilitating nucleophilic attack by the AMP. It appears that although this enzyme may be classified as forming a substrate bridge to give the enzyme–M^{2+}–substrate complex, an interaction of the cation with a group on the protein is also important. The role of the cation in adenylate kinase catalysis is both structural and electronic.

2. *Creatine Kinase*

Creatine kinase is the prototype of this group of enzymes; it forms a substrate bridge complex with the enzyme and utilizes a single divalent cation at the catalytic site. In the formation of the active enzyme–substrate complex that contains the metal ion, the metal binds to either ATP or ADP to form the ternary enzyme–substrate–cation complex. This complex and the binding of the second substrate or substrate analog have been characterized by PRR and high-resolution NMR studies (75) and by EPR investigations of bound Mn^{2+} (76, 77). The results indicate that the cation binds the α- and β-positions of ADP to form the bidentate complex and this complex will function as the phosphoryl acceptor for creatine phosphate. The cation appears to stabilize the Δ form of the M–ADP complex that is the active conformer in the active enzyme complex. Thus there appears to be a structural component to the formation of the metal–nucleotide complex where this enzyme is selective in terms of the conformation of the substrate complex. On binding of the creatine phosphate, the cation also binds to the phosphoryl group of the donor substrate and facilitates propinquity between the two substrates. The close proximity of the two substrates that is held by the cation also assists in partial charge neutralization. These two roles must facilitate the phosphoryl transfer as the catalytic event. The transition state for this phosphoryl transfer reaction has been suggested to appear much like the pentavalent phosphate in a displacement reaction, although analogous geometry for a dissociative reaction is also consistent with such geometry. The postulated configuration of the substrates is indicated in Fig. 5 (76).

3. *Hexokinase*

This enzyme catalyzes the phosphoryl transfer from MgATP to glucose or analogous sugars. Of specific interest in the study of the reaction mechanism of

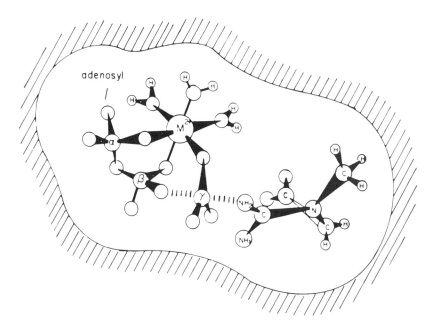

FIG. 5. A schematic view of the substrates in the transition state of the reaction catalyzed by creatine kinase. Coordination of Mn(II) to the *pro-S* oxygen at P_α is established by EPR; the conformation of creatine is based on the efficacy as substrates of conformationally restricted analogs of creatine. The β,γ-chelate ring of ATP is in the Λ configuration (76).

this enzyme is the observation that CrATP as the β,γ-isomer of the Λ diastereomer serves as a substrate for this reaction. The product is the β-monodentate MgADP complex. Thus it would appear that the cation serves both to aid in the formation of the optimal conformation of the nucleotide preferred by the enzyme and for electronic effects. Such results indicate that the cation serves to increase the electrophilicity of the γ-phosphate and to stabilize the ADP leaving group. Recent incisive kinetic studies using ^{18}O isotope effects indicate that there is an isotope effect of <1.0 ("inverse isotope effect") if [γ-^{18}O]ATP is used as the substrate. This indicates that this group undergoes a change in hybridization during the transition state of the reaction and suggests the formation of a metaphosphate (78). This is one of the first pieces of evidence to indicate that such a mechanism does occur in enzyme-catalyzed phosphoryl transfer eactions. Therefore, one of the roles of the metal ion is facilitation of P–O bond cleavage and stabilization of the metaphosphate transition state in this reaction. The function of the metal would appear to first form the β,γ-complex of ATP and the proper conformation then interacts and is stabilized on the enzyme. The M–ATP complex, in the presence of the glucose, changes structure to form the β-monodentate complex of M–ADP and facilitates the formation of metaphosphate, which will then move toward the activated form of the glucose. The glucose is activated by a base on the protein. This mechanism appears to differ from that of the analogous creatine kinase discussed above.

4. *Phosphoenolpyruvate Carboxykinase*

This enzyme catalyzes a reaction that appears to be very homologous to that of pyruvate kinase. One similarity other than the proposed mechanism is that this enzyme also requires two cations to elicit the most active form of the enzyme. Although calculations of kinetic activation data indicate that both GDP and metal–GDP (and GTP and metal–GTP) can both serve as substrates for the enzyme, the function of GDP_{free} as substrate has not been demonstrated, and if so, the metal–GDP complex is the more active form (*79*). Unlike pyruvate kinase, this enzyme does not require a monovalent cation. Kinetic activation studies indicate that the enzyme-bound metal facilitates the interaction of the substrates phosphoenolpyruvate and nucleotide with the enzyme, but not the substrate CO_2. High-resolution NMR studies of both phosphoenolpyruvate and the nucleotide have demonstrated that in contrast to the classic case of pyruvate kinase, the substrates appear to bind in the second coordination sphere of the bound cation. These results indicate a much different role in the activation processes. The activation appears to be modulated by metal-bound water molecules. This interaction appears to be with both the phosphate of phosphoenolpyruvate (*80*) and the γ-phosphate of GTP (*79*) [as indicated in Fig. 6 (*80*).] Thus, in the activation of phosphoenolpyruvate, the metal must increase the cationic character of the water molecule with which the phosphate of the substrate interacts,

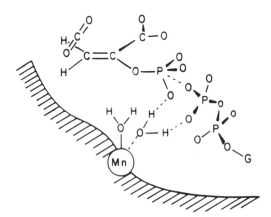

FIG. 6. Proposed mechanism for the phosphoenolpyruvate carboxykinase-catalyzed reaction. The orientation of the substrate with respect to enzyme-bound Mn^{2+} is shown. The second cation, which also takes part in the catalytic reactions, has not been included but is thought to form a β,γ-bidentate complex with GTP (*80*).

thus increasing the electrophilic character of this group. Since this charge effect is attenuated by the molecule of water, the effect of the metal is not as intense as in pyruvate kinase. The turnover number of this enzyme is about two orders of magnitude slower than that of the putative homologue, pyruvate kinase. The second cation appears to form a β,γ-bidentate complex with GTP but does not have a strong preference in conformation of this complex. The suggestion is that in the complex containing GDP and phosphoenolpyruvate, the second metal binds to the GDP as the β- or the α,β-complex and assists in stabilizing an active conformation of this potential nucleophile (69). As catalysis occurs, this cation begins to form a coordination complex with the incoming phosphoryl group, stabilizing the transition state by charge neutralization and by keeping the anionic portions of both substrates together.

5. *Pyruvate Kinase*

Pyruvate kinase is the prototype enzyme that forms a metal bridge complex with the enzyme and requires two divalent cations at the catalytic site (81). There is also a requirement for a monovalent cation at the catalytic site. The divalent cation that binds to the active site of the enzyme serves as a portion of the binding site for the substrate phosphoenolpyruvate and for the product ATP. The crystal structure of the cat muscle enzyme has been reported (82). In the formation of the ternary enzyme–metal–phosphoenolpyruvate complex, the substrate binds to the metal as an inner coordination sphere complex. In this case, the binding occurs at the phosphate group and apparently makes this center more electrophilic, facilitating nucleophilic attack by the metal–ADP complex as the second substrate. In this reaction, the roles of the metal ion on the protein appear to be as a cationic surface on the enzyme for specific phosphoenolpyruvate binding and in charge neutralization as part of the activation process. The phosphoryl transfer process yields primarily the Δ isomer of metal–ATP and the enolate of pyruvate. The enolate binds as a first coordination sphere complex with the enzyme and is stabilized as this complex. This complex, which appears to be transition-state-like, must be stabilized by the cation. This form of pyruvate can be activated to serve as the potential nucleophile for the reverse reaction by formation of such a cation–enolate. These complexes have been indicated as forming from NMR studies, but EPR studies have been more definitive as to the nature of the chelation structure of these complexes (83). The enzyme–bound cation interacts with approximately three ligands of the enzyme and directly with phosphoenolpyruvate or with the γ-phosphate of ATP. From studies of the selectivity of pyruvate kinase for the interaction of Co(III)ATP and Cr(III)ATP complexes, results indicate that the Δ form of the β,γ-bidentate complex of metal–ATP is the preferred conformer (Table IV). The metal ion that forms the metal–ADP complex serves as the second divalent cation requirement for this enzyme, and its role in the catalytic process appears to be to stabilize the active conformer of

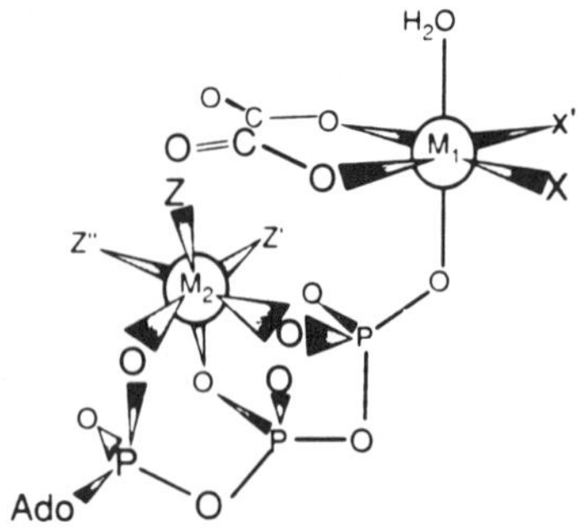

FIG. 7. Schematic drawing showing oxalate, ATP, and the two divalent cations bound at the active site of pyruvate kinase. M_1 is the divalent cation at site I and M_2 is the divalent cation at site II. Four oxygen ligands for M_1 have been identified by EPR experiments with ^{17}O labeling. Ligands X and X' are contributed from the protein. The α,β,γ-tridentate coordination scheme for M_2 was determined for Mn(II) at this site in one of the hybrid complexes formed in the presence of Cd(II). Ligands Z, Z', and Z" are likely water molecules given the facile binding of the aquo-$Cr^{III}ATP$ species at this site (83).

ADP at the catalytic site and for charge neutralization and stabilization of the product ATP. The two divalent cations have been shown to be in close proximity at the catalytic site and serve in the roles postulated (see Fig. 7). The monovalent cation is also found at the catalytic site, binding about 8 Å from the divalent cation (84, 85). In the presence of phosphoenolpyruvate but not ATP, a conformational change occurs that results in bringing the two cations 2 Å closer together. Binding and NMR studies indicate that this cation binds to the substrate via the carboxylate group to facilitate proper binding for catalysis, although this specific chelation has not been definitively demonstrated.

6. *Enolase*

Enolase catalyzes the trans dehydration of 2-phosphoglycerate to yield phosphoenolpyruvate and water; only a small free energy change (~ 1 kcal/mol) is associated with the reaction. The process is entropically driven and is readily reversible. This dimeric protein requires a divalent cation for activity and is rather promiscuous in that any one of about nine different cations can activate the enzyme (86). Depending upon the cation studied, the apoenzyme has either one or two metal binding sites per subunit. Metal ions such as Mg^{2+} and Mn^{2+} have one site per monomer, whereas Co^{2+} and Zn^{2+} will bind at two sites. In the presence of substrate there are two sites per subunit for all of the metal ions and, depending upon the pH, a third site is also induced. As the pH decreases, the third site is lost but not sites I and II. This third site is an inhibitory site, as the loss of this site parallels the loss of metal ion inhibition (87). The nature of this inhibition is not clear but may be due to the binding of the substrate at the phos-

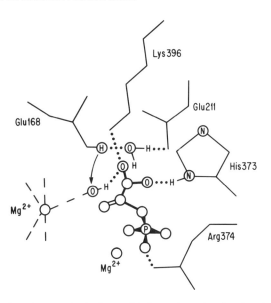

FIG. 8. Postulated mode of substrate binding at the active site of enolase (*91*).

phate site. The location of the substrate with respect to the metal Mn^{2+} at site I, using ^1H and ^{31}P NMR relaxation rate methods, has demonstrated that the substrate binds in the primary coordination sphere of the cation via the C-3 hydroxymethyl group (see Fig. 8). The results indicate that the cation stabilizes the leaving OH^- from the carbanionic intermediate to facilitate the second step of this reaction (*88*). The first step is proton removal of the C-2 proton by an active site base (*89*), presumably a glutamic acid. The first step may be partially rate determining, depending on the pH and the cation used. Thus, the cation at the catalytic site facilitates the stabilization of the cationic transition state and stabilizes the product by the formation of the $M(OH)^+$ as the leaving group. In the microscopic reverse, the activation of the nucleophilic water by being $M(OH)^+$ is the primary function of the cation. The role of the second cation has not been clarified. Its location appears to be greater than 10 Å from the cation at site I. Kinetic studies with mixed metal ions indicate that it is the cation at site I that modulates catalytic activity. Substitution at site II with the exchange-inert Co(III) yields enzyme that is still activated by Mg^{2+} or Mn^{2+} at about 85–90%, but there is only one exchangable cation site per monomer (*90*). The results suggest that the second cation serves no functional role in the catalytic process and may only serve to stabilize an active conformation.

   A recent analysis of crystal structure has attempted to add structural data to aid in understanding the mechanism (*91*). As mentioned above, the reaction involves dehydration of 2-phospho-D-glycerate to phosphoenolpyruvate, and the

H$^+$ and OH$^-$ removal steps must be assisted by catalytic groups. For catalytic activity, both a "structural" and "catalytic" metal must be bound. To date, the crystal structure only shows the position of the "structural" metal ion, which is coordinated to the hydroxyl group of the substrate. This ligation would be satisfactory for the mechanism proposed above, since it provides electrophilic assistance for removal of OH$^-$ in one direction, while serving to make water a good nucleophile in the opposite direction. The structural data must be viewed with caution in developing an overall structural understanding of the mechanistic events because the "catalytic" metal ion is absent from the structure and rearrangement of side chains along with shifts in the position of the bound substrate may occur when the full catalytic complex is formed.

7. Zn^{2+}-Metallopeptidases

The chemical mechanism of peptidases of this class has been understood for some time, but the precise role the Zn^{2+} plays has been altered slightly due to the determination of very accurate crystal structures with tightly bound transition-state analogs (92, 93). An analog that binds with an affinity of 3 pM has been cocrystallized with carboxypeptidase A, and the two phosphinyl oxygens bind to the Zn^{2+} in a manner that mimics the tetrahedral intermediate believed to be part of the reaction pathway. The Zn–O distances are 2.2 and 3.1 Å, which strongly implicates the metal ion in a dual role in catalysis, i.e., to activate Zn^{2+}-bound water to aid in nucleophilic attack on the peptide carbonyl and to bind and polarize the peptide carbonyl (Fig. 9). An identical mechanism is proposed for thermolysin, a Zn^{2+}-metallopeptidase, also from crystallographic analysis of bound transition-state analogs (94).

An analysis of phosphonamidate and phosphonate transition-state analogs for the thermolysin reaction was reported (95, 96) and the data were interpreted to suggest that one hydrogen bond could contribute ~4 kcal/mol binding to stabilize the formation of the tetrahedral "transition state." A reevaluation was re-

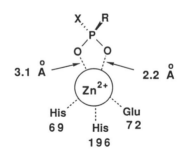

FIG. 9. Structure for a transition-state analog complexed to the Zn^{2+} at the active site of carboxypeptidase. Data taken from Kim and Lipscomb (94).

ported to include phosphinate analogs and the free energy changes were separated into ligation effects of the P–O⁻ group to the Zn^{2+} as well as hydrogen bonding to enzyme residues (97). From this more extensive analysis, the hydrogen bonding contribution was a more reasonable 1.4 kcal/mol and the ligation contribution was ~2.7 kcal/mol. The latter value is typical of metal ion–ligand binding energies and demonstrates that metal ion chelation of the transition state is an important factor in catalysis of this class of enzymes.

8.   Glutamine Synthetase

The bacterial enzyme uses two metal ions in substrate binding and catalysis, and a high-resolution crystal structure was reported a few years ago (98). An extensive kinetic analysis was just reported in which metal ion substitutions were made to explore the role of metal ions in individual catalytic reactions of a multiple-step reaction as well as to study the fate of allosteric modification on the catalytic behavior of the enzyme (99). Physiologically, Mg^{2+} is the activating metal ion and two ions are required for catalysis. The metal ions are about 6 Å apart as determined by EPR and NMR measurements (100, 101), and this was confirmed in the reported crystal structure. One metal ion is considered to bind and orient the carboxyl of glutamate to attack the γ-phosphoryl of ATP, while the other metal ion binds to ATP and facilitates ADP departure (leaving-group activation). Rapid-quench kinetic studies (99) revealed that in the physiologically active (unadenylylated) enzyme that product release, not chemistry, was rate limiting. Adenylylation (the physiological "off" switch) alters the reaction so that chemistry is now slowed and is rate limiting. The change in chemical activation steps is dramatically demonstrated by an examination of the internal equilibrium constants for the bound reactants and products; k_{int} is 2(with Mg^{2+}) or 5(with Mn^{2+}) for "active" (unadenylylated) enzyme and is altered to 0.1 by adenylylation (Fig. 10). Thus, phosphoryl transfer is now unfavorable in the adenylylated enzyme and provides an excellent chemical understanding of physiological regulation at the level of catalysis. Substrate binding is also altered by adenylylation. All these factors taken together begin to provide an understanding of the regulation of this complex enzyme at a chemical level. The "structure" of the putative tetrahedral intermediate formed by NH_3 attack on γ-glutamyl-P has been substantiated by kinetic (102) and spin-echo EPR (103) studies utilizing transition-state analogs of this intermediate. Because the nitrogen atom of the analog methionine sulfoximine binds to the metal ion at the substrate glutamate site, the role of this metal ion in stabilizing the negative charge that has developed in the tetrahedral intermediate seems reasonable. The metal ion at this site could also assist in the departure of phosphate in the final stage of the reaction by binding to the transition state involved in the C–O–P bond cleavage step. The chemical analogy for this last step can be found in the model reactions mentioned earlier in this article (7).

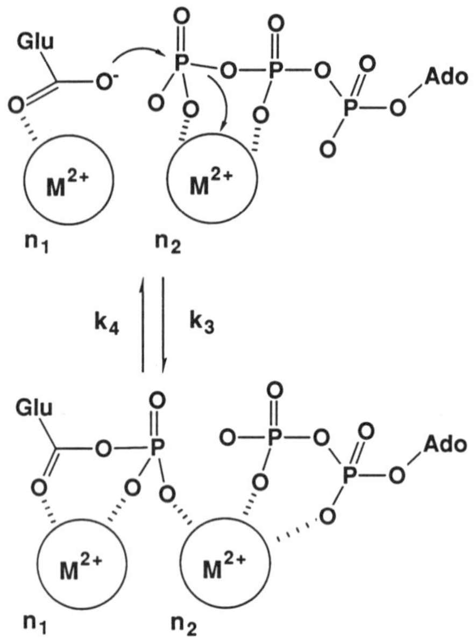

FIG. 10. Active site of glutamine synthetase. The two metal ions are designated n_1 and n_2 and represent their location in the crystal structure (98). The metal ions coordinate the substrates glutamate (n_1) and ATP (n_2) and are considered to be involved in the mechanism as shown. The n_1 site serves to orient the glutamate for attack on ATP and the n_2 metal ion aids in making ADP a good leaving group. The reaction shown represents the initial step in the overall mechanism. Subsequent steps involve attack of NH_3 on the γ-glutamyl-P intermediate.

9. Adenosine Deaminase

Adenosine deaminase is a key enzyme in purine metabolism. The reaction mechanism has been the subject of many kinetic studies, including the use of substrate analogs to explore the chirality of hydrolysis and the involvement of hydrogen bonding in transition-state stabilization (104). The crystal structure of the enzyme complexed with 6-hydroxy-1,6-dihydropurine ribonucleoside, a nearly ideal transition-state analog, was recently determined (105). A surprising result was that a stoichiometric amount of Zn^{2+} was bound complexed to the inhibitor and the enzyme; the enzyme had never been reported to be a metalloenzyme. The chirality of the bound inhibitor is R at the 6 position and the 6-hydroxyl group is coordinated to the metal ion. Buried at the bottom of a deep pocket is Zn^{2+}, which is coordinated by five atoms: the nitrogens of His-15, His-17, and His-214, the O-∂2 of Asp-295, and the O-6 (hydroxyl) of the inhibitor. The structure of the enzyme–Zn–inhibitor complex suggests a reaction pathway in which Zn^{2+} activates a water molecule for nucleophilic attack with the par-

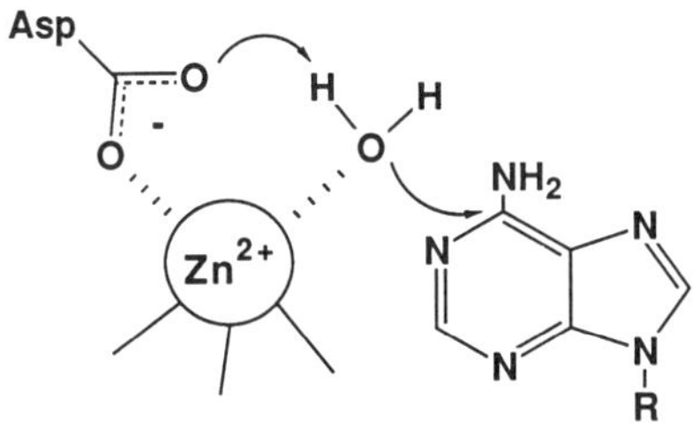

Fig. 11. The structure of the metal ion site in adenosine deaminase. The drawing is adapted from the data in Ref. *105*. The Zn^{2+} serves to activate a water molecule for nucleophilic attack while also orienting a carboxylate ligand to serve as a general base to deprotonate the water molecule.

ticipation of Asp-295, which acts as a general base, removing a proton from the water molecule (Fig. 11). The reaction may occur through a concerted S_N2 pathway (water attack/NH_3 departure) since there does not appear to be a proton donor appropriately placed to assist in departure of NH_3; however, the role of the metal ion would be unchanged. The presence of the metal ion at the active site alters the transition-state analysis and assignment of hydrogen bond contributions as reported previously (*104*), since contributions due to metal–inhibitor ligation must now be included in the overall analysis. The current status of protein–ligand discrimination using transition-state analogs has been presented (*106*).

REFERENCES

1. Rutter, W. J. (1961). "The Enzymes," 2nd Ed., Vol. 5, p. 341.
2. Sussman, F., and Weinstein, H. (1989). *Proc. Natl. Acad. Sci. U.S.A.* **86,** 7880.
3. Rashin, A. A., and Hönig, B. (1985). *J. Chem. Phys.* **89,** 5588.
4. Snyder, E. E., Buoscio, B. W., and Falke, J. J. (1990). *Biochemistry* **29,** 3937; Falke, J. J., Snyder, E. E., Thatcher, K. C., and Voertler, C. S. (1991). *Biochemistry* **30,** in press.
5. Vyas, N. K., Vyas, M. N., and Quiocho, F. A. (1988). *Science* **242,** 1290.
6. Shannon, R. D. (1976). *Acta Crystallogr., Sect. A: Cryst. Phys. Diffr., Theor. Gen. Crystallogr.* **A32,** 751.
7. Herschlag, D., and Jencks, W. P. (1990). *J. Am. Chem. Soc.* **112,** 1942.
8. Gray, H. B. (1980). *In* "Methods for Determining Metal Ion Environments in Proteins" (D. W. Darnall and R. G. Wilkins, eds.), Vol. 2, p. 1. Elsevier, Amsterdam.
9. Blumberg, W. D., and Peisach, J. (1971). *In* "Bioinorganic Chemistry" (R. Dessey, J. Dillard, and L. Taylor, eds.), p. 271. Am. Chem. Soc., Washington, D.C.
10. Mims, W. B., and Peisach, J. (1981). *In* "Biological Magnetic Resonance" (L. J. Berliner and J. Reuben, eds.), Vol. 3, p. 213. Plenum, New York.
11. Kevan, L. (1979). *In* "Time Domain Electron Spin Resonance" (L. Kevan and R. N. Schwartz, eds.), p. 279. Wiley (Interscience), New York.
12. McCracken, J., Peisach, J., and Dooley, D. M. (1987). *J. Am. Chem. Soc.* **109,** 4064.
13. Reed, G. H., and Leyh, T. S. (1980). *Biochemistry* **19,** 5472.

14. Kofron, J. L., Ash, D. E., and Reed, G. H. (1988). *Biochemistry* **27**, 4781.
15. Powers, L. (1982). *Biochim. Biophys Acta* **683**, 1.
16. Blumberg, W. E., Desai, P. R., Powers, L., Freedman, J. H., and Villafranca, J. J. (1989). *J. Biol. Chem.* **264**, 6029.
17. Villafranca, J. J. (1989). *In* "Methods in Enzymology" (N. J. Oppenheimer and T. L. James, eds.), Vol. 177, p. 403. Academic Press, New York.
18. Mildvan, A. S. (1988). *FASEB J.* **3**, 1705.
19. Villafranca, J. J. (1982). *In* "Methods in Enzymology" (D. L. Purich, ed.), Vol. 87, p. 180. Academic Press, New York.
20. Mildvan, A. S., and Gupta, R. K. (1978). *In* "Methods in Enzymology" (C. H. W. Hirs and S. N. Timasheff, eds.), Vol. 49, p. 322. Academic Press, New York.
21. Nowak, T. (1981). *In* "Spectroscopy in Biochemistry" (J. E. Bell, ed.), Vol. 2, p. 109.
22. Westheimer, F. H. (1968). *Acc. Chem. Res.* **1**, 70.
23. Herschlag, D., and Jencks, W. P. (1990). *Biochemistry* **29**, 5172.
24. Usher, D. A., Richardson, D., and Eckstein, F. (1970). *Nature (London)* **228**, 663; Usher, D. A., Erenrich, E. S., and Eckstein, F. (1972). *Proc. Natl. Acad. Sci. U.S.A.* **69**, 115.
25. Sheu, K-F. R., Richard, J. P., and Frey, P. A. (1979). *Biochemistry* **18**, 5548.
26. Richard, J. P., Prasher, D. C., Ives, D. H., and Frey, P. A. (1979). *J. Biol. Chem.* **254**, 4339.
27. Blätter, W. A., and Knowles, J. R. (1979). *J. Am. Chem. Soc.* **101**, 510.
28. Richard, J. P., Carr, M. C., Ives, D. H., and Frey, P. A. (1980). *Biochem. Biophys. Res. Commun.* **94**, 1052.
29. Richard, J. P., and Frey, P. A. (1978). *J. Am. Chem. Soc.* **100**, 7757.
30. Hansen, D. H., and Knowles, J. R. (1981). *J. Biol. Chem.* **256**, 5967.
31. Pollard-Knight, D., Potter, B. V. L., Cullis, P. M., Lowe, G., and Cornish-Bowden, A. (1982). *Biochem J.* **201**, 421.
32. Orr, G. A., Simmons, J., Jones, S. R., Chin, G. J., and Knowles, J. R. (1978). *Proc. Natl. Acad. Sci. U.S.A.* **75**, 2230.
33. Pluria, D. H., Schomberg, D., Richard, J. P., Frey, P. A., and Knowles, J. R. (1980). *Biochemistry* **19**, 325.
34. Lowe, G., and Potter, B. V. L. (1981). *Biochem. J.* **199**, 227.
35. Sheu, K.-F., Ho, H.-T., Nolan, L. D., Markovitz, P., Richard, J. P., Utter, M. F., and Frey, P. A. (1984). *Biochemistry* **23**, 1779.
36. Konopka, J. M., Lardy, H. A., and Frey, P. A. (1986). *Biochemistry* **25**, 5571.
37. Jarvest, R. L., Lowe, G., and Potter, B. V. L. (1981). *Biochem. J.* **199**, 427.
38. Tsai, M.-D., and Chang, T. T. (1980). *J. Am. Chem. Soc.* **102**, 5416.
39. Webb, M. R., and Trentham, D. R. (1980). *J. Biol. Chem.* **255**, 1775.
40. Bryant, F. R., Benkovic, S. J., Sammons, D., and Frey, P. A. (1981). *J. Biol. Chem.* **256**, 5965.
41. Jarvest, R. L., and Lowe, G. (1981). *Biochem. J.* **199**, 273.
42. Lowe, G., Cullis, P. M., Jarvest, R. L., Potter, B. V. L., and Sproat, B. S. (1981). *Philos. Trans. R. Soc. London Ser. B*, **293**, 75.
43. Miziorko, H., and Eckstein, F. (1984). *J. Biol. Chem.* **259**, 13037.
44. Arnold, J. P., Cheng, M.-S., Cullis, P. M., and Lowe, G. (1986). *J. Biol. Chem.* **261**, 1985.
45. Saini, M. S., Buchwald, S., VanEtten, F. L., and Knowles, J. R. (1981). *J. Biol. Chem.* **256**, 10456.
46. Jones, S. R., Kindman, L. A., and Knowles, J. R. (1978). *Nature (London)* **275**, 564.
47. Webb, M. R., and Trentham, D. R. (1981). *J. Biol. Chem.* **256**, 4884.
48. Lowe, G., and Potter, B. V. L. (1982). *Biochem. J.* **201**, 665.
49. Tsai, M.-D., and Chang, T. T. (1980). *J. Am. Chem. Soc.*, **102**, 5416.

50. Webb, M. R., Grubmeyer, C., Penefsky, H. S., and Trentham, D. R. (1980). *J. Biol. Chem.* **255**, 11637.
51. Webb, M. R., and Trentham, D. R. (1980). *J. Biol. Chem.* **255**, 8629.
52. Eccleston, J. F., and Webb, M. R. (1982). *J. Biol. Chem.* **257**, 5046.
53. Senter, P. D., Eckstein, F., and Kagawa, Y. (1983). *Biochemistry* **22**, 5514.
54. Gonzalez, M. A., Webb, M. R., Welsh, K. M., and Cooperman, B. S. (1984). *Biochemistry* **23**, 797.
55. Lowe, G., and Potter, B. V. L. (1981). *Biochem. J.* **199**, 693.
56. Blättler, W. A., and Knowles, J. R. (1980). *Biochemistry* **19**, 738.
57. Jones, J. P., Weiss, P. M., and Cleland, W. W. (1991). *Biochemistry* **30**, 3634.
58. Cleland, W. W., and Mildvan, A. S. (1979). *Adv. Inorg. Biochem.* **1**, 163.
59. Cornelius, R. D., and Cleland, W. W. (1978). *Biochemistry* **17**, 3279.
60. Merritt, E. A., Sundaralingam, M., Cornelius, R. D., and Clelend, W. W. (1978). *Biochemistry* **17**, 3274.
61. Dunaway-Mariano, D., and Cleland, W. W. (1980). *Biochemistry* **19**, 1496, 1506.
62. Pecoraro, V. L., Rawlings, J., and Cleland, W. W. (1984). *Biochemistry* **23**, 153.
63. Granot, J., Mildvan, A. S., Brown, E. M., Kondo, H. N., and Kaiser, E. T. (1979). *FEBS Lett.* **103**, 265.
64. Granot, J., Gibson, K. J., Switzer, R. L., and Mildvan, A. S. (1980). *J. Biol. Chem.* **255**, 10931.
65. Kramer, P., and Nowak, T. (1988). *J. Inorg. Biochem.* **32**, 135.
66. Eckstein, F., and Goody, R. S. (1976). *Biochemistry* **15**, 1685.
67. Burgers, P. M. J., and Eckstein, F. (1980). *J. Biol. Chem.* **255**, 8229.
68. Jaffe, E. K., and Cohn, M. (1978). *J. Biol. Chem.* **253**, 4823.
69. Lee, M. H., Goody, R. S., and Nowak, T. (1985). *Biochemistry* **24**, 7594.
70. Fry, D. C., Byler, M., Susi, H., Brown, E. M., Kuby, S. A., and Mildvan, A. S. (1988). *Biochemistry* **27**, 3588.
71. Pai, E. F., Sachsenheimer, W., Schirmer, R. H., and Schultz, G. E. (1977). *J. Mol. Biol.* **114**, 37.
72. Yan, H., Dahnke, T., Zhou, B., Nakazawa, A., and Tsai, M.-D. (1990). *Biochemistry* **29**, 10956.
73. Kalbitzer, H. R., Marquetant, R., Connolly, B. A., and Goody, R. S. (1983). Eur. J. Biochem. **133**, 226.
74. Yan, H., and Tsai, M.-D. (1991). *Biochemsitry* **30**, 5539.
75. Jarori, G. K., Ray, B. D., and Nageswara Rao, B. D. (1985). *Biochemistry* **24**, 3487.
76. Leyh, T. S., Sammons, R. D., Frey, P. A., and Reed, G. H. (1982). *J. Biol. Chem.* **257**, 15047.
77. Leyh, T. S., Goodhart, P. J., Nguyen, A. C., Kenyon, G. L., and Reed, G. H. (1985). *Biochemistry* **24**, 308.
78. Cleland, W. W. (1990). *FASEB J.* **4**, 2899.
79. Lee, M. H., and Nowak, T. (1984). *Biochemistry* **23**, 6506.
80. Duffy, T. H., and Nowak, T. (1985). *Biochemistry* **24**, 1152.
81. Gupta, R. K., Fung, C. H., and Mildvan, A. S. (1976). *J. Biol. Chem.* **251**, 2421.
82. Muirhead, H., Clayden, D. A., Barford, D., Lorimer, C. G., Fothergill-Gillmore, L. A., Schlitz, E., and Schmitt, E. (1986). *EMBO J.* **5**, 475.
83. Lodato, D. T., and Reed, G. H. (1987). *Biochemistry* **26**, 2243; Buchbinder, J. L., and Reed, G. H. (1990). *Biochemistry* **29**, 1799.
84. Nowak, T., and Mildvan, A. S. (1972). *Biochemistry* **11**, 2819.
85. Villafranca, J. J., and Raushel, F. M. (1982). *Fed. Proc.* **41**, 2961.

86. Brewer, J. M. (1981). *Crit. Rev. Biochem.* **11,** 209.
87. Lee, B. H., and Nowak, T. (1992). *Biochemistry* **31,** 2165.
88. Nowak, T., Kenyon, G. L., and Mildvan, A. S. (1973). *Biochemistry* **12,** 690.
89. Stubbe, J. A., and Abeles, R. H. (1980). *Biochemistry* **19,** 5505.
90. Lee, M. E., and Nowak, T. (1992). *Biochemistry* **31,** 2172; Lee, M. E. (1988). Ph.D. Thesis, University of Notre Dame, Notre Dame.
91. Lebioda, L., and Stec, B. (1991). *Biochemistry* **30,** 2817.
92. Matthews, B. W. (1988). *Acc. Chem. Res.* **21,** 333; Holden, H. M., Tronurd, D. E., Monzingo, A. F., Weaver, L. H., and Matthews, B. W. (1987). *Biochemistry* **26,** 8542.
93. Christianson, D. W., and Lipscomb, W. N. (1989). *Acc. Chem. Res.* **22,** 62.
94. Kim, H., and Lipscomb, W. N. (1990). *Biochemistry* **29,** 5546.
95. Bartlett, P. A., and Marlowe, C. K. (1987). *Science* **235,** 569.
96. Bash, P. A., Singh, U. C., Brown, F. K., Langridge, R., and Kollman, P. A. (1987). *Science* **235,** 574.
97. Grobelny, D., Goli, U. B., and Galardy, R. E. (1989). *Biochemistry* **28,** 4948.
98. Yamishita, M. M., Almassy, R. J., Janson, C. A., Cascio, D., and Eisenberg, D. (1989). *J. Biol. Chem.* **264,** 17681.
99. Abell, L. M., and Villafranca, J. J. (1991). *Biochemistry* **30,** 1413.
100. Balakrishnan, M. S., and Villafranca, J. J. (1978). *Biochemistry* **17,** 3531.
101. Ransom, S. C. (1984). Ph.D. Thesis, Pennsylvania State University, University Park.
102. Abell, L. M., and Villafranca, J. J. (1991). *Biochemistry* **30,** 6135.
103. Eads, C. D., LoBrutto, R., Kumar, A., and Villafranca, J. J. (1988). *Biochemistry* **27,** 165.
104. Jones, W., Kurz, L. C., and Wolfenden, R. (1989). *Biochemistry* **28,** 1242.
105. Wilson, D. K., Rudolph, F. B., and Quiocho, F. A. (1991). *Science* **252,** 1278.
106. Wolfenden, R., and Kati, W. M. (1991). *Acc. Chem. Res.* **24,** 209.

3

Phosphate Ester Hydrolysis

JOHN A. GERLT

Department of Chemistry and Biochemistry
University of Maryland
College Park, Maryland 20742

I. Introduction

Nearly 20 years have elapsed since the subject of phosphate ester hydrolysis has been discussed in a chapter in "The Enzymes" (*1*). During that time, meth-

95

THE ENZYMES, Vol. XX

ods were developed for the determination of the stereochemical consequences of both enzymic and nonenzymic hydrolyses of phosphate mono- and diesters, and high-resolution crystallographic studies have been completed on wild-type and site-directed mutant versions of phosphate mono- and diesterases. These advances, together with the application of more traditional methods of physical organic chemistry to the study of nonenzymic hydrolysis reactions, have permitted new insights into our understanding of enzyme-catalyzed phosphate ester hydrolysis. This chapter does not provide an exhaustive discussion of the progress made in the past two decades but emphasizes a selection of recent advances pertinent to our understanding of enzyme-catalyzed phosphate ester hydrolysis. The reader is referred to a number of recent reviews for more detailed discussions of some of the topics discussed in this chapter (2–8).

In this chapter I will examine recent evidence for and against two chemically distinct mechanisms with two chemically distinct intermediates in the hydrolyses of phosphate esters. Early experimental evidence for these mechanisms and intermediates was discussed in the previously cited chapter in this series (1), and, as such, this introduction will only briefly summarize the fundamental chemical and stereochemical differences between these mechanisms.

The first mechanism for a hydrolysis reaction at phosphorus is a nucleophilic or associative displacement of the leaving group. This mechanism is analogous to S_N2 displacement reactions at tetrahedral carbon and is often referred to as the $S_N2(P)$ mechanism. Unlike S_N2 reactions at carbon, $S_N2(P)$ reactions often involve the formation of a transiently stable intermediate in which five ligands are bonded to phosphorus. This species has trigonal bipyramidal geometry in which the two apical bonds are longer, and therefore weaker, than the three equatorial bonds. According to the principles put forth by Westheimer (9, 10), attacking nucleophiles and leaving groups occupy apical positions. However, the pentacoordinate phosphorane is capable of a ligand reorganization process, termed pseudorotation, in which the two apical ligands can become equatorial ligands and two of the equatorial ligands can become apical ligands. Thus, in $S_N2(P)$ reactions, "adjacent" attack of a nucleophile and departure of a leaving group is possible. Whereas the stereochemical consequence of an $S_N2(P)$ reaction in which pseudorotation does not occur is the expected inversion of configuration, the stereochemical consequences of reactions whose $S_N2(P)$ mechanisms involve pseudorotation are more complicated, since each pseudorotation step is the equivalent of an inversion of configuration. Thus, an $S_N2(P)$ reaction involving a single pseudorotatory process is retention: one inversion for the attack of the nucleophile and departure of the leaving group, and a second for the pseudorotation.

The second mechanism for a hydrolysis reaction at phosphorus involves the initial dissociative loss of the leaving group to generate the metaphosphate anion

(*11*). This reactive species is then captured by the nucleophile. This mechanism is analogous to S_N1 displacement reactions at tetrahedral carbon and is often referred to as the $S_N1(P)$ mechanism. Depending on the stability (lifetime) of the metaphosphate anion, the leaving group could have time to separate from the anion and be replaced by solvent so that it would be symmetrically solvated. Then, either of two solvent molecules could react with the metaphosphate anion to generate the hydrolysis product. If the mechanism were truly this dissociative, the stereochemical consequence of the $S_N1(P)$ reaction would be racemization. If the metaphosphate anion were so reactive that it would react with solvent as soon as the bond to the leaving group had been broken, the stereochemical consequence would be inversion of configuration; this variation of the $S_N1(P)$ mechanism has been termed "preassociative."

II. Methods for Syntheses and Configurational Analyses of Oxygen Chiral Phosphate and Thiophosphate Esters

In 1970, Eckstein and co-workers reported the first stereochemical study of an enzyme-catalyzed hydrolysis of a phosphate ester, the hydrolysis of the *endo* isomer of uridine 2′,3′-cyclic phosphorothioate (*endo*-cyclic UMPS) (*12*) by ribonuclease A (RNase A) (*13*). The hydrolysis of RNA catalyzed either by base or by RNase A proceeds by a two-step mechanism in which the 2′-hydroxyl group of a nucleotide unit within an RNA molecule acts as a nucleophile on the 3′-phosphodiester bond to displace the 5′-hydroxyl group of the neighboring nucleoside to form a 2′,3′-cyclic phosphate intermediate. RNase A then catalyzes the hydrolysis of this cyclic phosphate, mimicked by Eckstein's *endo*-cyclic UMPS, to yield the ultimate 3′-mononucleotide product.

The purpose of this investigation of RNase A was to determine whether pseudorotatory processes were important in the hydrolysis reaction. In the case of the reaction catalyzed by RNase A, no chemical evidence had been gathered that gave even a hint that the mechanism of the enzymic reaction involved the transient formation of a nucleotidylated enzyme intermediate. Thus, if the reaction were a simple, in-line $S_N2(P)$ reaction, the stereochemical course of the reaction would be inversion. However, if the $S_N2(P)$ reaction involved an initial attack of the water to form a trigonal bipyramidal intermediate whose breakdown to the 3′-mononucleotide would have to be preceded by a pseudorotation, the stereochemical course of the reaction would be retention. Thus, in this context, the observed *inversion* of configuration was used as early and important evidence against the involvement of pseudorotatory ligand reorganization processes of transient trigonal bipyramidal pentacovalent intermediates in enzyme-catalyzed reactions.

In view of the outcome of this experiment and the total agreement of a large body of subsequent chemical and stereochemical investigations of enzyme-catalyzed phosphoryl transfer reactions, including phosphate ester hydrolysis, pseudorotation of transient pentacovalent intermediates is now regarded as being unimportant in enzyme-catalyzed reactions. For this reason, the value of a stereochemical investigation of an enzyme-catalyzed phosphoryl transfer reaction is that it reliably provides an assessment of whether the reaction involves the formation of transient phosphorylated enzyme intermediates. An inversion of configuration is interpreted as a single $S_N2(P)$ displacement reaction on phosphorus (either associative or preassociative); a retention of configuration is interpreted as two successive $S_N2(P)$ displacement reactions, one to form a phosphorylated enzyme intermediate and a second to convert this intermediate to product. More involved interpretations invoking two or more even numbers of intermediates to explain an inversion of configuration or three or more odd numbers of intermediates to explain a retention of configuration are certainly possible (*14*), but no evidence has yet been obtained to invoke these complications. Racemization, the third possible stereochemical consequence of a hydrolysis reaction, would suggest the intermediacy of a *free* metaphosphate anion, but no example of racemization has yet been reported in an enzyme-catalyzed reaction. However, as detailed in Section III,A, considerable effort has been directed toward searching for appropriate substrates and experimental conditions in nonenzymic reactions that lead to the formation of free metaphosphate and metathiophosphate anions as evidenced by racemization of configuration.

Although the analytical techniques used in Eckstein's stereochemical study of the hydrolysis of *endo*-cyclic UMPS by RNase A might now be considered primitive in the context of more recently developed methodologies, this stereochemical experiment provided the intellectual basis and experimental design for many of the studies of both enzymic and nonenzymic hydrolyses of phosphate esters that have been performed in the succeeding years. Namely, a chiral substrate of known configuration is converted to a product of unknown configuration in solvent containing an oxygen isotope; the phosphorus in the reaction product, be it inorganic phosphate or a phosphate monoester, is converted to a diastereomeric center so that the chirality of the reaction product can be determined. Following Eckstein's seminal development of the use of chiral phosphorothioates to study the stereochemical consequences of enzyme-catalyzed phosphoryl and nucleotidyl transfer reactions, a number of laboratories independently developed methods for the chemical synthesis of phosphate monoesters and diesters chiral by virtue of oxygen isotope substitution and of phosphorothioate monoesters chirally labeled with two oxygen isotopes, so that the stereochemical consequences of phosphoryl transfer reactions, including phosphate ester hydrolysis, could be studied.

A. SYNTHESIS OF PHOSPHATE MONOESTERS CHIRAL
   BY VIRTUE OF OXYGEN ISOTOPES

Although phosphate monoesters chiral by virtue of oxygen isotope substitutions cannot be used in stereochemical studies of phosphate monoester hydrolysis (since there are only three stable isotopes of oxygen), they have been used profitably in studies of phosphoryl transfer reactions relevant to the question of the intermediacy of monomeric metaphosphate anion in phosphoryl transfer reactions (see Section III,A). The laboratories of Knowles and Lowe have reported general methods for the synthesis of phosphate monoesters chiral by virtue of oxygen isotope substitution, and these syntheses are summarized in this section.

1. *Knowles Synthesis*

The synthetic methodology developed by Knowles and co-workers (15) relies on the synthesis of a chiral cyclic phosphoramidate diester of the desired alcohol (or anhydride of the desired phosphoric acid, e.g., ADP in the synthesis of $[\gamma\text{-}^{16}O,^{17}O,^{18}O]$ATP). Knowles utilized ephedrine to provide both a chiral center required for separation of diastereomeric intermediates and convenient functional groups for the reactions that convert the neutral phosphoramidate diesters to the dianionic phosphate monoester product. In the usual execution of this synthesis, the ^{17}O, the least abundant stable isotope of oxygen that can be routinely obtained in enrichments no greater than approximately 60%, is efficiently incorporated in the essentially quantitative hydrolysis of PCl_5 to $P^{17}OCl_3$. The ^{18}O is later incorporated from $H_2^{18}O$, and the ^{16}O is derived from the ephedrine. The reader is referred to the original account by Knowles (15) for details of the synthesis (Fig. 1). Due to the epimeric composition of a synthetic intermediate, the configuration of the major phosphate monoester obtained from $P^{17}OCl_3$ is R_P; the enantiomeric or epimeric phosphate monoester having the S_P configuration is most conveniently and abundantly obtained by preparing $P^{18}OCl_3$ and subsequent incorporation of $H_2^{17}O$.

2. *Lowe Synthesis*

Simultaneously and independently, Cullis and Lowe developed a second general methodology for the synthesis of $^{16}O,^{17}O,^{18}O$-labeled chiral phosphate monoesters (16, 17). This synthesis relies upon the synthesis of a cyclic hydrobenzoin triester of the alcohol or phosphoric acid followed by hydrogenolysis to liberate the isotopically labeled monoester product (Fig. 2). Hydrobenzoin, chiral by virtue of stereospecific labeling with ^{16}O and ^{18}O, is the source of the two specified oxygen isotopes, and ^{17}O is derived from $H_2^{17}O$ via $P^{17}OCl_3$. The reader is referred to the articles by Cullis and Lowe for details of the synthesis.

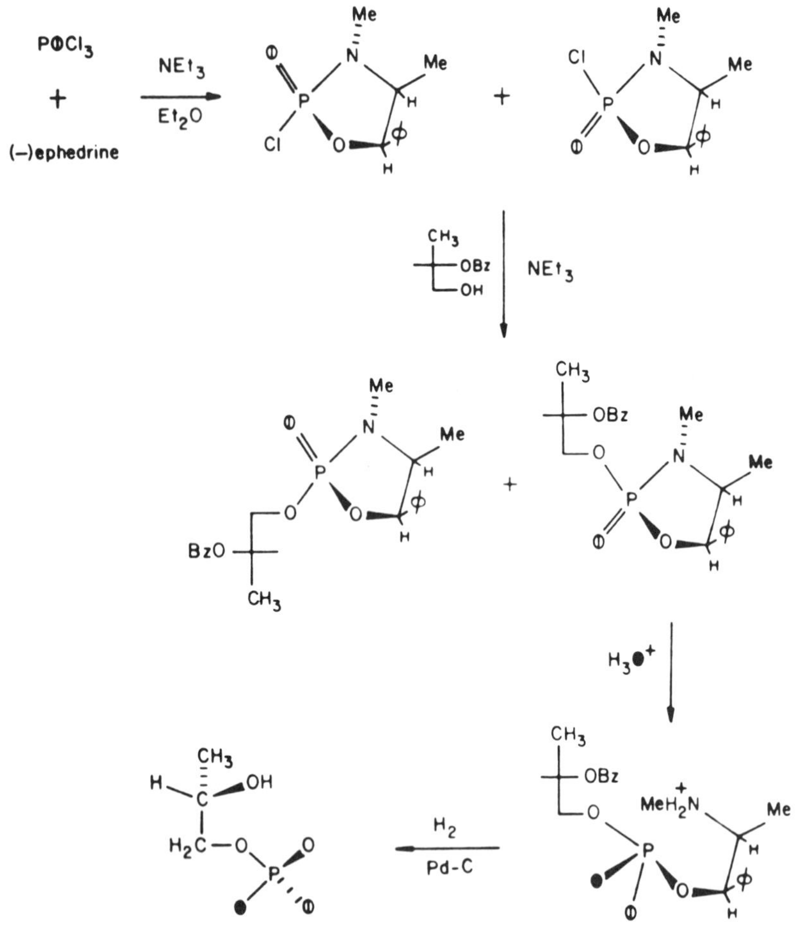

FIG. 1. Stereoselective synthesis of [1(R)-^{16}O,^{17}O,^{18}O]phospho-(S)-1,2-propane-diol. O, ^{16}O; Ø, ^{17}O; ●, ^{18}O. From Ref. 15.

The absolute configuration of a cyclic triester intermediate was initially assigned incorrectly (16), but was later corrected (18) when two independent studies of the stereochemical consequence of the reaction catalyzed by bovine cyclic 3',5'-nucleotide phosphodiesterase, using oxygen chiral substrates, obtained by completely independent procedures produced stereochemically opposite results (18–20). This procedure yields only the S_P enantiomer of the epimer of the phosphate monoester; the monoester having the opposite configuration must be obtained by exchanging the positions of two of the isotopes of oxygen.

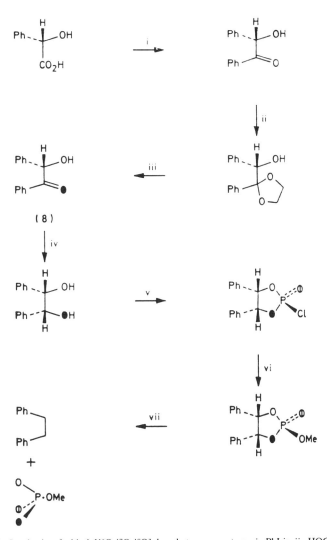

FIG. 2. Synthesis of chiral [^{16}O,^{17}O,^{18}O]phosphate monoesters. *i*, PhLi; *ii*, HOCH$_2$CH$_2$OH, *p*-CH$_3$C$_6$H$_4$SO$_3$H; *iii*, H$_2$[^{18}O]-dioxane-*p*-CH$_3$C$_6$H$_4$SO$_3$H; *iv*, LiAlH$_4$; *v*, P^{17}OCl$_3$, C$_5$H$_5$N; *vi*, Ch$_3$OH, C$_5$H$_5$N; *vii*, H$_2$, Pd. Ø, ^{17}O; ●, ^{18}O; Me, CH$_3$. From Ref. *17*.

B. SYNTHESIS OF THIOPHOSPHATE MONOESTERS CHIRAL
     BY VIRTUE OF OXYGEN ISOTOPES

Enzymes that hydrolyze phosphate monoesters cannot be stereochemically studied with oxygen chiral substrates unless they also catalyze a transphosphorylation reaction to an alcohol acceptor, e.g., alkaline and acid phosphatases. Thus, stereochemical studies of enzymes such as ATPases have utilized phosphorothioate analogs of nucleotides so that chiral thiophosphate can be generated as product and configurationally determined. In addition, a number of recent studies of the mechanisms of phosphate monoester hydrolysis have focused on the chemical mechanisms of phosphorothioate ester hydrolysis, since thiometaphosphate anion is believed to be more stable than the biologically potentially more relevant metaphosphate anion, which has questionable mechanistic significance due to its presumed high reactivity.

1. *Nucleoside Phosphorothioates*

Frey and Richard have described clever syntheses of the terminal phosphorothioate analogs of AMP, ADP, and ATP, designated AMPS, ADPβS, and ATPγS, respectively (*21*).

2. *General Syntheses of Phosphorothioate Monoesters*

Lowe, Cullis, and co-workers have described general methods for the synthesis of oxygen chiral phosphorothioate monoesters (see Figs. 3 and 4). The synthesis by Cullis is strictly analogous to the synthesis of chiral phosphate monoesters by Knowles except that ($-$)-ephedrine is thiophosphorylated with $PSCl_3$ (*22*). The synthesis due to Lowe of chiral phosphorothioate monoesters is strictly analogous to the synthesis of chiral phosphate monoesters except that $PSCl_3$ is used to thiophosphorylate the oxygen chiral hydrobenzoin (*23*).

C. SYNTHESIS OF PHOSPHATE DIESTERS CHIRAL BY VIRTUE
     OF OXYGEN ISOTOPES AND OF THIOPHOSPHATE DIESTERS

1. *Reaction of Phosphoranilidates with Carbonyl Compounds*

The laboratories of Stec (*24*) and the author (*26–28*) have reported that the nitrogen anions of phosphoramidates react with carbonyl compounds (benzaldehyde or CO_2) to incorporate an atom of oxygen bonded to phosphorus at the expense of the nitrogen substituent, and that this reaction proceeds with retention of configuration at phosphorus. This stereochemically useful reaction, known as the Wittig–Staudinger reaction, is not limited to carbonyl compounds but can also be used with thiocarbonyls or selenocarbonyls to produce chiral phosphorothioates or phosphoroselenates with retention of configuration at phosphorus (*27*). The primary application of this reaction has been in the preparation of both

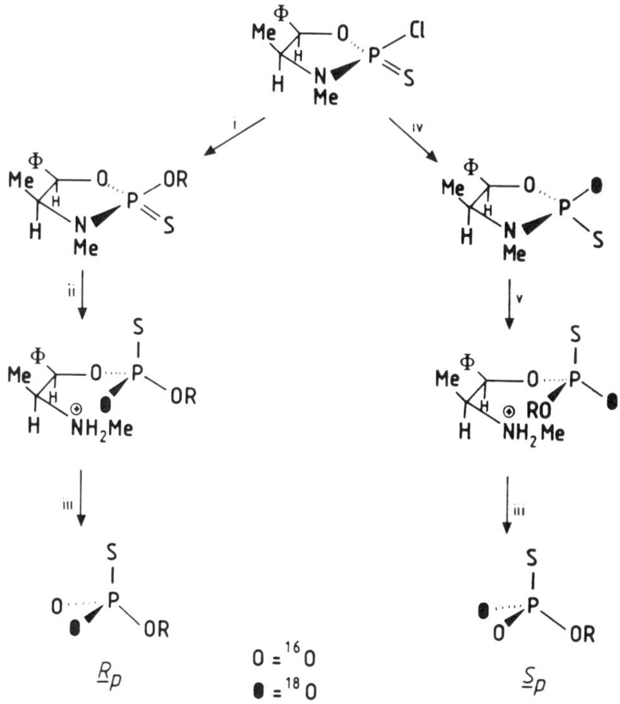

FIG. 3. Synthesis of isotopically chiral [$^{16}O,^{18}O$ (or ^{17}O)]thiophosphate monoesters of either R_P or S_P absolute configuration. *i*, RO$^{\ominus}$/ROH; *ii*, H$_2$^{18}O/CF$_3$CO$_2$H; *iii*, (CH$_3$)$_3$SiI or Na, liquid NH$_3$; *iv*, Li^{18}OH; *v*, ROH/CF$_3$CO$_2$H. O, ^{16}O; ●, ^{18}O; Me, CH$_3$. From Ref. *22*.

FIG. 4. Synthesis of enantiomeric [$^{16}O,^{17}O,^{18}O$]thiophosphates. (*i*) PSCl$_3$, C$_5$H$_5$N; (*iia*) H$_2$^{17}O; (*iib*) Na, liquid NH$_3$. Ø, ^{17}O; ●, ^{18}O. From Ref. *23*.

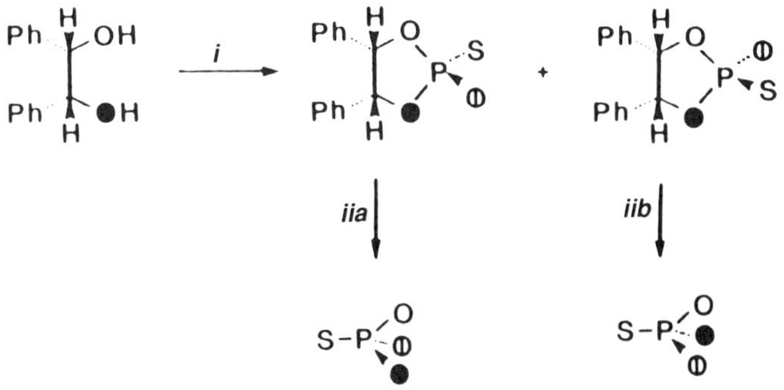

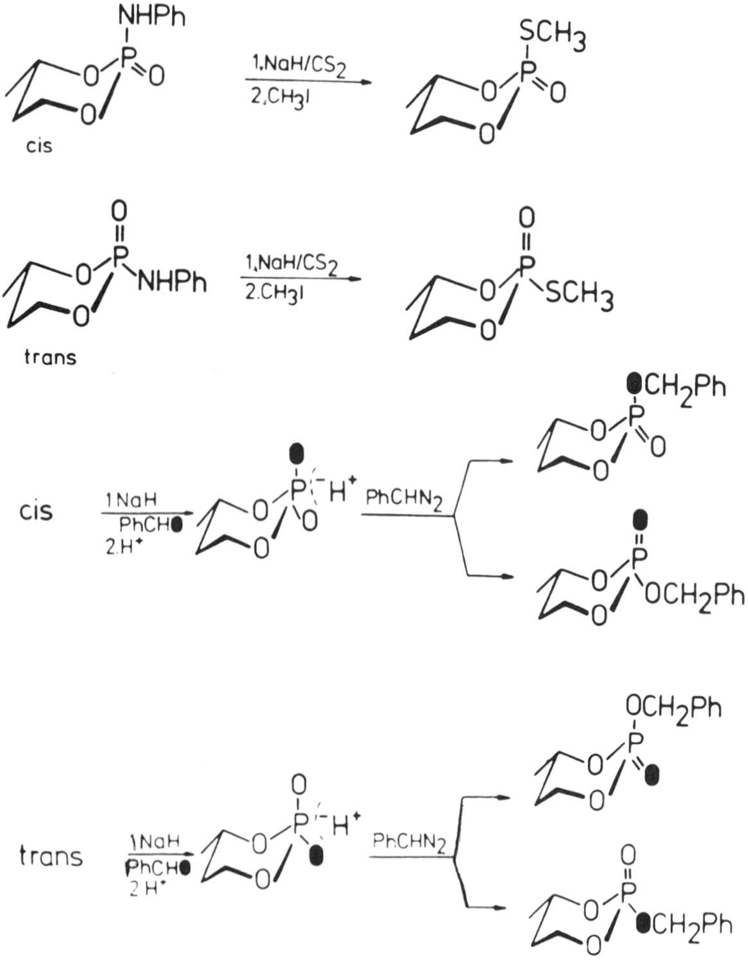

FIG. 5. Diastereomers used in the stereochemical studies on PN to PS conversion. ●, ¹⁸O. From Ref. 27.

acyclic and cyclic chiral phosphate diesters and phosphorothioate diesters. The reader is referred to the original literature for synthetic details (27) (Fig. 5).

D.   CONFIGURATIONAL ANALYSES OF OXYGEN CHIRAL PHOSPHATES

Methods based on the effects of the quadrupolar ^{17}O nucleus ($I = \frac{5}{2}$) and the ^{18}O nucleus ($I = 0$) on the ^{31}P NMR spectral properties of the isotopically labeled ^{31}P nucleus are now the only methods used to ascertain the configurations of tetrahedral phosphorus atoms in oxygen chiral phosphate esters. The chemical shift effect of the heavy ^{18}O nucleus on the NMR resonance of the directly bonded ^{31}P nucleus was first used to determine configurations of chiral diesters; later, the quadrupolar effects (line broadening) of ^{17}O on the directly bonded ^{31}P nucleus were combined with the effects of ^{18}O to determine the configurations of chiral monoesters. These methods are summarized in this section.

1. Effect of ^{18}O on ^{31}P NMR Chemical Shifts

Cohn and Hu (29), Lutz et al. (30), and Lowe and Sproat (31) independently reported that ^{18}O directly bonded to a ^{31}P nucleus caused an upfield chemical shift of the associated ^{31}P NMR resonance, the magnitude of which is related to the number of bonds between the ^{18}O and ^{31}P nuclei; the upfield chemical shift effect is approximately 0.02 parts per million (ppm) per single bond. Thus, by measuring the upfield chemical shift perturbation relative to the chemical shift of unlabeled phosphorus compound, the number of bonds between a ^{31}P nucleus and directly bonded ^{18}O nuclei can be counted.

The first application of this ^{18}O effect for determining the configuration of an oxygen chiral phosphate ester was the author's determination of the configuration of diastereomeric samples of cyclic [^{16}O,^{18}O]dAMP, the chiral substrate for studying the stereochemical consequences of the reverse reaction catalyzed by adenylate cyclase (formation of cyclic AMP from ATP), and of the hydrolysis reaction catalyzed by 3',5'-cyclic nucleotide phosphodiesterase (25) (see Fig. 6). The cyclic nucleotide structure provided a very convenient framework on which to demonstrate that ^{31}P NMR spectroscopy could be used to determine configuration, since the exocyclic oxygens on the chiral phosphorus atom are diastereomeric and are forced to occupy chemically nonequivalent axial and equatorial dispositions relative to the rigid and asymmetric six-membered ring. Thus, alkylation of one of the exocyclic oxygens would convert the rigid cyclic phosphodiester to the rigid cyclic axial and equatorial phosphotriesters, with the alkylation serving to introduce single- or double-bond character to the bonds between the phosphorus and the isotopically labeled exocyclic oxygens. Depending upon whether the ^{18}O nucleus is located in the axial or equatorial exocyclic position, the effect of the ^{18}O nucleus on the chemical shift of the ^{31}P nucleus would be different due to the varying number of bonds between the ^{18}O and ^{31}P nuclei.

2. Combined Effects of ^{17}O and ^{18}O on ^{31}P NMR Chemical Shifts

Whereas the ^{18}O effect on ^{31}P NMR chemical shifts is sufficient for determining the configurations of chirally labeled samples of prochiral phosphorus atoms in a diastereomeric environment, additional use of the quadrupolar effect of ^{17}O on ^{31}P NMR resonances is required for configurational analyses of oxygen chiral phosphate monoesters. The basic strategy for the configurational analyses of phosphate monoesters is the same, i.e., the enantiomeric center in the monoester must be converted to a diastereomeric center in a cyclic phosphodiester so that

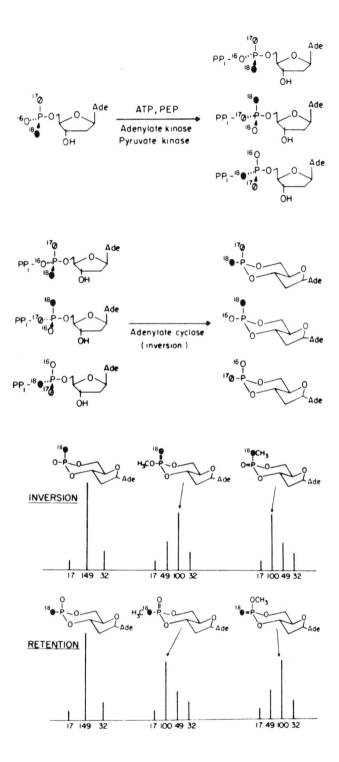

alkylation will yield diastereomeric triesters in which the number of bonds between the ^{18}O and ^{31}P nuclei can be counted. However, the function of the ^{17}O nucleus is to broaden the resonances having bonds between ^{17}O and ^{31}P nuclei so that the only species that will be visible by ^{31}P NMR spectroscopy are those with only ^{16}O and/or ^{18}O bonded to the ^{31}P nucleus (32, 33). If ^{17}O were available in 100% enrichment, the analysis would be identical to that already described for ^{16}O,^{18}O-labeled phosphodiesters; however, ^{17}O is not available routinely in enrichments exceeding 60%, so the analysis is complicated by the ^{16}O and ^{18}O, which "contaminate" the ^{17}O.

The first analysis of the configuration of an oxygen chiral phosphate monoester by ^{31}P NMR spectroscopy was accomplished in the laboratory of Knowles. Knowles and co-workers had previously used metastable ion mass spectroscopy to determine the configurations of [^{16}O,^{17}O,^{18}O]phosphate monoesters. A phosphatase that catalyzes a competing transphosphorylation reaction transfers the chiral phosphoryl group from the ester of unknown configuration to an acceptor chiral diol with retention of configuration; in this way, the transfer the chiral phosphoryl group to a chiral sample of 1,2-propanediol (34) or 1,3-butanediol (35) can be accomplished in a low but tolerable yield, but the desired effect of generating a diastereomeric center at phosphorus is achieved. The resulting monoester is then chemically cyclized by a reaction that proceeds by inversion of configuration and is accompanied by loss of one atom of oxygen from the chiral phosphoryl group. The resulting phosphodiesters are converted to phosphotriesters by alkylation, and the configuration of the chiral phosphorus center is established by ^{31}P NMR spectroscopy. The reader is referred to articles by Knowles for a detailed description of this procedure (34,35).

A conceptually similar approach can be used to determine the configurations of chiral phosphate monoesters of chiral alcohols without the necessity for transferring the chiral phosphoryl group to 1,3-butanediol. For example, in a number of stereochemical studies of hydrolysis reactions performed in the author's laboratory, 3'- or 5'-nucleotides were obtained as reaction products. In the case of samples of 5'-AMP, these can be enzymically converted to isotopically labeled cyclic 3',5'-nucleotides by initial pyrophosphorylation with adenylate kinase and pyruvate kinase, followed by enzymic cyclization with inversion of configuration by a bacterial adenylate cyclase (20) (see Figs. 6 and 7). In the case of samples of 3'-nucleotides or 5'-nucleotides other than 5'-AMP, the cyclization reactions can be accomplished chemically (23). In either case, following chemical con-

---

FIG. 6. (Top) Enzymic phosphorylation of the [^{16}O,^{17}O,^{18}O]dAMP to a mixture of three types of isotopically labeled dATP that could be cyclized (middle) enzymically with inversion of configuration to yield a mixture of three types of chiral cyclic dAMP. (Bottom) Idealized spectra for the hydrolysis of cyclic [^{17}O,^{18}O]dAMP in H_2^{16}O. From Ref. 20.

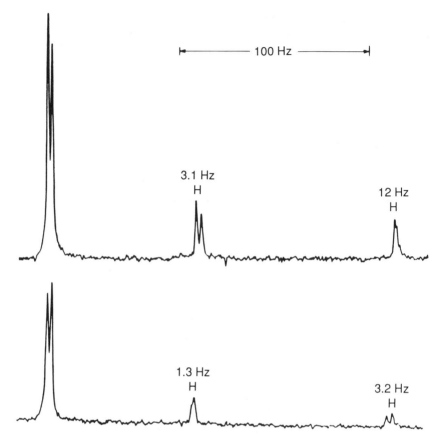

FIG. 7. 80.9 MHz ^{31}P NMR spectra of ethyl esters of cyclic 2'-deoxy-AMP prepared from equimolar mixtures of unlabeled cyclic 2'-deoxy-AMP and the cyclic 2'-[^{18}O]deoxy-AMP samples. (*Top*) ^{18}O-Labeled ester from the axial P-anilidate. The approximate chemical shift of the unlabeled diester is 3.0 ppm, that of the equatorial ester is 4.5 ppm, and that of the axial ester is 6.5 ppm. From Ref. *25*.

version to the diastereomeric triesters, ^{31}P NMR can be used to ascertain configuration (Fig. 8).

E.   CONFIGURATIONAL ANALYSIS OF OXYGEN CHIRAL THIOPHOSPHATE

As noted previously, the stereochemical course of the hydrolysis of a phosphate monoester can be studied only if an oxygen chiral phosphorothioate ester is used a substrate so that an enantiomer of ^{16}O,^{17}O,^{18}O-labeled chiral thiophosphate is the product. The laboratories of both Webb and Trentham (*36*) and of Tsai (*37*) developed methods for determining the configurations of the enantiomers of chiral thiophosphate; these analyses are based on the previously dis-

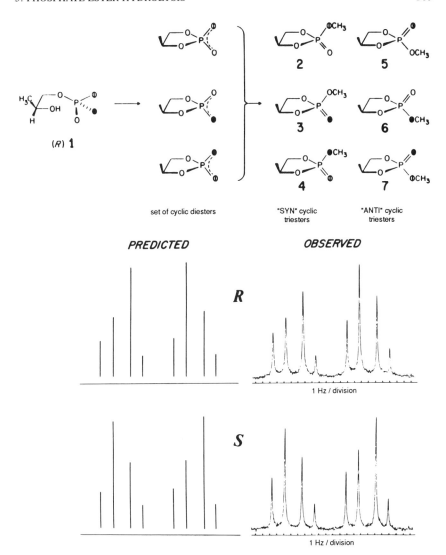

FIG. 8. (*Top*) The three cyclic diesters and six cyclic triesters that are derived from 1-(*R*)-[^{16}O,^{17}O,^{18}O]phosph-(*S*)-1,2-propane-diol by "in-line" ring closure and methylation. (*Bottom*) Predicted and observed ^{31}P NMR spectra of the mixtures of syn and anti cyclic triesters derived from labeled samples of 1-phospho-(*S*)-1,2-propane-diols that are *R* or *S* at phosphorus. From Ref. *34*.

cussed effects of ^{17}O and ^{18}O on ^{31}P NMR resonances. Using a series of enzymic reactions, the oxygen chiral thiophosphate is incorporated in reactions of known stereochemistry to the β-thiophosphoryl group of the S_P diastereomer of ATPβS, and the magnitudes of the ^{18}O perturbations on the resonances of the β- and γ-phosphorus atoms are measured (see Fig. 9).

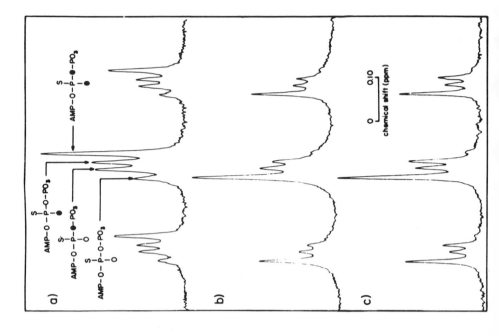

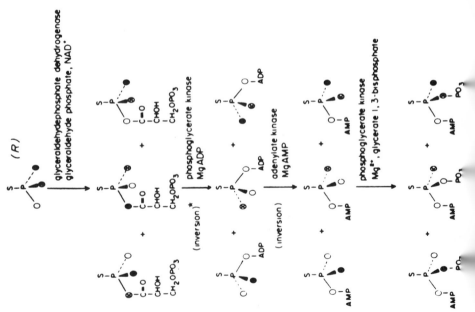

110

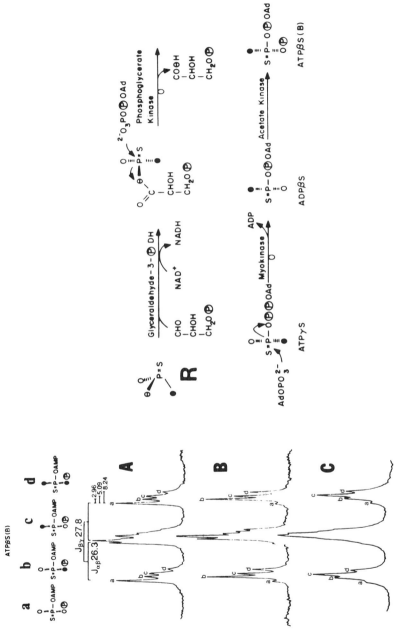

FIG. 9. Sequence of reactions used in the analysis of the enantiomers of the inorganic thiophosphate and ^{31}P NMR spectra of the β-phosphorus of (a) [β-^{18}O; $\beta\gamma$-^{18}O]ATPβS; ATPβS derived from the R enantiomer (b) or S enantiomer of inorganic [^{16}O,^{17}O,^{18}O]thiophosphate of Webb and Trentham (*top*) and Tsai (*bottom*). From Refs. *36* and *37*.

111

F. CONFIGURATIONAL ANALYSIS OF OXYGEN CHIRAL THIOPHOSPHATE MONOESTERS

As noted previously, studies of the mechanisms of phosphate monoester solvolysis have been extended to the mechanisms of the analogous phosphorothioate ester solvolysis because the thiometaphosphate anion is believed to be more stable than the metaphosphate anion. Thus, a general method based upon ^{31}P NMR spectroscopy for the configurational analysis of chiral thiophosphate monoesters (see Fig. 10) was described recently by Cullis and co-workers (38).

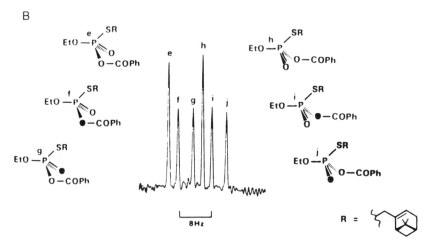

FIG. 10. ^{31}P NMR spectra of (A) the compound derived from R_P-4-nitrophenyl [^{16}O,^{18}O]thiophosphate and (B) the compound derived from racemic ethyl [^{16}O,^{18}O]thiophosphate. From Ref. 38.

III. Nonenzymic Phosphate Ester Hydrolysis

A. EVIDENCE FOR METAPHOSPHATE INTERMEDIATES

In recent years, perhaps the most intensely studied question concerning the mechanism of nonenzymic phosphate ester hydrolysis has been whether metaphosphate anion has sufficient stability to exist as a truly stable intermediate in solvolysis reactions. As was noted in Section I, one stringent test for the existence of a symmetrically solvated (and, therefore, "free") metaphosphate anion is whether the transfer of a chiral phosphoryl group is accompanied by racemization of configuration. The development of methods of syntheses and configurational analyses for oxygen chiral phosphate esters permitted stereochemical tests for the existence of "free" metaphosphate anion. In parallel with the search for metaphosphate anion, stereochemical studies directed toward determining whether the analogous thiometaphosphate anion can exist were also initiated. Thiometaphosphate anion has been proposed to be involved in the solvolyses of thiophosphate monoesters, since, like the solvolyses of phosphate monoesters, the dependence of reaction rate on pH is bell shaped and there is little sensitivity to the nucleophilicity of the solvent (39). The dependence of rate on pH has been interpreted as the monoprotonated ester generating the reactive species by a pre-equilibrium proton transfer from the phosphoryl group to the leaving group. Thiometaphosphate anion has been predicted to be more stable than metaphosphate anion since thiophosphate monoesters, both in the mono- and dianionic forms, solvolyze more rapidly than the analogous phosphate monoesters (40). Also, the effects of pressure on the hydrolysis of a thiophosphate monoester have been examined and compared to the effects of pressure on the hydrolysis of phosphate monoesters (41). The volume of activation for an associative mechanism is negative whereas that for a dissociative mechanism is positive. Whereas the volume of activation for the hydrolysis of the dianion of 2,4-dinitrophenyl phosphate is -4.8 cm^3 mol^{-1}, the volume of activation for the hydrolysis of the dianion of 2,4-dnitrophenyl thiophosphate is $+11$ cm^3 mol^{-1}. The observation of a symmetrically solvated thiometaphosphate anion, as evidenced by solvolysis of a chiral thiophosphate monoester with racemization of configuration, would be pertinent to the interpretation of results obtained in stereochemical studies of enzyme-catalyzed hydrolyses of phosphate monoesters, because chiral thiophosphate monoesters must be used as substrates. However, the observation of "free" thiometaphosphate anion could not be used to support the mechanistically significant occurrence of "free" metaphosphate anion in enzyme-catalyzed phosphoryl transfer reactions.

1. Solvolyses of Thiophosphate Monoesters

Cullis, Lowe, and co-workers have independently described reaction conditions that allow the formation of free thiometaphosphate anion, as evidenced by racemization of configuration.

The ethanolysis of the R_P diastereomer of 4-nitrophenyl [^{18}O]thiophosphate has been studied by Cullis and co-workers (42). The leaving group (4-nitrophenoxide) is a sufficiently good leaving group that protonation is not required for a solvolytic reaction; thus, the stereochemical consequences of solvolysis reactions involving both the monoanionic (where intramolecular proton transfer to the leaving group is possible) and dianionic (where intramolecular proton transfer is impossible) states of ionization can be and have been performed. The monoanion is solvolyzed in ethanol at 45°C to yield ethyl [^{18}O] thiophosphate, and the configuration of this product was found to be 60% S_P and 40% R_P, in contrast to the 95% R_P and 5% S_P configuration of the starting material. Under these reaction conditions, neither the reisolated substrate nor the authentic chiral ethyl thiophosphate provided any evidence for racemization. Under analogous solution conditions, except that the dianion of the chiral substrate was present, complete racemization of the transferred thiophosphoryl group was observed. Thus, the transfer of the chiral thiophosphoryl group from 4-nitrophenol or 4-nitrophenoxide to ethanol occurs with significant racemization and a small component of inversion, so a largely "free" thiometaphosphate anion must be involved in the solvolysis reaction. Similar observations were made when the monoanion was solvolyzed in water–ethanol mixtures and the ethyl thiophosphate product was configurationally analyzed.

Lowe and co-workers have reported studies of the hydrolyses of unlabeled deoxyadenosine 5′-β-thiodiphosphate (ADPβS) in H_2^{16}O as well as of both diastereomers of [β-^{17}O]ADPβS in H_2^{18}O (43). The hydrolysis of ADPβS is more rapid at pH 4.5, where the terminal thiophosphoryl group is monoanionic, than at pH 7.2, where this group is dianionic, thereby suggesting a mechanism involving thiometaphosphate. The chiral starting materials were synthesized by a procedure analogous to the procedure of Frey for preparing nucleoside phosphorothioates, and the configurational analysis of the chiral thiophosphate product was determined by a procedure developed in the laboratory of Lowe. At pH 4.5, the R_P diastereomer was observed to produce 73% S_P and 27% R_P [^{16}O, ^{17}O,^{18}O]thiophosphate, and the S_P diastereomer was observed to produce 75% R_P and 25% S_P [^{16}O,^{17}O,^{18}O]thiophosphate. Both the unreacted chiral ADPβS and authentic chiral thiophosphate were found to retain their configurational integrity under the reaction conditions. Thus, in water at a pH where intramolecular proton transfer is possible (pH 4.5 for 1.5 hr), ADPβS hydrolyzes with a significant amount of racemization of configuration. This demonstrates that in aqueous solution thiometaphosphate anion has sufficient stability to diffuse away from the AMP leaving group so that approximately 50% of the molecules of ADPβS that hydrolyze do so via a symmetrically solvated thiometaphosphate anion rather than via an associative [either $S_N2(P)$ or preassociative $S_N1(P)$] mechanism. At higher pH (pH 7.2 for 5.6 hr), hydrolysis of the chiral ADPβS proceeded with complete inversion of configuration and can be explained by a change to a mechanism that does not involve "free" thiometaphosphate anion.

In summary, Cullis and Lowe have demonstrated that the generation of stable thiometaphosphate in solution is possible. The next section describes analogous studies on probing for the existence of "free" metaphosphate anion, since it is this intermediate that has potential significance in enzyme-catalyzed phosphoryl transfer reactions.

2. Solvolysis of Phosphate Monoesters

Knowles has cogently summarized evidence obtained from a wide range of studies of solvolyses of phosphate monoesters that supports the participation of metaphosphate anion (35). The most recent evidence will be briefly summarized in this section before the results of stereochemical studies are discussed.

A very clever "three-phase test" for the detection of metaphosphate intermediates in phosphoryl transfer reactions has been described by Rebek and co-workers (44). The basis of this test is the use of two polymers suspended in solution. The donor polymer contains a potential precursor to metaphosphate anion, e.g., an acyl phosphate or a phosphoramidate, and the recipient polymer contains an acceptor nucleophile, e.g., an amine. After reaction and physical separation of the polymers, the recipient polymer is analyzed for covalently bound phosphate. Since very few of the phosphoryl groups to be transferred will be on the surface of the donor polymer, detection of significant transfer to the recipient polymer provides evidence for a diffusible intermediate, i.e., free metaphosphate anion. Significant transfer did occur in dioxane or acetonitrile suspensions of the polymers, thereby providing evidence for an intermediate. However, this test for diffusible and, therefore, relatively stable metaphosphate anion is compromised by the choice of solvent. Both dioxane and acetonitrile can provide unshared electron pairs for the highly electrophilic metaphosphate anion such that the actual species that migrates from the donor polymer to the recipient polymer may be a complex between metaphosphate anion and the solvent. Such a role for solvent has been investigated stereochemically, the results of which will be described later in this section.

Ramirez and Marecek have investigated the solvolyses of 2,4-dinitrophenyl phosphate in aprotic and protic solvents and described reaction conditions that are consistent with the involvement of "free" metaphosphate anion (45, 46). In particular, the dianion of the reactive phosphate monoester is capable of transferring its phosphoryl group to tert-butanol whereas the monoanion of the same ester is essentially unreactive in the same reaction. Since tert-butanol is unreactive as a nucleophile for steric reasons, the phosphorylation of this alcohol is considered to be a diagnostic test for the involvement of the highly reactive metaphosphate anion.

Gorenstein and co-workers have used primary ^{18}O isotope effects to study the structure of the transition state in the hydrolysis of 2,4-dinitrophenyl phosphate in which the bridging ester oxygen is labeled with ^{18}O (47). The results of 32 determinations were averaged to reveal an isotope effect of 1.0204 + 0.0044.

This seemingly small isotope effect must be compared to the isotope effect measured for the hydrolysis of the dibenzyl ester precursor to the isotopically labeled monoester [phosphate triesters can only hydrolyze by an $S_N2(P)$ mechanism], $1.0070 + 0.0038$, for proper interpretation. The fact that the hydrolysis of the monoester has a larger isotope effect was interpreted as demonstrating that substantial cleavage of the phosphate ester bond occurs in the transition state; this is in accord with the expectation based on an $S_N1(P)$ mechanism.

More recently, Cleland and co-workers have measured secondary ^{18}O isotope effects in the hydrolysis of glucose 6-phosphate (48). The rationale for this study was the expectation that an associative mechanism would require a decrease in bond order to the labeled oxygens as the transition state is approached ($\frac{4}{3}$ in the tetrahedral substrate and 1 in the trigonal bipyramidal transition state); in contrast, a dissociative mechanism would involve an increase in bond order ($\frac{8}{3}$ in metaphosphate). Thus, an $S_N2(P)$ mechanism would be accompanied by a normal isotope effect; an $S_N1(P)$ mechanism would be accompanied by an inverse isotope effect. Glucose 6-phosphate was prepared with ^{18}O substitution in all three nonbridging oxygens and ^{13}C in the C-1 position. The remote labeling with ^{13}C was utilized such that any rate effects in the hydrolysis of the oxygen-labeled phosphate monoester would be reflected in the isotope content at C-1. Both the product glucose and unreacted starting material can be enzymically degraded to liberate C-1 as CO_2, and the isotope content of the CO_2 can be accurately determined with an isotope ratio mass spectrometer. The cube root of the observed ^{18}O isotope effect, 1.013 ± 0.002, was taken to obtain the isotope effect for a single ^{18}O substitution, 1.0046. Since this study was conducted at pH 4.5, where glucose 6-phosphate exists largely as the monoanion, the isotope effect on the pK_a (1.0125) was also utilized to obtain the final value of 1.0004 (1.001 for three ^{18}O atoms) (49). While this value is not inverse, it also is not overwhelmingly normal. Thus, the conclusion is that the mechanism is either dissociative or the axial bond order in the transition state is very low (rather than unity) for a true $S_N2(P)$ mechanism. In either case, the transition states would be very similar, and this supports the belief that the monoanions of phosphate monoesters hydrolyze via a dissociative mechanism.

Jencks, Skoog, and Herschlag have carried out hallmark studies of the nature of the transition state in the phosphoryl group transfer reaction, including phosphate monoester hydrolysis.

Skoog and Jencks initially investigated the rates of transfer of the phosphoryl group from phosphorylated 3-methoxypyridine to a variety of substituted pyridines and amines that differ in pK_a and concluded that "free" metaphosphate anion is not involved as an intermediate (50). The basis for this experiment is the expectation that if metaphosphate anion is a true intermediate, a change in the rate-determining step should occur as the basicity of the acceptor pyridine or amine is varied. When the acceptor is less basic (nucleophilic) than 3-

methoxypyridine, the rate of reaction of the acceptor with the putative intermediate would be slower than the back reaction of 3-methoxypyridine with the intermediate; when the acceptor is more basic (nucleophilic) than 3-methoxypyridine, the rate of reaction of the acceptor with the putative intermediate would be faster than the back reaction. Thus, such a stepwise mechanism would predict a change in the rate-determining step, with the pK_a of 3-methoxypyridine being the breakpoint in the Brønsted plot. If the reaction were concerted with a single transition state, the Brønsted plot would be linear (or gently curved as the structure of the transition state was changed as a function of the pK_a of the acceptors). The Brønsted plot that was obtained was essentially linear with a slope $\beta_{nuc} = 0.17$. This rules out the occurrence of "free" metaphosphate anion and a preassociation mechanism in which bond breaking from the donor and bond making to the acceptor occur completely stepwise, since in each case the slope would be zero for basic acceptors. Instead, the results are consistent with a concerted reaction or a preassociation mechanism in which there is weak interaction with the nucleophile in both steps. A similar conclusion was reached by Bourne and Williams in concurrent studies (51).

Jencks and Jencks (52) have defined an interaction coefficient that relates β_{nuc} and β_{lg} for a reaction in which bond formation to the nucleophile (nuc) and bond breaking to the leaving group (lg) are coupled, as evidenced by a change in β_{nuc} as the leaving group is varied (or, analogously, a change in β_{lg} as the nucleophile is varied). Skoog and Jencks determined the interaction coefficient $p_{xy} = 0.013$ for the transfer of a phosphoryl group between N-phosphorylated pyridines and nitrogen nucleophiles. If either a truly dissociative mechanism (free metaphosphate) or a preassociation mechanism were operating, no coupling between the nucleophile and leaving group should be observed.

In a recent extension of these studies, Herschlag and Jencks have examined the reactions of phosphorylated pyridines with oxygen nucleophiles (53). Their studies are based on a kinetic distinction between preassociative and associative mechanisms, i.e., since the rate constant for diffusion that separates two species in an encounter pair is $\sim 10^{10}$ sec^{-1} and the frequency for a bond vibration is $\sim 10^{13}$ sec^{-1}, the most that a rate constant can be for a reaction occurring via a preassociation mechanism is 10^3. In the case of nitrogen nucleophiles studied by Skoog and Jencks, the observed variation in rate constants was 10^5. In particular, they have evaluated whether water behaves as other oxygen nucleophiles or whether it has anomalous properties that might allow the formation of metaphosphate in aqueous solution. As in the case of nitrogen nucleophiles, the reaction of phosphorylated pyridines with oxygen nucleophiles, including water, is characterized a linear Brønsted plot ($\beta_{lg} = 0.51$) and by a nonzero value for the interaction coefficient between β_{nuc} and β_{lg} ($p_{xy} = 0.013$). These observations are consistent with a concerted associative reaction mechanism.

The studies by Herschlag and Jencks also describe conditions that, in principle, would allow the transfer of a phosphoryl group from a phosphorylated pyridine to involve "free" metaphosphate (54). The nonzero value of the interaction coefficient p_{xy} allows calculation of pK_a values for the nucleophile and leaving group such that β_{nuc} and β_{lg} would be zero, i.e., no dependence of bond making and breaking on the leaving group and nucleophile, respectively. The values so obtained, $pK_{nuc} = -20$ and $pK_{lg} = -13$, are not compatible with reactions in aqueous solutions, so the occurrence of a stepwise mechanism involving metaphosphate can be discounted.

All of these studies are consistent with the solvolyses of phosphate monoesters involving some form of metaphosphate anion; however, on the basis of these experiments, the question of whether the anion exists symmetrically solvated and, therefore, can be considered to be a "free" intermediate in solution remains uncertain. Knowles, Cullis, and co-workers have used stereochemical techniques to examine the intermediacy of "free" metaphosphate anion in hydrolyses and alcoholyses of chiral phosphate monoesters. These experiments are important since they place significant constraints on the lifetime of the reactive intermediate and, therefore, clarify the large number of observations made in other laboratories on the question of whether metaphosphate anion is sufficiently stable that it can be considered a mechanistically significant intermediate.

The stereochemical consequences of the methanolyses of the monoanion of phenyl [$^{16}O,^{17}O,^{18}O$]phosphate, the dianion of 4-nitrophenyl [$^{16}O,^{17}O,^{18}O$] phosphate, and [$^{16}O,^{17}O,^{18}O$]phosphocreatine have been determined by Knowles and co-workers (35). The monoanion of phenyl phosphate behaves as a typical phosphate monoester in that its rate of hydrolysis is maximal at pH 4, where an intramolecular proton transfer is possible. The dianion of 4-nitrophenyl phosphate is highly reactive since protonation of the leaving group is not necessary. Finally, N-phosphoguanidines have been reported to be the most reactive phosphoryl compound (the chiral phosphocreatine can be enzymically synthesized from [γ-$^{16}O,^{17}O,^{18}O$]ATP). Thus, the solvolyses of all three of these compounds are believed to involve the participation of metaphosphate anion. The methanolysis of each of these compounds proceeds with *quantitative* inversion of configuration.

These stereochemical observations clearly do not support the intermediacy of a "free" metaphosphate anion. Instead, they support a preassociative mechanism in which the transfer of the phosphoryl group to the acceptor occurs only when it is properly positioned to receive the phosphoryl group; otherwise, the phosphoryl group reverts to starting material faster than it can diffuse from the leaving group to be symmetrically solvated. In view of the observations made by Jencks and Skoog, the methanolysis is likely to occur via a preassociative concerted mechanism involving a single transition state. Based on these experiments, it can be concluded that the *hydrolysis* of a phosphate monoester will also occur without the intermediacy of "free" metaphosphate anion, in contrast to the previously discussed ability to detect "free" thiometaphosphate anion in the hydrolysis of ADPβS. Thus, it is highly unlikely that metaphosphate anion is a mechanistically significant intermediate in enzyme-catalyzed phosphoryl transfer reactions.

Given the inability to implicate "free" metaphosphate anion in these studies, Cullis and Rous examined the stereochemical consequence of an alcoholysis reaction (55). When $[^{16}O,^{17}O,^{18}O]P^1$-$O$-ethyl-$P^1$-thiopyrophosphate is treated with methyl iodide in ethanol, the S-methyl derivative rapidly decomposes to form O-ethyl-S-methyl thiophosphate and ethyl $[^{16}O,^{17}O,^{18}O]$phosphate; this reaction is thought to proceed via a mechanism involving metaphosphate anion, but, as would be expected on the basis of the observations by Knowles, the stereochemical consequence is nearly quantitative inversion of configuration. In dichloromethane solution, the decomposition of the oxygen chiral substrate yields ethyl $[^{16}O,^{17}O,^{18}O]$phosphate that has suffered extensive racemization (approximately 70%). This stereochemical result can be rationalized by the intermediacy of "free" metaphosphate anion in the absence of nucleophile acceptors.

The laboratories of both Knowles and Cullis have described solvolysis conditions in which "free" metaphosphate anion can exist. Initially both laboratories investigated the possible stabilization of metaphosphate anion by acetonitrile, since this solvent was reported by Rebek et al. to allow a successful application of the three-phase text for "free" metaphosphate anion (44). The Harvard laboratory studied the reaction of phenyl $[^{16}O,^{17}O,^{18}O]$phosphate with tert-butanol in acetonitrile (56), and the Leicester laboratory studied the reaction of $[\beta$-$^{16}O,^{17}O,^{18}O]$ADP with 2-O-benzyl-(S)-1,2-propanediol in acetonitrile (57). In both cases, complete racemization was observed, and this can be explained by the complexation of metaphosphate anion by the acetonitrile solvent. Thus, the success of the three-phase test of Rebek et al. for metaphosphate presumably can be attributed to diffusion of an acetonitrile–metaphosphate anion complex rather than "free" metaphosphate anion.

The criterion given by Ramirez for the participation of metaphosphate in a solvolysis reaction was that the sterically hindered (and, therefore, nonnucleophilic) alcohol tert-butanol could be phosphorylated. Accordingly, the logical extension of this hypothesis to stereochemical studies searching for conditions favoring "free" metaphosphate anion was to conduct solvolysis reactions in neat tert-butanol. The solvolysis of 4-nitrophenyl $[^{16}O,^{17}O,^{18}O]$phosphate in tert-butanol was studied in the laboratory of Knowles and the tert-butyl $[^{16}O,^{17}O,^{18}O]$phosphate product was found to be completely racemic (56). This result can be explained only by the generation of a "free" metaphosphate anion in the sterically hindered solvent, which, once generated and symmetrically solvent, can react with the solvent to form the sterically hindered product. This conclusion is supported by complementary ^{31}P NMR experiments performed in the laboratory of Cullis. The reaction of $[\alpha,\beta\text{-}^{18}O]ADP$ in acetonitrile, acetonitrile–*tert*-butanol, and *tert*-butanol was studied by positional isotope exchange techniques in which scrambling of the bridging ^{18}O into a nonbridging position was studied (58). Such scrambling can occur only if metaphosphate anion is generated and sufficiently stable to allow rotation of the torsiosymmetric phosphoryl group in the $[^{18}O]AMP$. Thus, this technique also probes the lifetime of the metaphosphate anion that is generated, since the preassociative concerted mechanism previously discussed would not be expected to allow such rotation to occur, i.e., if the acceptor is not present, the cleavage of the P–O bond to liberate metaphosphate should not occur. Significant scrambling of the position of the ^{18}O in the unreacted ADP was detected in each solvent, with the scrambling detected in neat *tert*-butanol being in accord with the liberation of "free" metaphosphate anion (the scrambling detected in the presence of acetonitrile could be attributed to formation of an acetonitrile–metaphosphate anion complex).

Thus, even though "free" metaphosphate anion cannot be considered a mechanistically significant intermediate in enzyme-catalyzed phosphate monoester ester hydrolysis, since it is unlikely that an acceptor nucleophile will not be present to participate in a preassociative mechanism, it can exist in solution under appropriate solvent conditions when an acceptor nucleophile is unavailable.

B.   Evidence for Phosphorane Intermediates

As noted in Section I, the mechanism of an $S_N2(P)$ reaction can involve the formation of a transiently stable pentacovalent, trigonal bipyramidal intermediate that, depending on the exact reaction, can undergo pseudorotation before breakdown to product can be achieved. No evidence for pseudorotation has yet been obtained for enzyme-catalyzed reactions; however, many nonenzymic displacement reactions have been demonstrated to proceed by *retention* of configuration, and this is best explained by pseudorotation of a pentacovalent intermediate.

1.   *Stereoelectronic and Ring Strain Effects in Rates of Hydrolyses of Phosphate Esters*

The previously mentioned chapter in Volume VIII of "The Enzymes" summarized Westheimer's pioneering work on the importance of pentacoordinate intermediates and pseudorotation in the hydrolyses of five-membered ring cyclic phosphorus acid esters (1). That work sought to explain both the rapid hydrolyses of five-membered ring cyclic phosphates (hydrolyses of the five-membered ring cyclic phosphate triester methyl ethylene phosphate and of the cyclic phosphate diester ethylene phosphate are at least 10^6 times faster than acyclic analogs) and the interesting observation that hydrolysis of the cyclic triester proceeds with both endo- and exocyclic P–O bond cleavage (9, 10). The rapid rates were attributed to relief of ring strain in a trigonal bipyramidal pentacoordinate intermediate, and the occurrence of endo- and exocyclic cleavage was explained by the possibility for pseudorotation of the intermediate.

Gorenstein and co-workers have proposed that stereoelectronic effects are also important in the cleavage of P–O bonds to potential leaving groups in penta-coordinate species: when an oxygen lone pair is antiperiplanar to the P–O bond to a potential leaving group, the rate of cleavage of the bond is greater than the situation when oxygen lone pairs are not in this arrangement (59). While this effect was first recognized by Deslongchamps in the hydrolysis of acetals (60), several observations led Gorenstein to propose that these effects are important in the hydrolysis of phosphate esters.

Perhaps the initial suggestion that stereoelectronic effects might be important was the observation that the rates of hydrolysis of five-membered ring cyclic phosphate esters could not be explained entirely on the basis of relief of strain in the transition state. While thermochemical measurements of the ring strain in the methyl ethylene phosphate have yielded a range of values between 5.5 and 9 kcal/mol (61, 62), the enthalpy of hydrolysis of ethylene phosphate is more exothermic than that of the acyclic diethyl phosphate by only 4.6 kcal/mol (63). Since the activation energy for hydrolysis of the cyclic phosphate esters is approximately 10 kcal/mol lower than that for the acyclic esters, relief of ring strain apparently cannot entirely explain the rapid rates of hydrolysis; the remaining decrease in activation energy may, in part, be caused by stereoelectronic effects.

Gorenstein sought additional evidence for stereoelectronic effects by examining the rates of hydrolysis of conformationally constrained cyclic phosphotriesters in which antiperiplanar arrangements of electron pairs on heteroatoms would either be optimal for enhanced cleavage of specific P–O bonds or would be missing (59). The results of these studies have uniformly provided support for the stereoelectronic arguments. In these studies, Gorenstein has also sought to quantitate the magnitude of the contribution of the stereoelectronic effect to the observed rate acceleration in five-membered ring cyclic esters, and the conclusion is that the stereoelectronic effects contribute about 3 kcal/mol and ring strain contributes about 5 kcal/mol to the decrease in activation energies.

2. *Stereochemical Detection of Phosphorane Intermediates*

Stereochemical data exist for a large number of displacement reactions on neutral phosphoric acid and thiophosphoric acid esters and amides (64). Many of these occur with retention of configuration, especially when the displacement reactions occur on five-membered ring compounds, and this stereochemical course is explained by pseudorotation of a pentacoordinate intermediate. In contrast, only one stereochemical study has been performed on an anionic phosphate ester with the intention of detecting the participation of a phosphorane intermediate.

Knowles and co-workers have studied the intramolecular transfer of an oxygen

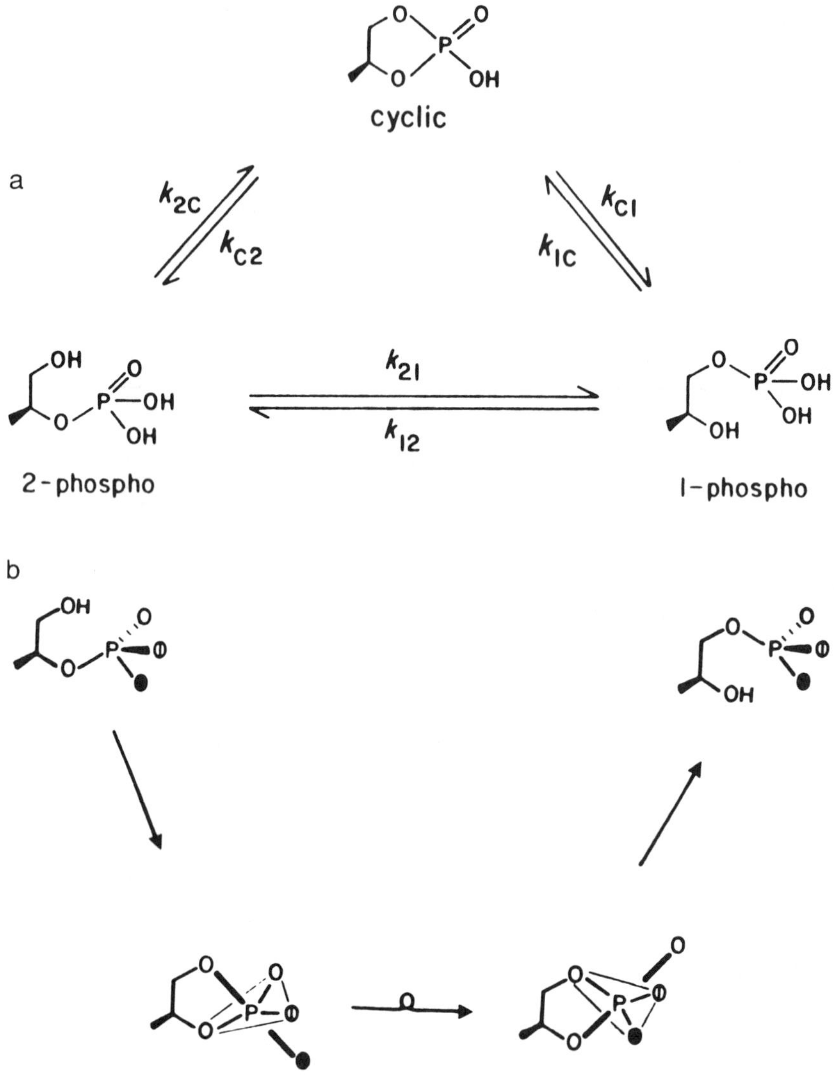

FIG. 11. Pathways for the isomerization of 2-phospho-1,2-propane-diol to 1-phospho-1,2-propane-diol via (a) a cyclic intermediate or (b) pseudorotation. From Ref. 35.

chiral phosphoryl group from the 2- to the 1-hydroxyl group of S-1,2-propanediol (Fig. 11) that occurs without the incorporation of solvent oxygen (65). Two mechanisms have been proposed for this intramolecular transfer, one involving the intermediate formation of the five-membered ring 1,2-cyclic phosphate and the second involving direct transfer via a pentacoordinate intermediate, which would necessarily undergo pseudorotation to accomplish the "adjacent associative" mechanism. In both mechanisms the stereochemical course of the transfer reaction would be retention, with the first mechanism being the result of two sequential inversions of configuration and the second being the result of formation and breakdown of a pentacoordinate species and pseudorotation. However, the first mechanism necessarily must be accompanied by incorporation of solvent oxygen while the second would involve no incorporation of solvent oxygen.

At 85°C in 0.5 N HClO$_4$, the secondary phosphate monoester of 1,2-propanediol isomerizes to the primary phosphate monoester with minimal hydrolysis to yield inorganic phosphate. The necessary detailed kinetic and solvent isotope incorporation analysis that was performed revealed that 36% of the intramolecular phosphoryl group transfer was direct and proceeded via the cyclic phosphorane intermediate. When the isomerization reaction was studied with oxygen chiral 2-phospho-2(S)-propanediol, the observed extent of retention of configuration was consistent only with a mechanism involving direct intramolecular transfer that was accompanied by retention of configuration. The rate of pseudorotation of the intermediate is sufficiently rapid to compete successfully with breakdown of the same intermediate to yield the 1,2-cyclic phosphate. Thus, in this enzymologically relevant example, pseudorotation of a pentacoordinate phosphorane intermediate can occur in aqueous solution.

3. Kinetic Detection of Phosphorane Intermediates

Breslow and co-workers have studied the mechanism of imidazole-catalyzed hydrolysis of both cyclic phosphate esters and of RNA (67–72). These studies are directed toward a more detailed understanding of the mechanism of the hydrolysis of RNA catalyzed by ribonuclease A (RNase A). In particular, recent studies by Breslow and co-workers have addressed the interesting and enzymologically pertinent question of the origin of the bell-shaped dependence of hydrolytic rate constant on pH in both enzymic and nonenzymic reactions.

Classically, the bell-shaped dependence of rate of the enzymic reaction on pH has been attributed to general acid and base catalysis by the two histidine residues in the active site, His-12 and His-119 (66). Support for this explanation based on the kinetic properties of a model system was first provided by an observation by Breslow and co-workers that β-cyclodextrin functionalized with two imidazole groups will catalyze the 1,2-cyclic phosphate of 4-tert-butylcatechol (67). The dependence of hydrolysis rate on pH mimics that of RNase A, and this behavior demonstrates that the presence of two imidazole functional groups on a nonionizable framework is the simplest kinetic mimic of the enzyme.

More recently, Breslow and co-workers have described a provocative observation that could suggest that the mechanism of the reaction catalyzed by RNase A involves a kinetically significant phosphorane intermediate whose reactivity is controlled by the active site histidine residues. Having devised a sensitive assay for measuring the rate of hydrolysis of phosphodiester bonds in RNA and oligo-nucleotides (68), the dependence of the imidazole-catalyzed hydrolysis of RNA polymers and dimers on both pH and imidazole concentration was studied (69, 70). The expected bell-shaped dependence of rate on pH was observed, suggesting roles for both general acid and general base catalysis. However, the dependence of rate on imidazole concentration was first order. The mechanism usually cited for the imidazole-catalyzed hydrolysis of RNA, i.e., the mechanism of the reaction catalyzed by RNase A, occurs in a single step, with imidazole catalyzing the abstraction of a proton from the 2′-hydroxyl group to facilitate nucleophilic attack on phosphorus, and imidazolium ion catalyzing the protonation of the leaving 5′-hydroxyl group. Such a mechanism predicts that the reaction should be second order in imidazole.

An alternate explanation for the bell-shaped dependence of rate on pH is a change in rate-determining step as the relative amounts of the basic imidazole and acidic imidazolium species are changed. If the hydrolysis of RNA catalyzed by imidazole involved two steps, with one requiring the basic form and the second requiring the acidic form, the dependence of rate on pH would also be bell shaped. Moreover, if the hydrolysis reaction involves two steps, a transiently stable, kinetically significant intermediate must lie on the reaction pathway. The most reasonable structure for such an intermediate would be a five-membered ring containing cyclic phosphorane.

Given these observations, Breslow has proposed that formation of the intermediate would involve imidazole-catalyzed abstraction of the proton from the 2′-hydroxyl group followed by protonation of the dianionic phosphorane resulting from attack of the alkoxide to regenerate the imidazole. The second step would involve imidazolium-catalyzed protonation of the 5′-hydroxyl leaving group followed by abstraction of a proton from the monoanionic phosphorane to regenerate imidazolium. Alternatively, the first step could involve initial imidazolium ion-catalyzed protonation of the phosphodiester linking in RNA followed by abstraction of the proton from the 2′-hydroxyl group, and the second step could involve imidazole-catalyzed removal of a proton from the monoanionic phosphorane intermediate followed by protonation of the leaving 5′-hydroxyl group. In either case, the mechanism involves the participation of both imidazole and imidazolium ion, but, in contrast to the concerted mechanism, these species catalyze two separate steps that catalyze the formation and breakdown of a phosphorane intermediate.

Recently, Anslyn and Breslow have examined the effectiveness of the three possible β-cyclodextrins functionalized with two imidazole groups (71). These substituents are located on the C-6 of the parent cyclodextrin. In the case when the imidazole functional groups are located on adjacent glucose moieties, the angle between the functional groups may average 51°; in the case when the two functional groups are separated by two intervening unmodified glucose residues, the angle between the functional groups may average 153°. These strikingly different locations of the imidazole functional groups were predicted to place severe geometric restrictions on the mechanism of cyclic phosphate ester hydrolysis. Interestingly and in confirmation of the hypothesis by Breslow that general acid catalysis is required to facilitate general base-catalyzed formation of a phosphorane intermediate, the bisfunctionalized cyclodextrin with imidazole groups on adjacent glucose units was the best catalyst and the modified cyclodextrin with the imidazole groups most widely separated was the poorest catalyst. Subsequent proton inventory experiments demonstrated that two protons were undergoing changing in bonding (72). This evidence provides further support for the intermediacy of kinetically significant phosphorane intermediates in the imidazole-catalyzed hydrolyses of RNA and five-membered ring cyclic phosphodiesters.

The key question is whether this interesting mechanistic complexity detected in the absence of RNase A is pertinent to the mechanism of the reaction catalyzed by RNase A. Of course, since the enzyme-catalyzed reaction can be considered to be an intramolecular reaction, concentration dependence of the reaction on the concentrations of His-12 and His-119 cannot be studied. Furthermore, experiments by Breslow do not reveal the steady-state concentration of the phosphorane intermediate, and it can be anticipated that this presumably reactive intermediate would not accumulate in the active site of RNase A, at least at physiological temperatures. However, the kinetic detection of such intermediates in the imidazole-catalyzed hydrolysis of RNA does suggest that the mechanism of the reaction catalyzed by RNase A should receive additional scrutiny.

IV. Enzymic Phosphate Ester Hydrolysis

A. ALKALINE PHOSPHATASE FROM *ESCHERICHIA COLI*

Alkaline phosphatase from *Escherichia coli* is a homodimeric enzyme that requires one Mg^{2+} and two Zn^{2+} ions for catalysis; the identical subunits contain 449 amino acids (73). A 2.8-Å X-ray structure for the enzyme is available (74), thereby allowing formulation of hypotheses regarding the functions of the metals ions and of active site residues in catalysis. The metal ions are clustered in the active site in the immediate vicinity of Ser-102, the residue that has been implicated in the formation of a covalent phosphorylated enzyme intermediate. Also, the guanidinium group of Arg-166 electrostatically interacts with the carboxylate of Asp-101 and has been proposed to properly position Ser-102, since the sequence Asp^{101}-Ser^{102}-Ala^{103} is conserved among a number of alkaline phosphatases; arsenate, a competitive inhibitor, binds between Ser-102 and the two Zn^{2+} ions, with the phosphate analog also being located within hydrogen bonding distance of Arg-166. In addition, since the gene for the enzyme has been cloned and sequenced, mutations of active site residues have been constructed by site-directed mutagenesis with the intention of testing mechanistic hypotheses.

The mechanism of an actual hydrolysis reaction catalyzed by this prototype phosphomonoesterase has never been studied stereochemically. This apparent omission is presumably explained by the very low catalytic efficiency of the enzyme toward phosphorothioate monoesters as compared to phosphate monoesters (75); certainly, chiral [^{17}O,^{18}O]phosphorothioate O-ester substrates already exist, and methodology is available for the configurational analysis of the chiral [^{16}O,^{17}O,^{18}O]thiophosphate that would be produced if the chiral substrate were hydrolyzed in H_2^{16}O. In fact, the low catalytic reactivity of phosphorothioate O-esters and the high reactivity of phosphorothioate S-esters has been explained by the enzyme utilizing nucleophilic catalysis (an associate mechanism) to achieve hydrolysis of the phosphate ester bond (40).

The first enzymic reaction to be studied with oxygen chiral phosphate monoesters was a reaction catalyzed by alkaline phosphatase (76). Alkaline phosphatase is well known to catalyze a transphosphorylation reaction in which the phosphoryl group of an alcohol donor is transferred to the primary hydroxyl group of diols (e.g., 1,2-propanediol or 1,3-butanediol). Knowles and co-workers synthesized phenyl [^{16}O,^{17}O,^{18}O]phosphate, whose configuration was assigned on the basis of the method of synthesis. This chiral phosphoryl group of this substrate was transferred to S-1,2-propanediol in less than 10% yield. Following ring closure, methylation, and analysis by ^{31}P NMR spectroscopy, the conclusion was reached that the transphosphorylation reaction had proceeded with *retention* of configuration at phosphorus. Because no experimental and stereochemical data on enzyme-catalyzed transfer reactions have demonstrated a requirement for pseudorotatory processes within the active site of an enzyme (so that a single-displacement reaction would proceed with retention of configuration) or the intermediacy of free metaphosphate anion (so that a single-displacement reaction would proceed with racemization of configuration), each enzyme-catalyzed displacement reaction at phosphorus is assumed to proceed with inversion of configuration. Thus, the stereochemical course of the reaction catalyzed by alkaline phosphatase is evidence that *two* displacement reactions at the chiral phosphoryl group necessarily occur during the reaction, and this is entirely consistent with a large body of chemical and biophysical data that have allowed characterization of a covalent phosphoryl enzyme intermediate (73). This stereochemical experiment, however, does not allow distinction between an associative $S_N2(P)$ mechanism and a preassociative $S_N1(P)$ mechanism.

Although the structure/reactivity studies conducted by Jencks and co-workers suggest that metaphosphate will not occur as an intermediate in aqueous reactions (54), and it is known that phosphorothioate esters undergo dissociative hydrolysis more rapidly than phosphate esters (40), Weiss and Cleland have recently utilized secondary ^{18}O isotope effects to probe the mechanism of the hydrolysis of glucose 6-[$^{18}O_3$]phosphate catalyzed by alkaline phosphatase (77). As noted earlier in this chapter, the expectation is that secondary ^{18}O isotope effects will be normal for an associative $S_N2(P)$ mechanism but inverse for mechanisms where the formation of metaphosphate is rate limiting (48). Isotope effects were measured both at pH 8, the pH optimum where phosphate dissociation is rate limiting, and at pH 6, where bond making and breaking are thought to be rate limiting. At the higher pH, the isotope effect per ^{18}O is 0.9994; at the lower pH, the isotope effect is 0.9982. The ratio between these numbers, 0.9988 ± 0.0006, is a small inverse isotope effect that can be interpreted as indicating that at the lower pH, where chemistry is rate limiting, the mechanism involves a dissociative contribution.

Specific active site mutations of alkaline phosphatase have been constructed in the laboratories of the late E. T. Kaiser and Kantrowitz. The work of Kaiser focused on Ser-102, with the major emphasis placed on the Cys substitution (S102C) (78). Unlike the wild-type enzyme, V_{max} for the S102C mutant was observed to depend on the pK_a of the leaving group. This observation suggested that the rate-determining step for the mutant enzyme was phosphorylation of the active site thiol nucleophile rather than product dissociation. Further evidence for this hypothesis was provided by the finding that Tris buffer does not increase the overall rate of substrate turnover as found for the wild-type enzyme; again, this can be most readily explained by the phosphorylation of the enzyme being the rate-determining step. The stereochemical course of the transphosphorylation reaction catalyzed by S102C is also retention (79).

More recently Kantrowitz and co-workers have examined the effects of substitutions for Arg-166 on catalysis (80). The alanine and serine substitutions, R166A and R166S, respectively, were constructed and found to retain similar and significant levels of catalytic activity. In the absence of Tris base, the V_{max} values of the mutant enzymes were approximately 30-fold lower than wild type; in the presence of Tris base, the V_{max} values differed only threefold. The primary effect of these mutations appears to be on binding of substrate and the product inorganic phosphate, and through an altered binding of substrate to the mutant enzymes the substitutions may also affect V_{max}. In the laboratory of Kantrowitz, the D101A mutant has also been constructed and characterized (recall that Asp-102 interacts with Arg-166); under some conditions this substitution was found to increase the catalytic activity of the enzyme, thereby negating the idea that this residue is essential for maintenance of a productive geometry in the active site.

B. 3',5'-CYCLIC NUCLEOTIDE PHOSPHODIESTERASE

Stereochemical studies have been performed on phosphohydrolases, which are considerably less well structurally characterized than alkaline phosphatase from

E. coli. One such enzyme and the first phosphohydrolase that was subjected to stereochemical scrutiny with both an oxygen chiral (*20*) and a chiral phosphorothioate (*81*) substrate was the 3′,5′-cyclic nucleotide phosphodiesterase obtained from bovine heart. The results of these studies were important in determining whether stereochemical studies employing phosphorothioates yield mechanistically reliable results, given the fact that phosphorothioates are considerably less reactive to nucleophilic attack given the lower electronegativity of the sulfur substituent (*40*). The demonstration that the enzyme-catalyzed hydrolyses of both the natural phosphate ester substrate and the phosphorothioate substrate analog occur with the same stereochemical consequence (inversion) was especially gratifying, given an initial report that sulfur substitution did, in fact, reverse the stereochemical outcome (*19*).

The stereochemical experiment employing the S_P diastereomer of cyclic AMPS, the phosphorothioate analog of 3′,5′-cyclic AMP, was performed by Stec and co-workers (*81*). The substrate was synthesized from the precursor phosphoroanilidate by reaction with CS_2. The hydrolysis was performed in $H_2^{18}O$ to generate one of the diastereomers of [$^{16}O,^{18}O$]AMPS as product. In view of the techniques that were subsequently developed for the identification of the configuration of chiral [$^{16}O,^{18}O$]AMPS (i.e., determination of the magnitude of the ^{18}O perturbation on the S_P diastereomer of ATPαS that was generated in the first step of the configurational analysis), the configurational analysis was laborious; however, the hydrolysis reaction was found to proceed with *inversion* of configuration. At the time of this experiment, 1979, some uncertainty existed as to the importance of pseudorotatory processes in ring-opening reactions involving six-membered ring cyclic phosphodiesters such as cyclic AMP. Thus, two mechanisms were proposed to be consistent with the stereochemical result: a single displacement with direct, in-line displacement of the 3′-hydroxyl group to yield the product, and the formation of an adenylated enzyme intermediate with retention of configuration (displacement accompanied by pseudorotation) followed by hydrolysis of the intermediate with inversion of configuration. On the basis of a stereochemical study of the nonenzymic hydrolysis of cyclic [$^{17}O,^{18}O$]dAMP in which inversion of configuration was observed for formation of both the oxygen chiral 3′- and 5′-dAMP products, the former mechanism is preferred (*82*).

The stereochemical experiment employing the R_P diastereomer of cyclic [$^{17}O,^{18}O$]dAMP as substrate was reported in 1981 by Gerlt and co-workers (*20*). This substrate was also synthesized from a precursor phosphoroanilidate by reaction with $C^{18}O_2$. The hydrolysis reaction was performed in $H_2^{16}O$ to generate one of the diastereomers of [$^{16}O,^{17}O,^{18}O$]dAMP as product. The configurational analysis of this material was performed with very high stereochemical integrity by first using adenylate kinase, pyruvate kinase, a trace of ATP, and phosphoenolpyruvate to convert the acyclic nucleotide to a mixture of three iso-topomers at the α-phosphorus of dATP. This isotopically labeled sample of $[^{16}O, ^{17}O, ^{18}O]$dATP was converted (with inversion of configuration) to a mixture of the three possible types of doubly oxygen-labeled cyclic dAMP using the adenylate cyclase from *Brevibacterium liquefaciens* as catalyst. The ^{31}P NMR method described in Section II,D,2 was used to determine that the absolute configuration of the cyclic $[^{16}O, ^{18}O]$dAMP present in this mixture was R_P. When the structures of the R_P diastereomers of cyclic $[^{17}O, ^{18}O]$dAMP and $[^{16}O, ^{17}O, ^{18}O]$dAMP are compared, the conclusion is reached that the enzymic hydrolysis of the labeled cyclic dAMP had occurred with inversion of configuration. [An independent stereochemical study of the reaction of oxygen chiral cyclic AMP had concluded that the reaction occurred with retention of configuration (*18*), but this result was later corrected (*19*).]

These comparative studies constituted the first example of an enzyme-catalyzed hydrolysis reaction whose stereochemical course was unaffected by sulfur substitution. At the time these experiments were performed, the stereochemical courses of the reactions catalyzed by glycerol kinase (*83, 84*) and by the bacterial adenylate cyclase (*85, 86*) had already been compared in the laboratories of Knowles and Gerlt, respectively, and these were also found to be unaffected by the sulfur substitution. A number of other comparisons of this type have been made, and in no case were the stereochemical consequences of the reactions studied with chiral phosphate esters and the chiral thiophosphate analogs found to differ. This agreement suggests that the necessary use of oxygen chiral thiophosphate monoesters to study the stereochemical course of phosphomonoesterases will provide pertinent results for ascertaining whether phosphorylated intermediates are involved in the reaction mechanism.

C. STAPHYLOCOCCAL NUCLEASE

In contrast to the lack of detailed structural information for the 3′,5′-cyclic nucleotide phosphodiesterase, staphylococcal (or micrococcal) nuclease (SNase), an extracellular nuclease produced by *Staphylococcus aureus,* is well characterized. SNase is an extraordinarily efficient catalyst for the endo- and exonucleolytic hydrolysis of single-stranded DNA and RNA, with the rate acceleration for DNA being approximately 10^{15} relative to the uncatalyzed rate; the final products of the reaction are 3′-mononucleotides. The sequence of 149 amino acids that constitute the enzyme has been determined both by classical degradation procedures and by sequence analysis of the cloned gene. A highly refined 1.65-Å X-ray structure determined in the presence of Ca^{2+} and the competitive inhibitor thymidine 3′,5′-bisphosphate (pdTp) was recently completed (*87*); this structure differs only slightly from a less refined 1.5-Å structure that was reported in 1979

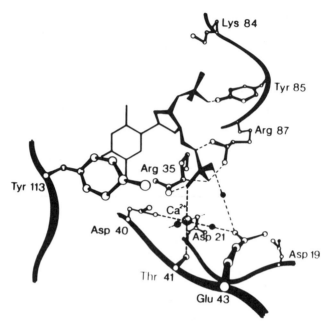

FIG. 12. Active site of wild-type staphylococcal nuclease with bound active site ligands Ca²⁺ and thymidine 3′,5′-diphosphate. Reproduced with permission from Ref. 88.

(88) (Fig. 12). The X-ray structure shows the positions of amino acids in the active site that can be used to formulate a mechanism. The Ca²⁺ ion, held in place by coordination to the β-carboxylates of Asp-21 and Asp-40 and to the carbonyl oxygen of Thr-41, is coordinated to the 5′-phosphate group of the competitive inhibitor. This phosphate group is also coordinated to the guanidinium functional groups of Arg-35 and Arg-87. If the assumption is made that the position of the bound inhibitor mimics the position of bound substrate, and since SNase produces 3′-mononucleotides, the position of the scissile P–O bond is between the 5′-phosphate group and its 5′-ester oxygen. Thus, the Ca²⁺ ion and the guanidinium groups of both arginines can be proposed to be electrophilic catalysts that facilitate the direct attack of water on the phosphorus to effect P–O cleavage. The γ-carboxylate group of Glu-43 is also present in the vicinity of the 5′-phosphate, and in the X-ray structure it is hydrogen bonded to two water molecules, one that is in the inner coordination sphere of the Ca²⁺ and a second that serves as a bridge between the carboxylate group and the 5′-phosphate group of the bound inhibitor. Thus, the γ-carboxylate group of Glu-43 can be proposed to be a general basic catalyst that facilitates the direct attack of water on the phosphorus.

Our laboratory has determined the stereochemical course of a hydrolysis re-action catalyzed by SNase (89), namely, the hydrolysis of the R_P diastereomer of thymidine 5'-(4-nitrophenyl-[^{17}O,^{18}O]phosphate) in H_2^{16}O. This substrate is hydrolyzed to thymidine and 4-nitrophenyl [^{16}O,^{17}O,^{18}O]phosphate. The con-figuration of the chiral phosphate ester was determined using the procedure de-scribed by Knowles and co-workers (76) in which alkaline phosphatase was used to transfer the chiral phosphoryl group to S-1,2-propanediol with inversion of configuration. Continuing the procedure for the configurational analysis (Section II,D,2) prescribed by Knowles and co-workers (34), the configuration of the chiral phosphate monoester was found to be S_P. This change in absolute configu-ration reveals that the stereochemical course of the hydrolysis reaction is inver-sion. This stereochemical outcome of the reaction is consistent with the mecha-nism proposed on the basis of the X-ray structure.

A second mechanism consistent with the proposed mechanism is that the car-boxylate of Glu-43 acts as a nucleophile to displace the 5'-hydroxyl group of the phosphodiester to form a transient nucleotidylated enzyme intermediate (acyl phosphate ester), which is converted to the phosphomonoester product by attack of a water nucleophile on the carboxylate carbon, i.e., formation but not break-down of the intermediate involves nucleophilic attack on the chiral phosphorus atom. This mechanism has been shown to be incorrect by carrying out the hy-drolysis of a substrate in H_2^{18}O with a molar excess of enzyme (single-turnover conditions) and determining that the product phosphomonoester is quantitatively labeled with ^{18}O (90); if the acyl phosphate ester intermediate were mechanisti-cally significant, the oxygen of the monoester product in this experiment should not have been labeled with ^{18}O, since the oxygen atom would have been derived from the unlabeled carboxylate group of Glu-43. Thus, in this case, a nucleoti-dylated enzyme intermediate is not mechanistically relevant.

The putative general base catalyst, Glu-43, and electrophilic catalysts, Arg-35 and Arg-87, have been specifically mutated to Asp (91, 92) and Lys (93) residues, respectively, in the author's laboratory to assess the roles of these amino acid residues in catalysis. Other amino acids were also introduced at these positions, but the present discussion will briefly outline the results obtained with the "conservative" substitutions. As mentioned previously, the rate acceleration characteristic of the SNase-catalyzed hydrolysis of DNA is approximately 10^{15}. The Asp substitution for Glu-43 (E43D) decreased the catalytic efficiency ap-proximately 10^3, and the Lys substitutions for Arg-35 (R35K) and Arg-87 (R87K) decreased the catalytic efficiency approximately 10^4 and 10^5, respec-tively. While such decreases in catalytic efficiency have been used to describe quantitatively the roles of various active site residues in catalysis, such interpre-tation is clearly unwarranted in the case of these active site mutants of SNase. The melting temperatures of all three of these mutant enzymes differ significantly from that of wild type, and one- and two-dimensional NMR spectroscopies reveal significant changes in chemical shifts and interresidue nuclear Overhauser effects (93). The suggestion that these differences from wild-type properties are the result of conformational changes produced by the amino acid substitutions has been verified crystallographically in the case of E43D (94); a highly refined 1.7-Å structure of E43D reveals significant changes from the structure of the wild-type protein in the positions of the backbone near the active site as well as in the positions of side chains both within the active site and up to 15 Å removed from the active site. This situation underscores the difficulties that can be encountered in the use of mutagenesis techniques to describe mechanisms.

D. DEOXYRIBONUCLEASE I

DNase I from bovine pancreas hydrolyzes double-stranded DNA to yield oligonucleotides with 5′-phosphoryl ends (95). The enzyme requires divalent metal ions for activity, with Ca^{2+} and either Mg^{2+} or Mn^{2+} producing optimal activity. Although classical peptide sequencing suggested that the enzyme contained 254 amino acids in a defined sequence, a recent X-ray analysis to 2.0 Å resolution revealed that a repeated tripeptide had been missed in the sequence analysis (96). The X-ray structure of DNase I was determined in the presence of Ca^{2+} and in both the presence and the absence of pdTp (also a competitive inhibitor of this enzyme as well as SNase), so an active site region could be identified and a mechanism could be proposed. The high-resolution X-ray structure provides an interesting active site environment in which general base catalysis can facilitate hydrolysis of the phosphodiester backbone of DNA with inversion of configuration. Like the active site of SNase, the presumed catalytically essential Ca^{2+} is found to be coordinated to a phosphate group of the pdTp. A water molecule is located adjacent to the position of the bound inhibitor and Ca^{2+}, appropriately so if it is to be the nucleophile that directly attacks the phosphorus atom. Interestingly, this water molecule is within hydrogen bonding distance of His-131, and, in turn, His-131 is within hydrogen bonding distance of Glu-75 (Fig. 13). No carboxylate group within the active site is sufficiently close to the position of the bound pdTp to act as a nucleophile to form an acyl phosphate ester intermediate. Thus, the mechanism of the reaction catalyzed by DNase I can be hypothesized to be analogous to that of trypsin, chymotrypsin, and other serine- and cysteine-dependent proteases: the nucleophile (water in the case of DNase I and an active site serine or cysteine in the case of the proteases) is activated with respect to attack on the substrate (phosphodiester in the case of DNase I and a peptide bond in the case of the proteases) by hydrogen bonding to an active site histidine residue; the essential histidine residue is properly positioned with respect to the hydroxylic substrate by hydrogen bonding to an active site carboxylate group.

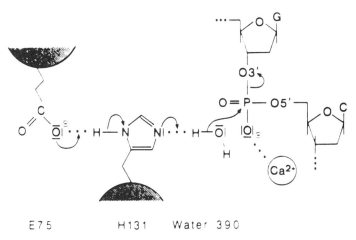

E75   H131   Water 390

FIG. 13. Active site of DNase I. From Ref. *96*.

Prior to the X-ray analysis, the stereochemical course of the hydrolysis of the S_P diastereomer of thymidine 3'-(4-nitrophenyl [$^{17}O,^{18}O$]phosphate) 5'-(4-nitrophenyl phosphate) in $H_2{}^{16}O$ was determined (*26*). This synthetic oligonucleotide analog is not a good substrate, and the S_P diastereomer of thymidine 3'-[$^{16}O,^{17}O,^{18}O$]phosphate 5'-(4-nitrophenyl phosphate) was obtained as product (recall that on double-stranded DNA the oligonucleotide products contain 5'-phosphate groups). The simplest explanation for this inversion of configuration is that the enzyme catalyzes the direct attack of water on the phosphodiester backbone of double-stranded DNA, presumably by general basic catalysis.

In view of the apparent convergent evolution of mechanism in the serine and cysteine protease family, it is interesting that two phosphodiesterases that require Ca^{2+} for catalytic activity by virtue of presumed electrophilic catalysis via direct coordination to the anionic phosphoryl oxygens of the substrate have evolved conceptually similar (general basic catalysis) but structurally distinct solutions to the problem of phosphodiester hydrolysis.

E.   PHOSPHODIESTERASES FROM SNAKE VENOM AND BOVINE SPLEEN

We have also studied the stereochemical outcomes of the reactions catalyzed by the phosphodiesterase from snake venom (*99*) and from bovine spleen (*101*). The former enzyme catalyzes the hydrolysis of esters of 5'-nucleotides (free 3'-hydroxyl group), and the latter enzyme catalyzes the hydrolysis of esters of 3'-nucleotides (free 5'-hydroxyl group). No structural information of mechanistic consequence is available for these enzymes.

The stereochemical course of the reaction catalyzed by the enzyme from snake venom has been studied in four laboratories, with two studies employing chiral phosphorothioate substrates (97, 98) and two employing oxygen chiral substrates (99, 100). The study performed in our laboratory (99) employed the S_P diastereomer of thymidine 5'-(4-nitrophenyl [$^{17}O,^{18}O$]phosphate) as substrate, and after hydrolysis in $H_2^{16}O$ the R_P enantiomer of thymidine 5'-[$^{16}O,^{17}O,^{18}O$]phosphate was isolated and configurationally analyzed by the methods described in Section II,D,2. These structures for the substrate and product indicate that the hydrolysis reaction catalyzed by this enzyme proceeds with retention of configuration at phosphorus. The same conclusion was reached in the laboratory of Lowe using a mixture of three chiral isotopomers of [α-$^{16}O,^{17}O,^{18}O$]ATP as substrate (100). The laboratories of Eckstein and of Benkovic independently utilized the S_P diastereomers of phenyl esters of a phosphorothioate (98) and a phosphonothioate (97), respectively, as substrates in their stereochemical investigations. (Note that this enzyme is stereospecific with respect to the configuration of the phosphorothioate substrate and prefers the S_P diastereomer. This stereospecificity has been used to assign the absolute configurations of nucleoside phosphorothioate esters of unknown configuration.)

The stereochemical course of the reaction catalyzed by phosphodiesterase from bovine spleen was determined by using the S_P diastereomer of thymidine 3'-(4-nitrophenyl [$^{17}O,^{18}O$]phosphate) as substrate in $H_2^{16}O$ (101). The R_P diastereomer of thymidine 3'-[$^{16}O,^{17}O,^{18}O$]phosphate was isolated and configurationally characterized. This hydrolysis reaction also proceeds with retention of configuration at phosphorus. In this case, neither diastereomer of thymidine 3'-(4-nitrophenyl phosphorothioate) was a substrate for the enzyme from spleen. Instead, the enzyme catalyzed a transnucleotidylation reaction: 4-nitrophenol was first displaced to form a 3'-nucleotidylated enzyme intermediate, and this intermediate was then intercepted by the 5'-hydroxyl group of a second substrate molecule to generate an internucleosidic 3',5'-phosphorothioate linkage. The stereochemical consequence of this nucleotidyl transfer reaction has been found to be the anticipated retention of configuration based on this hypothesis (102).

The stereochemical consequences of the reactions catalyzed by these two phosphodiesterases are the opposite of those determined for the reactions catalyzed by SNase and DNase I. The structural and mechanistic divergence already noted for the two phosphodiesterases, which require Ca^{2+} for activity and probably involve general base-assisted attack of water on the substrate, is expanded on by the finding that the phosphodiesterases from snake venom and spleen utilize mechanisms that have retention as their stereochemical outcomes. Clearly, the stereochemical and structural studies of phosphodiesterases reported to date reveal mechanistic complexity that contrasts with the stereochemical (and mechanistic) uniformity that has been discovered for all of the kinases studied to date. The diversity in the phosphodiesterases is perhaps even more surprising when it is realized that the substrates for all of the enzymes that have been stereochemically characterized are nucleotide esters. In the kinases, transfers of the γ-phosphoryl group of ATP to carboxylate, hydroxyl, amino, and guanidino functional groups all proceed with inversion of configuration, even though the pK_a values of the accepting functional groups vary over a wide range.

F. CLASS I APURINIC/APYRIMIDINIC SITE ENDONUCLEASES

Until recently, all enzymes that catalyze the cleavage of phosphate ester bonds were assumed to be phosphohydrolases, i.e., the cleavage reaction occurs by hydrolysis of a phosphodiester bond to generate ends containing a phosphate monoester and a hydroxyl group. This assumption has now been shown to be incorrect for enzymes that introduce strand breaks on the 3'-side of aldehydic abasic sites (also referred to as apurinic, apyrimidinic, or simply AP sites) during DNA repair. Despite the title of this chapter, the mechanism of the strand cleavage reaction will be briefly described.

One pathway for the repair of damaged bases in DNA (e.g., uracil, thymine glycol, dihydrothymine, and urea) involves the excision of the damaged nucleotide unit and its replacement with the undamaged precursor specified by the coding information on the opposite strand (103). In some cases, the excision of the damaged nucleotide occurs first by the action of a glycosylase to remove the damaged base followed by strand cleavage on the 3'-side of the damaged nucleotide unit; both of these reactions can be catalyzed by the same enzyme molecule. The sugar–phosphate residue at the nick can then be removed by the action of an enzyme such as exonuclease III, producing a gap that can be repaired by a DNA polymerase and DNA ligase.

The first suggestion that the strand cleavage reaction catalyzed endonuclease III from *E. coli*, UV endonuclease V from bacteriophage T4, and UV endonuclease from *Micrococcus luteus* did not proceed by a hydrolytic mechanism was that the sugar–phosphate end could not be removed by the action of the 3' → 5' exonuclease activity of DNA polymerase I from *E. coli* (104). Later, the observations were made that strand cleavage reactions catalyzed by these enzymes are accompanied by abstraction of ^{3}H from the 2-position of the abasic site (105, 106) and that the sugar–phosphate end that was enzymically released did not have the chromatographic properties of deoxyribose 5-phosphate (105, 107), the product expected from a phosphohydrolase activity. The suggestion was made that the strand cleavage reaction was β-elimination of the nucleotide unit esterified to the O-3 position of the abasic site.

We have directly identified the sugar product of the strand cleavage reaction catalyzed by endonuclease III and UV endonuclease V as the trans-α,β-unsaturated aldehyde obtained by elimination of the 3'-phosphate (108, 109); with the determination that these endonucleases stereospecifically abstract the 2-*pro-S* hydrogen, the stereochemical course of the β-elimination reaction is syn

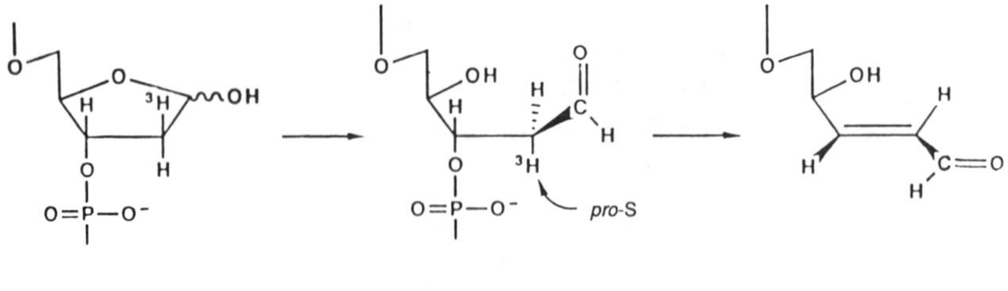

syn β-elimination

FIG. 14. Stereochemical course of the β-elimination reaction catalyzed by endonuclease V from bacteriophage T4 and by endonuclease III from *Escherichia coli*. From Ref. *108*.

(Fig. 14). Whereas a mixture of anomeric hemiacetals constitutes the major contribution to the structure of an aldehydic abasic site (*111*), the syn stereochemical course indicates that the strand cleavage occurs from an acyclic species such as the aldehyde or an activated acyclic species derived from the aldehyde, i.e., an imine (*110*).

In this case, cleavage of the phosphate ester bond occurs by an unusual reaction mechanism; however, the structure of the substrate allows a facile β-elimination reaction that is characterized by a lower activation energy barrier than attack of a nucleophile on an anionic phosphate ester.

REFERENCES

1. Benkovic, S. J., and Schray, K. J. (1972). "The Enzymes," 3rd Ed., Vol. 8, Chap. 6.
2. Knowles, J. R. (1980). *Annu. Rev. Biochem.* **49**, 877.
3. Frey, P. A. (1982). *New Compr. Biochem.* **3**, 201.
4. Gerlt, J. A., Coderre, J. A., and Mehdi, S. (1983). *Adv. Enzymol.* **55**, 201.
5. Lowe, G. (1983). *Acc. Chem. Res.* **16**, 244.
6. Eckstein, F. (1985). *Annu. Rev. Biochem.* **54**, 367.
7. Frey, P. A. (1989). *Adv. Enzymol.* **62**, 119.
8. Thatcher, G. R. J., and Kluger, R. (1989). *Adv. Phys. Org. Chem.* **25**, 100.
9. Westheimer, F. H. (1968). *Acc. Chem. Res.* **1**, 70.
10. Westheimer, F. H. (1980). *In* "Rearrangements in Ground and Excited States" (P. deMayo, ed.), Vol. 2, p. 229. Academic Press, New York.
11. Westheimer, F. H. (1981). *Chem. Rev.* **81**, 313.
12. Eckstein, F., and Gindl, H. (1968). *Chem. Ber.* **101**, 1670.
13. Usher, D. A., Richardson, D. I., and Eckstein, F. (1970). *Nature (London)* **228**, 663.
14. Spector, L. B. (1980). *Proc. Natl. Acad. Sci. U.S.A.* **77**, 2626.
15. Abbott, S. J., Jones, S. R., Weinman, S. A., Bockhoff, F. M., McLafferty, F. W., and Knowles, J. R. (1979). *J. Am. Chem. Soc.* **101**, 4323.
16. Cullis, P. M., and Lowe, G. (1978). *J. Chem. Soc., Chem. Commun.* p. 512.
17. Cullis, P. M., and Lowe, G. (1981). *J. Chem. Soc., Perkin Trans. 1*, p. 2317.

18. Cullis, P. M., Lowe, G., Jarvest, R. L., and Potter, B. V. L. (1981). *J. Chem. Soc., Chem. Commun.* p. 245.
19. Jarvest, R. L., Lowe, G., and Potter, B. V. L. (1980). *J. Chem. Soc., Chem. Commun.* p. 1145.
20. Coderre, J. A., Mehdi, S., and Gerlt, J. A. (1981). *J. Am. Chem. Soc.* **103**, 1872.
21. Richard, J. P., and Frey, P. A. (1982). *J. Am. Chem. Soc.* **104**, 3476.
22. Cullis, P. M., Iagrossi, A., and Ross, A. J. (1986). *J. Am. Chem. Soc.* **108**, 7869.
23. Arnold, J. R. P., Bethell, R. C., and Lowe, G. (1987). *Bioorg. Chem.* **15**, 250.
24. Baraniak, J., and Stec, W. J. (1987). *J. Chem. Soc., Perkin Trans. 1*, p. 1645.
25. Gerlt, J. A., and Coderre, J. A. (1980). *J. Am. Chem. Soc.* **102**, 4531.
26. Mehdi, S., and Gerlt, J. A. (1984). *Biochemistry* **23**, 4844.
27. Stec, W. J. (1983). *Acc. Chem. Res.* **16**, 411.
28. Gerlt, J. A., Mehdi, S., Coderre, J. A., and Rogers, W. O. (1980). *Tetrahedron Lett.* **21**, 2385.
29. Cohn, M., and Hu, A. (1978). *Proc. Natl. Acad. Sci. U.S.A.* **75**, 200.
30. Lutz, O., Nolle, A., and Staschewski, D. (1978). *Z. Naturforsch. A: Phys. Chem. Kosmophys.* **A33**, 380.
31. Lowe, G., and Sproat, B. S. (1978). *J. Chem. Soc., Chem. Commun.* p. 565.
32. Tsai, M. D. (1979). *Biochemistry* **18**, 1468.
33. Lowe, G., Potter, B. V. L., Sproat, B. S., and Hull, W. E. (1979). *J. Chem. Soc., Chem. Commun.* p. 733.
34. Buchwald, S. L., and Knowles, J. R. (1980). *J. Am. Chem. Soc.* **102**, 6601.
35. Buchwald, S. L., Friedman, J. M., and Knowles, J. R. (1984). *J. Am. Chem. Soc.* **106**, 4911.
36. Webb, M. R., and Trentham, D. R. (1980). *J. Biol. Chem.* **255**, 1775.
37. Tsai, M. D. (1980). *Biochemistry* **19**, 5310.
38. Cullis, P. M., Misra, R., and Wilkins, D. J. (1987). *Tetrahedron Lett.* **28**, 4211.
39. Domanico, P., Mizrahi, V., and Benkovic, S. J. (1986). In "Mechanisms of Enzymatic Reactions: Stereochemistry" (P. A. Frey, ed.), p. 127. Elsevier, New York.
40. Breslow, R., and Katz, I. (1968). *J. Am. Chem. Soc.* **90**, 7376.
41. Ramirez, F., Marecek, J., Minore, J., Srivastava, S., and le Noble, W. J. (1986). *Am. Chem. Soc.* **108**, 348; Burgess, J., Blundell, N., Cullis, P. M., Hubbard, C. D., and Misra, R. (1988). *J. Am. Chem. Soc.* **110**, 7900.
42. Cullis, P. M., Misra, R., and Wilkins, D. J. (1987). *J. Chem. Soc., Chem. Commun.* p. 1594.
43. Harnett, S. P., and Lowe, G. (1987). *J. Chem. Soc., Chem. Commun.* p. 1416.
44. Rebek, J., Gavina, F., and Navarro, C. (1978). *J. Am. Chem. Soc.* **100**, 8113.
45. Ramirez, F., and Marecek, J. F. (1979). *Tetrahedron* **35**, 1581.
46. Ramirez, F., and Marecek, J. F. (1980). *Tetrahedron* **36**, 3151.
47. Gorenstein, D. G., Lee, Y.-G., and Kar, D. J. (1977). *J. Am. Chem. Soc.* **99**, 2264.
48. Weiss, P. M., Knight, W. B., and Cleland, W. W. (1986). *J. Am. Chem. Soc.* **108**, 2761.
49. Knight, W. B., Weiss, P. M., and Cleland, W. W. (1986). *J. Am. Chem. Soc.* **108**, 2759.
50. Skoog, M. T., and Jencks, W. P. (1984). *J. Am. Chem. Soc.* **106**, 7597.
51. Bourne, N., and Williams, A. (1984). *J. Am. Chem. Soc.* **106**, 7591.
52. Jencks, D. A., and Jencks, W. P. (1977). *J. Am. Chem. Soc.* **99**, 9948.
53. Herschlag, D., and Jencks, W. P. (1989). *J. Am. Chem. Soc.* **111**, 7579.
54. Herschlag, D., and Jencks, W. P. (1989). *J. Am. Chem. Soc.* **111**, 7587.
55. Cullis, P. M., and Rous, A. J. (1985). *J. Am. Chem. Soc.* **107**, 6721.
56. Friedman, J. M., Freeman, S., and Knowles, J. R. (1988). *J. Am. Chem. Soc.* **110**, 1268.
57. Cullis, P. M., and Rous, A. J. (1986). *J. Am. Chem. Soc.* **108**, 1298.
58. Cullis, P. M., and Nicholls, D. (1987). *J. Chem. Soc., Chem. Commun.* p. 783.
59. Gorenstein, D. G. (1987). *Chem. Rev.* **87**, 1047.

60. Deslongchamps, P. (1983). "Stereoelectronic Effects in Organic Chemistry." Pergamon, Oxford.
61. Kumamoto, J., Cox, J. R., and Westheimer, F. H. (1956). *J. Am. Chem. Soc.* **78,** 4858.
62. Kaiser, E. T., Panar, M., and Westheimer, F. H. (1963). *J. Am. Chem. Soc.* **85,** 602.
63. Gerlt, J. A., Westheimer, F. H., and Sturtevant, J. M. (1975). *J. Biol. Chem.* **250,** 5059.
64. Hall, C. R., and Inch, T. D. (1980). *Tetrahedron* **36,** 2059.
65. Buchwald, S. L., Pliura, D. H., and Knowles, J. R. (1984). *J. Am. Chem. Soc.* **106,** 4916.
66. Richards, F. M., and Wyckoff, H. W. (1971). "The Enzymes," 3rd Ed., Vol. 4, Chap. 24.
67. Breslow, R., Doherty, J. B., Guillot, G., and Lipsey, C. (1978). *J. Am. Chem. Soc.* **100,** 3227.
68. Corcoran, R., LaBelle, M., Czarnik, A. W., and Breslow, R. (1985). *Anal. Biochem.* **144,** 563.
69. Breslow, R., and LaBelle, M. (1986). *J. Am. Chem. Soc.* **108,** 2655.
70. Ansyln, E., and Breslow, R. (1989). *J. Am. Chem. Soc.* **111,** 4473.
71. Anslyn, E., and Breslow, R. (1989). *J. Am. Chem. Soc.* **111,** 5972.
72. Anslyn, E., and Breslow, R. (1989). *J. Am. Chem. Soc.* **111,** 8931.
73. Coleman, J. E., and Gettins, P. (1983). *Adv. Enzymol.* **55,** 381.
74. Sowadski, J. M., Handschumacher, M. D., Murthy, H. M. K., Kundrot, C., and Wyckoff, H. W. (1985). *J. Mol. Biol.* **186,** 417.
75. Neumann, H. (1968). *J. Biol. Chem.* **243,** 4671.
76. Jones, S. R., Kindman, L. A., and Knowles, J. R. (1978). *Nature (London)* **275,** 564.
77. Weiss, P. M., and Cleland, W. W. (1989). *J. Am. Chem. Soc.* **111,** 1928.
78. Ghosh, S. S., Bock, S. C., Rokita, S. E., and Kaiser, E. T. (1986). *Science* **231,** 145.
79. Butler-Ransohoff, J. E., Kendall, D. A., Freeman, S., Knowles, J. R., and Kaiser, E. T. (1988). *Biochemistry* **27,** 4777.
80. Chaidaroglou, A., Brezinski, D. J., Middleton, S. A., and Kantrowitz, E. R. (1988). *Biochemistry* **27,** 8338.
81. Burgers, P. M. J., Eckstein, F., Hunneman, D. H., Baraniak, J., Kinas, R. W., Lesiak, K., and Stec, W. J. (1979). *J. Biol. Chem.* **254,** 9959.
82. Mehdi, S., Coderre, J. A., and Gerlt, J. A. (1983). *Tetrahedron* **39,** 3483.
83. Blättler, W. A., and Knowles, J. R. (1979). *Biochemistry* **18,** 3927.
84. Pliura, D. H., Schomburg, D., Richard, J. P., Frey, P. A., and Knowles, J. R. (1980). *Biochemistry* **19,** 325.
85. Gerlt, J. A., Coderre, J. A., and Wolin, M. S. (1980). *J. Biol. Chem.* **255,** 331.
86. Coderre, J. A., and Gerlt, J. A. (1980). *J. Am. Chem. Soc.* **102,** 6594.
87. Loll, P., and Lattman, E. E. (1989). *Proteins* **5,** 189.
88. Cotton, F. A., Hazen, E. E., and Legg, M. J. (1979). *Proc. Natl. Acad. Sci. U.S.A.* **76,** 2551.
89. Mehdi, S., and Gerlt, J. A. (1982). *J. Am. Chem. Soc.* **104,** 3223.
90. Hibler, D. W., Stolowich, N. J., Mehdi, S., and Gerlt, J. A. (1986). *In* "Mechanisms of Enzymatic Reactions: Stereochemistry" (P. A. Frey, ed.), p. 101. Elsevier, New York.
91. Hibler, D. W., Stolowich, N. J., Reynolds, M. A., Gerlt, J. A., Wilde, J. A., and Bolton, P. H. (1987). *Biochemistry* **26,** 6278.
92. Wilde, J. A., Bolton, P. H., Dell'Acqua, M., Hibler, D. W., Pourmotabbed, T., and Gerlt, J. A. (1988). *Biochemistry* **27,** 4127.
93. Pourmotabbed, T., Dell'Acqua, M., Gerlt, J. A., Stanczyk, S. M., and Bolton, P. H. (1990). *Biochemistry* **29,** 3677.
94. Loll, P., and Lattman, E. E. (1990). *Biochemistry* **29,** 6866.
95. Laskowski, M. (1971). "The Enzymes," 3rd Ed., Vol. 4, p. 289.
96. Suck, D., and Oefner, C. (1986). *Nature (London)* **321,** 620.
97. Bryant, F. R., and Benkovic, S. J. (1979). *Biochemistry* **18,** 2825.

98. Burgers, P. M. J., Eckstein, F., and Hunneman, D. H. (1979). *J. Biol. Chem.* **254,** 7476.
99. Mehdi, S., and Gerlt, J. A. (1981). *J. Biol. Chem.* **256,** 12164.
100. Jarvest, R. L., and Lowe, G. (1981). *Biochem. J.* **199,** 447.
101. Mehdi, S., and Gerlt, J. A. (1981). *J. Am. Chem. Soc.* **103,** 7018.
102. Uznanski, B., Niewiarowski, W., and Stec, W. J. (1986). *J. Biol. Chem.* **261,** 592.
103. Myles, G. M., and Sancar, A. (1989). *Chem. Res. Toxicol.* **2,** 197.
104. Bailly, V., and Verly, W. G. (1985). *FEBS Lett.* **178,** 223.
105. Bailly, V., and Verly, W. G. (1987). *Biochem. J.* **242,** 565.
106. Bailly, V., Sente, B., and Verly, W. G. (1989). *Biochem. J.* **259,** 751.
107. Kim, J., and Linn, S. (1988). *Nucleic Acids Res.* **16,** 1135.
108. Mazumder, A., Gerlt, J. A., Rabow, L., Absalon, M. J., Stubbe, J., and Bolton, P. H. (1989). *J. Am. Chem. Soc.* **111,** 8029.
109. Mazumder, A., Gerlt, J. A., Absalon, M. J., Stubbe, J., Withka, J., and Bolton, P. H. (1990). *Biochemistry* **29,** 1119.
110. Manoharan, M., Ransom, S. C., Mazumder, A., Gerlt, J. A., Wilde, J. A., Withka, J. A., and Bolton, P. H. (1988). *J. Am. Chem. Soc.* **110,** 1620.
111. Wilde, J. A., Bolton, P. H., Mazumder, A., Manoharan, M., and Gerlt, J. A. (1989). *J. Am. Chem. Soc.* **111,** 1894.

4

Nucleotidyltransferases and Phosphotransferases: Stereochemistry and Covalent Intermediates

PERRY A. FREY

Institute for Enzyme Research, Graduate School, and
Department of Biochemistry, College of Agricultural and Life Sciences
University of Wisconsin—Madison
Madison, Wisconsin 53705

THE ENZYMES, Vol. XX

I. Introduction

Phosphotransferases and nucleotidyltransferases were last reviewed in this series 15 years ago in Volumes VIII and IX. At that time a major mechanistic question was whether these enzymes catalyze their reactions by single-displacement
or double-displacement mechanisms. The two mechanisms differed chemically
with respect to whether the phosphoryl or nucleotidyl group is transferred directly between two substrates, or whether the group transfer is mediated by a
nucleophilic group of the enzyme in a two-step mechanism via a covalent phosphoenzyme or nucleotidyl–enzyme.

For a simple Bi Bi reaction, Eq. (1) describes the overall course of group
transfer, in which A and B symbolize the group donor and receptor, respectively,
and P symbolizes the phosphoryl or nucleotidyl group.

$$A–P + B \rightleftharpoons A + B–P \tag{1}$$

In the single-displacement or direct transfer mechanism, the donor and acceptor
substrates are bound at adjacent subsites in the active center of the enzyme, and
the group is transferred directly from the donor substrate to the acceptor, as in
Eq. (2).

$$E + B + A–P \rightleftharpoons E·A–P·B \rightleftharpoons E·A·B–P \rightleftharpoons B–P + A + E \tag{2}$$

The donor and acceptor can be bound in compulsory order or random order, the
main point being that both must be bound at adjacent sites before group transfer
can occur. This mechanism is kinetically indistinguishable from one in which an
additional covalent phosphoryl–enzyme or nucleotidyl–enzyme exists and con-
nects the two ternary complexes, as in Eq. (3). In this mechanism the enzyme
mediates group transfer by nucleophilic catalysis, utilizing an enzymic nucleophile as the catalytic functional group.

$$E–N·A–P·B \rightleftharpoons E–N–P·A·B \rightleftharpoons E–N·A·B–P \tag{3}$$

The one kinetic mechanism that requires nucleophilic catalysis by group-
transferring enzymes is the Ping-Pong pathway, which is defined for phosphotransferases by Eqs. (4a) and (4b).

$$E-N + A-\mathbf{P} \rightleftharpoons E-N\cdot A-\mathbf{P} \rightleftharpoons E-N-\mathbf{P} + A \qquad (4a)$$

$$E-N-\mathbf{P} + B \rightleftharpoons E-N-\mathbf{P}\cdot B \rightleftharpoons E-N + B-\mathbf{P} \qquad (4b)$$

In this pathway the group donor binds to the active site and reacts with an enzymic nucleophile, transferring the group to the nucleophile, and the first product dissociates. In the second step the acceptor binds to the active site, and the phosphoryl or nucleotidyl group is then transferred to the acceptor from the enzymic nucleophile.

The classical methods for distinguishing the mechanisms of Eqs. (2), (3), (4a), and (4b) are well known. The mechanisms of Eqs. (2) and (3) involve sequential kinetic pathways, which give converging lines in double-reciprocal plots of initial rates versus substrate concentrations, with one substrate varied and the other held constant (1). The mechanism of Eqs. (4a) and (4b) can be distinguished from those of Eqs. (2) and (3) by kinetic analysis, since it is a Ping-Pong Bi Bi kinetic pathway and will give the parallel double-reciprocal plots that characterize Ping-Pong pathways (1). This mechanism is also characterized by the observation of certain exchange reactions, for example, the exchange of radiochemically labeled A (A*) into A–P in the absence of B, at rates that are compatible with the Ping-Pong kinetic pathway. Furthermore, the intermediate E–N–**P** can generally be isolated and characterized. An intermediate such as that in Eq. (3) may or may not be isolable.

In practice, the classical methods alone have proved to be inadequate to settle the question of whether enzymes utilize nucleophilic catalysis to catalyze phosphoryl and nucleotidyl group transfer reactions. Since the publication of Volumes VIII and IX of "The Enzymes," stereochemical analysis of these reactions has given the most important mechanistic information on the question of whether nucleophilic catalysis and covalent phosphoryl–enzyme or nucleotidyl–enzyme intermediates are involved in the mechanisms. Acetate kinase, phosphoglycerate kinase, adenosine kinase, cAMP-dependent protein kinases, and hexokinase, as well as many other phosphotransferases, have in the past been proposed to catalyze phospho transfer by mechanisms involving covalent phosphoenzymes as intermediates. Several have also been proposed to catalyze their reactions by Ping-Pong kinetic pathways. However, all are now known to catalyze direct phospho transfer between ATP and their substrates, and in all cases some of the most direct and convincing evidence for this mechanism is the fact that phospho transfer proceeds with inversion of configuration at phosphorus. Moreover, in several cases in which apparent Ping-Pong kinetics has been invoked as the major evidence in support of the assignment of a double-displacement mechanism, a more complete kinetic analysis has shown that the kinetics is sequential. In several cases in which evidence had been presented for phosphorylation of the enzyme by ATP, later studies showed either that the observation of a [^{32}P]phosphoenzyme could be attributed to adventitious phosphorylation of an enzymic group by [γ-^{32}P]ATP or to phosphorylation of a contaminating bound substrate by [γ-^{32}P]ATP. Similar phosphorylations were also observed when other high-energy phosphorylating agents such as acetyl phosphate, a substrate for acetate kinase, were incubated with their cognate enzymes. In the case of nucleoside phosphotransferase, the sequential catalytic pathway originally postulated was first brought into question by the observation of overall retention of configuration at phosphorus in a chiral [^{18}O]thiophospho transfer. A more complete kinetic analysis revealed that the reaction proceeds by a specialized Ping-Pong kinetic pathway, and the phosphoenzyme was also isolated.

II. Stereochemistry of Phospho and Nucleotidyl Transfer

Stereochemical information is important in the analysis of most reaction mechanisms. This is true for all substitution reactions at atoms with substituents in tetrahedral array, such as saturated carbon and tetravalent phosphorus. Enzymic substitution at phosphorus in phosphoric esters and phosphoanhydrides is not an exception to the rule. There are experimental complications, however, in that all naturally occurring biological phosphates have two or more chemically equivalent oxygens, so that none has chirally substituted phosphorus. Inasmuch as an asymmetric arrangement of substituents is required for stereochemical analysis, P-chiral substrates for stereochemical studies of phosphotransferases and nucleotidyltransferases must be synthesized with sulfur or heavy isotopes of oxygen as substituents in an asymmetric array.

The possibilities are exemplified by consideration of the ATP molecule:

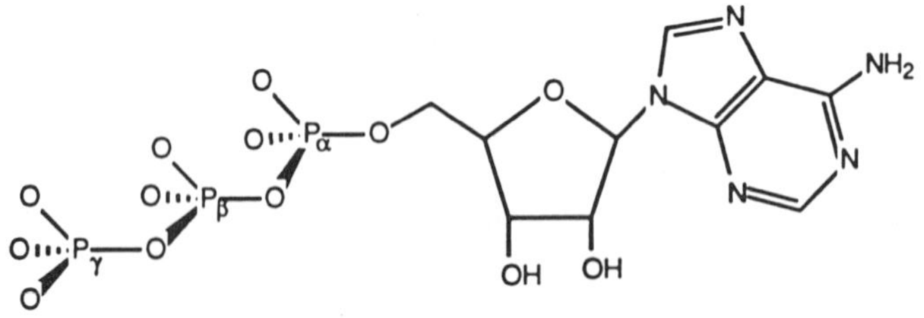

Each phosphorus is either a prochiral or a proprochiral center that can be made chiral by substitution of S, ^{17}O, or ^{18}O for O at P$_\alpha$, P$_\beta$, the substitution of S and ^{17}O or ^{18}O for two atoms of ^{16}O at P$_\gamma$, and the substitution of ^{18}O and ^{17}O for two atoms of ^{16}O at P$_\gamma$. All of these can be and have been synthesized.

The methodologies for synthesis and configurational analysis of P-chiral biological phosphates are extensively reviewed elsewhere and will not be repeated here (2–7). Most and perhaps all biological phosphates can be synthesized with a chiral phosphorus, and the configuration about chiral P in any enzymic product can probably be determined by spectroscopic and chromatographic methods.

Two kinds of information about nucleotidyltransferases and phosphotransferases are obtained by use of substrates or substrate analogs with chiral P. The stereochemical course of phosphoryl transfer and nucleotidyl transfer gives important information about the reaction mechanism. If inversion of configuration at phosphorus is observed, it may be concluded that an uneven number of displacements at phosphorus occurs in the reaction mechanism. If retention of configuration at phosphorus is observed, it may be concluded that the mechanism entails an even number of displacements at phosphorus. Inversion corresponds to the single-displacement mechanism of Eq. (2), and retention indicates a mechanism such as that of Eq. (3) or Eqs. (4a) and (4b).

Before proceeding to consider a few examples of the use of stereochemical analysis, the rules for assigning stereochemical symbols to centers of chirality at phosphorus will be presented. The R,S system of Cahn, Ingold, and Prelog is adapted to chiral P centers (8). The following exception to these rules eliminates a few potential ambiguities and other difficulties, but it does not alter the basic principles on which the system is based. In most phosphate anions the negative charges are delocalized, resulting in fractional bond orders between phosphorus and oxygen, and because symbols can be assigned without reference to bond orders in this series of compounds, negative charges and multiple bonds are ignored in assigning the symbols R or S. With this exception, the rules are otherwise applied as originally formulated. The assignments are made as follows: (1) The substituents to a chiral P are ordered in accord with the priority rules that, in general, give increasing priority to substituents according to increasing atomic masses of the atoms most proximal to the chiral center. (2) The chiral center is viewed from the side opposite the lowest priority substituent. (3) The symbol is assigned as R (or R_P if there are other chiral centers in the molecule) when the remaining substituents appear in a clockwise array in order of decreasing priority, and it is assigned as S (or S_P) when the remaining substituents appear in counterclockwise array in order of decreasing priority. The symbols R and S refer to *rectus* and *sinister,* respectively. Some common substituents to phosphorus in biological phosphates are, in order of decreasing priority, $S > OPO_3X > OPO_3$. The family of substituents OPO_3X are ordered according to the priority rules for the X group. Examples of X in decreasing priority are $PO_3Y > PO_3. >$ aryl $>$ secondary alkyl $>$ primary alkyl, etc., where the substituent Y can be any other substituent. The priorities for heavy isotopes increase in order of atomic masses, but these priorities are applied only as required to make an unambiguous assignment. Thus, the isotopic mass is sometimes used only to determine from which side the chiral P must be viewed in order to assign a symbol, and it is otherwise ignored in applying the priority rules. In other cases, the isotopic masses must be considered at every stage of the assignment process, but they are used only as needed to make an assignment. The symbols used in the following sections in reference to enzymic substrates for stereochemical analysis show how these rules are applied to the assignment of configurational symbols.

A few specific cases of stereochemical analysis exemplify the use of stereochemistry to analyze phospho transfer mechanisms. UDPglucose pyrophosphorylase catalyzes the interconversion of (R_P)-UTPαS and (S_P)-UDPαS-glucose, as shown in reaction (5) (9, 10).

$$PP\!-\!O\!\cdots\!P\!-\!O\!-\!Urd \;+\; Glc\text{-}1\text{-}P \;\xrightleftharpoons{\;Mg^{2+}\;}\; O\!\cdots\!P\!-\!O\!-\!Urd \;+\; PP_i \quad (5)$$

The configuration about P is inverted in this reaction, which must, therefore, proceed by a mechanism involving an uneven number of displacements at P, most likely one in this case. Adenylate kinase catalyzes reaction (6), a $[^{18}O]$ thiophospho transfer from (R_P)-$[\gamma\text{-}^{18}O_2]$ATPγS to AMP to form (S_P)-$[\beta\text{-}^{18}O]$ ADPβS (11).

$$ADP\!-\!{}^{18}O\!-\!P\!\cdots\!{}^{18}O \;+\; AMP \;\xrightleftharpoons{\;Mg^{2+}\;}\; ADP\text{-}^{18}O \;+\; {}^{18}O\!\cdots\!P\!-\!O\!-\!AMP \quad (6)$$

Again, the configuration is inverted. Nucleoside diphosphate kinase catalyzes the same transfer, but to a nucleoside diphosphate rather than to AMP, and with retention of configuration rather than with inversion (10). The mechanism of action of adenylate kinase involves a single displacement at P and that of nucleoside diphosphate kinase involves a double displacement at P via an intermediate phosphoenzyme. Although alkaline phosphatase is not classified as a phosphotransferase, it catalyzes transphosphorylation via the same phosphoenzyme that is the intermediate in the phosphatase reaction. This enzyme catalyzes reaction (7), a phosphoryl transfer, with retention of configuration at P (12).

Many other similar or analogous stereochemical studies have been carried out on over 60 enzymes in recent years. A compilation of results has recently appeared (7). Very recent papers have reported inversion for cAMP-dependent protein kinase-catalyzed phosphorylation of a peptide (*13*) and inversion for phosphofructo-2-kinase-catalyzed phosphorylation of fructose 6-phosphate (*14*). In each case inversion of configuration was interpreted to mean that the reaction proceeds by a mechanism involving a single displacement on phosphorus. The stereochemical test has for some time been accepted as a definitive indicator of whether phospho transfer reactions proceed in even or uneven numbers of displacement steps. In enzymic reactions the simplest and most probable mechanism giving inversion is a single displacement, in which the group is transferred directly from the group donor to the group acceptor, both of which are bound in adjacent subsites at the active site of the enzyme. The simplest and most probable mechanism giving retention of configuration is a double displacement, in which the enzyme mediates group transfer by nucleophilic catalysis via a covalent phosphoenzyme or nucleotidyl–enzyme.

The second major type of stereochemical information that can be obtained about phosphotransferases and nucleotidyltransferases is the coordination structure of nucleotide–metal complexes as they are bound at the active sites of enzymes. Two of the simplest coordination complexes of MgATP are shown below to exemplify the stereochemical difference. These are two stereoisomers differing in screw sense in the coordination ring.

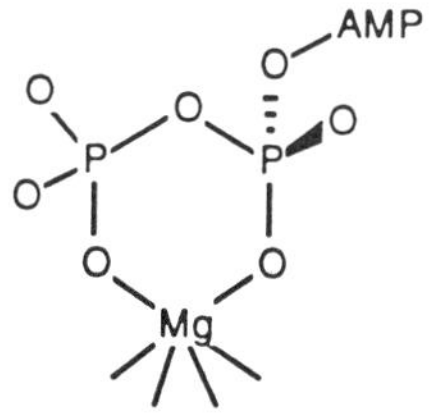

Δ Mg(β,γ)ATP 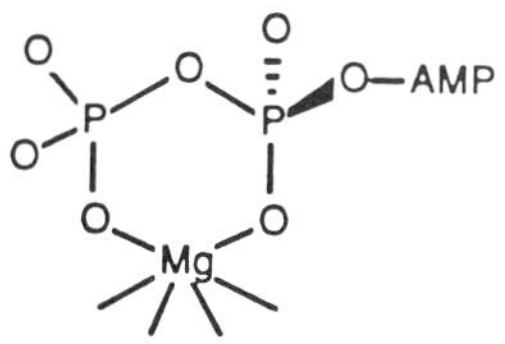   Λ Mg(β,γ)ATP

The isomer designated as Δ is the one that, when viewed from the side of the ring from which AMP is projected, shows the closest pathway from Mg to AMP in clockwise rotation. The isomer is designated as Λ when the shortest path from Mg to AMP is in the counterclockwise direction. The two isomers shown are the β,γ-bidentate isomers. There are 15 other possible monodentate, bidentate, and tridentate isomers of MgATP, all of which are in rapid coordination exchange equilibrium with one another in aqueous solution.

Only one coordination isomer is likely to be the true substrate in the active site of an enzyme. The reason for expecting this to be so is that it is almost certain that only one transition state is involved in a given step of an enzymic reaction, and the structure of this transition state will be strongly influenced by enzymic binding interactions. These binding interactions will manifest themselves in the ground states of complexes such as E·MgATP. They will tend to destabilize MgATP and to facilitate the formation of the transition state. It is unlikely that such binding interactions would equally facilitate the formation of one transition state from two different ground-state structures of MgATP in the complex E·MgATP. Therefore, only a single isomer is expected to be productively bound at the active site.

Since the coordination isomers of MgATP are rapidly interconverted in solution, and no spectroscopic method can distinguish the isomers when bound at the active sites of enzymes, it is not as yet possible to determine which is bound at the active site of an enzyme. However, three techniques are available for determining the structures of analogs of MgATP that are preferentially utilized as substrates in place of MgATP (or MgADP). These methods are outlined in the following paragraphs.

Coordination exchange-inert metal nucleotide complexes have been synthesized, their structural and stereoisomers have been separated by chromatographic and enzymic methods, and their structures have been determined by X-ray crystallography and correlated to their circular dichroism and ^{31}P NMR spectra. The pure isomers have been tested as substrates for enzymes in place of MgATP or MgADP, and from the results the structures of the enzyme-bound and active isomers have been deduced. The most widely used complexes of this type have been Cr(III)–aquo complexes and Co(III)–ammine complexes such as those shown below.

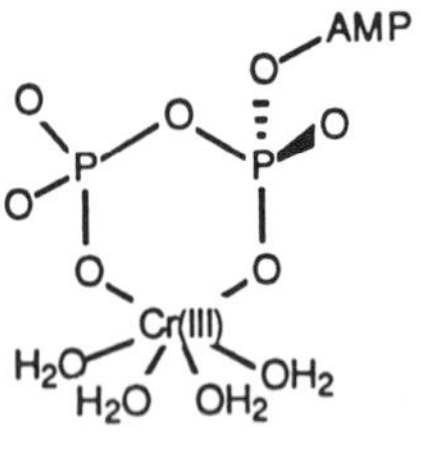

Δ Cr(III)(H$_2$O)$_4$ATP 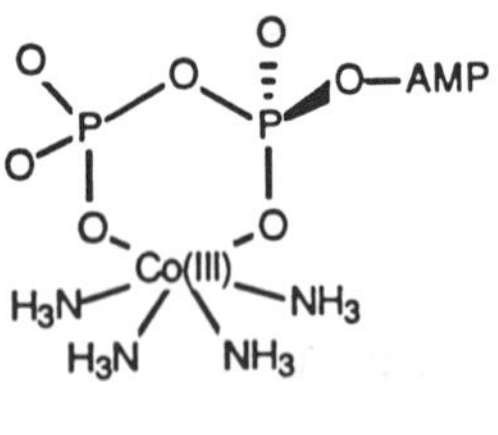 Λ Co(III)(NH$_3$)$_4$ATP

The syntheses and uses of these and other analogous complexes is reviewed elsewhere (15–17). The complex Λ Co(III)(NH$_3$)$_4$ATP is a substrate for hexokinase, whereas the Δ isomer is not, which suggests that the bidentate complex Λ MgATP is the natural substrate (19). The complex Δ Co(III)(NH$_3$)$_4$ATP is a substrate for phosphoribosyl pyrophosphate synthetase (20). The results of many similar studies are compiled in the review by Eckstein (21).

A second method for studying the coordination structures of active Mg–nucleotide complexes is to use the P-chiral nucleoside thiotriphosphates and various divalent metal ions as substrates. The divalent metal ions form coordination exchange-labile complexes with ATPαS and ATPβS, but the various metal ions preferentially form coordination bonds with either S or O. In general, the coordination patterns follow the rules for hard and soft acids and bases, wherein hard metals preferentially bind O and soft metals preferentially bind S (21). Thus, in MgATPβS, magnesium preferentially binds to oxygen at P_β, whereas in CdATPβS, cadmium preferentially binds to sulfur at P_β. These facts can allow the coordination geometry at P_β and P_α in active complexes to be deduced.

The first and clearest example of the use of nucleoside phosphorothioates to determine coordination geometry at an active site was the work of Jaffe and Cohn on hexokinase (22). They found that (R_P)-ATPβS is a far better substrate with Mg(II) as the activating cation than with Cd(II); however, (S_P)-ATPβS is a far better substrate with Cd(II) than with Mg(II) as the metal ion. That is, the favored complex depended on both the metal ion and the configuration at P_β of ATPβS, and a change in metal ion from Mg(II) to Cd(II) led to a change in selectivity for the configuration at P_β. The structures of the preferred substrates were thereby deduced as those shown below. The other isomers, Mg(S_P)-ATPβS and Cd(R_P)-ATPβS, were less active or practically inactive. By the simplest interpretation of the data, both active complexes were the Λ-screw sense isomers shown here, in which Mg is coordinated to oxygen in one and Cd is coordinated to sulfur in the other.

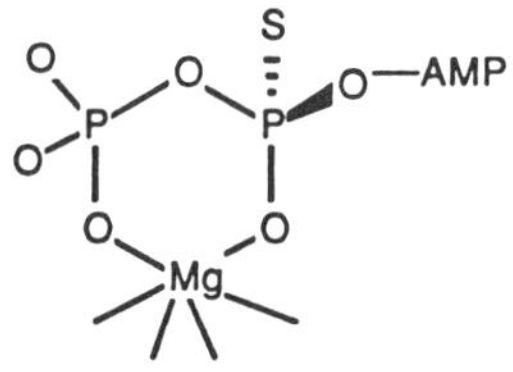

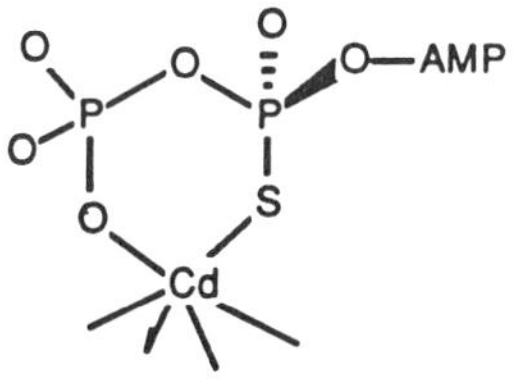

Λ Mg(R_p)-ATPβS Λ Cd(S_p)-ATPβS

It was concluded on this basis that the active complex of MgATP is the Λ-screw sense isomer.

The coordination structures of enzyme-bound manganese nucleotides can in favorable cases be determined by analysis of electron paramagnetic resonance (EPR) spectra of Mn(II) coordinated to ^{17}O-labeled nucleotides. When the nucleotide is stereospecifically labeled with ^{17}O at one diastereotopic position of a prochiral center, either oxygen can in principle be bound to Mn(II) in the coordination complex in an enzymic site. When the coordination bond is between Mn(II) and ^{17}O, the EPR signals for Mn(II) are broadened and attenuated, owing to unresolved superhyperfine coupling between the nucleus of ^{17}O and the unpaired electrons of Mn(II) (23). No such effect is possible with ^{16}O, which has no nuclear spin. The effect is observable in samples in which all the Mn(II) is specifically bound in one or two defined complexes of the nucleotide with the enzyme. Thus the complex Mg(S_P)-[α-^{17}O]ADP bound at the active site of creatine kinase in a transition-state analog complex exhibits the broadening effect, whereas the complex $Mn(R_P)$-$[\alpha$-$^{17}O]ADP$ does not and is indistinguishable from the same complex formed with unlabeled ADP (24).

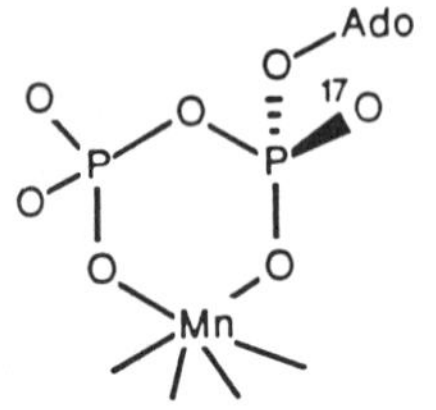

Δ Mn(R_p)-$[\alpha$-$^{17}O]ADP$ 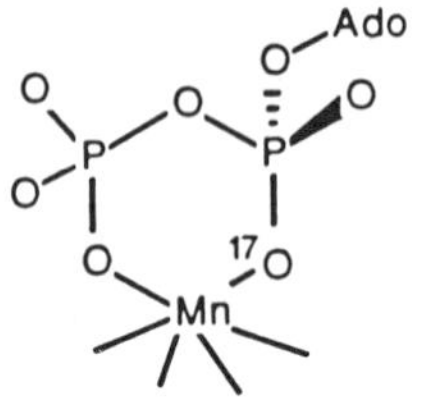   Δ Mn (S_p)-$[\alpha$-$^{17}O]ADP$

Therefore, the complex bound at the active site must be the Δ isomer shown above. Similarly, (S_P)-$[\beta$-$^{17}O]ATP$ complexed with Mn(II) at the active site of creatine kinase with an unreactive substrate analog elicits the effect, whereas (R_P)-$[\beta$-$^{17}O]ATP$ does not. Therefore, the coordination stereochemistry of MnATP at the active site of creatine kinase is fully defined as that shown below (25).

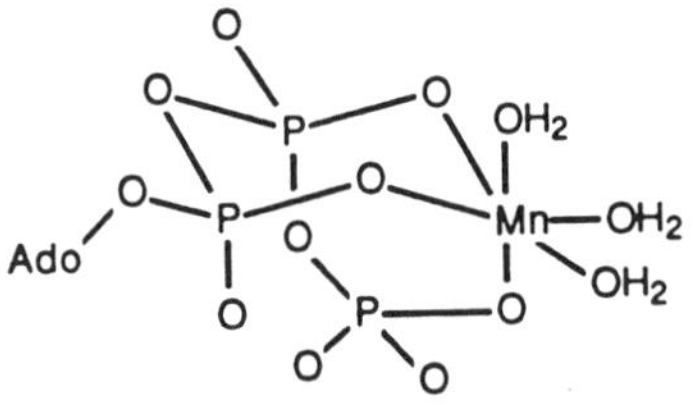

It is also known from experiments with $[\gamma$-$^{17}O]ATP$ and ^{17}O-labeled transition-state analogs of metaphosphate monoanion that P_γ is coordinated to Mn(II) in the active complex and that the remaining three ligands to Mn(II) are water molecules (23). The stereochemical results by this technique are in agreement with those obtained using the epimers of ADPαS and ADPβS as substrates and various metal ions as activators (21).

The coordination structures of manganese nucleotides at enzymic active sites are especially good models for magnesium nucleotides, since they are comparably reactive as substrates. The coordination exchange-inert complexes are somewhat less reactive, but no conflicting results have been obtained to date by these three methods.

III. Nucleotidyltransferases

Nucleotidyltransferases are all enzymes that catalyze the simple transfer of a nucleoside phosphoryl group from a donor substrate to an acceptor. They differ from ATP-dependent synthetases such as DNA ligase and acyl-CoA synthetases and aminoacyl-tRNA synthetases in that they catalyze only a single group transfer and do not catalyze the joining of two molecules. The nucleotidyltransferases include those enzymes known as pyrophosphorylases, polynucleotide polymerases, and various cyclases such as adenylyl cyclase. The mechanisms of action of representative examples of each of these enzymes will be discussed in the following sections.

A. PYROPHOSPHORYLASES

Most nucleotidyltransferases utilize nucleoside triphosphates as nucleotidyl donor substrates and produce pyrophosphate as a product. The reactions are reversible, so that many of the enzymes are known by their common name, i.e., as pyrophosphorylases. A few nucleotidyltransferases utilize nucleoside diphosphates as nucleotidyl group donors and produce inorganic phosphate as a product. The pyrophosphorylases catalyze the formation of nucleoside diphosphate compounds in reactions of the magnesium complexes of nucleoside triphosphates, such as ATP, UTP, dTTP, GTP, and CTP, with phosphomonoesters, such as sugar 1-phosphates, nucleoside phosphates, and phospholipids. Molecules such as NAD^+, FAD, GDPmannose, and CDPdiglycerides are produced in these reactions, which follow the general form of Eq. (8).

$$
\begin{array}{c}
PP-O-\overset{\overset{\displaystyle O}{\|}}{\underset{\underset{\displaystyle O^-}{|}}{P}}-O-\text{Nucleoside} \\
+ \\
R-O-\overset{\overset{\displaystyle O}{\|}}{\underset{\underset{\displaystyle O^-}{|}}{P}}-O^-
\end{array}
\quad\rightleftharpoons\quad
\begin{array}{c}
R-O-\overset{\overset{\displaystyle O}{\|}}{\underset{\underset{\displaystyle O^-}{|}}{P}}-O-\overset{\overset{\displaystyle O}{\|}}{\underset{\underset{\displaystyle O^-}{|}}{P}}-O-\text{Nucleoside} \\
+ \\
\mathbf{PP_I}
\end{array}
\tag{8}
$$

By the criteria of steady-state kinetic patterns and stereochemistry, these enzymes appear to catalyze their respective reactions by similar or closely related mechanisms. The steady-state kinetic mechanisms are of the sequential type, and in all cases so far investigated the reactions proceed with inversion of configuration at P_α of the nucleoside triphosphate. Thus, these reactions proceed via ternary complexes of enzyme·NTP·ROPO$_3{}^{2-}$, and the nucleotidyl transfer is a one-step transfer directly from the NTP to the acceptor, that is, by a single displacement at P_α of NTP.

A specific example is UDPglucose pyrophosphorylase, which catalyzes reaction (9).

$$\text{MgUTP} + \text{glucose 1-P} \rightleftharpoons \text{UDPglucose} + \text{MgPP}_i \tag{9}$$

The properties of this enzyme from various sources are reviewed in Volume VIII of this series (26). The kinetic pathway followed by the UDPglucose pyrophosphorylase from yeast and other species, including calf liver, is the ordered Bi Bi pathway shown in Eq. (10), in which the nucleotidyl donor substrate must bind first to its site on the enzyme and the acceptor substrate must bind second to form a ternary complex (27, 28).

$$
\begin{array}{ccccc}
\text{UTP} & \text{Glc-1P} & & \text{PP}_i & \text{UDP-Glc} \\
\downarrow & \downarrow & & \uparrow & \uparrow \\
\hline
\text{E} & \text{E·UTP} & (\text{E·UTP·Glc-1-P} \rightleftharpoons \text{E·UDP-Glc·PP}_i) & \text{E·UDP-Glc} & \text{E}
\end{array}
\qquad (10)
$$

The chemical reaction then takes place to form a product-containing ternary complex, from which the products dissociate in reverse order, that is, with the nucleotide product dissociating last. This pathway is clearly established by the results of steady-state kinetic analyses carried out in several laboratories. The reaction proceeds with inversion of configuration at P_α of the nucleotide substrate in both reaction directions, as shown in Eq. (5) (9, 10). Therefore, the interconversion of ternary complexes proceeds in an uneven number of displacement steps. In the absence of any other evidence for the involvement of nucleophilic catalysis by the enzyme, it is almost certain that the UMP transfer occurs in a single displacement between bound substrates.

UDPglucose pyrophosphorylase from calf liver exhibits the property of binding its uridylyl donor substrates very tightly (28). The purified enzyme binds added radiochemically labeled uridine nucleotides, and the intact nucleotides remain noncovalently bound to the enzyme after gel permeation chromatography. The bound uridine nucleotides do not spontaneously dissociate but rapidly exchange with free nucleotides. The exact mechanism by which this exchange occurs is not known. It may be that the binding mechanism for nucleotides is a multistep process in which two substrate moieties, say the nucleoside and the polyphosphate moieties, bind separately to different subsites in sequential steps. In this case it is possible for two nucleotides to bind transiently to separate subsites and facilitate the exchange, as illustrated by the following binding scheme (Scheme I).

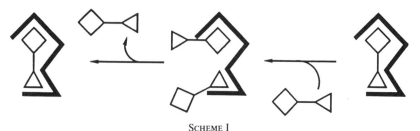

SCHEME I

Alternatively, the exchange may be related to the multimeric nature of the enzyme. It may be that the binding mechanism is defined by extreme anticooperative binding, such that in the multimeric enzyme only half the binding sites may be occupied, while exchange is allowed via transient species that never accumulate. Such a binding scheme based on a dimeric structure is illustrated in Scheme II, where UDPR is either UTP or UDPglucose.

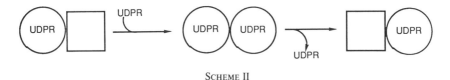

SCHEME II

The intermediate form with the binding sites in both subunits occupied does not accumulate, but it is the only form that can release a bound nucleotide at a significant rate. UDPglucose pyrophosphorylase exists in several multimeric forms, including dimers, tetramers, and octamers (26). In view of the fact that the calf liver enzyme tightly binds up to one uridine nucleotide per subunit (28), it may be that the former binding mechanism is operative. Either mechanism is consistent with the compulsory ordered binding of substrates, as revealed by the kinetics.

B.   GALACTOSE-1-PHOSPHATE URIDYLYLTRANSFERASE

An exceptional nucleotidyltransferase is galactose-1-P uridylyltransferase (UTP–hexose-1-phosphate uridylyltransferase), which catalyzes reaction (11).

$$\text{UDPglucose} + \text{galactose 1-P} \rightleftharpoons \text{UDPgalactose} + \text{glucose 1-P} \qquad (11)$$

This enzyme is not a pyrophosphorylase, but it catalyzes a similar reaction utilizing a nucleotide sugar as the uridylyl group donor. Reaction (11) is an important step in the Leloir pathway for galactose metabolism. Galactose 1-P is produced from galactose by the action of galactokinase, and galactose 1-P is converted to UDPgalactose by galactose-1-P uridylyltransferase in reaction (11). UDPgalactose is equilibrated with UDPglucose by UDPgalactose 4-epimerase. Glucose 1-P produced in reaction (11) is metabolized as glucose, allowing galactose to be catabolized as efficiently as glucose. Galactosemia is a metabolic disease of humans in which galactose-1-P uridylyltransferase is inactive.

Galactose-1-P uridylyltransferase from *Escherichia coli* has been purified and its molecular properties characterized (29). It was found to have a M_r of 80,000 and to contain two subunits with a M_r of about 40,000. Active site studies showed the subunits to be functionally identical, so that it could be assumed that they are structurally identical (30). Nucleotide sequence analysis of the cloned gene *galT* confirmed the size and, by translation, provided the amino acid sequence of the protein (*31, 32*). The amino acid sequences of the yeast and human enzymes were also derived from the nucleotide sequences of the respective genes (*33*).

Galactose-1-P uridylyltransferase is an exceptional nucleotidyltransferase because of its reaction mechanism. The kinetic mechanism is Ping-Pong Bi Bi, and the chemical reaction pathway involves a uridylyl–enzyme (UMP–enzyme), as shown in Eq. (12) (*34*).

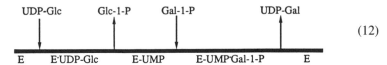

$$ \tag{12} $$

The enzyme catalyzes all of the exchange reactions that are implied by the kinetic pathway at rates that are required by the overall kinetics (*30, 35*). These are the exchange of ^{14}C between UDP[^{14}C]galactose and galactose 1-P in the absence of glucose 1-P and the exchange of ^{14}C between UDP[^{14}C]glucose and glucose 1-P. These exchanges reflect the independent reactions of UDPglucose and UDPgalactose with the free enzyme and the independent reactions of glucose 1-P and galactose 1-P with the uridylyl–enzyme [E–UMP in Eq. (12)]. The enzyme also catalyzes the exchange of ^{14}C between [*uracil*-^{14}C]UDPglucose and UDPgalactose in the absence of both galactose 1-P and glucose 1-P. This exchange is also predicted by Eq. (12), given the reversibility of the reaction. The uridylyl group in the intermediate is bonded to N-3 of a histidine residue (*36*). The evidence supporting this mechanism is reviewed in detail elsewhere (*37*).

The stereochemistry of the galactose-1-P uridylyltransferase reaction is known for each displacement step. The overall reaction proceeds with retention of configuration at P_α as shown by Eq. (13), which shows a double-displacement mechanism with inversion of configuration in each displacement (*10*).

$$ \tag{13} $$

This is the general interpretation that applies when overall retention is observed. In the case of galactose-1-P uridylyltransferase, the configuration at phosphorus of the uridylyl–enzyme is known from stereochemical analysis of the isolated intermediate (*38*).

Inasmuch as the isolated intermediate was stereochemically pure and inverted in configuration at phosphorus, its identification as the true reaction intermediate was supported by the stereochemical analysis. The question of the authenticity of isolated intermediates has always been problematical, since the isolation process could allow for or even lead to isomerization of a metastable intermediate to a more stable form. In the case of the uridylyl–enzyme, it was possible that the UMP group might have undergone migration to a second nucleophilic group of the enzyme during the isolation process. Indeed, this possibility was later accented by the discovery of two histidine residues in the active site of this enzyme. Any isomerization would have entailed stereochemical consequences, however, and would have led to loss of stereochemical purity in the isolated intermediate.

The question of which histidine residue is the nucleophilic catalyst has recently been investigated in this laboratory (39). From the translated nucleotide sequence of the E. coli gene galT, the sequence positions of 15 histidine residues were located. The importance of each histidine to the activity of this enzyme was then determined by site-directed mutagenesis. By changing the code for each histidine in turn to a code for asparagine, 15 mutant genes were generated, in each of which a specific code for histidine had been changed to one for asparagine. The change of histidine to asparagine was considered to be moderately conservative from a structural standpoint, because histidine and asparagine have similar H bonding capabilities and space requirements. Mutation to glutamine might have been equally conservative, but this would have entailed a two-base change for each mutant, and so would have been experimentally more problematical. Assays of the mutant enzymes for activity showed that 13 were active and 2 were completely inactive. The two inactive mutants were H164N and H166N. In addition to showing no detectable activity in turnover assays, the purified proteins failed to undergo uridylylation by UDPglucose within 2 hr by a sensitive assay procedure under conditions in which wild-type enzyme reacted within milliseconds. These experiments identified His-164 and His-166 as essential residues in the active site. The active site sequence was found to contain the triad His-Pro-His, which has also been found in the yeast and human enzymes (33).

The nucleophilic histidine is now known to be His-166 from examination of the catalytic properties of the two other specific mutant proteins, H166G and H164G. The rationale is illustrated in Fig. 1, which shows at the top how the uridylyl–enzyme reacts with glucose 1-P to form UDPglucose. The specific mutants lack the methylene imidazole side chain of either His-164 or His-166. One of the two mutants, H166G, accepts uridine 5′-phosphoimidazolate (UMP-Im) as a uridylyl donor substrate with glucose 1-P as the acceptor to form UDPglucose (39a). This is illustrated at the bottom of Fig. 1. The mutant H166G does not catalyze the reaction of UDPgalactose with glucose 1-P to produce UDPglu-

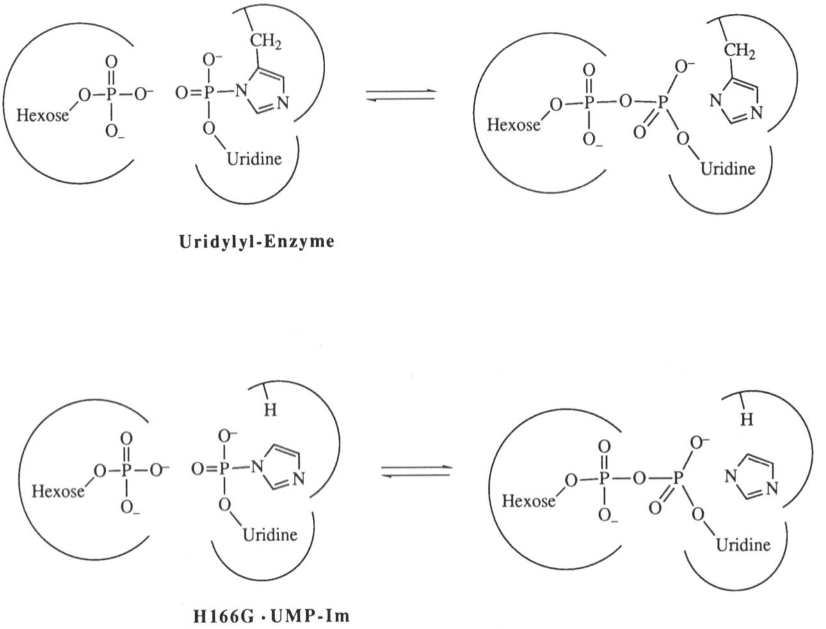

Uridylyl-Enzyme

H166G · UMP-Im

FIG. 1. The reactions of the uridylylgalactose-1-P uridylyltransferase and the mutant H166G. At the top, the uridylyl–enzyme reacts with glucose 1-P to form UDPglucose. Shown at the bottom is the reaction of UMP-Im with glucose 1-P to form UDPglucose and imidazole, catalyzed by the mutant H166G.

cose and galactose 1-P; the wild-type galactose-1-P uridylyltransferase does not catalyze the reaction of UMP-Im with glucose 1-P to form UDPglucose and imidazole. The mutant H166G also catalyzes the reverse of the reaction at the bottom of Fig. 1, the cleavage of UDPglucose by imidazole to form glucose 1-P and UMP-Im. The mutant H166G catalyzes its reaction by a sequential kinetic mechanism, analogous to the adenylate kinase and UDPglucose pyrophosphorylase mechanisms. This protein, H166G, is an enzyme in its own right, a previously unknown UDPhexose synthase. It is a newly engineered enzyme, with a k_{cat} about 1/250th that of galactose-1-P uridylyltransferase. The mutant H164G exhibits no known catalytic activities and is structurally unstable.

C. DNA AND RNA POLYMERASES

Nucleic acid polymerases catalyze nucleotidyl group transfers to the growing 3′-hydroxyl ends of polynucleotides according to Eq. (14).

$$(14)$$

This reaction is chemically similar to other nucleotidyl transfers, but it is complicated by the fact that the substrate is or rapidly becomes a large polymer, by the fact that the recognition of Base$_2$ of the nucleotidyl donor is provided by the respective templates rather than by the enzymes themselves, and by the fact that the enzymes must slide along the templates to carry out the polymerizations. The properties and functions of these enzymes are reviewed in Volume X of this series (40–44).

The DNA and RNA polymerase reactions, as well as the reverse transcriptase and polynucleotide phosphorylase reactions, proceed with inversion of configuration at P_α of the nucleoside triphosphate (45–50). Thus, an uneven number of displacements at phosphorus is involved in the chemical reaction mechanism, and the stereochemistry provides no evidence for the involvement of a covalent nucleotidyl–enzyme as an intermediate on the catalytic pathway. No other evidence for such an intermediate is available. Therefore, it must be concluded that the physicochemical requirements for nucleotidyl group transfer, substrate recognition, and movement along the template are derived from binding interactions between the enzyme and its template and substrate rather than through nucleophilic catalysis. This is also true of polynucleotide phosphorylase and other nucleotidyltransferases that catalyze reactions of polynucleotides (51, 52).

In contrast to nucleic acid polymerases, polynucleotide processing enzymes often act by mechanisms that involve covalent polynucleotide enzymes as compulsory intermediates (53, 54). The covalent linkages are through phosphodiesters comprising an enzymic nucleophile, usually the phenolic group of tyrosine, and a nucleotidyl moiety of the nucleic acid. These enzymes are not classified as nucleotidyltransferases, but they catalyze nucleotidyl group transfer as the basic reaction in isomerization processes. Examples are topoisomerases and site-specific recombinases. These enzymes utilize the enzymic nucleophile to cleave the polynucleotide in such a way as to preserve the energy of the covalent bond in the form of a new covalent bond between the enzyme and the polynucleotide. The covalent linkage is also probably used to anchor one end of the polymer during the course of an isomerization reaction. The phosphodiester bond is later restored to the polynucleotide by a rejoining reaction accompanied by cleavage of the fragment from the enzymic nucleophile subsequent to the isomerization process. The isomerizations carried out by these enzymes involve complex physical and chemical reorganizations of the polymer chains in processes such as recombination and supercoil relaxation. The mechanisms by which these processes are carried out are poorly understood. It is possible that more than one covalent intermediate might be involved in some of these reactions. Stereochemical analysis would give information about the number of displacements on phosphorus, but such information is not available.

D.   ADENYLYL CYCLASE

Adenylyl cyclase catalyzes an intramolecular adenylyl group transfer of ATP to form 3′,5′-cyclic AMP (cAMP) according to reaction (15).

$$\text{(15)}$$

3′,5′-Cyclic nucleotides are second messengers for a variety of hormonal processes. The hormones bind to cell membranes and stimulate the activities of the intramembrane cyclases, which produce cyclic nucleotides from nucleoside 5′-triphosphates and release them inside the cells, where they in turn regulate cellular metabolism. Hence their designation as second messengers.

Adenylyl and guanylyl cyclases catalyze their respective cyclization reactions with inversion of configuration at P_α of ATP or GTP (55–59). There is no other evidence for the involvement of a nucleotidyl–enzyme in these reactions. Therefore, the mechanism apparently does not involve nucleophilic catalysis by the enzyme, but instead proceeds with intramolecular nucleophilic displacement of MgPP$_i$ from P_α of the substrate by the 3′-hydroxyl group. This forms cAMP or cGMP and MgPP$_i$ in an in-line reaction with a single displacement at phosphorus.

It is interesting that, in contrast to nearly all nucleotidyltransferase reactions, most nonenzymic reactions that are used for synthesizing phosphoanhydrides and phosphodiesters are best carried out in nucleophilic solvents or in reaction media containing nucleophilic catalysts. In one case stereochemical evidence has been advanced as evidence for nucleophilic catalysis by pyridine used as the solvent in the synthesis of a thiophosphoanhydride (60). While many of these nonenzymic reactions proceed at slow rates without nucleophilic catalysis, the catalysts increase the rates and enhance the yields. However, enzymes catalyze very similar reactions at high rates by single-displacement mechanisms, with nucleophilic catalysis playing no role. Exactly how enzymes catalyze these reactions is not known. It seems likely that charge neutralization through binding interactions with metal ions and lysine-ε-ammonium ions and arginine-guanidinium ions may be important in binding the phosphoryl and nucleotidyl groups for transfer reactions.

IV. Phosphotransferases

The phosphotransferases hexokinase, adenylate kinase, phosphoglycerate kinase, nucleoside diphosphate kinase, phosphofructokinase, pyruvate kinase, creatine kinase, arginine kinase, glycerol kinase, protein kinases, and others were reviewed in Volumes VIII and IX. Evidence cited in nearly all of these reviews supported the involvement of covalent phosphoenzymes as catalytic intermediates. Ping-Pong kinetics had been reported from some laboratories for most of these enzymes, but those claims had been challenged by other laboratories. Sequential kinetic pathways seemed to be the rule for the phosphotransferases, with the notable exception of nucleoside diphosphate kinases. However, the question of the involvement of phosphoenzymes as components of ternary complexes remained unsettled and controversial. In recent years this question has been resolved in favor of the absence of a phosphoenzyme species in the catalytic pathways for phosphotransferases, again with the notable exceptions of a few cases, such as nucleoside diphosphate kinase and nucleoside phosphotransferase, in which Ping-Pong kinetics was found to be involved.

The most decisive and generally accepted single point of evidence on this question is the stereochemistry of the phospho transfer process. In all reactions so far investigated, with the exception of nucleoside diphosphate kinase and nucleoside phosphotransferase, the phospho transfer proceeds with inversion of configuration at phosphorus (7). This stereochemistry all but excludes mechanisms involving an even number of phospho transfer steps in the mechanism, because there is no other known chemical basis for configurational retention in phospho transfer. The pseudorotation mechanism can lead to retention of configuration at phosphorus in a single displacement, but this mechanism is known only for certain special cases of nonenzymic phosphoryl ester transfer and is not known for any enzymic reaction (61).

The mechanisms of action of a few phosphotransferases discussed in the following sections have been especially controversial. Stereochemical analysis has in some instances resolved outstanding questions of mechanism and in other cases reopened mechanistic questions previously thought to be settled. These studies point up the importance of making complete mechanistic studies before reaching firm conclusions. The functional significance of double-displacement mechanisms has also emerged from considerations of the instances in which these mechanisms have evolved.

A. ACETATE KINASE

Acetate kinase of *E. coli* catalyzes reaction (16), the reversible phosphorylation of acetate by ATP to form acetyl-P and ADP.

$$\text{acetyl-P} + \text{MgADP} \rightleftharpoons \text{acetate} + \text{MgATP} \tag{16}$$

Reaction (16) is written in the thermodynamically favored direction, which is the reverse of the phosphorylation of acetate. However, the enzyme is known to be required for the activation of acetate in intermediary metabolism as a means of generating acetyl-P, which then reacts as a substrate for phosphotransacetylase (phosphate acetyltransferase) and is thereby converted to acetyl-CoA and utilized as a source of carbon and energy.

The mechanism of action of acetate kinase from *E. coli* has been particularly controversial. Early kinetic and chemical experiments appeared to show that reaction (16), as catalyzed by this enzyme, proceeds by a double-displacement mechanism involving a covalent phosphoenzyme. The controversy has been discussed in detail elsewhere (*62*). Briefly stated, the evidence for the Ping-Pong pathway and an intermediate phosphoenzyme was as follows: (1) Steady-state kinetic data appeared to be consistent with a Ping-Pong kinetic pathway, because apparently parallel lines were found for double-reciprocal plots. (2) The enzyme could be phosphorylated by either [^{32}P]acetyl-P or [γ-^{32}P]ATP. (3) The enzyme was also found to catalyze the exchange reactions that characterize the Ping-Pong kinetic pathway. These included the exchange of [^{14}C]acetate into acetyl-P in the absence of added ADP [reaction (16a)] and the exchange of [^{14}C]ADP into ATP in the absence of added acetate [reaction (16b)].

$$\text{acetyl-P} + [^{14}\text{C}]\text{acetate} \rightleftharpoons [^{14}\text{C}]\text{acetyl-P} + \text{acetate} \tag{16a}$$

$$\text{ATP} + [^{14}\text{C}]\text{ADP} \rightleftharpoons [^{14}\text{C}]\text{ATP} + \text{ADP} \tag{16b}$$

(4) The [^{32}P]phosphoenzyme was isolated and chemically characterized and shown to be an enzymic glutamyl phosphate. Therefore, the evidence in favor of the double-displacement mechanism appeared to be powerful.

Despite the facts cited above, which are undisputed, the mechanism of reaction (16) is no longer believed to involve a double displacement or a phosphoenzyme. The apparent Ping-Pong kinetics, the rates of the exchange reactions (16a) and (16b), and the isolated phosphoenzyme fail to meet the most rigorous criteria for the double-displacement mechanism. And the exchange reaction (16a) is not completely independent of ADP. It is now known from a more complete kinetic analysis carried out in both reaction directions that the kinetics is sequential and that the apparent parallel lines in one direction result from particular values of kinetic parameters and the conditions of rate measurements (63). Furthermore, the rates of the exchange reactions are too slow to reflect the operation of a Ping-Pong pathway. The relationship between the maximum exchange rates for reactions (16a) and (16b) and the maximum forward and reverse overall rates is that the sum of the reciprocals of the maximum exchange rates must equal the sum of the reciprocals of the overall maximum forward and reverse rates [Eq. (17)] (63).

$$V_{exch\ 1}^{-1} + V_{exch\ 2}^{-1} = V_f^{-1} + V_r^{-1} \tag{17}$$

This relationship is not satisfied by the maximum exchange and overall rates of the acetate kinase reactions. Therefore, the reaction does not follow a simple Ping-Pong kinetic pathway.

Despite the fact that reaction kinetics cannot be Ping-Pong, the enzyme is phosphorylated by substrates and the phosphoenzyme reacts with ADP to form ATP. Is this process a part of the mechanism of reaction (16)? The stereochemical analysis of phospho transfer argues strongly against the importance of the phosphoenzyme in the reaction mechanism. The phospho transfer proceeds with inversion of configuration at phosphorus, which is consistent with a single displacement at phosphorus and inconsistent with a double-displacement mechanism (64). Therefore, the phosphoenzyme does not appear to be on the main catalytic pathway.

The catalytic pathway is best described as a random binding kinetic mechanism involving the formation of the ternary complex E·acetyl-P·ADP, with direct phosphoryl group transfer between enzyme-bound substrates to form the product ternary complex E·acetate·ATP. The formation and decomposition of these ternary complexes involve only noncovalent binding interactions of the enzyme with the substrates and products. The stereochemistry is inconsistent with a mechanism in which the phosphoryl group is transferred to an enzymic nucleophile as a step in the interconversion of the ternary complexes. The case of acetate kinase is one in which the stereochemical course of phosphoryl group transfer essentially discredited a double-displacement mechanism that had been reasonably well supported by other evidence.

The reaction of acetyl-P with acetate kinase to form a phosphoenzyme either is an adventitious side reaction of the catalytic process or reflects some other function of acetate kinase in the E. coli cell. The latter is almost certainly correct. Acetate kinase is now known to interact with the sugar phosphate transport system (PTS) of E. coli and Salmonella typhimurium (65). This system transports sugars across the plasma membrane; in the process it phosphorylates the sugars and releases sugar phosphates from the inner membrane surface inside the cell. Phosphoenolpyruvate is the best known and most thoroughly studied phosphoryl group donor for the PTS system. Phosphoenolpyruvate phosphorylates Enzyme I (EC 2.7.3.9; phosphoenolpyruvate–protein phosphotransferase) of this system (66). Phospho-Enzyme I then transfers the phosporyl group to the protein histidine (HPr), which in turn transfers the phosphoryl group to the protein III^{Glc}. Phospho-III^{Glc} is the immediate phospho donor substrate for sugars being transported, and the enzyme II^{Glc} catalyzes the phospho transfer to sugars such as glucose. Acetate kinase also catalyzes the phosphorylation of Enzyme I by ATP. In the presence of the other components of the PTS system, but in the absence of phosphoenolpyruvate, acetate kinase catalyzes the phosphorylation of III^{Glc}. Phospho-III^{Glc} also transfers the phosphoryl group to acetate kinase in the presence of HPr and Enzyme I. The biochemical importance of phosphoryl-acetate kinase seems to reside in this transport function rather than in the activation of acetate for intermediary metabolism, as in the reverse of reaction (16). Thus, acetate kinase has at least two important functions in the *E. coli* cell: (1) the activation of acetate, which involves direct phospho transfer between enzyme-bound substrates, and (2) the transport of sugars across the membrane, which involves the phosphoenzyme form of acetate kinase in the phosphorylation of Enzyme I in the PTS system of sugar phosphate transport.

B.   PHOSPHOGLYCERATE KINASE

Phosphoglycerate kinase catalyzes reaction (18), the formation of ATP and 3-phosphoglycerate from 1,3-diphosphoglycerate and ADP.

$$
\begin{array}{c}
O \diagdown C \diagup OPO_3^{2-} \\
| \\
H-C-OH \\
| \\
CH_2OPO_3^{2-}
\end{array}
+ ADP \underset{Mg^{2+}}{\rightleftharpoons} ATP +
\begin{array}{c}
O \diagdown C \diagup O^- \\
| \\
H-C-OH \\
| \\
CH_2OPO_3^{2-}
\end{array}
\qquad (18)
$$

This is a late step of the glycolytic pathway, thus the enzyme is important in many species (67). The reaction is closely analogous to reaction (16) catalyzed by acetate kinase, and, as for acetate kinase, a double-displacement mechanism for phosphotransfer has been advanced and supported by apparently direct evidence (68).

The evidence supporting the double-displacement mechanism for phosphoglycerate kinase was less extensive than that advanced for the acetate kinase reaction. The strongest evidence was the formation of a [^{32}P]phosphoenzyme by reaction of the purified enzyme with either [γ-^{32}P]ATP or 1,3,[1-^{32}P]diphosphoglycerate. The ^{32}P was removed from the enzyme by incubation with 3-phosphoglycerate or ADP. Reaction of the ^{32}P-labeled phosphoenzyme with hydroxylamine also removed the label as $^{32}P_i$, suggesting that, as with acetate kinase, the ^{32}P was bound as an acyl phosphate. Phosphoglycerate kinase also was found to catalyze the exchange reactions expected for a double-displacement mechanism.

All steady-state kinetic studies of this enzyme from various sources indicated that the kinetics is sequential and the substrate-binding and product-releasing steps involve the formation and decomposition of ternary complexes (63, 67). No evidence for a Ping-Pong kinetic pathway was found. This meant that no free phosphoenzyme such as one formed in a Ping-Pong kinetic pathway could be involved in the mechanism. Again, the kinetics could not rule out a double-displacement mechanism, but the kinetics gave no evidence of such a mechanism. Nevertheless, the phosphorylation of the enzyme by a substrate could readily be observed, and it did not seem reasonable to dismiss this as a side reaction.

The mechanistic issue is now regarded as resolved by the results of two studies. Phospho transfer catalyzed by phosphoglycertate kinase proceeds with inversion of configuration at phosphorus, which is inconsistent with a double-displacement mechanism and consistent with direct transfer of the phosphoryl group from the donor substrate to the acceptor within the ternary complex (69). Second, the [^{32}P]phosphoenzyme is not a covalent phosphoenzyme. The enzymes from yeast and horse liver contain 3-phosphoglycerate even after extensive purification (70). This substrate can and does react with [γ-^{32}P]ATP to become phosphorylated, and the resulting 1,3-[1-^{32}P]diphosphoglycerate remains bound to the protein, even after passage through a gel permeation column. ADP removes the ^{32}P by reacting with the bound 1,3-[1-^{32}P]diphosphoglycerate to form [γ-^{32}P]ATP. Hydroxylamine releases ^{32}Pi by reaction with enzyme-bound 1,3-[1-^{32}P]diphosphoglycerate, forming the hydroxamate of 3-phosphoglycetate, without inactivating the enzyme (70). In as much as hydroxylamine does not inactivate the enzyme, it cannot be reacting with an enzymic acyl [^{32}P]phosphate, because such a reaction would lead to the conversion of the enzymic carboxyl group to a hydroxamate.

One type of evidence often presented in support of double-displacement mechanisms involving covalent enzyme–substrate intermediates is the observation of exchange reactions of the type discussed above for galactose-1-P uridylyltransferase and acetate kinase. The exchange reactions (19a) and (19b), where PGA is 3-phosphoglycerate and DPG is 1,3-diphosphoglycerate, are catalyzed by phosphoglycerate kinase (68, 70).

$$ATP + [^3H]ADP \rightleftharpoons [^3H]ATP + ADP \qquad (19a)$$
$$[^{14}C]PGA + DPG \rightleftharpoons PGA + [^{14}C]DPG \qquad (19b)$$

These exchange reactions can support the existence of a double-displacement mechanism via a covalent intermediate *if cosubstrates are absent*. Reaction (19a) could proceed if ATP phosphorylates the enzyme and the phosphoryl group is

captured by [³H]ADP. Reaction (19b) could similarly occur if DPG phosphorylates the enzyme and the phosphoryl group is captured by [¹⁴C]PGA. In the case
of phosphoglycerate kinase, the exchange reactions are very slow, orders of
magnitude slower than the overall reaction. Given that the kinetic mechanism
is sequential, the slow rates of the exchange reactions do not by themselves
rule out a double displacement, because in the absence of a free covalent intermediate one could not expect the exchange reactions to be fast. However, the
exchange reactions must be interpreted in another way. Given the presence of
3-phosphoglycerate, reaction (19a) simply represents the operation of the overall
reaction, since all substrates are present. Reaction (19b) is much more difficult
to explain. The enzyme is known to contain 3-phosphoglycerate, but it is not
known to contain ADP, which would be required to complete the reaction mixture. It is possible that the enzyme is contaminated with ADP as well. Alternatively, the complex coupling system used to assay for this exchange reaction
may have been contaminated with ADP or ATP. In any case, the fact that the
phosphoenzyme is really a complex of the enzyme with 1,3-diphosphoglycerate
adequately accounts for the presence of ^{32}P in the enzyme after reaction with
[γ-^{32}P]ATP or 1,3-[1-^{32}P]diphosphoglycerate.

Both acetate kinase and phosphoglycerate kinase were found to catalyze exchange reactions of the type normally associated with Ping-Pong kinetics and
double-displacement mechanisms involving covalent intermediates. In both cases,
^{32}P-labeled phosphoenzymes could be isolated, a covalent species in the case of
acetate kinase and a tight complex of enzyme and 1,3-diphosphoglycerate in the
case of phosphoglycerate kinase. However, in both cases these indicators of the
double-displacement mechanism have been explained in other ways. In both
cases the stereochemical course of phospho transfer and the steady-state kinetics
have supported the single-displacement mechanism, with direct transfer of the
phosphoryl group between bound substrates of ternary enzyme–substrate complexes. As with many other enzymes, the stereochemical test has proved to be a
highly reliable indicator of reaction mechanism.

C. NUCLEOSIDE PHOSPHOTRANSFERASE

The enzyme nucleoside phosphotransferase catalyzes reaction (20), the transfer of a phosphoryl group from one nucleoside 5'-phosphate to another nucleoside.

$$N_1MP + N_2 \rightleftharpoons N_1 + N_2MP \tag{20}$$

This activity is reported to be widespread in tissues, including liver. The enzyme
has been purified from plant sources such as carrots and barley sprouts (71–74).
The purified enzyme is activated by metals such as Mg^{2+} and is fairly specific

for nucleoside 5'-phosphates, with a distinct preference for purine nucleosides, especially purine deoxynucleosides. The enzyme also catalyzes the hydrolysis of nucleoside 5'-phosphates. Thus, it is both a nucleoside phosphotransferase and a phosphatase with selectivity for nucleoside 5'-phosphates. The phosphatase activity is inhibited by nucleosides.

Nucleoside phosphotransferase catalyzes thiophosphoryl group transfer as well as phosphoryl group transfer, and thiophospho transfer proceeds with overall retention of configuration at phosphorus (75). This fact first suggested that the enzyme may catalyze phospho transfer by a double-displacement mechanism. The steady-state kinetics is consistent with the Ping-Pong catalytic pathway of Eq. (21), which differs from the simplest Ping-Pong mechanism by allowing for hydrolysis of the intermediate phosphoenzyme in the absence of a nucleoside to accept the phosphoryl group (74).

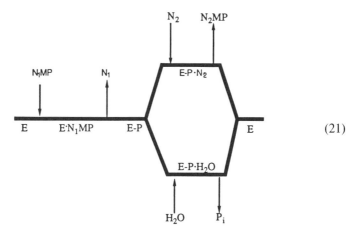 (21)

The hydrolytic lability of the phosphoenzyme intermediate prevented its isolation in an active form; however, the ^{32}P-labeled phosphoenzyme could be isolated in good yield in a denatured form by precipitation of the protein with acid at low temperature in the presence of a ^{32}P-labeled substrate (74). Further chemical analysis indicated that the intermediate is an acyl phosphate, with the phosphoryl group bonded to an enzymic carboxyl group (76).

In early studies the kinetics appeared to be sequential because of the convergence of double-reciprocal plots (72, 77). It was not entirely clear at the time that the phosphatase activity was an intrinsic part of the phosphotransferase activity, and the kinetics was not pursued further. Efforts to isolate an active phosphoenzyme failed, presumably due to its hydrolysis in the absence of a nucleoside (73, 77). The stereochemical analysis showing retention of configuration at phosphorus reopened the mechanistic question and stimulated further studies that

eventually clarified the Ping-Pong kinetics and led to the isolation and characterization of the phosphoenzyme.

D. ADENOSINE KINASE

Adenosine kinase is one of a family of nucleoside kinases that are widely
found in animal tissues, microorganisms, and plants. This enzyme catalyzes
reaction (22), the phosphorylation of the 5′-hydroxyl group of adenosine by
MgATP.

$$Ado + MgATP \rightleftharpoons AMP + MgADP \tag{22}$$

The enzyme from beef liver catalyzes thiophospho transfer with ATPγS as the
phosphodonor, and the reaction of (R_P)-[γ-$^{18}O_2$]ATPγS proceeds with inversion
of configuration at phosphorus (78). This means that the catalytic pathway includes an uneven number of phospho transfer steps, almost certainly a single
step in this case.

As is known for other nucleoside kinases, the steady-state kinetics of the
adenosine kinase reaction is complex owing to regulatory effects. Adenosine and
AMP apparently bind at a regulatory site, where they modulate activity, as well
as at the active site, where they act as substrates. These interactions complicate
the kinetics, but a careful analysis shows that the basic kinetic pathway is sequential and involves the compulsory formation of ternary complexes (79).
Thus, the kinetics is consistent with the stereochemistry and suggests that the
phospho transfer is a direct, one-step displacement between substrates bound at
the active site in a ternary complex. Complications introduced into the mechanistic analysis of this enzyme by the adventitious phosphorylation of the protein
by ATP have been discussed elsewhere (7).

V. ATP-Dependent Synthetases and Ligases

Synthetases and ligases are not classified as nucleotidyltransferases and phosphotransferases, but those that utilize ATP to activate a substrate for ligation
catalyze nucleotidyl transfers or phospho transfers in activation steps. The
mechanistic questions regarding these steps are analogous but not identical to
those for the simple transferases. The chemical aspects of the adenylyl and phospho transfer steps should be the same for ATP-dependent synthetases and ligases
as for the simpler transferases. However, enzymic catalysis frequently involves
factors other than purely chemical rate enhancement of a single step. In the case
of ATP-dependent synthetases and ligases, these factors include the requirement
to bind several substrates in close proximity or the involvement of two or more
sites in a polymer. In a few cases the complications imposed by the interactions

of the enzyme with additional substrates or reaction sites are overcome by the formation of covalent nucleotidyl–enzymes or phosphoenzymes as reaction intermediates. In other cases no such intermediates are involved. A few enzymes of each type are briefly discussed here as more complex examples of enzymes that catalyze nucleotidyl or phospho transfer, but are not themselves classified as transferases.

A.   DNA and RNA Ligases

DNA ligase and RNA ligase catalyze phosphodiester bond formation between 5'-phosphate and the 3'-hydroxyl ends of DNA or RNA. DNA ligases are required for DNA replication and RNA ligases are required for certain RNA splicing reactions. The function and overall reaction mechanism for DNA ligases are reviewed in Volume X of this series (80). The chemical mechanistic pathway for DNA ligase is outlined in reactions (23a)–(23c).

$$\text{E-Lys} + \text{MgATP} \rightleftharpoons \text{E-Lys-AMP} + \text{MgPP}_i \qquad (23a)$$

$$(23b)$$

$$(23c)$$

These enzymes are not classified as nucleotidyltransferases, although they catalyze nucleotidyl group transfers in the course of activating the 5'-phosphoryl groups for the ligation process. The activation mechanism involves a covalent adenylyl–enzyme as an intermediate and a double displacement on P_α of ATP (or NAD$^+$). The chemical mechanism of the RNA ligase reaction is similar. The stereochemistry of these reactions is known for RNA ligase and is consistent with the mechanism as formulated above (81, 82).

From a chemical standpoint the activation phases of the DNA and RNA ligase reactions [reactions (23a) and (23b)] are analogous to the pyrophosphorylase reactions discussed above, and the ligase phases [reaction (23c)] are analogous to the reactions catalyzed by the DNA and RNA polymerases. Both the pyrophosphorylases and the polymerases catalyze single displacements at phosphorus; however, the DNA and RNA ligases catalyze double displacements in the pyrophosphorylase phase and form covalent adenylyl–enzyme intermediates. The reasons for the differing reaction mechanisms are not fully understood. The involvement of adenylyl–enzymes as intermediates should not reflect a specific chemical requirement for nucleophilic catalysis in the enzymic cleavage of ATP, because the pyrophosphorylases and polymerases function well by single-displacement mechanisms. The involvement of nucleophilic catalysis and adenylyl–enzymes in the nucleic acid ligase reactions may reflect special requirements for the accessibility of separate substrate-binding subsites to an activated adenylyl group.

B. GLUTAMINE SYNTHETASE

The mechanism of action of glutamine synthetase has been extensively studied, and both the bacterial and mammalian enzymes were reviewed in Volume X of this series (*83, 84*). Many experimental approaches have been applied to the question of the chemical activation of the γ-carboxyl group of glutamate and its reaction with ammonia. All of these experiments supported the stepwise chemical mechanism outlined in reactions (24a) and (24b) (*83*),

$$ \text{(24a)} $$

$$ \text{(24b)} $$

with the intermediate γ-glutamyl phosphate remaining enzyme bound throughout the reaction. Much chemical evidence supporting this mechanism has been generated.

Inasmuch as γ-glutamyl phosphate is unstable in solution, cyclizing rapidly with displacement of P_i by the α-amino group to form 5-oxoproline, the evidence implicating it as an intermediate is indirect. Because of its instability, γ-glutamyl phosphate can neither be isolated and characterized nor synthesized and used as a substrate in place of glutamate and ATP. Indeed, among the most direct evidence for the formation of γ-glutamyl phosphate at the active site of glutamine synthetase is the fact that the enzyme catalyzes the formation of 5-oxoproline, ADP, and P_i at a slow rate from glutamate and ATP in the absence of NH_3 (*83*). Cyclization at the active site is presumably prevented in the presence of NH_3 by the rapid reaction of this substrate with γ-glutamyl phosphate to form glutamine.

To examine the kinetic competence of γ-glutamyl phosphate formation at the active site of glutamine synthetase, the technique of positional isotope exchange (PIX) was introduced to studies of the mechanism of phospho transfer (85). This experiment circumvented the problem of the instability of the intermediate, which complicated direct kinetic experiments. In the PIX experiment, the rate at which glutamine synthetase catalyzes reaction (25) was measured.

$$
\begin{array}{c}
{}^{18}\text{O}\quad\quad\text{O}\\
\|\quad\quad\|\\
{}^{18}\text{O}-\text{P}-{}^{18}\text{O}-\text{P}-\text{O}-\text{AMP}\\
|\quad\quad|\\
{}^{18}\text{O}\quad\quad\text{O}
\end{array}
\quad\rightleftharpoons\quad
\begin{array}{c}
{}^{18}\text{O}\quad\quad{}^{18}\text{O}\\
\|\quad\quad\|\\
{}^{18}\text{O}-\text{P}-\text{O}-\text{P}-\text{O}-\text{AMP}\\
|\quad\quad|\\
{}^{18}\text{O}\quad\quad\text{O}
\end{array}
\qquad (25)
$$

This reaction consisted of an isomerization of ^{18}O-labeled ATP in the presence of glutamate, in which the oxygen bridging P_β and P_γ of ATP underwent scrambling between bridging and nonbridging positions. Isotopic equilibration resulted from the reversibility of reaction (24a) at the active site of the enzyme and from the torsional motion of the β-phosphoryl group in enzyme-bound MgADP. The scrambling proceeded in the presence but not in the absence of glutamate, and it was catalyzed by the enzyme at a rate that corresponded to the overall rate of the glutamine synthetase reaction. The results indicated that reaction (24a) proceeds at a rate that is comparable to the overall rate of the glutamine synthetase reaction.

The PIX experiment has been extensively applied to problems of the type presented by glutamine synthetase, where the putative intermediates are too unstable to study directly and information on their presence and rates of formation is needed. For tests of kinetic competence this technique can be superior to more direct experiments, in which isolated or synthesized intermediates are mixed with enzymes in order to construct active complexes and the rates of reactions are measured. In such experiments the rates are rarely if ever found to be kinetically competent, since the putative intermediate must bind to an enzyme form that is never generated in a catalytic cycle, and the rate at which the intermediate complex is formed will be less than the overall rate. In the PIX experiment only the normal catalytic reaction is used to generate the desired complexes. The main drawback to the PIX experiment is that it depends on the torion rate in an intermediate such as MgADP. When the torsion rate is slow, e.g., owing to metal complexation, the experiment fails.

Phospho transfer in the overall glutamine synthetase reaction proceeds with inversion of configuration at phosphorus (86). Only reaction (24a) of the chemical pathway involves phosphoryl group transfer, since the subsequent reaction (24b) occurs with carbon–oxygen bond cleavage. The mechanism of reaction (24a) could in principle involve either a single displacement and direct phospho transfer from ATP to the γ-carboxylate of glutamate, or a double displacement via a covalent phosphoenzyme. The stereochemical course indicates the former mechanism, a single displacement and direct transfer.

C. SUCCINYL-COA SYNTHETASE

Succinyl-CoA, which is produced in the tricarboxylic acid cycle, is normally converted to succinate by the action of succinyl-CoA synthetase in catalyzing reaction (26).

$$+ \text{ GTP } + \text{ CoASH} \qquad\qquad\qquad\qquad + \text{ GDP } + \text{ P}_i \qquad (26)$$

The enzyme from *E. coli* utilizes ATP and ADP as substrates in place of the guanine nucleotides utilized by the mammalian enzymes. The mechanism of this reaction is reviewed in Volume X of this series (*87*). It is an interesting mechanism in that it involves an intermediate phosphoenzyme, with the phosphoryl group bonded to a histidine imidazole ring. The chemical reaction pathway consists of reactions (27a)–(27c), in which succinyl phosphate is an enzyme-bound intermediate.

$$\text{E–His } + \text{ GTP } \rightleftharpoons \text{ E–His–P } + \text{ GDP} \qquad\qquad (27a)$$

$$\text{E–His–P } + \text{ succinate } \rightleftharpoons \text{ E–His·succinyl-P} \qquad\qquad (27b)$$

$$\text{E–His·succinyl-P } + \text{ CoASH } \rightleftharpoons \text{ E–His } + \text{ succinyl-CoA } + \text{ P}_i \qquad (27c)$$

These equations describe the chemical events but do not represent the kinetic pathways of substrate binding and product release. The kinetic pathway is one that requires the formation of quaternary complexes within which the chemical events take place. The kinetic pathway for the enzyme from *E. coli*, outlined in Eq. (28), is sequential and partially random with no Ping-Pong component. The intermediate E–His–P is involved in the interconversion of the quaternary complexes.

$$(28)$$

D. AMINOACYL-tRNA SYNTHETASES

In the activation of amino acids for protein biosynthesis, the aminoacyl-tRNA synthetases catalyze their ligation as acyl esters to the 3-hydroxyl ends of their cognate species of tRNA. The chemical activation mechanism requires ATP and occurs in two steps, the activation of the amino acid by reaction with ATP to form an aminoacyl adenylate in reaction (29a), and the transfer of the activated aminoacyl group to the 3'-hydroxyl end of tRNA in reaction (29b) (*88*).

$$E + R\!-\!\underset{\overset{|}{^+NH_3}}{CH}\!-\!CO_2^- + ATP \rightleftharpoons E\cdot R\!-\!\underset{\overset{|}{^+NH_3}}{CH}\!-\!\overset{\overset{O}{\|}}{C}\!-\!O\!-\!AMP + PP_i \quad (29a)$$

$$E\cdot R\!-\!\underset{\overset{|}{^+NH_3}}{CH}\!-\!\overset{\overset{O}{\|}}{C}\!-\!O\!-\!AMP + tRNA^{aa} \rightarrow E + E\cdot R\!-\!\underset{\overset{|}{^+NH_3}}{CH}\!-\!\overset{\overset{O}{\|}}{C}\!-\!O\!-\!tRNA^{aa} \quad (29b)$$

The nucleotidyl transfer step is reaction (29a), which proceeds with inversion of configuration at phosphorus in all of the aminoacyl-tRNA synthetase reactions so far studied [for amino acids (aa) Phe, Ile, Tyr, and Met] (*89–92*). Stereochemical inversion shows that the nucleotidyl transfer mechanism involves an uneven number of substitutions on phosphorus. Since no other evidence of an adenylyl–enzyme can be found, aminoacyl activation most likely occurs by a single-displacement mechanism, with direct transfer of the AMP group from ATP to the carboxylate group of the amino acid within the enzyme–amino acid–ATP complex.

Acetyl-CoA synthetase from mammalian tissues and yeast catalyzes the reaction of acetate with ATP and CoA to form acetyl-CoA by a chemical mechanism similar to that of the aminoacyl-tRNA synthetases. The catalytic pathway is similar to that of reactions (29a) and (29b), substituting acetate for the amino acid and CoA for tRNA (*93*). The activation of acetate via the intermediate acetyl adenylate also occurs with inversion of configuration at P_α of ATP (*94*). Thus, as for aminoacyl-tRNA synthetases, acetyl-CoA synthetase appears to catalyze the activation of acetate by a single-displacement mechanism.

E. FORMATION OF PHOSPHOENOLPYRUVATE

Phosphoenolpyruvate can be produced from pyruvate in bacteria by reactions (30) and (31), catalyzed, respectively, by phosphoenolpyruvate synthetase and pyruvate phosphate dikinase (*95, 96*).

$$pyruvate + ATP \rightleftharpoons phosphoenolpyruvate + AMP + P_i \quad (30)$$

$$pyruvate + ATP + P_i \rightleftharpoons phosphoenolpyruvate + AMP + PP_i \quad (31)$$

Reaction (30) is favored as written for the formation of phosphoenolpyruvate by utilization of the energy of two phosphoanhydride bonds to phosphorylate pyruvate. Reaction (31) is energetically unfavored as written and would lead to the decomposition of phosphoenolpyruvate. This may be its function in some cells or under certain conditions (95). However, in the presence of inorganic pyrophosphatase to catalyze the hydrolysis of pyrophosphate, reaction (31) can also produce phosphoenolpyruvate.

The mechanisms of these reactions are interesting in that they include both phosphoenzymes and pyrophosphoenzymes as intermediates. The chemical catalytic pathways follow reactions (32a)–(32c).

$$\text{E-His} + \text{ATP} \rightleftharpoons \text{E-His-}\mathbf{PP} + \text{AMP} \tag{32a}$$

$$\text{E-His-}\mathbf{PP} + \text{H}_2\text{O(P}_i\text{)} \rightleftharpoons \text{E-His-}\mathbf{P} + \text{P}_i(\text{PP}_i) \tag{32b}$$

$$\text{E-His-}\mathbf{P} + \text{pyruvate} \rightleftharpoons \text{E-His} + \text{phosphoenolpyruvate} \tag{32c}$$

In reaction (32b) the phosphoryl acceptor substrate is water in the case of phosphoenolpyruvate synthetase and inorganic phosphate in the case of pyruvate phosphate dikinase. The evidence supporting these pathways is extensively reviewed elsewhere (95, 96). Briefly stated, this evidence includes the observation of Ping-Pong kinetics, the observation of relevant exchange reactions that occur at rates compatible with the overall reaction rates and Ping-Pong kinetics, labeling experiments demonstrating that P_β of ATP appears in phosphoenolpyruvate, the isolation of phosphoenzymes, and, in the case of pyruvate phosphate dikinase, the isolation of the pyrophosphoenzyme. The hydrolytic lability of the pyrophosphorylated form of phosphoenolpyruvate synthetase prevents its being isolated as such, but there is little doubt that it is transiently formed.

According to reactions (32a)–(32c) there are two substitutions at P_β of ATP and one substitution at P_γ in the overall formation of phosphoenolpyruvate. Therefore, the stereochemical course of the overall reaction should involve retention of configuration at P_β and inversion at P_γ. These are, in fact, the stereochemical consequences at P_β and P_γ, thus the stereochemistry confirms reactions (32a)–(32c) as the catalytic pathway (97).

Phosphoenolpyruvate is produced from oxaloacetate in animals and plants by reaction (33), which is catalyzed by phosphoenolpyruvate carboxykinase.

In plants ATP is used in place of GTP or ITP in animals. Either Mn^{2+} or a mixture of Mn^{2+} and Mg^{2+} is required as a cofactor to facilitate both the phospho transfer and the decarboxylation. The kinetic binding pathway is sequential via the ternary complex E·oxaloacetate·MgGTP in the forward direction, with an additional metal presumably coordinated to oxaloacetate (*98, 99*). This reaction proceeds with inversion of configuration at phosphorus, suggesting that P_γ of GTP or ITP is transferred directly to enolpyruvate, which is produced in the decarboxylation of oxaloacetate (*100, 101*). The overall reaction probably occurs in two steps at the active site, with the decarboxylation of oxaloacetate leading the phospho transfer, as in reactions (34a) and (34b).

(34a)

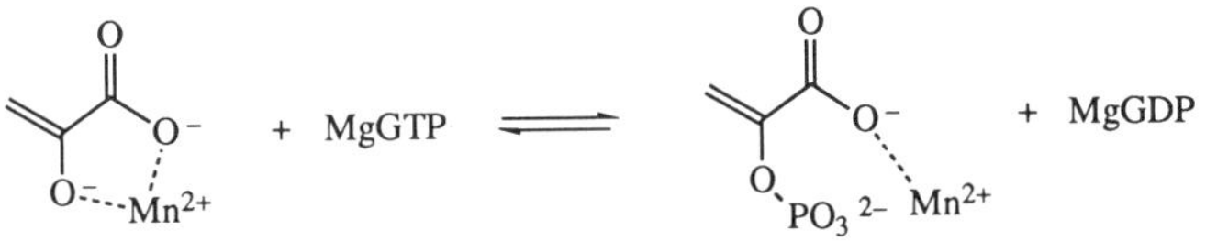

(34b)

VI. Importance of Covalent Intermediates

A few nucleotidyltransferases and phosphotransferases act by mechanisms that include a covalent nucleotidyl–enzyme or phosphoenzyme as a compulsory intermediate. A few more ATP-dependent synthetases act by such mechanisms. However, the overwhelming majority of these enzymes catalyze phosphoryl or nucleotidyl group transfer by mechanisms that entail the direct, one-step transfer of the nucleotidyl group or phosphoryl group between two substrates bound in adjacent subsites at the active site.

The existence and importance of the covalent intermediates that appear in the few such reactions in which they are involved can be rationalized in a few ways. Most can be rationalized on the basis of the principle of *economy in the evolution of binding sites*. In the cases of the pyrophosphoenzymes and phosphoenzymes that are intemediates in the phosphoenolpyruvate synthetase and pyruvate phosphate dikinase reactions, the same principle applies; however, additional interactions are probably more important. In one instance of a phosphoenzyme, the succinyl-CoA synthetase reaction, the role of the phosphoenzyme is unclear. But even in that case an attractive rationale can be advanced as a hypothesis.

A. PHOSPHOTRANSFERASES AND NUCLEOTIDYLTRANSFERASES

The principle of economy in the evolution of binding sites accounts for the few double-displacement reactions in this class. This principle may be stated as follows: An enzyme contains within its active center the minimum number of substrate binding sites required to catalyze the reaction for which it is fitted by evolution. The importance of this principle in the evolution of these transferases is most obvious when one compares those reactions that proceed by the double-displacement mechanism with chemically similar reactions that occur by single-displacement mechanisms. Consider the paired reactions (35a) and (35b), (36a) and (36b), and (37a) and (37b).

$$\text{UDPGlc} + \text{Gal 1-P} \rightleftharpoons \text{UDPGal} + \text{Glc 1-P} \qquad \text{(retention, Ping-Pong)} \qquad (35a)$$

$$\text{MgUTP} + \text{Glc 1-P} \rightleftharpoons \text{UDPGlc} + \text{MgPP}_i \qquad \text{(inversion, sequential)} \qquad (35b)$$

$$\text{MgATP} + \text{MgNDP} \rightleftharpoons \text{MgADP} + \text{MgNTP} \qquad \text{(retention, Ping-Pong)} \qquad (36a)$$

$$\text{MgATP} + \text{AMP} \rightleftharpoons \text{MgADP} + \text{ADP} \qquad \text{(inversion, sequential)} \qquad (36b)$$

$$\text{AMP} + \text{dAdo} \rightleftharpoons \text{Ado} + \text{dAMP} \qquad \text{(retention, Ping-Pong)} \qquad (37a)$$

$$\text{MgATP} + \text{Ado} \rightleftharpoons \text{MgADP} + \text{AMP} \qquad \text{(inversion, sequential)} \qquad (37b)$$

Reactions (35a) and (35b) are catalyzed by galactose-1-P uridylyltransferase and UDPglucose pyrophosphorylase, respectively; reactions (36a) and (36b) are catalyzed by nucleoside diphosphate kinase and adenylate kinase, respectively; and reactions (37a) and (37b) are catalyzed by nucleoside phosphotransferase and adenosine kinase, respectively.

The reactions listed above are chemically paired; that is, the two reactions of a pair are chemically similar. These paired reactions show three additional correlations: (1) In each pair one of the reactions is kinetically Ping-Pong and proceeds with retention of configuration at phosphorus, whereas the other is kinetically sequential and proceeds with inversion. (2) A covalent nucleotidyl–enzyme or phosphoenzyme is an intermediate in each Ping-Pong reaction, whereas there is no covalent intermediate in any of the kinetically sequential reactions. (3) In each Ping-Pong reaction the phospho-group acceptor substrates on the left and right sides of the equation are structurally similar, whereas in each sequential reaction the phospho-group acceptors are structurally dissimilar.

Of the possible kinetic mechanisms that can support an enzymic bisubstrate group transfer reaction, the one-site Ping-Pong Bi Bi kinetic pathway requires the simplest active center. This is because the Ping-Pong pathway can operate with fewer substrate binding sites than any sequential kinetic mechanism. The Ping-Pong pathway is, therefore, favored by the principle of economy in the evolution of binding sites whenever it is a practical mechanistic option. In the Ping-Pong pathway the two group acceptor substrates are never bound to the enzyme at the same time [see Eqs. (4a) and (4b)]. The condition that one acceptor must dissociate from the enzyme before the other can bind is one that cannot be violated in a Ping-Pong Bi Bi pathway, and it is most simply satisfied when the two acceptors bind at the same enzymic binding site or subsite. Since each group donor includes within its structure one of the potential acceptors, the basic binding site in an enzyme that operates by a one-site Ping-Pong pathway accommodates both

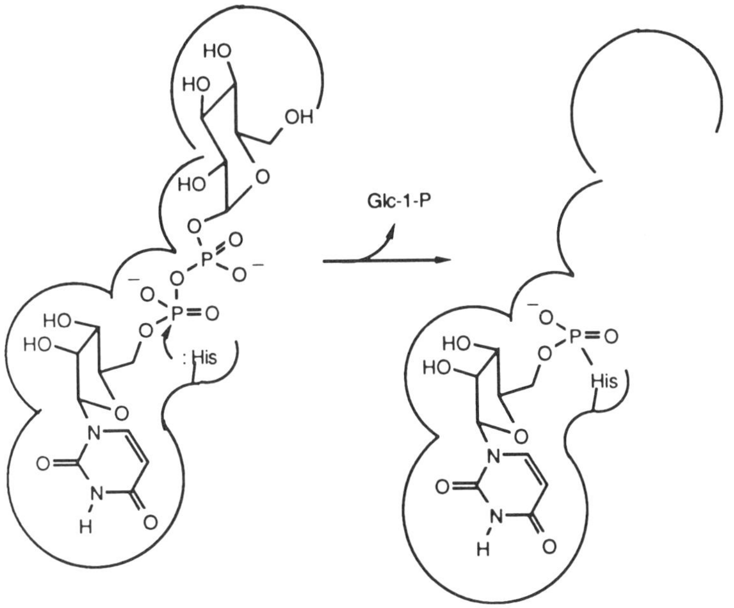

FIG. 2. A conceptualization of the substrate binding site in galactose-1-P uridylyltransferase.

group donor substrates. A subsite within this binding site is the binding locus for both acceptors, as illustrated in Fig. 2 for the galactose-1-P uridylyltransferase. Shown is a model of a conceptualization of the binding site for UDPglucose in this enzyme. The binding site can be thought of as consisting of subsites for UMP and glucose 1-P, and the subsite for glucose 1-P can also bind galactose 1-P because the two are sterically and electrostatically similar and have similar binding requirements. However, glucose 1-P and galactose 1-P cannot both bind at the same time. If a nucleophilic group is present and has access to P_α, it can cleave glucose 1-P away by forming a bond to P_α. Glucose 1-P can then dissociate and be replaced by galactose 1-P, which reacts with the uridylyl–enzyme and forms UDPgalactose. In this way, the enzyme can function with only a single binding site for a nucleotide sugar. The function of the nucleophilic histidine is to preserve the bond energy to the uridylyl group during the interchange between glucose 1-P and galactose 1-P binding at the acceptor subsite.

It seems likely that the structural similarity between glucose 1-P and galactose 1-P, which allows them to bind to the same site, has been a decisive factor in the evolution of this enzyme and its reaction mechanism. The principle of economy in the evolution of binding sites is nicely obeyed by the evolution of a single binding site for the two uridylyl group acceptors. The inclusion through evolution of a nucleophilic group for covalently binding the transferred group, UMP

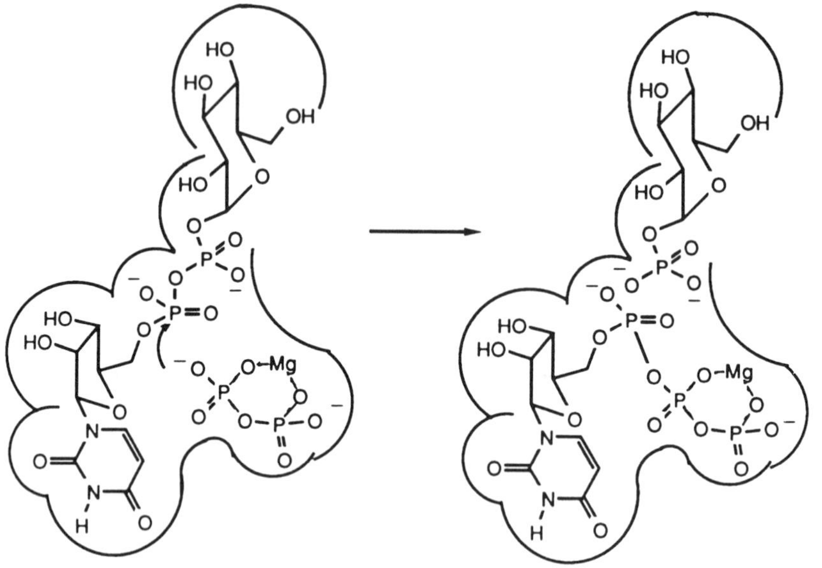

FIG. 3. A conceptualization of the substrate binding sites in UDPglucose pyrophosphorylase.

in this case, seems less demanding than the creation of separate binding sites for glucose 1-P and galactose 1-P.

The evolution of UDPglucose pyrophosphorylase could not have led to a single binding site for the uridylyl group acceptors glucose 1-P and PP_i because these acceptors are sterically and electrostatistically very different and could not be accommodated by any one binding subsite. Economy in the evolution of binding sites, therefore, dictated the evolution of separate sites for binding PP_i and glucose 1-P. Such sites are conceptualized in Fig. 3, which illustrates the substrate binding requirements for this enzyme. With separate binding sites for the two acceptor substrates in place, both can bind simultaneously, and the need for a nucleophile to mediate uridylyl group transfer is obviated. The uridylyl group is simply transferred from the donor substrate directly to the acceptor substrate bound in the adjacent site.

The rationale for the different mechanisms of the other paired reactions [(36a) and (36b) and (37a) and (37b)] is exactly the same. This pattern is followed by all enzymes of this class. That is, nucleophilic catalysis and covalent intermediates, which are necessary and required for the Ping-Pong kinetic pathway to operate, are found only in those cases involving Ping-Pong kinetics. No phosphotransferase or nucleotidyltransferase is known to catalyze phosphoryl or nucleotidyl group transfer by a sequential kinetic pathway via a double-displacement mechanism. However, it is not certain that in all cases of double displacements the two acceptors bind to the same site. Glucose-1,6-bisphosphate synthase catalyzes reaction (38)

$$\begin{array}{l} \text{glucose 1-phosphate} \\ + \text{ 1,3-diphosphoglycerate} \end{array} \rightleftharpoons \begin{array}{l} \text{glucose 1,6-biphosphate} \\ + \text{ 3-phosphoglycerate} \end{array} \qquad (38)$$

by a Ping-Pong kinetic mechanism via an intermediate phosphoenzyme (*102*). The two acceptors are sterically and electrostatically somewhat different and may or may not share a common binding site. The rule that double displacements and phosphoenzymes are limited to cases of Ping-Pong kinetics is not violated, however.

It must be concluded that, from a purely chemical catalytic standpoint, enzymes do not require nucleophilic covalent catalysis to mediate phospho-group transfer. The most common mechanism is the single displacement with direct transfer between substrates bound at adjacent sites. Nucleophilic catalysis is limited to those cases in which the bond energy to the phospho group must be maintained during some other process, such as the interchange of acceptor substrates in a Ping-Pong pathway. We shall see in the following section that ATP-dependent synthetases are more complex, and that phosphoenzymes can have more complex roles in a few of those reactions.

B. ATP-DEPENDENT SYNTHETASES

Covalent nucleotidyl–enzymes or phosphoenzymes are intermediates in a few reactions catalyzed by ATP-dependent synthetases, although most of these enzymes catalyze phospho-group transfer directly between bound substrates. However, the importance of covalent intermediates in the synthetase reactions is more complex than in the simpler transferase reactions. In all cases the covalent intermediates conserve bond energy to phospho groups as in the transferases, but they often have other essential functions as well. These other functions are not well understood, but reasonable hypotheses can be advanced.

DNA ligase catalyzes the joining of $3'$-hydroxyl and $5'$-phosphate ends of DNA strands to form a phosphodiester linkage in reactions (23a)–(23c). The first step is the reaction of a lysine-ε-NH_2 group at the active site with ATP (or NAD^+) to form an adenylyl–enzyme (E–Lys–AMP) and PP_i in reaction (23a). This reaction occurs in the absence of nicked DNA. The adenylyl–enzyme then binds nicked DNA and catalyzes ligation in reactions (23b) and (23c). Thus, the adenylylation of the enzyme and DNA strand ligation occur in separate steps and the reaction follows Ping-Pong kinetics (*80*).

It is not known whether magnesium pyrophosphate and the $3'$-hydroxyl group of a nick in DNA share a common acceptor binding site of DNA ligase, in analogy to the acceptors in reactions (35a), (36a), and (37a) discussed above. It seems unlikely that a limited binding site could accommodate such diverse acceptors, but there might be a degree of steric overlap between the two binding sites. Alternatively, there may be no spatial overlap in the binding sites for acceptors. In either case there must be a mechanism for both acceptors to interact chemically with the adenylyl group. This could be accomplished with separate sites if the adenylyl group possesses mobility and can interact with either site. The covalent attachment of AMP to the ε-amino group of lysine in the intermediate offers the possibility of a significant degree of mobility. The structure of lysine allows for torsion about five bonds in the tetramethylene system. It is conceivable that the active site lysine accepts the adenylyl group at the MgATP binding site and then allows the adenosine 5'-phosphoramide to rotate into a neighboring subsite, where it activates the 5'-phosphate in the ligation reaction. This reorientation is illustrated in principle by Scheme III, in which the two orientations are related by rotation about a single bond in the tetramethylene system of lysine.

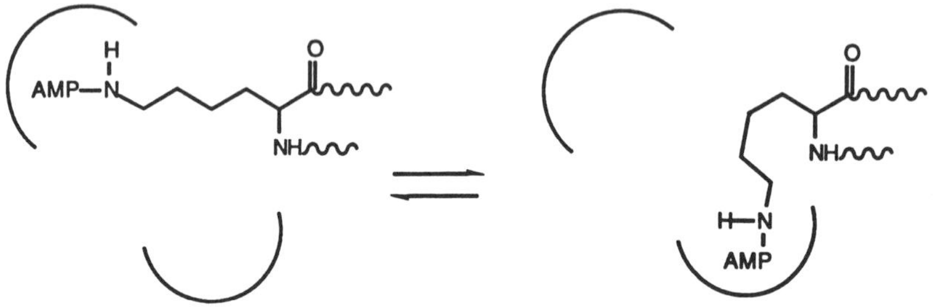

SCHEME III

This is one of the simplest of a number of reorientation mechanisms that are allowed by the potential rotational mobility in the side chain of lysine. In this mechanism the function of the covalent intermediate is both to maintain the bond energy to the AMP group and to provide for structural mobility within the active site.

The importance of the phosphoenzyme in the mechanism of action of succinyl-CoA synthetase in reactions (27a)–(27c) is also unknown. The mechanisms of action of aminoacyl-tRNA synthetases and of acyl-CoA synthetases do not include covalent enzymic intermediates. The fact that succinyl-CoA synthetase involves succinyl phosphate as the activated substrate, whereas the others involve acyl adenylates, does not explain the difference. There is no chemical catalytic basis for the mechanisms of the formation of these intermediates to vary in this way. Moreover, acetate kinase produces acetyl phosphate without the intermediate formation of a phosphoenzyme, so that at least acetate kinase has the capacity to catalyze direct phosphorylation of a carboxylate group.

The phosphoenzyme may have a transport function within the quaternary enzyme–substrate complex [E·ATP·Succ·CoA in Eq. (28)]. The active site of a quaternary complex is likely to be sterically crowded. In this complex the placement of the succinate carboxylate group is crucial; it must be near enough to the γ-phosphoryl group of ATP to accept a phosphoryl group and, after it is phosphorylated, it must be near enough to the sulfhydryl group of CoA to react and form a CoA ester. It is possible that the problem of providing for the proximity of the succinate carboxylate group to both the γ-phosphate and CoA is overcome in this case by allowing for motion by the γ-phosphoryl group. Such motion is possible in a phosphohistidyl residue by torsion about either bond to the β-methylene group. The following illustration shows how rotation of the phosphoimidazole group about one bond can physically translate the reactive phosphoryl group. Such a function of histidine could allow ATP and succinate to react over a significant distance of separation within the quaternary complex.

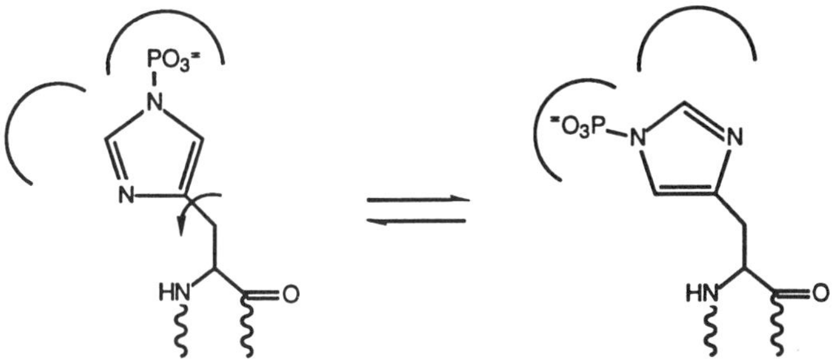

Phosphoenolpyruvate synthetase and pyruvate phosphate dikinase catalyze the formation of phosphoenolpyruvate utilizing the energy of two phosphoanhydride bonds of ATP [reactions (30) and (31)]. The mechanisms of action of these enzymes include two covalent intermediates, a pyrophosphoenzyme and a phosphoenzyme. These intermediates have a special chemical catalytic function in the formation of phosphoenolpyruvate that is not known to be important for any other enzyme. A basic mechanistic problem that must be solved in these reactions is how to utilize the energy of two phosphoanhydride bonds in a single phosphorylating enzyme form that has sufficient energy to phosphorylate pyruvate to phosphoenolpyruvate. The free energy for hydrolysis of a phosphoanhydride is -8 kcal/mol, whereas that for phosphoenolpyruvate is -13 kcal/mol. Thus, a single phosphoanhydride bond cannot phosphorylate pyruvate, whereas two phosphoanhydride bonds can provide sufficient free energy for this process. The mechanistic problem is to focus this energy into a single phosphoryl group.

The problem of focusing phosphorylation free energy is elegantly solved in the mechanisms of action of phosphoenolpyruvate synthetase and pyruvate phosphate dikinase. Phosphohistidyl residues are generated at the active sites of these enzymes by pyrophosphorylation of active site histidines, followed by hydrolysis or phosphorolysis to phosphohistidines. The resulting phosphohistidyl–enzymes are poised at a sufficiently high energy to phosphorylate pyruvate. Exactly how this is arranged cannot be known without detailed structural information about the active site. It can be expected, however, based in part on what is known about the basis for the free energy of hydrolysis of phosphoenolpyruvate and in part on certain binding properties of pyruvate phosphate dikinase, that the phosphohistidyl–enzymes can stabilize enolpyruvate through binding interactions at the active site. The major part of the free energy of hydrolysis of phosphoenolpyruvate is attributable to the ketonization of pyruvate in reactions (39a) and (39b). The hydrolysis of phosphoenolpyruvate

$$\underset{\text{H}_2\text{C}=\text{C}-\text{CO}_2^-}{\overset{\text{PO}_3^{2-}}{|}} + \text{H}_2\text{O} \rightleftharpoons \underset{\text{H}_2\text{C}=\text{C}-\text{CO}_2^-}{\overset{\text{OH}}{|}} + \text{HOPO}_3^{2-} \tag{39a}$$

$$\underset{\text{H}_2\text{C}=\text{C}-\text{CO}_2^-}{\overset{\text{OH}}{|}} \rightleftharpoons \underset{\text{H}_3\text{C}-\text{C}-\text{CO}_2^-}{\overset{\text{O}}{\|}} \tag{39b}$$

to phosphate and enolpyruvate in reaction (39a) should involve a free energy release of 3 to 5 kcal/mol, which is fairly typical of phosphomonoesters. The ketonization of enolpyruvate is a highly favored process that releases the remaining 8 to 10 kcal/mol of free energy. It is clear that the phosphohistidyl–enzyme can phosphorylate pyruvate if it can stabilize bound pyruvate in its enol form through binding interactions. That this is the case is indicated by the fact that the phosphoenzyme of pyruvate phosphate dikinase binds oxalate, much more tightly than does the resting enzyme (*103*). As shown,

enolpyruvate   oxalate

oxalate is a structural analog of the enolpyruvate dianion. Thus, there is good reason to believe that the phosphoenzyme stabilizes the enolate of pyruvate through binding interactions that are less optimal in the resting enzyme. Exactly how the phosphorylation of histidine potentiates the development of such binding interactions will require specific knowledge of the three-dimensional structure of the active site in both the resting enzyme and the phosphoenzyme.

VII. Transition State for Phospho Transfer

In nonenzymic substitution reactions of phosphoric esters and anhydrides, it is generally recognized that a diversity of mechanisms is allowed and that the structure of the substrate and the reaction conditions determine the transition state for reaction with a particular nucleophile (*104, 105*). The extreme cases are generally described as the *dissociative* and *associative* substitution mechanisms. The fully dissociative mechanism entails the formation of monomeric metaphosphate monoanion as a discrete intermediate and was first formulated by F. H. Westheimer, who pioneered the physical organic chemistry of the hydrolysis of phosphate esters (*106, 107*). This mechanism is depicted in Eq. (40) and is possible only for phosphomonoesters with good leaving groups, examples of which are shown.

$$(40)$$

It is possible that the metaphosphate anion (PO_3^-) is never fully formed; an extended structure–function study indicates that the reacting nucleophile participates in P–O bond cleavage (*108*). Moreover, the alcoholyses of P-chiral phenyl[$^{17}O,^{18}O$]phosphate and dinitrophenyl[$^{17}O,^{18}O$]phosphate proceed with inversion of configuration in acetonitrile (*109*). Alcoholysis of P-chiral *p*-nitrophenyl[$^{17}O,^{18}O$]phosphate in *tert*-butanol proceeds with racemization, however, indicating that the planar metaphosphate is an intermediate in the sterically hindered alcohol (*110*). Alcoholysis in dichloromethane of a P-chiral phosphoanhydride with an extraordinarily good leaving group also proceeds with racemization at phosphorus, indicating the intermediate formation of metaphosphate (*111*).

The physical organic chemistry of the solvolysis of phosphomonoesters indicates that the reaction either involves an extended transition state with little bonding to PO_3^- in a concerted displacement of the leaving group, as illustrated,

or it involves the formation of PO_3^- as a discrete intermediate as in reaction (40), depending on the substrate, the attacking nucleophile, and the reaction conditions. In either case, bond cleavage is very important in the transition state, which has less bonding to phosphorus than either the substrate or the product.

The presence of two negative charges on the departing phosphoryl group [see Eq. (40)] is thought to be the driving force for the expulsion of good leaving groups.

Phosphodiesters and phosphotriesters cannot undergo substitution reactions by dissociative mechanisms. Thus, the dibenzyl phosphate monoanion is much less reactive in hydrolysis than the benzyl phosphate monoanion (107).

The reason for the low reactivity of dibenzyl phosphate is that it cannot expel benzyl alcohol to form metaphosphate. Therefore, the solvolysis must proceed by nucleophilic attack of hydroxide on the dibenzyl phosphate monoanion or attack of water on the neutral, acidic dibenzylphosphoric acid.

Associative mechanisms may include the formation of discrete trigonal bipyramidal addition intermediates. In the case of the hydrolysis of five-membered ring cyclic phosphodiesters and phosphotriesters, the pentavalent adducts certainly are intermediates (61). The five-membered ring cyclic esters undergo hydrolysis in acid or base much faster than noncyclic esters such as dibenzyl phosphate because the five-membered rings are strained, and this strain is relieved by the formation of trigonal bipyramidal adducts. An example of the intermediate formation of a pentavalent intermediate is the alkaline hydrolysis of ethylene phosphate shown in reaction (41).

$$\tag{41}$$

In unstrained substrates it is possible that a discrete adduct does not form, but that the trigonal bipyramidal species is the high point, that is, the transition state, on the energy level–reaction coordinate profile.

Very little is known about the transition states for enzymic nucleotidyl and phospho transfer reactions. Inasmuch as the mechanisms of the nonenzymic solvolyses of phosphate esters establish the possibilities for comparable enzymic reactions, either associative or dissociative mechanisms can be considered. However, the dissociative mechanism is thought to be possible only for phospho transfer reactions and not for nucleotidyl transfers, because metaphosphate cannot be formed from diesters. A dissociative mechanism for a diester would entail the formation of a monomeric metaphosphate monoester, a species that can exist but which has not been observed in solvolysis reactions. The difficulty with the dissociative mechanism for phosphodiesters may be that a monoanion cannot provide enough driving force to expel a leaving group under solvolytic conditions.

Therefore, the nucleotidyltransferases are thought to catalyze substitution at phosphorus by associative mechanisms. However, no mechanistic information bearing on this question is available from studies on the enzymes themselves.

The question of whether phosphotransferases catalyze substitution at phosphorus by associative or dissociative mechanisms remains unresolved (*104, 105*). It has been pointed out that it is not clear how an enzyme could catalyze the dissociative mechanism, apart from simply bringing substrates together. Of course bringing substrates together is one of the most important catalytic processes for enzymes. Apart from this, however, the dissociative mechanism is not facilitated by general acid–base or nucleophilic catalysis. It should not be facilitated by charge neutralization through ionic bonding of phosphate esters with metal ions and enzymic guanidinium groups of arginine or ammonium groups of lysine. In fact, to the extent that charge neutralization diminishes negative charge on the phosphate group in the Michaelis complex, the dissociative mechanism should be suppressed, since this mechanism depends on concentrated negative charge to provide the driving force for the formation of metaphosphate. The associative mechanism, in contrast to the dissociative mechanism, might be facilitated by charge neutralization. Recent evidence, however, shows that metal complexation does not increase the associative character of the transition state for nonenzymic transfer of the phosphoryl group to substituted pyridines (*112*). Therefore, it appears that in metal complexes the phosphate groups sense the full negative charges and retain their normal reactivities.

Much more research will be required to address the question of the true nature of the transition states for enzymic substitution at phosphorus. The catalytic reaction pathways are now well understood with respect to the question of double-displacement and single-displacement mechanisms. Information about the nature of the transition states will require much more information about the structures of active sites, as well as many more structure–function studies on the enzymes for comparison with mechanistic information about comparable nonenzymic reactions.

ACKNOWLEDGMENT

The author's research in the field of enzymic substitution at phosphorus is supported by Grant No. GM 30480 from the National Institute of General Medical Sciences.

REFERENCES

1. Cleland, W. W. (1970). "The Enzymes," 3rd Ed., Vol. 2, p. 1.
2. Frey, P. A. (1982). *New Comp. Biochem.* **3**, 201.
3. Frey, P. A. (1982). *Tetrahedron* **38**, 1541.

4. Buchwald, S. L., Hansen, D. E., Hassett, A., and Knowles, J. R. (1982). In "Methods in Enzymology," Vol. 87, p. 301.
5. Frey, P. A., Richard, J. P., Ho, H.-T., Brody, R. S., Sammons, R. D., and Sheu, K.-F. (1982). In "Methods in Enzymology" (D. L. Purich, ed.), Vol. 87, p. 213. Academic Press, New York.
6. Gerlt, J. A., Coderre, J. A., and Mehdi, S. (1983). Adv. Enzymol. Related Top. Mol. Biol. 55, 291.
7. Frey, P. A. (1989). Adv. Enzymol. Related Areas Mol. Biol. 62, 119.
8. Cahn, R. S., Ingold, C. K., and Prelog, V. (1966). Angew. Chem., Int. Ed. Engl. 5, 385.
9. Sheu, K.-F. R., and Frey, P. A. (1978). J. Biol. Chem. 253, 3378.
10. Sheu, K.-F. R., Richard, J. P., and Frey, P. A. (1979). Biochemistry 18, 5548.
11. Richard, J. P., and Frey, P. A. (1978). J. Am. Chem. Soc. 100, 7757.
12. Jones, S. R., Kindman, L. A., and Knowles, J. R. (1978). Nature (London) 275, 564.
13. Ho, M., Bramson, H. R., Hanson, D. E., Knowles, J. R., and Kaiser, E. T. (1988). J. Am. Chem. Soc. 110, 2680.
14. Kountz, P. D., Freeman, S., Cook, A. G., Rafaat El-Maghrabi, M., Knowles, J. R., and Pilkis, S. J. (1988). J. Biol. Chem. 263, 16069; Kountz, P. D., Freeman, S., Cook, A. G., Rafaat El-Maghrabi, M., Knowles, J. R., and Pilkis, S. J. (1988). J. Biol. Chem. 15, 8116.
15. Cleland, W. W. (1982). In "Methods in Enzymology" (D. L. Purich, ed.), Vol. 87, p. 159. Academic Press, New York.
16. Villafranca, J. J. (1982). In "Methods in Enzymology" (D. L. Purich, ed.), Vol. 87, p. 180. Academic Press, New York.
17. Cleland, W. W., and Mildvan, A. S. (1979). Adv. Inorg. Biochem. 1, 163.
18. Deleted in proof.
19. Cornelius, R. D., and Cleland, W. W. (1978). Biochemistry 17, 3279.
20. Li, T. M., Mildvan, A. S., and Switzer, R. L. (1978). J. Biol. Chem. 253, 3918.
21. Eckstein, F. (1985). Annu. Rev. Biochem. 54, 367.
22. Jaffe, E. K., and Cohn, M. (1978). J. Biol. Chem. 253, 4823.
23. Reed, G. H., and Leyh, T. S. (1980). Biochemistry 19, 5472.
24. Leyh, T. S., Sammons, R. D., Frey, P. A., and Reed, G. H. (1982). J. Biol. Chem. 257, 15047.
25. Frey, P. A., Iyengar, R., Smithers, G. W., and Reed, G. W. (1986). "Biophosphates and Their Analogues—Synthesis, Structure, Metabolism and Activity" (K. S. Bruzik and W. J. Stec, eds.), p. 267. Elsevier, Amsterdam.
26. Turnquest, R. L., and Hansen, R. G. (1973). "The Enzymes," 3rd Ed., Vol. 8, p. 51.
27. Tsuboi, K. K., Fukunaga, K., and Petricciani, J. C. (1969). J. Biol. Chem. 244, 1008.
28. Gillett, T. A., Levine, S., and Hansen, R. G. (1971). J. Biol. Chem. 246, 2551.
29. Saito, S., Ozutsumi, M., and Kurahashi, K. (1960). J. Biol. Chem. 242, 2362.
30. Wong, L.-J., Sheu, K.-F. R., Lee, S.-L., and Frey, P. A. (1977). Biochemistry 16, 1010.
31. Lemaire, H. G., and Muller-Hill, B. (1986). Nucleic Acids Res. 14, 7705.
32. Cornwell, T. L., Adhya, S. L., Reznikoff, W. S., and Frey, P. A. (1987). Nucleic Acids Res. 15, 8116.
33. Reichardt, J. K. V., and Berg, P. (1988). Mol. Biol. Med. 5, 107.
34. Wong, L.-J., and Frey, P. A. (1974). Biochemistry 13, 3889.
35. Wong, L.-J., and Frey, P. A. (1974). J. Biol. Chem. 249, 2322.
36. Yang, S.-L. L., and Frey, P. A. (1979). Biochemistry 18, 2980.
37. Frey, P. A., Wong, L.-J., Sheu, K.-F., and Yang, S. L. (1982). In "Methods in Enzymology" (D. L. Purich, ed.), Vol. 87, p. 20. Academic Press, New York.
38. Arabshahi, A., Brody, R. S., Smallwood, A., Tsai, T.-C., and Frey, P. A. (1986). Biochemistry 25, 5583.

39. Field, T. L., Reznikoff, W. S., and Frey, P. A. (1989). *Biochemistry* **28**, 2094.
39a. Kim, J., Ruzicka, F., and Frey, P. A. (1990). *Biochemistry* **29**, 10590.
40. Kornberg, T., and Kornberg, A. (1974). "The Enzymes," 3rd Ed., Vol. 10, p. 119.
41. Loeb, L. A. (1974). "The Enzymes," 3rd Ed., Vol. 10, p. 174.
42. Temin, H. M., and Mizutani, S. (1974). "The Enzymes," 3rd Ed., Vol. 10, p. 211.
43. Chambon, P. (1974). "The Enzymes," 3rd Ed., Vol. 10, p. 261.
44. Chamberlin, M. J. (1974). "The Enzymes," 3rd Ed., Vol. 10, p. 333.
45. Brody, R. S., and Frey, P. A. (1981). *Biochemistry* **20**, 1245.
46. Burgers, P. M. J., and Eckstein, F. (1979). *J. Biol. Chem.* **254**, 6889.
47. Brody, R. S., Adler, S., Modrich, P., Stec, W. J., Leznikowski, Z. L., and Frey, P. A. (1982). *Biochemistry* **21**, 2570.
48. Romaniuk, P. J., and Eckstein, F. (1982). *J. Biol. Chem.* **257**, 7684.
49. Burgers, P. M. J., and Eckstein, F. (1978). *Proc. Natl. Acad. Sci. U.S.A.* **75**, 4798.
50. Yee, D., Armstrong, V. W., and Eckstein, F. (1979). *Biochemistry* **18**, 4116.
51. Bartlett, P. A., and Eckstein, F. (1982). *J. Biol. Chem.* **257**, 8879.
52. Burgers, P. M. J., and Eckstein, F. (1979). *Biochemistry* **18**, 450.
53. Horowitz, D. S., and Wang, J. C. (1987). *J. Biol. Chem.* **262**, 5339.
54. Pargellis, C. A., Nunes-Duby, S. E., Moitoso se Vargas, L., and Landy, A. (1988). *J. Biol. Chem.* **263**, 7678.
55. Gerlt, J. A., Coderre, J. A., and Wolin, M. J. (1980). *J. Biol. Chem.* **255**, 331.
56. Coderre, J. A., and Gerlt, J. A. (1980). *J. Am. Chem. Soc.* **102**, 6594.
57. Eckstein, F., Romaniuk, P. J., Heideman, W., and Storm, D. R. (1981). *J. Biol. Chem.* **256**, 9118.
58. Van Pelt, J. E., Iyengar, R., and Frey, P. A. (1986). *J. Biol. Chem.* **261**, 15995.
59. Senter, P. D., Eckstein, F., Mulsch, A., and Bohme, E. (1983). *J. Biol. Chem.* **258**, 6741.
60. Richard, J. P., and Frey, P. A. (1983). *J. Am. Chem. Soc.* **105**, 6605.
61. Westheimer, F. H. (1981). *In* "Rearrangements in Ground and Excited States" (P. deMayo, ed.), Vol. 2, p. 229. Academic Press, New York.
62. Purich, D. L. (1982). "Methods in Enzymology" (D. L. Purich, ed.), Vol. 87, p. 1. Academic Press, New York.
63. Janson, C. A., and Cleland, W. W. (1974). *J. Biol. Chem.* **249**, 2567.
64. Blattner, W. A., and Knowles, J. R. (1979). *Biochemistry* **18**, 3927.
65. Fox, D. K., Meadow, N., and Roseman, S. (1986). *J. Biol. Chem.* **261**, 13498.
66. Meadow, N. D., Kukuruzinska, M. A., and Roseman, S. (1985). *In* "The Enzymes of Biological Membranes" (A. Martonosi, ed.), 2nd ed., p. 523. Plenum, New York.
67. Scopes, R. K. (1973). "The Enzymes," 3rd Ed., Vol. 8, p. 335.
68. Walsh, C. T., and Spector, L. B. (1971). *J. Biol. Chem.* **246**, 1255.
69. Webb, M. R., and Trentham, D. R. (1980). *J. Biol. Chem.* **255**, 1775.
70. Johnson, P. E., Abbott, S. J., Orr, G. A., Semeriva, M., and Knowles, J. R. (1976). *Biochemistry* **15**, 2893.
71. Brunngraber, E. F., and Chargaff, E. (1967). *J. Biol. Chem.* **242**, 4834.
72. Brunngraber, E. F., and Chargaff, E. (1970). *J. Biol. Chem.* **245**, 4825.
73. Rogers, R., and Chargaff, E. (1972). *J. Biol. Chem.* **247**, 5448.
74. Prasher, D. C., Carr, M. C., Ives, D. H., Tsai, T.-C., and Frey, P. A. (1982). *J. Biol. Chem.* **257**, 4931.
75. Richard, J. P., Prasher, D. C., Ives, D. H., and Frey, P. A. (1979). *J. Biol. Chem.* **254**, 4339.
76. Stelte, B., and Witzell, H. (1986). *Eur. J. Biochem.* **155**, 121.
77. Grivell, A. R., and Jackson, J. F. (1976). *Biochem. J.* **155**, 571.
78. Richard, J. P., Carr, M. C., Ives, D. H., and Frey, P. A. (1980). *Biochem. Biophys. Res. Commun.* **94**, 1052.

79. Hawkins, C. F., and Bagnara, A. S. (1987). *Biochemistry* **26,** 1982.
80. Lehman, I. R. (1974). "The Enzymes," 3rd Ed., Vol. 10, p. 237.
81. Bryant, F. R., and Benkovic, S. J. (1982). *Biochemistry* **21,** 5877.
82. Harnett, S. P., Lowe, G., and Tansley, G. (1985). *Biochemistry* **24,** 7446.
83. Meister, A. (1974). "The Enzymes," 3rd Ed., Vol. 10, p. 699.
84. Stadtman, E. R., and Ginsburg, A. (1974). "The Enzymes," 3rd Ed., Vol. 10, p. 755.
85. Midelfort, C. F., and Rose, I. A. (1976). *J. Biol. Chem.* **251,** 5881.
86. Bethell, R. C., and Lowe, G. (1988). *Biochemistry* **27,** 1125.
87. Bridger, W. A. (1974). "The Enzymes," 3rd Ed., Vol. 10, p. 581.
88. Soll, D., and Schimmel, P. R. (1974). "The Enzymes," 3rd Ed., Vol. 10, p. 489.
89. Connolly, B. A., Eckstein, F., and Grotjahn, L. (1984). *Biochemistry* **23,** 2026.
90. Harnett, S. P., Lowe, G., and Tansley, G. (1985). *Biochemistry* **24,** *2908*.
91. Lowe, G., Sproat, B. L., Tansley, G., and Cullis, P. (1983). *Biochemistry* **22,** 1229.
92. Langdon, S. P., and Lowe, G. (1979). *Nature (London)* **281,** 320.
93. Londesborough, J. C., and Webster, Jr., L. T. (1974). "The Enzymes," 3rd Ed., Vol. 10, p. 469.
94. Midelfort, C. F., and Sarton-Miller, I. (1978). *J. Biol. Chem.* **254,** 6889.
95. Cooper, R. A., and Kornberg, H. J. (1974). "The Enzymes," 3rd Ed., Vol. 10, p. 631.
96. Goss, N. H., and Wood, H. G. (1982) *In* "Methods in Enzymology" (D. L. Purich, ed.), Vol. 87, p. 51. Academic Press, New York.
97. Cook, A. G., and Knowles, J. R. (1985). *Biochemistry* **24,** 51.
98. Miller, R. S., and Lane, M. D. (1968). *J. Biol. Chem.* **243,** 6041.
99. Jomain-Baum, M., and Schramm, V. L. (1978). *J. Biol. Chem.* **253,** 3648.
100. Sheu, K.-F., Ho, H.-T., Nolan, L. D., Markovitz, P., Richard, P., Utter, M. F., and Frey, P. A. (1984). *Biochemistry* **23,** 1779.
101. Konopka, J. M., Lardy, H. A., and Frey, P. A. (1986). *Biochemistry* **25,** 5571.
102. Wong, L.-J., and Rose, I. A. (1976). *J. Biol. Chem.* **251,** 5431.
103. Michaels, G., Milner, Y., and Reed, G. H. (1975). *Biochemistry* **14,** 3213.
104. Benkovic, S. J., and Schray, K. J. (1971). "The Enzymes," 3rd Ed., Vol. 8, p. 201.
105. Knowles, J. R. (1981). *Annu. Rev. Biochem.* **49,** 877.
106. Butcher, W. W., and Westheimer, F. H. (1955). *J. Am. Chem. Soc.* **77,** 2420.
107. Kumamoto, J., and Westheimer, F. H. (1955). *J. Am. Chem. Soc.* **77,** 2515.
108. Skoog, M. T., and Jencks, W. P. (1983). *J. Am. Chem. Soc.* **105,** 3356.
109. Buchwald, S. L., Friedman, J. M., and Knowles, J. R. (1984). *J. Am. Chem. Soc.* **106,** 4911.
110. Friedman, J. M., Freeman, S., and Knowles, J. R. (1988). *J. Am. Chem. Soc.* **110,** 1268.
111. Cullis, P., and Rous, A. J. (1985). *J. Am. Chem. Soc.* **107,** 6721.
112. Herschlag, D., and Jencks, W. P. (1987). *J. Am. Chem. Soc.* **109,** 4665.

5

Glycosidases and Glycosyltransferases

GREGORY MOOSER

University of Southern California
School of Dentistry
Los Angeles, California 90089

THE ENZYMES, Vol. XX

I. Introduction

New insights into glycosyltransferase and glycosidase catalysis have emerged over the past decade. The distinction between the two classes (glycosyltransferases, EC 2.4; glycosidase hydrolases, EC 3.2) has become less clear as structure and mechanism similarities emerge among the most intensely studied examples of each group. Many glycosyltransferases catalyze substrate hydrolysis, and some previously considered strict glycoside hydrolases have been found to catalyze transfer to carbohydrate acceptors. Recent analyses of gene and protein sequence data bases have revealed extensive homology among active site peptides of endo- and exo-α-D-glucosidases, α-D-glucosyltransferases, and α-D-glucanotransferases, and a number of high-resolution X-ray structure analyses of glycosidases and other carbohydrate-binding proteins have shed additional light on reaction mechanisms.

In spite of an increasingly diverse range of reported glycosidases and glycosyltransferases, only a small number have been studied in great detail, primarily hen egg-white lysozyme and *Escherichia coli lacZ* β-galactosidase. The enzymes have provided a wealth of detail on catalytic mechanisms and correlations with solution chemistry (*1–6*). Most mechanism data come from enzymes that catalyze the reaction with retention of anomer configuration; that is, reactions in which the glycosyl donor and product have the same anomeric form. As expected, but not necessarily required (*7, 8*), the enzyme reactions commonly proceed through a discrete glycosyl–enzyme intermediate. Two significant amino acid groups are generally involved: a carboxylate that stabilizes the glycosyl–enzyme and an amino acid that serves as a general acid catalyst to protonate the glycoside oxygen.

A general mechanism consistent with many of the data is shown in Fig. 1 for a hypothetical β-glucosidase that cleaves a β-glucoside bond and liberates β-D-glucose. The carboxylate anion can participate in two ways, by electrostatic stabilization of the developing oxocarbonium ion (upper pathway, Fig. 1) or by covalent nucleophilic attack of the reaction center (lower pathway, Fig. 1). Unless an acceptor rapidly traps the glycosyl cation, the potential to collapse to the covalent form is great given the short ion distances and extremely brief lifetime of the oxocarbonium ion (*9, 10*). When a covalent complex does form, either by direct nucleophilic attack or via an oxocarbonium ion, the reaction reduces to a typical double-displacement mechanism involving two sequential anomer inver-

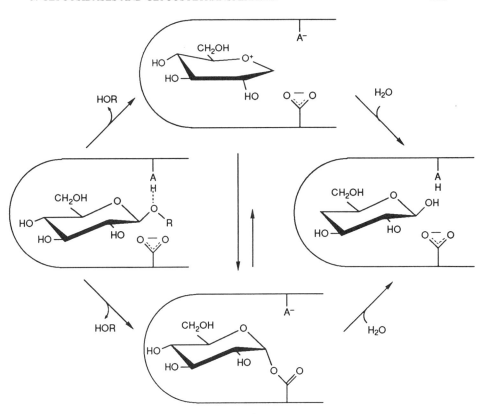

FIG. 1. Reaction intermediates of a hypothetical β-glucosidase proceeding with retention of anomeric configuration. The upper pathway passes through an oxocarbonium ion transition state and the lower pathway passes through a covalent α-glucosyl intermediate.

sions as described by Koshland (*11*). First, as in the lower pathway of Fig. 1, the substrate β-glucosyl moiety is transferred and inverted to form the α-anomeric glycosyl intermediate; in a second inversion, the α-glucosyl intermediate is transferred to an acceptor, resulting in a β-anomeric product.

An equilibrium between the covalent and noncovalent forms of the intermediate has been considered for several glycosidases (*2*, *12–14*). The much greater stability of the covalent complex would bias the equilibrium in that direction, but the actual productive form of the transition state in transfer of the glycosyl intermediate to an acceptor cannot be generalized; both noncovalent and covalent mechanisms are relevant: evidence supports lysozyme and baker's yeast β-glucosidase catalysis through a noncovalent oxocarbonium transition state (*1*, *15*, *16*), and E. coli lacZ β-galactosidase catalysis through either a covalent or noncovalent glycosyl–enzyme, depending on the substrate (*13*, *14*).

II. Lysozyme

The extensive information available on hen egg-white lysozyme serves as an appropriate foundation to introduce the principles illustrated in Fig. 1. The description below is prefatory to all discussions that follow and will be expanded on in relation to other enzymes.

A. INTRODUCTION

Solution chemistry relevant to acetal hydrolysis came together with glycosidase catalysis in the mid-1960s with the report of the three-dimensional structure of hen lysozyme, the first enzyme crystal structure resolved. The basic principles, proposed from detailed model building and structure comparisons of the native enzyme and enzyme–inhibitor complexes, remain intact today (*17–21*).

Lysozymes are *N*-acetylmuramidases that cleave the *N*-acetylmuramic-β-1, 4-*N*-acetylglucosamine (MurNAc-GlcNAc) bond common to polysaccharides found in gram-positive bacterial cell walls. Lysozymes are widely distributed among organisms and commonly grouped into three classes: chicken type, which includes the enzymes from hen and tortoise egg white and human secretions; goose type, found in goose eggs; and phage type, found in bacteriophage T4. Among types, there are modest variations in specificity, substantial variations in amino acid sequence, and both common and distinct three-dimensional features (*22–24*). The hen egg-white enzyme is the most thoroughly studied, but important comparative data on the goose- and phage-type enzymes have been reported as well. The 129-amino acid primary structure of the hen enzyme folds into a slightly elongated form, with a deep groove separating two major protein lobes. The groove houses the active site, which is composed of six saccharide-binding subsites (A–F). Oligosaccharides bind with the nonreducing end extending toward subsite A. The transition-state forms in subsite D, with scission occurring at the GlcNAc-MurNAc glycoside bond that straddles subsites D and E (*1*).

Much detail has been derived from the crystal structure of a nonproductive enzyme–(GlcNAc)$_3$ complex in which the inhibitor occupies subsites A–C (*1, 19, 24*). Saccharide interaction at subsite D has been studied indirectly through model building and theoretical analyses (*19, 25*), and directly from the resolved three-dimensional lysozyme complexes with (GlcNAc)$_4$ (*24*) and the tetrasaccharide transition-state analog (GlcNAc)$_4$-δ-lactone (*26*) at subsites A–D, and MurNAc-GlcNAc-MurNAc at subsites B–D (*27*). Saccharide interactions at subsites E and F were first predicted through model building (*19, 20*) and later from the crystal structure of GlcNAc-MurNAc bound to turkey lysozyme (*28*).

A summary of general features is described below as an introduction to later topics. Excellent and extensive reviews on lysozyme structure–function relationships have recently appeared that emphasize a structural perspective (*24*) and a mechanism perspective (*6*).

B. AMINO ACIDS DIRECTLY INVOLVED IN CATALYSIS

The catalytic mechanism of hen egg-white lysozyme utilizes Asp-52 in electrostatic stabilization of the developing carbonium ion and Glu-35 as a general acid proton donor to the glycoside oxygen. The significance of the residues has been clearly established from several experimental directions. X-Ray structure analysis shows Glu-35 and Asp-52 in close proximity to the sessile bond. The surrounding environment of the two carboxylic amino acids differ: Glu-35 is located in a hydrophobic pocket, which elevates the pK_a to ~6 and preserves the proton at catalytic pH until transfer to the glycoside oxygen; Asp-52 is in a polar hydrophilic environment with normal pK_a near 4.5 and is ionized at catalytic pH, consistent with its role in electrostatic stabilization of the oxocarbonium ion (*29–31*). The significance of Asp-52 and Glu-35 is further supported by an impressive history of chemical modification and site-directed mutagenesis studies (see Section VII,A,2).

General acid catalysis of acetals through proton transfer to the leaving-group oxygen is well documented through analysis of model organic compounds, particularly reactions involving intramolecular catalysis by proton transfer from a carboxylic acid (*2, 3, 32*). Two important conclusions can be extrapolated and applied to lysozyme. First, bond breaking must precede proton transfer to the leaving-group (glycoside) oxygen. Initially, proton transfer is unfavorable, but as bond breaking proceeds, the oxygen builds negative charge, which favors protonation and leaving-group departure. Second, a negative charge on the leaving-group oxygen is not a sufficient requisite for acid catalysis but requires that the carboxylic acid hydrogen bond to the leaving-group oxygen. The two features are consistent with the three-dimensional structure of hen lysozyme: Glu-35 is within hydrogen bond distance of the glycoside oxygen (*1*), and Asp-52 is aligned to stabilize the developing carbonium ion. As bond breaking proceeds in the transition state, the negative charge builds on the glycoside oxygen, followed by proton transfer and an orders-of-magnitude increase in catalytic efficiency (*6, 33*).

As previously noted, lysozyme Asp-52 can participate in catalysis through noncovalent stabilization of the oxocarbonium ion transition state and/or covalent nucleophilic attack of the glycoside reaction center. The experimental distinction of these routes can be difficult. With nucleophilic catalysis, for example, some oxocarbonium ion formation likely precedes the covalent complex so that the reaction flux in Fig. 1 would proceed through the upper pathway, then collapses to the covalent complex before glycosyl transfer to water or another acceptor. The critical distinction between the two mechanisms is not necessarily a carbonium ion transition state but the state of the intermediate in which bond scission occurs and the productive form in transfer to an acceptor. The distinction is a few angstroms in distance and the electron structure of the atoms. Kinetic isotope effects (KIEs) provide an approach to examine the differences.

C.   KINETIC ISOTOPE EFFECTS AND TRANSITION-STATE STRUCTURE

Glycosyl–enzyme structure is most subtly explored through kinetic analysis of the effects caused by isotope substitutions. The value of kinetic isotope effects lies in the fact that an uncharged neutron does not affect the basic electronic character of the transition state, but does affect force constants associated with vibrational modes. α-Deuterium KIEs, where deuterium substitution is at the carbon of the sessile C–O bond, are widely used because the enzyme reaction rate is affected differently in the two relevant mechanisms. Progression from the usual saccharide tetrahedral carbon at C-1 to a trigonal oxocarbonium ion transition state involves a change in electron hybridization, unlike the nucleophilic reaction where the hybridization remains tetrahedral. Thus, k_H/k_D ratios of 1.1 to 1.4 are indicative of a substantial abundance of an oxocarbonium ion transition state, while values close to 1.00 suggest no change in hybridization and a nucleophilic mechanism (34, 35). Caution in interpretation is needed, however, because at least two confounding situations can occur: (1) nucleophilic reactions at times exhibit both normal and inverse KIEs on the order of 20%, and (2) a rate-determining step independent of the chemical reaction, such as a protein conformational change, can mask kinetic information relevant to the transition state (36, 37). The lysozyme α-deuterium KIE accompanying hydrolysis of a p-nitrophenyl-β-D-glycoside reported by Dahlquist et al. is 1.11 (15), consistent with substantial carbonium ion character in the transition state and an S_N1 reaction; a secondary α-tritium KIE based on hydrolysis of the trisaccharide chitotriose is of comparable magnitude (38).

The potential confounding ambiguities inherent in α-deuterium KIEs can often be resolved by analysis of oxygen-18 leaving-group KIEs (39, 40). Isotopic oxygen-18 substitution at the glycoside oxygen reports changes in the force constants of the cleaved C–O bond, which can provide quantitative analysis of the minimum degree of bond scission in the overall rate-determining step of the reaction (calculated from V_{16}/V_{18} KIE) and the first irreversible step of the reaction [calculated from $(V/K)_{16}/(V/K)_{18}$ KIE].

Oxygen-18 leaving-group KIEs are approximately 10-fold more subtle than α-deuterium KIEs and require measurements with comparably increased precision (41); significant values are a few percent above 1.00 and require measurements within a fraction of a percent accuracy. The V_{16}/V_{18} KIE, measured by Rosenberg and Kirsch in the reaction of lysozyme with a p-nitrophenyl-β-D-glycoside substrate (42), was large (1.0467 $\pm$ 0.0015) and close to the theoretical limit, indicating nearly complete bond scission at the first irreversible step of the reaction. Furthermore, since proton addition to the glycoside oxygen lowers the KIE, and no decrease was apparent, scission must be well advanced of proton addition. The conclusions strongly support an S_N1 reaction with significant oxocarbonium ion character, consistent with the α-deuterium KIEs and the mechanism initially proposed from the enzyme crystal structure (19, 20).

D. ACTIVE SITE

Saccharide binding to hen lysozyme subsites A–F involves both hydrogen bonds and nonpolar force interactions. The latter is not commonly considered relevant to protein–carbohydrate binding but is an important factor contributing to specificity and affinity. Nonpolar regions occur in oligosaccharides at localized clusters of C–H groups oriented away from the pyranose-ring hydroxyls. In hen lysozyme, for example, a hydrophobic interaction between the (GlcNAc)₃ pyranose at subsite B and Trp-62 causes a 17° rotation that narrows the binding site and aligns hydrogen protein bonds to the carbohydrate (1, 24).

Saccharide affinity differs at each subsite, with the weakest binding at subsite D, where the transition state develops. The highest affinity sites are on either side of D (subsites C and E) and become progressively weaker toward peripheral subsites A and F (1, 43). With the exception of the catalytic subsite D, where the binding free energy is close to zero with GlcNAc and positive with MurNAc, affinities correlate well with the number of hydrogen bonds, hydrophobic interactions, degree of solvent access, and subsite depth. The highest affinity site, subsite C, has an extensive hydrogen-bonded network in which the saccharide contributes six hydrogen bonds from the reducing-end GlcNAc of (GlcNAc)₃ (1). As extensive as this appears, it falls short of the near-perfect hydrogen-bonded complex of L-arabinose and L-arabinose-binding protein, where all non-anomeric hydroxyls donate one and accept one or two hydrogen bonds (44) (see Section VII,C). In the (GlcNAc)₃–enzyme complex, the weaker affinity subsites A and B contribute only one hydrogen bond each and are more exposed to solvent than subsite C, which lies in a deeper, more hydrophobic environment.

The catalytic site, subsite D, is unique. The tetrasaccharide transition-state analog (GlcNAc)₄-δ-lactone (in which the lactone in subsite D approximates the half-chair conformation of a carbonium ion transition state) binds more tightly than the chair-conformation ligand (45). Model building suggests that the more planar half-chair structure avoids steric impingement by Asp-52, Trp-108, and the acetamido group from the neighboring saccharide in site C, and aligns Asp-52 and Glu-35 with the sessile glycoside bond.

Analysis by Kelly *et al.* (*27*) of subsites B–D in the enzyme complex with MurNAc-GlcNAc-MurNAc clearly shows an alternative arrangement at subsite D in which the pyranose chair conformation is accommodated with slight repositioning of amino acids relative to the bound transition-state analog, albeit with the chair conformation less imbedded than the half-chair. The ability of the catalytic site to accommodate both forms might depict the protein structure of a Michaelis complex in one case and the transition-state structure in the other case (*27*).

III. Glycosidase Reversible Inhibitors

Inhibition studies have substantially contributed toward unraveling the glycosidase and glycosyltransferase mechanisms. Reversible inhibitors have been particularly valuable in confirming a transition-state structure and exploring the active site environment [see reviews by Legler (*46*), Truscheit *et al.* (*47*), and Lalégerie *et al.* (*4*)], and irreversible inhibitors have been useful in mechanism studies and identification of active site functional amino acids (*4, 48*). The latter group is discussed in other sections relative to specific enzymes, and is emphasized in Section VII,A.

Examples of reversible inhibitors are presented primarily in the context of β-galactosidase because of the broad range of inhibition studies with this enzyme. The compounds are divided into four groups: (1) inhibitors that mimic the glycosyl oxocarbonium ion transition state; (2) 5-amino-5-deoxy-D-glycopyranoses (and the 1,5-dideoxyimino derivative) or nojirimycin (and deoxynojirimycin); and (3) D-glycosylamines and substituted D-glycosylamines. A final discussion includes the natural product acarbose, a strong inhibitor or α-amylases, and other α-D-glucosidases.

A. Transition-State Analogs

The capacity of an enzyme to bind selectively the substrate transition state or to tap substrate binding energy to distort the substrate into the transition state involves mechanisms capable of producing significant catalytic power (*49, 50*). In β-galactosidase, for example, several particularly strong inhibitors mimic the planar trigonal structure of the galactosyl carbonium ion (**II**). Three inhibitors considered here bind β-galactosidase with an affinity substantially greater than D-galactopyranose (**I**); two are clearly transition-state analogs—lactones and the furanose form of L-ribose and D-lyxose; a third inhibitor, D-galactal, is actually a very slow-reacting substrate but has some structural features appropriate to this category. The reaction of glycals is introduced here and is presented in greater depth in Section IV.

Sugar lactones are potent inhibitors of a number of glycosidases (4, 51), including mammalian (52, 53) and E. coli (54) β-galactosidases. Huber and Brockbank (54) tested inhibition of eight sugar lactones and found that they all inhibited E. coli β-galactosidase more strongly than did the parent sugar. The most effective was D-galactonolactone (54, 55), of which D-galactono-1,5-lactone (III) rather than the 1,4-lactone was responsible for inhibition. ^{13}C NMR analysis (54) and the crystal structure of other aldono-1,5-lactones (56) show that the inhibitor has a half-chair conformation with a planar trigonal structure at C-1, approximating the D-galactose carbonium ion.

L-Ribose and D-lyxose in the furanose form have hydroxyl positions comparable to galactose C-3 and C-4, the most significant positions in galactose binding (117). But the strong inhibition of these monosaccharides is due to the envelope-type conformation, similar to the half-chair of the lactones. The furanose form is actually in lower abundance in solution than the pyranose of both compounds (57), but it is the former that is inhibitory based on a correlation between the magnitude of inhibition and furanose concentration at different temperatures (54). The approximation of the structure to the D-galactose oxocarbonium imparts a 10^3 to 10^4 decrease in inhibition constant compared to D-galactose (54).

Glycals, such as D-galactal (1,2-dideoxy-D-lyxo-hex-1-enopyranose) (IV) (58), in part fit the requirements of an oxocarbonium transition state since the structure at C-1 is trigonal and planar. However, the similarity breaks down at the double bond, which, compared to the oxocarbonium ion, is positioned on the opposite side of C-1 (*52*). Glycals are slow-reacting substrates that hydrate at the double bond to form the 2-deoxy sugar (or 2-deoxyglycoside, depending on the nucleophile acceptor) (*59, 60, 85*). Hehre *et al.* (*86*) suggest that the strong D-glucal inhibition of α- and β-glucosidase is not actually due to the parent compound, but results from strong binding of the glycal transition state (either a deoxy-D-glucosyl cation or a covalent acylal).

In addition to implication of glycal inhibition on glycosidase transition state, the slow, readily measurable association and dissociation rates of the D-galactal reaction with β-galactosidase have been used to confirm an enzyme conformation change important in β-galactosidase catalysis. Viratelle and Yon (*61*) found that the D-galactal reaction is highly dependent on Mg^{2+}, where, in the absence of the ion, the D-galactal K_m increased some 2000-fold. Comparison of Mg^{2+} (*105*) and D-galactal binding to the Mg^{2+}-free enzyme indicates that a D-galactal-induced conformation change was similar if not identical to a conformation change induced by Mg^{2+} (*61*).

Glycals have been successfully used to tag active site amino acids. Denaturation of *Aspergillus wentii* β-glucosidase A_3 in the presence of D-glucal traps the hydrated product bound as an aspartate ester (*62*). The residue is apparently the same β-glucosidase aspartic acid labeled with the site-directed irreversible inhibitor conduritol B epoxide (*63, 64*).

An analogous *E. coli* β-galactosidase–galactal complex has been trapped in a reaction with radiolabeled D-galactal. The lability of the complex, however, prevented identification of the associated amino acid (*65*), although, as with *A. wentii* β-glucosidase A_3, the carboxyl of β-galactosidase Glu-461, labeled with conduritol C *cis*-epoxide, is a reasonable candidate (*66, 67*).

B.   5-AMINO-5-DEOXYGLYCOPYRANOSES: NOJIRIMYCIN AND DERIVATIVES

Replacing the pyranose ring oxygen with nitrogen produces extremely potent inhibitors of many glycosidases. The glucose analog, which is the antibiotic nojirimycin (*68*), and its 1-deoxy derivative (deoxynojirimycin) have inhibition constants as much as 10^5 smaller than glucose with several α- and β-glucosidases (*51, 69–71*). Similar inhibition enhancement is observed with α- and β-galactosidase and the respective D-galactose analogs [D-*galacto*-nojirimycin (**V**) and 1-deoxy-D-*galacto*-deoxynojirimycin (**VI**)] (*71*), α- and β-mannosidase and the respective D-mannose analogs (*70*), and α-L-fucosidase and the respective L-fucose analogs (*72, 73*). As with glycals, inhibition in many cases is very slow, on the order of minutes, for both formation and dissociation of the tight enzyme–inhibitor complex (*69, 74*). However, unlike the case with glycals, no reaction occurs and the slow steps can only be interpreted as protein conformational changes.

Details of the inhibition mechanism are not clear (*4, 72*), but Hanozet *et al.* (*69*) have introduced an interesting hypothesis based on the reaction of intestinal sucrase with deoxynojirimycin. They find that the extremely high affinity of unprotonated deoxynojirimycin for intestinal sucrase is almost fully lost in the protonated form. It was reasoned that the inhibition is a two-step process in which the unprotonated inhibitor binds and induces a protein conformational change (*75*), followed by protonation and electrostatic interaction with an active site carboxylate. Thus, the tight enzyme–protonated-inhibitor complex forms only after unprotonated inhibitor induces a protein conformational change.

C. GLYCOSYLAMINES

As with the 5-amino derivatives, substituting an amino group for the anomeric oxygen greatly enhances ligand affinity for a number of glycosidases (*4, 46, 76*). With *E. coli* β-galactosidases, the base properties of the nitrogen substantially contribute to the affinity, which can be enhanced still further by substitution with a hydrophobic aglycon that takes advantage of the hydrophobic glucose subsite on β-galactosidase (*77, 78*). Hydrophobic N-substituted β-D-galactosylamines with a relatively basic nitrogen such as *N*-heptyl-D-galactosylamine and *N*-benzyl-D-galactosylamine have inhibition constants in the low nanomolar range, approximately 10^3-fold lower than β-D-galactosylamine.

The N-substituted D-galactosylamine inhibition of β-galactosidase is enhanced by Mg^{2+} (*78, 79*). The metal ion functions to maintain alignment of a β-galactosidase acid catalyst that protonates the galactoside oxygen (*80*). Loeffler *et al.* (*79*) demonstrated with ^1H NMR and pH-dependency analyses that inhibition by cationic *N*-galactosylamines (e.g., β-1-D-galactopyranosyl pyridinium) likely results from electrostatic interaction of the deprotonated acid catalyst with the cationic nitrogen $(Gal-Pyr)^+ X^-$. Similarly, strong inhibition of uncharged N-substituted galactosylamines (e.g., *N*-heptyl-D-galactosylamine) results from acid catalyst proton donation to the bound inhibitor $(Gal-NH_2R)^+$ X^- (*78*). The consistency, however, does not hold for all substituted galactosylamines. Inhibition by uncharged D-galactosylpiperidine should, but does not, decrease at pH 10, where the β-galactosidase active site acid catalyst is fully deprotonated (*81*).

D. ACARBOSE

In a search for microbial α-glucosidase inhibitors, researchers at Bayer AG (Wuppertal, Germany) isolated and characterized a homologous series of compounds referred to as pseudooligosaccharides. Different homologues vary in affinity and specificity. The strongest pseudooligosaccharide inhibitor of many glucosidases is acarbose (**VII**) (*47*). All members share a pseudodisaccharide core composed of hydroxymethylconduritol linked to 4-amino-4,6-dideoxy-D-glucose (two moieties at the nonreducing end). Derivatives differ in the number of α-1,4-linked glucose residues at either end of the core; acarbose has no glucose residues linked to the hydroxymethylconduritol and two glucose residues linked through the anomeric carbon of dideoxylaminoglucose.

Acarbose inhibition is at least in part related to the protonated nitrogen substituting for the glycoside oxygen (47, 82) and the half-chair conformation of the conduritol segment, which likely binds in the position of the glucosyl transition state (4). However, strong α-amylase and sucrase inhibition requires the full complex; neither the hydroxymethylconduritol nor the aminodeoxyglucose segments are independent of the core (47).

Recently, Goldsmith et al. (83) analyzed the X-ray crystal structure of acarbose bound to glycogen phosphorylase a. Acarbose is a weak inhibitor of phosphorylase glycogenolysis and is noninhibitory in the direction of glycogen synthesis. The acarbose glycogenolysis $K_{i(app)}$ is 26 mM, which is on the order of the maltopentaose K_m (40 mM) (83). The values contrast with the stronger acarbose binding complexes with some α-amylases and glucoamylase, where inhibition constants are in the low micromolar to high nonomolar range (4, 47). Structure analysis of acarbose bound to phosphorylase a shows the inhibitor bound exclusively at the glycogen storage site and with no protein interaction with the conduritol moiety.

It is worth noting that acarbose has been applied to reduce free glucose production in the intestine by inhibiting intestinal brush border α-glucosidases and pancreatic amylase. In diabetes, the decreased rate of di- and oligosaccharide digestion improves control of glucose metabolism by reducing the rate of glucose absorption, postprandial blood glucose surge, and associated insulin and triglyceride response (84)

IV. Glycosidase Catalytic Flexibility: Catalysis of Substrates with Very Small Aglycon

Glycals as well as another class of nonglycoside substrates, glycosyl fluorides, have been used to substantiate unique glycosidase and glycosyltransferase reaction processes. The unusual activity of glycals and glycosyl fluorides results in part from the small aglycon, small particularly with glycals, which can be considered to possess an internal glycosyl bond in the C-1=C-2 double bond [see, for example, galactal (IV)]. Hydration of glycals can be equated with glycoside hydrolysis; both involve hydroxide attack at one end of the bond and proton addition at the other. For glycals this entails hydroxide attack at C-1 and proton addition at C-2 to form the 2-deoxy sugar.

Evidence of stereochemically unique results first emerged from analysis of β-

galactosidase protonation of D-galactal as reported by Lehmann and Zieger (85) and α- and β-glucosidase hydration of D-glucal as described by Hehre et al. (86). In recent years, Hehre and colleagues developed an impressive array of data demonstrating that, under select conditions, glycosidases and glycosyltransferases catalyze reactions that appear to violate what were considered stereochemical limitations. For example, several apparently stereospecific inverting glycosidases were found to catalyze hydrolysis of both α- and β-glycosyl fluoride. Bacillus pumilus β-xylosidase, which normally cleaves β-D-xylosidic bonds with inversion, hydrolyzes β-D-xylosyl fluoride to α-D-xylose, as expected, and hydrolyzes α-D-xylosyl fluoride with *retention* of configuration, i.e., the α-anomer is the product of both α- and β-D-xylosyl fluoride (87). It is pertinent to the following discussion that α-D-xylosyl fluoride hydrolysis also generates small amounts of an intermediate transfer product, 4-O-β-D-xylopyranosyl-D-xylopyranose. In addition to β-xylosidase, analogous characteristics are observed with sweet potato β-amylase (88); *Rhizopus* glucoamylase and *Arthrobacter* glucodextranase (89); and yeast, rabbit kidney (90), and *Trichoderma reesei* trehalase (91), all of which are inverting glycosidases.

The proposed mechanism of α- and β-glycosyl fluoride hydrolysis (88) is based on the assumption that catalysis requires a general acid and anionic base on opposite faces of the substrate-reactive center. The assumption is fully consistent with the operative mechanism of lysozyme, modified to catalyze a reaction with anomeric inversion. To continue the example of β-xylosidase, hydrolysis of the usual anomeric substrate (β-D-xylosyl fluoride) involves water attack at the reactive carbon from the *si* face or below C-1. The general acid, situated above C-1 on the *re* face, protonates the fluoride (or glycoside oxygen) as it departs from the β-position (87, 92) [Eq. (1)].

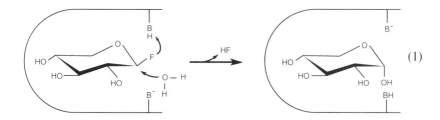

$$\tag{1}$$

Hydrolysis of the stereochemically "incompatible" substrate, α-D-xylosyl fluoride, generates the anomeric retained α-D-xylose by a mechanism that is somewhat more complex. The reaction is a two-step process that involves free xylose, and requires that the general acid and base switch roles. Following formation of the α-D-xylosyl fluoride Michaelis complex, free xylose (rather than water) attacks from the *re* face with the aid of what is now the general base; α-linked fluoride is displaced with assistance from proton donation at the *si* face.

The transfer generates 4-*O*-β-D-xylopyranosyl-D-xylopyranose. The disaccharide is a reactive substrate and rapidly hydrolyzes to α-D-xylose [Eq. (2)] (*87*).

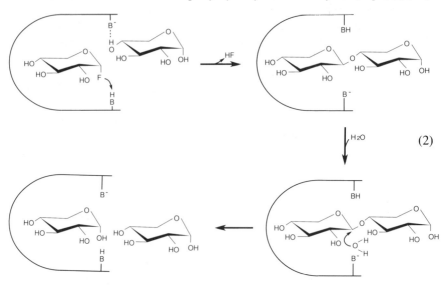

$$(2)$$

In this manner, both β- and α-D-xylosyl fluoride generate α-D-xylose, in the former instance by stereochemical inversion and in the latter by stereochemical retention. The series of events is reflected in each of the inverting glycosidases listed above. The reactions are unique not only because they involve hydrolysis of both (albeit truncated) substrate anomers, but the enzymes also catalyze an uncommon nonhydrolytic condensation reaction using the activated glycosyl fluoride.

Catalytic flexibility is also seen in the stereochemistry of glycosidase hydration of glycals. Substrate anomeric specificity is moot here because of the C-1=C-2 double bond, but C-1 is a prochiral center and hydration of D-glucal, for example, generates anomeric 2-deoxy α- or β-D-glucose. Thus, the stereochemistry of the *de novo* chiral carbon can be compared with the known stereochemical course of glycoside substrate hydrolysis. Also of interest is the stereochemistry of proton addition to glycal C-2. If glycoside hydrolysis and glycal hydration are simply viewed as two examples of glycosylation reactions, then general acid catalysis and protonation of the glycoside oxygen are comparable to protonation of the glycal double bond at C-2.

Glycal hydrations (and related enolic substrates) have been examined with β-galactosidase (*85, 93*), a number of α- and β-glucosidases (*86, 94–98*), exo-α-glucanases (*94–96*), trehalase (*97, 98*), exo- and endo-cellulases (*99*), and β-amylase (*100*). The enzymes represent both retentive and inverting glycosidases. Reactions are generally assayed in D_2O and the stereochemical progress is monitored by NMR spectroscopy.

Determining the anomeric configuration of the hydrated glycal is not always possible because mutarotation outpaces the slower hydration reactions. In all cases that have been analyzed, —OH (or —OD) addition to the prochiral center produced the expected anomer (that normally released in the hydrolytic reaction of a given enzyme). The course of proton addition across the double bond, however, was not as uniformly consistent. The analogy of glycal hydration and glycoside hydrolysis predicts that proton (deuteron) addition to the glycal double bond would occur from the same face as proton addition to a glycoside oxygen—that is, above the plane of the reactive center (*re* face) with β-glycosidases and below the plane (*si* face) with α-glycosidases. This, however, did not always hold. For example, while D-glucal hydration by three inverting *exo-α*-glucanases (*Arthrobacter* glucodextranase and *Rhizopus* and *Paecilomyces* glucoamylases) resulted in proton addition from below the *si* face, as expected (*94*), D-glycal protonation by another inverting *exo-α*-glucanase, β-amylase, developed from above the *re* face (*100*), opposite that expected. Furthermore, all retentive glycosidases, including several α- and β-glucosidases (*86, 94*), cellobiases (*99*), and β-galactosidase (*85*), were found to protonate the glycal double bond from the face opposite protonation of an exocyclic glycoside oxygen.

Based on the results with glycosyl fluorides, glycals, and related enolic analogs, glycosidase and glycosyltransferase catalysis is in part highly predictable: without exception and independent of substrate, each enzyme generates a single product anomeric configuration. In contrast, proton donation is substrate dependent and not limited to a single mechanism for a given enzyme. Hehre and coworkers propose that glycosidase catalysis (or possibly all glycosylase catalysis) has two phases: the generation of product anomeric configuration is a "conserved" phase and the exclusive domain of the enzyme; proton donation is a more "plastic" phase, subject to influence by the substrate (*94*).

V. β-Galactosidase

A. INTRODUCTION

The most thoroughly studied β-galactosidase is the *E. coli lacZ* gene product, which, after expression, is a tetramer of 116,353-Da identical subunits. The primary structure of the subunit has been determined by Fowler and Zabin (*101*) and essentially confirmed by the gene sequence (*102*). Each subunit binds a single Mg^{2+} (*103*), which is not mandatory for activity but significantly enhances the reaction rate for most (*104–106*) but not all (*107*) substrates. Mn^{2+} can substitute for Mg^{2+}, but other divalent metal ions are questionable (*108, 109*). Mn^{2+} actually has a slightly higher affinity than Mg^{2+} ($K_D = 1.1 \times 10^{-8} M$ compared to $2.8 \times 10^{-7} M$, respectively) and binds with a high degree of positive cooperativity (*109, 110*).

β-Galactosidase is primarily specific for the nonreducing end of O-, N-, and S-β-D-galactopyranosides (*104, 111*). Hydrolysis occurs with retention of configuration at C-1 (*112, 113*). The active site is composed of distinct glucose and galactose subsites (*114, 115*). The glucose subsite on the free enzyme has a higher affinity for hydrophobic ligands than for glucose (*104, 116*), but glucose affinity increases dramatically in a conformation change that accompanies formation of a galactosyl–enzyme intermediate (*114*). Galactose hydroxyl positions 3 and 4 are critical for binding at the galactose subsite and for catalysis; position 6 and possibly position 5 are important in binding but not in catalysis, and position 2 appears to have no influence on affinity (*117*). Like many glycosidases, β-galactosidase catalyzes transfer reactions to acceptors other than water (*104, 118*) and, notably, catalyzes the isomerization of lactose [galactosyl-β-D-(1,4)-glucopyranose] to allolactose [galactosyl-β-D-(1,6)-glucopyranose], the primary inducer of the *E. coli lac* operon (*118–120*).

B. GALACTOSYL–ENZYME

Independent lines of evidence establish a galactosyl–enzyme intermediate [Eq. (3)] in β-galactosidase catalysis [rather than direct displacement, as in Eq. (4)]. Analyses of deuterium and oxygen-18 kinetic isotope effects provide evidence of the form, abundance, and galactosyl transfer mechanism related to intermediates.

Several kinetic approaches can be used to test for the presence of a glycosyl–enzyme intermediate (*7, 8*). Some glycosylases, such as sucrose phosphorylase (*121*), *Bacillus subtilis* levansucrase (*122*), and oral streptococcal glucosyltransferases (*123*), have a Ping-Pong or modified Ping-Pong kinetic pattern, including isotope exchange half-reactions, which strongly implicates a glycosyl–enzyme intermediate. A galactosyl intermediate is also supported by studies using substrates that vary in rate-determining glycosylation or deglycosylation. This has been valuable with β-galactosidase since galactosyl donor and acceptor substrates are abundant. Water, methanol, ethanol, and other alkyl alcohols are all suitable acceptors (*124, 125*), and several chromophoric arylgalactosides, particularly nitrophenylgalactosides (*81, 126, 127*) and galactosylpyridinium salts (*128*) that vary in aglycon leaving-group potential, are excellent galactosyl donors.

$$E + G{-}X \underset{k_2}{\overset{k_1}{\rightleftharpoons}} E{\cdot}G{-}X \overset{k_3}{\rightarrow} E{\cdot}G + X \overset{k_4}{\rightarrow} E + G \tag{3}$$

$$E + G{-}X \underset{k_2}{\overset{k_1}{\rightleftharpoons}} E{\cdot}G{-}X \overset{k_3}{\rightarrow} E + G + X \tag{4}$$

Mechanisms (3) and (4) theoretically can be distinguished by using substrates with rate-determining glycosyl transfer to solvent. With this restriction, the aglycon influences k_{cat} in Eq. (4) but not in Eq. (3), since the latter reduces to the rate of degradation of E·G common to all substrates. However, the practical application of the approach has a significant limitation; ambiguity in the rate-determining step of any one of a range of glycoside substrates required to test the hypothesis will complicate interpretation of the data, and such assurances can be difficult to obtain.

Fortunately, the limitation is relatively easily overcome by measuring the ratio of products in the presence two competing glycosyl acceptors [Eq. (5)]. Here, if a common glycosyl–enzyme intermediate exists, the *ratio* of products is constant, independent of the leaving group and independent of the rate-determining step.

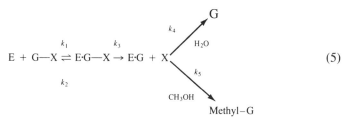

$$E + G\!-\!X \underset{k_2}{\overset{k_1}{\rightleftharpoons}} E\!\cdot\!G\!-\!X \overset{k_3}{\longrightarrow} E\!\cdot\!G + X \tag{5}$$

Stokes and Wilson (*129*) used the approach to established the participation of a galactosyl–enzyme in β-galactosidase solvolysis. They found the enzyme reaction in 0.247 M methanol and 0.171 M ethanol produced a constant ratio of phenol:alkyl-β-galactoside, with eight arylgalactoside substrates in the former case and four in the latter.

The intermediate has been confirmed through presteady-state "burst" experiments that capitalize on reactions in which degradation of a common intermediate is rate determining. Here, a rapid presteady-state release of leaving group occurs in the initial enzyme turnover before the reaction encounters the bottleneck of the rate-determining step. The slowest of the common β-galactosidase substrates with rate-determining degradation of galactosyl–enzyme is too rapid to monitor presteady-state events under conventional conditions (*13*). As an alternative, Fink and Angelides (*130*) cooled a reaction of β-galactosidase and *o*-nitrophenyl-β-D-galactopyranoside to $-22°C$ in 50% dimethyl sulfoxide. Even though galactosyl–enzyme degradation is not rate determining for *o*-nitrophenyl-β-D-galactopyranoside at 25°C (*81*), the decreased temperature reduces the degalactosylation rate to near zero while galactosyl–enzyme formation remains significant. The result is a negligible steady-state rate preceded by a presteady-state burst of *o*-nitrophenol equivalent to the concentration of enzyme. Burst kinetics have also been demonstrated by Deschavanne *et al.* (*131*) using an *E. coli* β-galactosidase expressed with a point mutation that dramatically reduces the rate of o- and p-nitrophenyl-β-D-galactopyranoside hydrolysis by lowering k_4 to a rate significantly slower than k_3 [Eq. (3)].

C. KINETIC ISOTOPE EFFECTS

As discussed in the description of hen lysozyme, KIEs can detect subtle changes that reflect the character of the reaction transition state. The results with lysozyme are clear: α-deuterium KIEs, oxygen-18 leaving-group KIEs, X-ray crystal structure analysis, and the binding of transition-state analogs all support an oxocarbonium ion transition state and an S_N1 reaction. β-Galactosidase KIE results are far less clear. Results from different experimental directions do not fully agree. In the arguments that follow, the role of Mg^{2+} in a protein conformational change that aligns the general acid catalyst is considered as an explanation for some of the contradictions.

Galactosyl–enzyme degradation is at least partly rate determining in hydrolysis of very reactive (more acidic leaving group) arylgalactosides such as 2,4- and 3,5-dinitrophenylgalactopyranoside, and formation of the complex is rate determining for slower substrates such as phenyl- and p-nitrophenylgalactopyranoside (81, 126, 127, 132). The distinction is important since the latter substrates more directly report characteristics during glycoside bond scission in the chemical step of the reaction.

The α-deuterium KIEs of reactive arylgalactoside substrates compared to less reactive substrates differ; the former have a large KIE ($k_H/k_D \sim 1.25$) (127, 133), implicating significant carbonium character in the transition state (34, 134), and the latter are close to 1.0 (127, 133), suggesting a nucleophilic reaction, or a rate-determining protein conformational change that would mask the effects of deuterium substitution (80).

Mg^{2+} serves a structural rather than a catalytic role in β-galactosidase (103), and its dynamics are fundamental to the argument. Comparison of Mg^{2+}–enzyme and apoenzyme galactoside hydrolysis implicates a role in conformational alignment of the general acid catalyst that protonates the galactoside oxygen. Two kinetic studies defend this. First, log k_{cat} or log k_{cat}/K_m appears independent of aglycon pK_a in the presence of Mg^{2+} and linearly dependent on pK_a in the absence of Mg^{2+} (133). Since acid assistance tends to override and eliminate the effects of aglycon pK_a, the results suggest acid assistance is functional with Mg^{2+} but not without it. The second series of experiments is a comparison of aryl- and pyridiniumgalactoside substrates where the cationic structure of the latter disallows leaving-group acid assistance. The results again suggest Mg^{2+} is associated with general acid catalysis: pyridinium substrates are unaffected by addition of Mg^{2+} while arylgalactoside rates increase 10-fold (80, 107, 128). The relation between Mg^{2+} and acid assistance indirectly supports a rate-determining protein conformation change as an explanation for an S_N1 reaction with a ^2H KIE near 1.0 for slow arylgalactoside substrates.

A different approach to explore the chemical step and develop evidence to distinguish S_N1 and S_N2 reaction mechanisms is analysis of the reaction in the absence of acid catalysis (and absence of the potential complication of a conformation change). Jones *et al.* (*50*) compared spontaneous and β-galactosidase-catalyzed hydrolysis of galactosylpyridinium salts and developed substantial support for bond cleavage as an S_N1 reaction. Spontaneous galactosylpyridinium hydrolysis was independent of pH between 2 and 8 and occurred with a positive entropy of activation, attributes characteristic of S_N1 reactions. The α-secondary ^2H KIE of β-D-galactopyranosylpyridinium was 1.16, which is borderline for an S_N1 reaction but within the commonly low range for glycopyranosides (*15*). These data compare favorably with α-secondary ^2H KIEs for the β-galactosidase-catalyzed reaction: 1.17 ± 0.03 and 1.136 ± 0.040 for the Mg^{2+}-free enzyme (*107*) and holoenzyme (*128*), respectively. Furthermore, both spontaneous hydrolysis of galactosylpyridinium substrates (*50*) and enzymic hydrolysis (with and without Mg^{2+}) (*80, 107, 128*) exhibit a linear correlation between log k_{cat} and aglycon pK_a, with a slope reasonably consistent with an S_N1 reaction (*50*) and inconsistent with an S_N2 reaction (*135*). There is, thus, substantial evidence that hydrolysis of this class of substrates proceeds by an S_N1 mechanism.

Rosenberg and Kirsch (*14*) analyzed the reaction from a different perspective using ^{18}O leaving-group KIEs and developed results supporting an S_N2 reaction, in conflict with the above conclusions. As noted previously, ^{18}O leaving-group KIEs provide quantitative analysis of the minimum degree of galactoside bond scission in the overall rate-determining step of the reaction (calculated from the V_{16}/V_{18} KIE) and the first irreversible step of the reaction [calculated from the $(V/K)_{16}/(V/K)_{18}$ KIE] (*39, 40*).

The ^{18}O leaving-group KIEs for the reactive substrate, 2,4-dinitrophenyl-β-D-galactopyranoside, where rate-determining steps occur after bond scission, exhibit results consistent with the steady-state kinetics: the V_{16}/V_{18} KIE is not significantly different from 1.00 (1.002 ± 0.009) (galactoside–oxygen bond scission does not occur in the rate-determining step), and the $(V/K)_{16}/(V/K)_{18}$ KIE is significantly greater than 1.00 (1.030 ± 0.003) (bond scission, at least in part, occurs at the first irreversible step). The latter result is also exhibited with the slow-reacting substrate, p-nitrophenyl-β-D-galactopyranoside, where rate-determining events occur prior to or at bond scission (*81, 126*) [$(V/K)_{16}/(V/K)_{18}$ KIE $= 1.014 \pm 0.003$].

Of particular importance, however, is the V_{16}/V_{18} for slow-reacting substrates, which reports the degree of glycoside bond scission in the chemical step. The magnitude is significant, 1.022 ± 0.002, which, based on solution chemistry, extrapolates to a minimum of $52 \pm 7\%$ bond scission in the rate-determining step. This requires that generation of the galactosyl–enzyme intermediate, not a protein conformation change, is rate determining. Therefore, the α-secondary ^{2}H KIE near 1.0 (*133*) likely reflects events that occur in bond cleavage and an S_N2 reaction.

D. Mechanism Summary

β-Galactosidase catalysis coordinates an active site carboxylate stabilizing a developing carbonium ion that can collapse to a covalent form. In the presence of Mg^{2+} and most substrates, a general acid catalyst protonates the leaving-group exocyclic oxygen.

The generation of an oxocarbonium galactosyl–enzyme is evidenced by the high affinity of β-galactosidase for carbonium-like transition-state analogs and by the hydrolysis of galactosides under conditions that eliminate general acid catalysis (*4, 13, 50*). The transition state in transfer to an acceptor, however, may be the carbonium ion or a covalent complex. The very short lifetime and instability of a glycosyl oxocarbonium (*9, 10*), and the unlikely prospect of maintaining a stabilized ion pair without collapse, suggest an equilibrium between the covalent and noncovalent forms (*2*). All data are consistent with this proposition (*13, 14, 136*).

Sinnott and Souchard (*133*) maintain that while the covalent acylal would be in greater abundance than the galactosyl cation of an oxocarbonium $\rightleftharpoons$ acylal equilibrium, the latter is nonproductive and all galactosyl transfer occurs through the cation as an S_N1 reaction. However, for this to be acceptable, the α-secondary ^{2}H KIE near 1.0 with slow-reacting arylgalactoside substrates must be explained. As noted above, there is substantial indirect evidence suggesting that the ^{2}H KIE reflects a rate-determining conformation change related to a Mg^{2+}-dependent alignment of an acid catalyst. The conclusion is consistent with an apparent random distribution of reaction rates as a function of aglycon acidity and a ^{2}H KIE above 1.0 only when acid assistance (and a Mg^{2+}-dependent conformation change) is not operative (*50, 80, 107, 128, 133*).

Rosenberg and Kirsch (*14*) contend that the conclusion is not strictly true when analyzing Mg^{2+}–enzyme hydrolysis of slow-reacting arylgalactosides where bond scission is rate determining. They suggest that the relation between log k_{cat}/K_m and aglycon pK_a is not random, but appears so due to steric factors of meta- and ortho-substituted substrates. When only para-substituted derivatives are considered, the correlation is reasonably good. This is consistent with the V_{16}/V_{18} leaving-group KIE for p-nitrophenylgalactoside (in which galactosyl–enzyme formation is rate determining), which predicts that bond scission contributes at least $52 \pm 7\%$ to the rate-determining step (*14*). The α-secondary ^{2}H KIE near 1.0 then must reflect a significant degree of galactosyl transfer through an S_N2 reaction. In an equilibrium analogous to that illustrated in Fig. 1 (Section I), hydrolysis can proceed through either a covalent or noncovalent galactosyl–

enzyme intermediate, depending on the relative ability of the acceptor to compete with the enzyme for the carbonium ion.

E. ACTIVE SITE AMINO ACIDS

The amino acid serving to stabilize the transition state was first identified as Glu-461 by Legler and Herrchen using the site-directed irreversible inhibitor conduritol C epoxide (66, 67). The significance has since been firmly established through kinetic analysis of Glu-461–β-galactosidase mutations (136).

Application of site-directed mutagenesis to establish significance of individual amino acids in catalysis has become a common tool in enzymology, although relatively rare with glycoside hydrolases and glycosyltransferases, which are limited to date to studies on β-galactosidase (136–140), lysozyme (141), and invertase (142).

The β-galactosidase study is an excellent example of the power of site-directed mutagenesis. Huber, Miller, and colleagues prepared and examined five Glu-461–β-galactosidase substitutions (Asp, Gly, Gln, His, and Lys) (136, 139). All substitutions had k_{cat} values less than 0.3% of the wild-type enzyme except the His-461 mutation, which was approximately 6%. For most of the substitutions it was possible to quantify K_m, k_{cat}, K_s, and rates of galactosylation and degalactosylation for three substrates, and K_i values for three inhibitors. Different enzyme kinetic properties resulting from different amino acid substitutions confirm that Glu-461 is directly involved in catalysis and contributes to active site structure stability. Heat inactivation at 55°C occurred more rapidly with each amino acid substitution compared to the wild-type enzyme, except for the structurally conservative Gln substitution, which was only moderately affected.

Glu-461 substitutions greatly reduced both galactosylation and degalactosylation rates. The glutamate carboxyl is well known for stabilization of the oxocarbonium ion transition state, but, in addition, examination of the effects of Glu-461 substitution on substrates with different leaving-group potentials (i.e., aglycon acidity) showed that Glu-461 contributes to galactosylation and degalactosylation in ways that are not necessarily related to stabilization of the transition state. As expected, the galactosylation rates of all five mutations increased with increasing leaving-group acidity; the tested substrate with the best leaving group (o-nitrophenyl-β-D-galactoside) had a relatively high galactosylation rate, much higher than solution hydrolysis, even with the cationic Lys-461 substitution, which would tend to destabilize the transition state. Acid assistance alone cannot account for the high rate because the Lys-461 mutation was essentially ineffective in catalyzing galactosylation of lactose. Thus, for these amino acid substitutions acting on substrates with a good leaving group, factors other than acid catalysis must be relevant.

The His-461 mutation had a large k_{cat}, representing nearly 6% of the native enzyme activity with o-nitrophenyl-β-D-galactoside. This high value was almost exclusively due to a rapid degalactosylation rate. This is best accounted for and offers confirmation of the potential of both a covalent and a noncovalent β-galactosidase intermediate (*13*, *14*). The His 3'-nitrogen, like the carboxylate oxygen, is a good nucleophile and occupies a position nearly identical to the carboxylate oxygen. Thus, His-461 may form a covalent galactosyl–His complex capable of relatively rapid degalactosylation (*136*).

The β-galactosidase residue considered responsible for acid catalysis in glycoside bond cleavage, Tyr-503, has also been subject to site-directed mutagenesis. Initial evidence for the role of Tyr-503 was indirect. Two site-directed irreversible inhibitors, N-bromoacetyl-β-D-galactopyranosylamine (*143*) and β-D-galactopyranosylmethyl-p-nitrophenyltriazene (*144*), inactivate β-galactosidase by alkylating a methionine, which, in the latter case, was shown to be Met-502. However, Naider *et al.* (*143*) eliminated methionine from consideration on showing that norleucine substitution did not significantly affect catalysis. Because of the location of Met-502 and the β-galactosidase pH dependency, it was hypothesized that the proton donor may be the adjacent residue, Tyr-503 (*144*). The assignment was consistent with the results of Ring *et al.* (*145*), who demonstrated that m-fluorotyrosine substitution lowered the alkaline side pK_a by 1.5 pH units, equivalent to the difference between tyrosine and the fluoro derivative. Finally, Tyr-503 site-directed mutagenesis showed that conservative substitution with phenylalanine resulted in greatly reduced activity (*137*, *138*)

VI. Sucrose Glycosyltransferases

A. INTRODUCTION

The sucrose glycosyltransferases discussed in this section—sucrose phosphorylase, *B. subtilis* levansucrase, *Streptococcus* glucosyltransferases, and *Leuconostoc* dextransucrase—share sucrose as a source of free energy in glucosyl or fructosyl transfer to an acceptor [-6600 cal/mol (*146*)]. The high specificity of the enzymes for sucrose has limited the variety of available substrates and consequently the extent of kinetic and mechanism analysis, although Chambert *et al.* (*122*) performed a remarkable series of kinetic experiments to establish the levansucrase kinetic mechanism.

1. *Sucrose Phosphorylase and Levansucrase*

Sucrose phosphorylase (sucrose:orthophosphate α-D-glucosyltransferase, EC 2.4.1.7) is distinguished as the first example of an experimentally trapped covalent enzyme intermediate complex, in this case a glucosyl–enzyme linked to a glutamic or aspartic acid carboxyl (*147*). The approach has since been success-fully applied to other enzymes in this group.

Sucrose phosphorylase catalyzes reversible glucosyl transfer from sucrose to orthophosphate, although other acceptors, including AsO_4^{3-}, water, fructose, and a number of other ketoses, are equally suitable. Donor specificity is more strict and, aside from the substrate/product pair of sucrose and α-D-glucosyl phos-phate, only α-D-glucosyl fluoride has been well characterized (*12, 148–151*).

Doudoroff *et al.* (*152*) proposed a sucrose phosphorylase covalent glucosyl–enzyme based on enzyme catalysis of exchange between phosphate and α-D-glucosyl phosphate, a proposition that was later strengthened by the observation of exchange between fructose and sucrose (*153*), a modified Ping-Pong steady-state kinetic pattern (*121*), and trapping a glucosyl–enzyme complex (*147*). The collective data appear to comprise a compelling argument for covalent catalysis, but it is recognized that the results fall short of unambiguously discriminating between a covalent and a noncovalent intermediate (*12*).

Bacillus subtilis levansucrase (sucrose:2,6-β-D-fructan 6-β-D-fructosyltrans-ferase, EC 2.4.1.10) catalyzes fructosyl transfer from sucrose to levan (*154*). In the absence of a fructosyl acceptor, the primary reaction is sucrose hydrolysis, although a limited amount of self-initiated levan synthesis occurs as well (*155*). As with sucrose phosphorylase, acceptor specificity is broad; a number of sac-charides and other nucleophiles are suitable fructosyl acceptors (*154–158*). The complete amino acid sequence of the approximately 50-kDa enzyme has been determined by both protein (*159*) and gene (*160*) sequence analyses. The three-dimensional structure at 3.8 Å reveals a rod- or ellipsoid-shaped protein with a length some four times the diameter (*161*).

Chambert, Dedonder, and colleagues (*122, 162*) deduced the levansucrase kinetic mechanism through an extensive series of studies. One complication of levansucrase (and a potential complication of other polymer synthases) is that a moderately sized levan acceptor is indistinguishable from levan as a product. All polysaccharide substrate/product terms are kinetically relevant, resulting in non-linear double-reciprocal plots for many common mechanisms (*122*). Nonethe-less, levansucrase steady-state kinetics of fructosyl transfer from sucrose to levan (and the reverse), fructosyl transfer from levan to water, fructosyl transfer from sucrose to water, and fructose isotope exchange between sucrose and levan at equilibrium support a Ping-Pong mechanism (*122*). As expected with Ping-Pong kinetics, levansucrase catalyzes the half-reaction of isotope exchange between fructose and sucrose. The partial reaction is of particular value and has been exploited as a means to avoid the complication of levan substrate/product inhi-bition (*162*). The accumulated steady-state kinetic and isotope exchange results were sufficient to assign values to each rate constant in the forward and reverse reactions, with levan and water as fructosyl acceptors. As with sucrose phosphorylase, a levansucrase-denatured fructosyl–enzyme complex has been trapped (*163*). In the native state, thermodynamic analysis established that the high free energy of the sucrose glycosidic bond is preserved in the fructosyl–enzyme, whereas the free energy change from sucrose to intermediate is only -460 ± 200 cal/mol (*162*).

2. *Streptococcus and Leuconostoc Sucrose Glycosyltransferases*

Another extensively studied family of sucrose glucosyltransferases includes the extracellular enzymes from *Leuconostoc* and *Streptococcus* spp. Hehre (*164*) first indentified dextransucrase (sucrose: 1,6-α-D-glucan 6-α-D-glucosyltransferase, EC 2.4.1.5) in a culture supernatant of *Leuconostoc mesenteroides* and initiated structure and mechanism characterization. The enzyme became of practical interest because of the widespread application of dextran as a plasma support during World War II (*165*). *Leuconostoc* dextransucrase is sucrose induced (*166*), unlike the streptococcal enzymes (*167*), although expression of the latter varies considerably with culture conditions (*168, 169*).

Streptococcal glucosyl- and fructosyltransferases have received recent attention because of their role in initiation of dental caries (*170–174*). A number of oral streptococci secrete sucrose glucosyltransferases, including representative strains of *Streptococcus sanguis*, *Streptococcus salivarius*, *Streptococcus mitis*, and the four bacteria of the mutans group, *Streptococcus mutans*, *Streptococcus sobrinus*, *Streptococcus cricetus*, and Streptococcus rattus (*175–177*). The cariogenicity of the bacteria, particularly those of the mutans group, is associated with synthesis from sucrose of dental plaque polysaccharides, which adhere to smooth enamel surfaces and enhance bacterial colonization of teeth (*170, 172, 178–181*). In addition to promoting adherence, the polysaccharides stabilize the colony environment and serve as a carbohydrate reserve that can be tapped by secreted extracellular glycanases (*182–184*).

a. *Nomenclature.* Oral streptococci secrete several sucrose glucosyltransferases, which vary among species and strain (*176, 185–187*). The enzymes are commonly divided into two broad categories: those that synthesize water-soluble α-(1,6)-D-glucans (dextran), and those that synthesize water-insoluble α-1,3-linked D-glucans. Included in this family are enzyme variations in product glucan branching, glucan size, and degree of dependence on exogenous glucan acceptor. Oral streptococci also produce at least two extracellular β-fructosyltransferases: *S. salivarius* cultures produce levansucrase (sucrose: 2,6-β-D-fructan 6-β-D-fructosyltransferase) (*188, 189*) and the *S. mutans* serotypes secrete an inulosucrase (sucrose: 2,1-β-D-fructan 1-β-D-fructosyltransferase, EC 2.4.1.9) (*190–193*).

The terms used here, GTase-S and GTase-I, will denote the streptococcal glucosyltransferases that synthesize water-soluble glucans and water-insoluble glucans, respectively. Distinctions between enzymes that vary in degree of product branching or size will be noted when required. The *Leuconostoc* enzyme will be termed dextransucrase.

   b. *Reaction Characteristics.* Only a few sucrose substitutes have been identified for streptococcal and *Leuconostoc* glucosyltransferases. These include 4-O-α-D-galactopyranosylsucrose (*194*), α-D-glucopyranosyl-α-D-sorbofuranoside (*195*), and α-D-glucopyranosyl fluoride (*196, 197*). *p*-Nitrophenyl-α-D-glucopyranoside is a much weaker substrate (possibly due to solubility limitations), but is still capable of transferring glucose to a suitable acceptor (*198*). In contrast to donor specificity, acceptor specificity is broad (*165, 177, 199*). A large number of *Leuconostoc* dextransucrase carbohydrate acceptors have been characterized, of which maltose and isomaltose are the most effective (*200, 201*). The enzymes have highest affinity for native glycans that contribute to a tendency to aggregate in the presence of polysaccharide products (*202–206*).

   Streptococcal and *Leuconostoc* glycosyltransferase-catalyzed reactions are analogous to levansucrase and sucrose phosphorylase. The reaction proceeds with retention of α-anomeric configuration, with bond scission between the glycoside oxygen and anomeric carbon (*207*). The fundamental *Leuconostoc* dextransucrase and streptococcal GTase reactions are glucosyl transfer from sucrose to glucan [Eq. (6)], glucosyl transfer from sucrose to water [Eq. (7)], and isotope exchange between sucrose and fructose [Eq. (8)].

$$\text{Suc} \;+\; \text{Glc}_n \;\rightarrow\; \text{Glc}_{n+1} \;+\; \text{Fru} \tag{6}$$

$$\text{Suc} \;\rightarrow\; \text{Glc} \;+\; \text{Fru} \tag{7}$$

$$\text{Suc} \;+\; {}^*\text{Fru} \;\rightleftharpoons\; {}^*\text{Suc} \;+\; \text{Fru} \tag{8}$$

However, with one exception (*208*), the enzymes do not catalyze the reverse reactions of either sucrose synthesis from fructose and glucan or the second partial reaction of glucose exchange between glucans. Both *Leuconostoc* dextransucrase and *Streptococcus* GTase-S catalyze a very slow glucosyl transfer between glucans (*209*), but the reaction is too sluggish to be representative of the second half-reaction.

   There is indirect structure and kinetic evidence suggesting that the absence of or extremely slow GTase-S reverse reaction may result from a required sucrose-induced protein conformational change; GTase has distinct sucrose and glucan-binding domains (*210–212*), and the binding of one substrate is known to influence the other—sucrose and fructose affinity increase approximately sixfold in the presence of glucan (*213*). The characteristics are consistent with a hinge mechanism (*214*) or type of induced-fit model (*215*) in which sucrose binding energy is tapped to align substrate domains to form the active catalytic site. Thus, in the absence of sucrose, the active site would not form and catalysis does not occur.

In some cases, sucrose hydrolysis is the exclusive GTase-S reaction in the absence of added acceptor, and the hydrolytic reaction progressively partitions to glucan or sucrose synthesis with added glucan or fructose, respectively (*213*). The enzyme has a high affinity for glucan; half-saturation of 10-kDa polysaccharide is 0.4 μg/ml, or 40 nM. Trace glucan contamination in commercial sucrose (*216*) is sufficient to initiate some steady-state glucan synthesis (*213*) even in the absence of an exogenous glucan acceptor. This does not discount the property of several glycosyltransferases that catalyze rapid self-initiated glucan synthesis (*185, 217, 218*).

Steady-state kinetic analysis of *S. sobrinus* GTase-S glucosyl transfer from sucrose to glucan or sucrose to water, fructose product inhibition, and isotope exchange between sucrose and fructose established a hybrid rapid-equilibrium random/Ping-Pong mechanism (*213*). The reaction flux of glucosyl transfer from sucrose to glucan proceeds as a sequential process with formation of an enzyme–sucrose–glucan ternary complex, while glucosyl transfer from sucrose to fructose occurs as a nonsequential reaction that does not allow concurrent binding of sucrose and acceptor fructose. The hybrid mechanism is analogous to that deduced for acetate kinase (*219, 220*), arginine kinase (*221*), and phosphoenolpyruvate carboxykinase (*222*).

B. Glycosyl–Enzyme

Sucrose phosphorylase, *B. subtilis* levansucrase, and streptococcal and *Leuconostoc* glucosyltransferases catalyze glycosyl transfer with retention of anomeric configuration, and in each case catalysis proceeds by a Ping-Pong or modified Ping-Pong kinetic mechanism (*121, 122, 213*). Both characteristics implicate a distinct glycosyl–enzyme intermediate. A very small kinetic term can mask a ternary complex and possible direct displacement, but this is highly unlikely considering the isotope exchange reaction. For example, in the GTase-S partial reaction of isotope exchange between fructose and sucrose, both reactants bind to overlapping sites based on obvious structural similarities and reasonably high affinity for the enzyme (*213*). Thus, product fructose must dissociate from the active site before acceptor fructose binds, leaving an activated glucosyl–enzyme mediating the transfer.

There is no direct evidence on the form of the glycosyl–enzyme intermediate. Nonetheless, stabilization of a noncovalent oxocarbonium for the period between product fructose release and acceptor binding may not be realistic for slow reactions, given the extremely short lifetime of a glycosyl oxocarbonium ion (*9*). Levansucrase and particularly GTase-S are quite slow enzymes, with sucrose hydrolysis k_{cat} of 48 sec^{-1} (*122*) and 9.1 sec^{-1} (*213*), respectively. Thus, the carbonium ion may well collapse to a more stable covalent complex or develop an equilibrium between the two forms. Nucleophilic catalysis is consistent with available data as suggested by Chambert and Gonzy-Treboul (*162*) and Fu and Robyt (*223*). Nonetheless, no result clearly answers the question of the productive glycosyl–enzyme transition state that precedes product formation.

Two approaches have been used with *Leuconostoc* and *Streptococcus* glucosyltransferases to trap a glycosyl–enzyme: a method based on chemically trapping a denatured glycosyl–enzyme through denaturing a steady-state reaction of enzyme and sucrose, analogous to experiments with sucrose phorphorylase and levansucrase; and a method based on stabilizing a native glucosyl–enzyme bound to a liquid chromatography resin.

In the latter case, stabilizing a native glycosyl–enzyme requires reaction conditions that dramatically reduce the deglycosylation rate. For example, the defructosylation half-life of levansucrase fructosyl–enzyme with water as an acceptor is 14 msec (*122*), and while less complete kinetic data are available, the maximum GTase-S deglucosylation half-life (based on k_{cat}) or glucosyl–enzyme transfer to water is less than 76 msec (*213*). Thus, accumulation of a stable intermediate requires deglycosylation rate reductions of 10^4- to 10^5-fold. A reduction of this magnitude generally requires thermal trapping at subzero temperatures (*224*) or use of alternative substrates that very slowly deglycosylate. Mayer and colleagues used solvents at 4°C to preserve a native GTase, which, at elevated temperature, transferred glucose to maltose, fructose, and water acceptors (*225–227*).

C. CHEMICALLY TRAPPED GLYCOSYL–ENZYME

Enzyme intermediates trapped by chemical modification can provide pertinent details about the enzyme active site and catalytically significant amino acids that directly reflect on the reaction mechanism. However, when the chemical modification is irreversible, demonstrating kinetic relevance by intermediate transfer along the remainder of the reaction pathway at a rate consistent with catalysis is not possible. Thus, distinguishing an authentic covalent intermediate from a collapsed form of a glycosyl–cation is not possible.

Voet and Abeles (*147*) captured a covalent glucosyl–enzyme of sucrose phosphorylase by two methods: acid quenching a steady-state reaction of sucrose and enzyme, and reacting sucrose with sodium periodate-modified enzyme. The acid-quenched complex is irreversibly denatured, but the periodate-modified form retains some, albeit limited, catalytic activity (20,000-fold reduction). Specific glucosyl substitution at the active site is clear from the absolute requirement of sucrose and active enzyme, a mole ratio of glucose to enzyme near 1, and glucose linkage to the enzyme exclusively as the β-anomer (*12, 147*). The latter is consistent with retention of glucose anomeric configuration and a double-displacement mechanism involving two successive anomer inversions (*11*). The sucrose phosphorylase-trapped complex is extremely base labile, with a half-life at pH 6.0 of 80 min. This, plus methanolysis product analysis (*147*) and carbodiimide-mediated labeling, implicates a glucosyl linkage to a carboxyl of aspartic or glutamic acid at the enzyme active site (*228*).

Characteristics of acid-quenched levansucrase (*163*) and *S. sobrinus* GTase-S (*123*) glycosyl–enzymes are similar to sucrose phosphorylase. Stoichiometry is consistent with substitution exclusively at the enzyme active sites, and both complexes are base labile; GTase glucosyl–enzyme has a half-life at pH 7.0 of 32 min (*123*), comparable to sucrose phosphorylase glucosyl–enzyme, with a half-life extrapolated from the first component of a biphasic first-order decay of 60 min at the same pH (*147*); the levansucrase fructosyl–enzyme is slightly more stable, with an equivalent half-life approximately 3.5 pH units higher (*163*).

Mild alkaline hydrolysis of *S. sobrinus* GTase-S glucosyl–enzyme releases glucose exclusively as the β-anomer. This contrasts with α-D-glucose liberated during sucrose hydrolysis by native enzyme. The difference can be attributed to the character of acylals, which can hydrolyze at the acetal carbon or the ester carbon with respective anomeric retention or inversion (*229, 230*). The native enzyme hydrolyzes sucrose as an acetal so that released α-D-glucose translates to an active site carboxyl linked as the β-anomer or ion paired to a carbonium ion from the *re* face. The denatured covalent complex, without the active site structure to cleave at the acetal carbon, hydrolyzes the β-linked D-glucose at the ester carbon, with release of the retained β-anomer.

Characteristics of sucrose phosphorylase, levansucrase, and GTase-trapped glycosyl–enzymes affirm intermediate stabilization through an active site carboxyl (*147, 163, 231*). Furthermore, a general acid catalyst likely assists leaving-group departure (*12*). An imidazolium of histidine has been suggested as the acid catalyst of levansucrase (*163*) and dextransucrase based on photooxidation (*232*) and diethyl pyrocarbonate (*223*) amino acid modifications. The second-order rate of native levansucrase fructosyl–enzyme defructosylation (assuming 1 M water) is on the same order (40 $\pm$ 3 M^{-1} min^{-1}) as the spontaneous nucleophilic catalysis of denatured fructosyl–enzyme (67 $\pm$ 7 M^{-1} min^{-1}) (*163*). The reactions are not fully comparable but are consistent with a covalent intermediate.

VII.   Glycosidase and Glycosyltransferase Structure

The primary structure of a large number glycosidases and glycosyltransferases is known from the gene sequence and in a few instances directly from the amino acid sequence. Conserved segments among functionally related groups have been identified, some of which are clearly associated with structure or catalysis and others which are conserved for as-yet unexplained reasons. The homology spanning the broadest evolutionary scale is found in the active site of α-glucosidases, primarily those enzymes catalyzing glucosyl transfer with retention of anomeric configuration. More recently, homology among active sites in one of at least two structural classes of fructosylases has been documented (*233, 234*).

A. IDENTIFICATION OF ACTIVE SITE AMINO ACIDS AND PEPTIDES

Unambiguous identification of catalytic amino acids usually requires confirmation from several experimental directions. There are three common approaches: (1) crystal structure analysis of enzyme–ligand complexes, (2) selective amino acid modification through site-directed irreversible labeling and site-directed mutagenesis, and (3) exploiting the catalytic mechanism by trapping a covalent complex [as with the sucrose glycosyltransferases (Section VI,C)] or more commonly with mechanism-based inhibitors.

Each method has been used successfully in the study of glycosidases; some approaches are more reliable than others, but Purich (*8*) emphasizes that all methods require confirmation because of the reactive quality of the reagents or intermediates, and the potential to sequester artifacts. Once a catalytic amino acid has been clearly identified, homology alignments can serve to localize active site segments in other enzymes.

1. *Mechanism-Based Irreversible Inhibitors*

Some of the more reactive irreversible inhibitors have been variously called suicide substrates, k_{cat} inhibitors, and mechanism-based inhibitors. These compounds are relatively innocuous substrate analogs until converted by the enzyme to highly reactive products capable of covalent attachment at the active site. Since the enzyme mechanism is involved, the covalently conjugated amino acid is often directly involved in catalysis. There has recently been increased interest in mechanism-based glycosidase inhibitors because of their value in studying the reaction mechanism (*4, 48*) and in their potential therapeutic application (*47*).

It is convenient to divide the inhibitors into two categories. The first is represented by reagents that form highly reactive products after catalytic conversion. These include glycosylmethyl aryltriazenes (**VIII**) (*235, 236*), conduritol epoxides (**IX**) (*46*), 1′,1′-difluoroalkylglycosides (**XII**) (*237*), and two recently introduced aziridines derived from conduritol (**X**) (*238*) and deoxynojirimycin (**XI**) (*239*).

The second group consists of substrates that have relatively rapid glycosylation rates but very slow or nonexistent deglycosylation rates, which results in terminating the reaction at the glycosyl–enzyme intermediate at the first enzyme turnover [see Eq. (3), where $k_3 > k_4$ and k_4 is small]. If deglycosylation is slow but finite, the complex can be more permanently trapped by denaturing the protein during catalysis. Glycals can be considered in this category (*86*) (see Section III), as can 2-deoxy-2-fluoroglycosides (**XIII**) (*48, 240*).

Glycosylmethyl aryltriazenes (**VIII**) owe their reactivity to potential decomposition on protonation of the aniline nitrogen. While the mechanism is not fully defined (*235*), decomposition presumably leads to generation of a reactive glycosylmethyl carbonium ion or a diazonium. If the proton is derived from the enzyme's general acid, there is a reasonable possibility of covalent attachment to this residue. The relevant glycopyranosylmethyl aryltriazenes have successfully and irreversibly inhibited α- and β-galactosides, β-glucosidase, α-xylosidase, and α-L-arabinofuranosidase. The most extensively examined is inhibition of the *E. coli lacZ* β-galactosidase Met-502 (*144*). This is not the enzyme acid catalyst that was later identified as the adjacent residue, Tyr-503 (*137, 138, 144*) (see Section V,E).

Conduritol epoxides are derived from inositol isomers, where, for example, conduritol B epoxide (**IX**) approximates the glucose hydroxyl orientation and conduritol C epoxide approximates galactose. In addition, specificity can be dictated by the epoxide group, which can be oriented on either face of the pyranose ring; enzyme inactivation commonly occurs only with epoxide alignment consistent with the enzyme anomeric specificity. Conduritol epoxides are quite stable near neutral pH but become highly reactive on proton transfer to the epoxide oxygen. The glycosidase inhibition mechanism, when successful, proceeds by protonation from the active site general acid and capture by the carboxylate nucleophile. Thus, conduritol B epoxide has been used to label an *A. wentii* β-glucosidase active site aspartate presumed associated with the glucosyl intermediate (*64*), and conduritol C epoxide labels the *lacZ* β-galactosidase catalytic nucleophile, Glu-461 (*66, 67*).

An analogous mechanism that relies on proton transfer and nucleophilic attack is apparently operative with aziridine derivatives of conduritol and deoxynojirimycin. Preliminary studies show both compounds are effective inhibitors: the conduritol B aziridine (**X**) irreversibly inactivates *Alcaligenes faecalis* β-glucosidase and yeast α-glucosidase (*238*),and the *galacto*deoxynojirimycin aziridine (**XI**) is an extremely potent inhibitor of green coffee bean α-galactosidase, but not yeast α-glucosidase, jack bean α-mannosidase, or bovine β-galactosidase (239).

Difluoroalkylglucosides (**XII**), also recently reported, are designed around a different concept. The difluoroalkyl aglycon is stable when conjugated as a glycoside, but the α,α-fluoroalkyl alcohol, which is released at the active site by enzyme hydrolysis, spontaneously loses HF to generate an acid fluoride that can react with an active site nucleophile. The glucoside with R = CHFCl successfully inhibited yeast α-glucosidases with psuedo-first-order kinetics, but reacted only as a substrate with intestinal sucrase–isomaltase (237).

2-Deoxy-2-fluoroglycosides (**XIII**) are recently conceived mechanism-based inhibitors designed on the premise that substitution of electronegative fluorine at C-2 will destabilize the transition state and reduce the rates of glycosylation and/or deglycosylation. If the glycosylation rate is significant and the deglycosylation rate is extremely slow, the glycosyl–enzyme complex will accumulate. Results of initial studies are very positive; 8 of 11 diverse glycosidases were inactivated by the 2-fluoroglycosyl fluorides, suggesting excellent potential for identification of catalytically significant active site amino acids and associated peptides and examination of the native enzyme intermediate (48, 240).

2. *Glycosidase Active Site Carboxylates*

X-Ray structure analysis of hen egg-white lysozyme complex with oligosaccharides provided the first evidence of a carboxylate role in glycosidase catalysis and the foundation for the mechanism of glycoside bond scission (20, 21). Comparisons of the native enzyme and the oligosaccharide–enzyme complex not only aligned two carboxylic acids (Glu-35 and Asp-52) with the sessile glycoside bond, but also implicated the distortion of the pyranose ring to approximate the transition state. Based on the polarity of the environment, Asp-52 was likely ionized and ion paired with the transition and Glu-35 was un-ionized and in position to donate a proton to the glycoside oxygen (19, 29). The presumptive evidence of the role of Asp-52 has been confirmed by extensive studies involving affinity labeling, biochemical modification, and site-directed mutagenesis.

Before the introduction of site-directed mutagenesis, chemical modification was the predominant means of altering a specific amino acid; the earliest report with lysozyme was from Parsons and Raftery (241), who prepared the ethyl ester derivative of Asp-52 by reaction with triethyloxonium fluoroborate; the modified enzyme lost catalytic function but not substrate affinity. An ethyleneimine reaction product of Asp-52 has also been prepared with similar effects on catalysis (242). Sharon and co-workers modified and then regenerated Asp-52 to eliminate concern that inactivation results from experimental manipulation rather than specific amino acid modification (243, 244). Thus, Asp-52 was first esterified with an epoxypropyl-β-glycoside derivative of di-(N-acetyl-D-glucosamine), then reduced to homoserine or hydrolyzed to return the free aspartate. Both the ester and homoserine derivatives lost activity but the regenerated aspartate was fully active.

Highly conservative modifications such as replacement of aspartate with glutamate or asparagine are possible with site-directed mutagenesis. Notably, however, the first demonstration was from Imoto and co-workers, who used a chemical approach that involved selective esterification followed by ammonolysis to produce Asn-52 (*245*). The derivative lost 97 to 99 % activity (depending on the pH) and had a twofold to threefold reduction in substrate affinity. The identical derivative, prepared by site-directed mutagenesis where unambiguous and complete substitution is assured, had comparable reductions in catalysis and substrate affinity (*141*).

In addition to lysozyme, a number of modifications can be drawn upon as evidence of a carboxyl role in glycosidases. Carboxylic amino acids associated with catalysis have been labeled in intestinal sucrase–isomaltase (*246*), bitter almond (*247*), *A. wentii* (*64*) and calf liver (*248*) β-glucosidases, soybean β-amylase (*249*), and yeast invertase (*233*) using site-directed irreversible inhibitors. The β-galactosidase active site nucleophile, Glu-461, was identified by labeling with the galactose analog, conduritol C epoxide (*67*). Five amino acid substitutions of Glu-461 (Asp, Gln, Gly, His, and Lys) were subsequently prepared by site-directed mutagenesis (*136*). These data were discussed in Section V,D.

As outlined in Section IV,C., the catalytic mechanism has also been exploited to stabilize and isolate a transient glycosyl–enzyme intermediate in three glycosyltransferases that depend on sucrose as the glycosyl donor: sucrose phosphorylase (*147*), *B. subtilis* levansucrase (*163*), and *S. sobrinus* α-glucosyltransferases (*231*). Each was prepared by denaturing a steady-state reaction of radiolabeled substrate and enzyme. All were covalently bound to a carboxyl group at the enzyme active site. Only sucrose phosphorylase has been renatured, and in this case only to a limited degree. Sucrose phosphorylase and streptococcal GTase (both anomer-retentive α-glucosyltransferases) are linked to the active site carboxylic amino acid as the expected β-glucosyl anomer.

B. GLYCOSIDASE AND GLYCOSYLTRANSFERASE SEQUENCE HOMOLOGY

The discovery and subsequent detailed analysis of sequence homology among α-amylases revealed six highly conserved segments (*250*). Sequence homology of some segments, particularly those containing residues directly involved in catalysis, extends to other α-glucosidases as well. For example, region 4, which contains the carboxylic amino acid involved in stabilizing the glycosyl intermediate, is found in α-glucosyltransferases, α-transglucanosylases, α-glucosamylases, and other α-glucosidases in addition to α-amylases (Table I, Section A). Region 6, which includes the general acid catalyst that donates a proton to the glycoside oxygen, also is homologous with peptide segments in α-glucosylases other than amylases. The remaining conserved amylase regions (1, 2, 3, and 5) lack amino acids directly involved in catalysis but include residues associated with oligosaccharide binding; these have some (but more limited) homology with other α-glucosylases (250).

Region 4 deserves further comment since participation of a carboxylate in stabilizing the transition state in glycoside bond cleavage is thoroughly documented and, to date, without exception. The relevant sequence in a large group of α-glucosidases and transferases is conserved across a broad evolutionary range, spanning prokaryotes, eukaryotes, plants, and animals. The conservation is most apparent in enzymes that catalyze the reaction with retention of α-anomeric configuration, although weak homology is also found among α-glucoamylases, which catalyze anomer inversion (250, 251).

The broad base of active site peptide analyses derived from labeling the active site carboxyl allows sequence comparisons and division of enzymes into related groups. Four groups are listed in Table I, each keyed to an active site carboxyl identified in at least one enzyme in each group. For the present it is assumed that all of the sequences are aligned with the carboxylic amino acid that stabilizes the glycosyl–enzyme intermediate. In cases where evidence is limited, the assignments might be viewed with caution; multiple independent experimental confirmations are needed for unambiguous conclusions. The assignment is clear for the α-glucosidases and transferases in Table I, Section A, where the relevant amino acid and sequence are based on site-directed irreversible inhibition (246), functional labeling (231), X-ray structure analysis (262), and site-directed mutagenesis (252). There is less evidence for the relevant β-glucosidase sequences since the assignment is based exclusively on site-directed irreversible inhibition (64, 247). However, added confidence that the conduritol B epoxide-labeled carboxylic amino acid in A. wentii β-glucosidase A_3 is associated with stabilizing the glucosyl intermediate comes from analogy with conduritol C epoxide labeling of a Glu-461 in E. coli β-galactosidase (67), which has been confirmed as a catalytically significant amino acid (136).

In Table I, active site carboxyls that have been specifically identified are shown in boldface type. The α-glucosidases and transferases in Section A, all catalyze glucosylation with retention of anomeric configuration except for the glucoamylases, wherein the sequence conservation is weakest. Inclusion of glucoamylase in this group is supported by sequence homology with other segments of α-amylase primary structure (250). Many additional sequences could have been included with the α-glucosidases and transferases; several can be found in Svensson (250).

Section C lists four β-amylases (like α-amylases) catalyze cleavage of α-1,4-glucosidic bonds but with inversion rather than retention of anomeric configura-

TABLE I

GLUCOSIDASES, FRUCTOFURANOSIDASES, AND RELATED TRANSFERASES

Enzyme	Source	Sequence location [a]
A. α-Glucosidases and transferases		
Isomaltase	Rabbit intestine	499 Y D G L W I **D** M N E V [b]
Sucrase	Rabbit intestine	1388 F D G L W I **D** M N E P [b]
Glucosyltransferase-I	Streptococcus mutans	445 F D S I R V **D** A V D N [c]
Glucosyltransferase-S	Streptococcus mutans	459 F D S I R V **D** A V D N [d]
Taka-amylase A	Aspergillus oryzae	200 I D G L R I **D** T V K H [e]
α-Amylase	Porcine pancreas	191 V A G F R I **D** A S K H [f]
α-Amylase	Barley type B	197 L D G W R F **D** F A K G [g]
Pullulanase	Clostridium thermohydrosulfuricum	623 A D D G W R L **D** V A N E [h]
Isoamylase	Pseudomonas amyloderamosa	494 V D D G F R F **D** L A S V [i]
Acid α-glucosidase	Human lysosome	511 F D D G M W I **D** M N E P [j]
Maltase	Saccharomyces carlsbergensis	208 V D D G F R I **D** T A G L [k]
Glucanotransferase	Klebsiella pneumonia	247 V D A I R I **D** A I K H [l]
Glucan branching enzyme	Escherichia coli	398 I D A L R V **D** A V A S [m]
Glucoamylase	Aspergillus niger	175 Y D — L W E E **E** V D G S [n]
B. β-Amylases		
β-Amylase	Soybean	180 G L G P A G **E** L R Y P [o]
β-Amylase	Sweet potato	181 G C G A A G **E** L R Y P [p]
β-Amylase	Barley	178 G L G P A G **E** M R Y P [q]
β-Amylase	Bacillus polymyxa	192 S G G P S G **E** L R Y P [r]
C. β-Glucosidases		
β-Glucosidase	Aspergillus wentii	— Z G F V M S **D** W A A H [s]
β-Glucosidase	Kluyveromyces fragilis	219 D G M L M S **D** W F G T [t]
β-Glucosidase	Candida pelliculosa	295 Q G F V M T **D** W — G A [u]
β-Glucosidase	Bitter almond	— Z Z G V F G **D** S (ABBPZ) [v]
D. β-Fructofuranosidases		
β-Fructofuranosidases		
Yeast invertase	Saccharomyces cerevisiae	36 N K G W M N **D** P N G L [w]
Levanase	Bacillus subtilis	43 E A N W M N **D** P N G M [x]
Sucrase	Bacillus subtilis	37 P V G L L N **D** P N G V [y]

220

Sucrase	*Zymomonas mobilis*	37	L	T	S	W	M	N	D	P	N	G	L	[z]
Sucrase	*Streptococcus mutans*	41	K	T	G	L	L	N	D	P	N	G	F	[aa]
Sucrase	*Vibrio alginolyticus*	45	L	T	S	W	M	N	D	P	N	G	L	[bb]

[a] Sequence homologies are keyed to an active site carboxylic amino acid that stabilizes a glycosyl–enzyme intermediate. Sequence numbers include the signal peptide where available. Amino acids shown in boldface type are residues identified as catalytically significant. References list the sequence source followed by the reference for the identification of the residue that is indicated in boldface type.

[b] Sucrase/isomaltase, rabbit intestine (246, 257).

[c] Sucrose 3-α-glucosyltransferase, *S. mutans* (258) and *Streptococcus downei* (210) which have identical sequences in this peptide segment; aspartic acid was labeled in the *Streptococcus sobrinus* enzyme (231).

[d] Sucrose-6-α-glucosyltransferase, *S. mutans* (259) and *S. downei* (260); aspartic acid was labeled in the *S. sobrinus* enzyme (231).

[e] Taka-amylase A, *A. oryzae* (261); catalytic role of Asp-206 supported by X-ray structure analysis and homology with other α-glucosidases, although Asp-297 is considered as well (262).

[f] α-Amylase, porcine pancreas (263); X-ray structure analysis supports Asp-197 as a catalytic residue (264).

[g] α-Amylase, barley type B (265).

[h] α-Amylase–pullulanase, *C. thermohydrosulfuricum* (266).

[i] Isoamylase, *P. amyloderamosa* (267).

[j] α-Glucosidase (acid maltase), human lysosome (268).

[k] Maltase, *S. carlsbergensis* (269).

[l] Cyclomaltodextrin–glucanotransferase, *K. pneumonia* (270).

[m] Glycogen branching enzyme, *E. coli* (271).

[n] Glucoamylase G2, *A. niger* (251, 272).

[o] β-Amylase, soybean (249, 273).

[p] β-Amylase, sweet potato (274).

[q] β-Amylase, barley (275).

[r] β-Amylase, *B. polymyxa* (276).

[s] β-Glucosidase A₃, *A. wentii* (64).

[t] β-Glucosidase, *K. fragilis* (277).

[u] β-Glucosidase, *C. pelliculosa* (278).

[v] β-Glucosidase, bitter almond (247).

[w] Invertase, *S. cerevisiae* (233, 234).

[x] Levanase, *B. subtilis* (279).

[y] Sucrase, *B. subtilis* (280).

[z] Sucrase, *Z. mobilis* (281).

[aa] Sucrase, *S. mutans* (282).

[bb] Sucrase, *V. alginolyticus* (283).

tion. Section D lists four β-glucosidases, including three homologous fungal β-glucosidases; an aspartic acid in the *A. wentii* enzyme is known from labeling with conduritol B epoxide as noted above. It has been suggested that several β-glucosidases and β-glucanases have sequence homology with the peptide segment surrounding the functional active site anion of hen egg-white lysozyme, Asp-52 (*253, 254*). There is no apparent homology with bitter almond β-glucosidase (*247*).

Table I (*257–283*) lists six β-fructofuranosidases, which have homology with the active site peptide surrounding Asp-42 in yeast invertase (*233, 234*). Two fructosyltransferases, *S. mutans* fructosyltransferase and *B. subtilis* levansucrase, have extensive sequence homology but are not homologous with the invertase group (*255*).

For comparison with the sequences listed in Table I, the analogous active site region of hen egg-white lysozyme (*256*) is

46   N T D G S T D Y G I L

and *E. coli lacZ* β-galactosidase (*102*) is

455   I W S L G N E S G H G

C. TERTIARY STRUCTURE

A number of lysozymes and lysozyme–inhibitor complexes have been resolved (*24*) and have provided the most detailed glimpse of molecular events in glycosidase bond cleavage. General structure features of lysozyme were described in Section II,A; a very limited number of other glycosidase structures have been resolved, but several nonglycosidase carbohydrate-binding proteins have been reported, which broadens the base for comparison of structure motifs. Quiocho has presented an excellent review on this subject (*284*).

Carbohydrate–protein interactions are dominated by hydrogen bond forces with significant contributions from hydrophobic interactions. The extensive hydration of carbohydrates in aqueous solution is essentially replaced by hydrogen bonds on binding to the protein. Hydrophobic interactions with the protein are contributed by nonpolar clusters of C–H atoms on carbohydrates. High-affinity binding is generally characterized by an extensive network of hydrogen bonds with solvent access shielded by sites buried in the protein surface or protected by an apolar environment.

1. L-*Arabinose-Binding Protein*

These principles are well illustrated by the 1.7-Å analysis of the L-arabinose complex with L-arabinose-binding protein shown in Fig. 2 (*44*). The extensively refined structure is the most detailed example of a hydrogen bond network at a

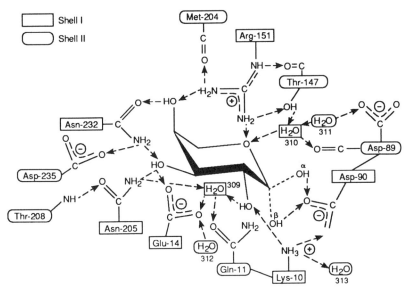

FIG. 2. Schematic representation from the extensively refined structure analysis of L-arabinose complexed with the L-arabinose-binding protein. Two levels of hydrogen bonds stabilizing the structure are shown. The residues in shell I hydrogen bond to L-arabinose and to shell II. Adapted from Quiocho and Vyas (44) with permission from *Nature*.

carbohydrate binding site. The arabinose-binding protein is a 33-kDa monomer with a single arabinose binding site formed in a cleft between two large globular domains. The site tightly binds both the α- and β-anomers with a dissociation constant of approximately $4 \times 10^{-7} M$ (285).

The structure analysis resolved three levels of noncovalent bonding at the L-arabinose binding site; the outer two levels (shells I and II) are shown in Fig. 2. The structure includes two integrated water molecules, Wat-309 and Wat-310, that donate and accept hydrogen bonds from within shell I. Two remarkable features of shell I bonding to L-arabinose are worth noting. First, all hydroxyls, with the exception of the anomeric hydroxyl, participate as both hydrogen bond donors and acceptors; in each instance, the acceptor and donor take the form NH → OH → O. In two instances, L-Arabinose O-3 and the ring O-5 hydroxyl oxygens accept two NH-derived hydrogen bonds (from Asn-205 and Asn-232 in the former case; from Arg-151 and Wat-310 in the latter case). This detailed network establishes the protein specificity and contributes the bulk of the binding affinity. The second notable feature is the facile accommodation of either α- or β-L-arabinose at the binding site. Both anomers form identical hydrogen bonds, including the anomeric hydroxyl, where the equatorial α-anomer and the axial β-anomer can align with Asp-90 atom OD2. Nonpolar interactions are also present but not shown in Fig. 2; C-3–H, C-4–H, and C-5–H form a hydrophobic cluster by virtue of the proximity of these atoms and absence of polar hydroxyls punctuating the space. The apolar cluster aligns with indole ring of Trp-16.

2. α-Amylases and Cyclodextrin Glycosyltransferase

In addition to lysozyme, three-dimensional structure information on glyco-sidases and transferases includes a low-resolution image of *B. subtilis* levan-sucrase, which shows the 50-kDa enzyme folding into an elongated ellipsoid with approximate dimensions of 26 × 32 × 117 Å (*161*). The remaining solved structures, all within the α-amylase homology group, include a preliminary analysis of barley malt α-amylase 2 (*286*), *Aspergillus oryzae* α-amylase (Taka-amylase A; TAA) at 3 Å resolution (*262, 287, 288*), porcine pancreatic α-amylase (PAA) at 5 Å (*289*) and 2.9 Å (*264*), and *Bacillus circulans* cyclo-dextrin glycosyltransferase (CGT) at 3.4 Å (*290*). The enzymes all catalyze cleav-age of α-1,4-D-glucoside bonds and have substantial sequence homology and common architectural features. PPA (496 residues) and TAA (478 residues) to-pology includes three domains, as suggested by Buisson *et al.* (*264*). Domain A is located at the N-terminus and accounts for more than half of the structure. Within domain A is an $(\alpha/\beta)_8$-barrel composed of a cylindrical sheet of eight parallel β-strands surrounded by a eight α-helices. The pattern is found in a growing number of enzymes (*291*), including triose-phosphate isomerase (*292*), glucose-6-phosphate isomerase (*293*), and pyruvate kinase (*294*). Domain A is not continuous in the primary structure; a central portion forms domain B, which extends between the third β-strand and third α-helix of the α/β barrel. Domain B contains a long loop and one (TAA) or two (PPA) antiparallel β-sheets. Do-main C is a globular segment at the C-terminus that folds into an eight-stranded antiparallel β-sheet. Domains A and C are connected by a single peptide chain. The larger enzyme, CGT (about 670 residues), has analogous A, B, and C domains with added C-terminal D and E domains both folded into antiparallel β-sheet topologies.

The α-amylase active site lies in a cleft formed at the carboxyl end of the domain A parallel β-barrel and domain B. The active site gains added stability from an essential calcium ion that bridges the cleft with ligands to asparagine and histidine on domain A and two aspartic acids on domain B (*264*). Glucose binding sites have been located on TAA by difference Fourier analysis based on an enzyme–isomaltose complex. The very slow-reacting maltotriose was appar-ently hydrolyzed in the crystal, which left only maltose at the active site. Model building generated the seven binding subsites shown in Fig. 3 (*262*). Bond scission occurs between subsites 4 and 5. Based on homology with related α-glucosidases, Asp-206 (Asp-197 in PPA) appears to be the best candidate for

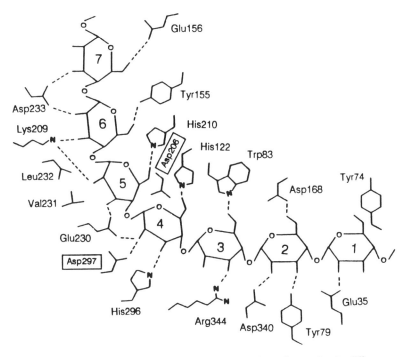

Fig. 3. Substrate binding site on Taka-amylase A deduced from electron density difference maps with the enzyme–maltose complex and model building. The seven saccharide binding sites are numbered. Presumed catalytic amino acids Asp-206 and Asp-297 surround the sessile glycoside bond. Glu-230 is considered as a possible catalytic amino acid as well because of its proximity to the reaction center. Adapted from Matsuura et al. (262) with permission from J. Biochem (Tokyo).

the carboxylate nucleophile (264), although Matsuura et al. considered Asp-297 in that role (262). Asp-297 (Asp-300 in PPA) may be more appropriately assigned to the general acid catalyst (262, 264), although these hypotheses require confirmation.

ACKNOWLEDGMENT

The unpublished research reported from this laboratory was supported by a grant from the National Institutes of Health (DE 03739).

REFERENCES

1. Imoto, T., Johnson, L. N., North, A. C. T., Phillips, D. C., and Rupley, J. A. (1972). "The Enzymes," 3rd Ed., Vol. 7, p. 665.

 2. Dunn, B. M., and Bruce, T. C. (1973). *Adv. Enzymol.* **37,** 1.
 3. Fife, T. H. (1975). *Adv. Phys. Org. Chem.* **11,** 1.
 4. Lalégerie, P., Legler, G., and Yon, J. M. (1982). *Biochimie* **64,** 977.
 5. Sinnott, M. L. (1984). *In* "The Chemistry of Enzyme Action" (M. I. Page, ed.), p. 389. Elsevier, New York.
 6. Kirby, A. J. (1987). *Crit. Rev. Biochem.* **22,** 283.
 7. Jencks, W. P. (1987). "Catalysis in Chemistry and Enzymology," p. 42. Dover, New York.
 8. Purich, D. L. (1982). *In* "Methods in Enzymology" (D. L. Purich, ed.) Vol. 87. p. 3. Academic Press, New York.
 9. Young, P. R., and Jencks, W. P. (1977). *J. Am. Chem. Soc.* **99,** 8238.
10. Sinnott, M. L., and Jencks, W. P. (1980). *J. Am. Chem. Soc.* **102,** 2026.
11. Koshland, D. E. (1954). *In* "Mechanism of Enzyme Action" (W. E. McElroy and B. Glass, eds.), p. 608. Academic Press, New York.
12. Mieyal, J. J., and Abeles, R. H. (1970). "The Enzymes," 3rd Ed., Vol. 7, p. 515.
13. Sinnott, M. L. (1978). *FEBS Lett.* **94,** 1.
14. Rosenberg, S., and Kirsch, J. F. (1981). *Biochemistry* **20,** 3189.
15. Dalhquist, F. W., Rand-Meir, T., and Raftery, M. A. (1968). *Proc. Natl. Acad. Sci. U.S.A.* **61,** 1194.
16. Rosenberg, S., and Kirsch, J. F. (1981). *Biochemistry* **20,** 3196.
17. Blake, C. C. F., Koenig, D. F., Mair, G. A. North, A. C. T., Phillips, D. C., and Sarma, V. R. (1965). *Nature (London)* **206,** 757.
18. Johnson, L. N., and Phillips, D. C. (1965). *Nature (London)* **206,** 761.
19. Blake, C. C. F., Johnson, L. N., Mair, G. A., North, A. C. T., Phillips, D. C., and Sarma, V. R. (1967). *Proc. R. Soc. London Ser. B* **167,** 378.
20. Phillips, D. C. (1966). *Sci. Am.* **215,** 78.
21. Phillips, D. C. (1986). "Protein Structure and Function." Oxford Univ. Press, London and New York.
22. Grutter, M. G., Weaver, L. H., and Matthews, B. W. (1983). *Nature (London)* **303,** 828.
23. Weaver, L. H., Grutter, M. G., Remmington, S. J., Gray, T. M., Isaacs, N. W., and Matthews, B. W. (1985). *J. Mol. Evol.* **21,** 97.
24. Johnson, L. N., Cheetham, J., McLaughlin, P. J., Acharya, K. R., Barford, D., and Phillips, D. C. (1988). *Curr. Top. Microbiol. Immunol.* **139,** 81.
25. Levitt, M. (1974). *In* "Peptides, Polypeptides, and Proteins" (E. R. Blout, F. A. Bovrey, M. Goodman, and N. Lotan, eds.), pp. 99–113. Wiley, New York.
26. Ford, L. O., Johnson, L. N., Machin, P. A., Phillips, D. C., and Tjian, R. J. (1974). *J. Mol. Biol.* **88,** 349.
27. Kelly, J. A., Sielecki, A. R., Sykes, B. D., James, M. N. G., and Phillips, D. C. (1979). *Nature (London)* **282,** 875.
28. Sarma, R., and Bott, R. (1977). *J. Mol. Biol.* **113,** 555.
29. Kuramitsu, S., Ikeda, K., and Hamaguchi, K. (1977). *J. Biochem. (Tokyo)* **82,** 585.
30. Parsons, S. M., and Raftery, M. A. (1972). *Biochemistry* **11,** 1630.
31. Parsons, S. M., and Raftery, M. A. (1972). *Biochemistry* **11,** 1633.
32. Capon, B. (1969). *Chem. Rev.* **54,** 407.
33. Kirby, A. J. (1980). *Adv. Phys. Org. Chem.* **17,** 183.
34. Kirsch, J. F. (1977). *In* "Isotope Effects in Enzyme Catalyzed Reactions" (W. W. Cleland, M. H. O'Leary, and D. B. Northrop, eds.), p. 100. Univ. Park Press, Baltimore, Maryland.
35. Hogg, J. L. (1978). *In* "Transition States in Biochemical Processes" (G. D. Gandour and R. L. Schowen, eds.), p. 201. Plenum, New York.
36. Craze, G. A., Kirby, A. J., and Osborne, R. (1978). *J. Chem. Soc., Perkin Trans. 2* p. 357.
37. Knier, B. L., and Jencks, W. P. (1980). *J. Am. Chem. Soc.* **102,** 6789.

38. Smith, L. E. H., Mohr, L. H., and Raftery, M. A. (1973). *J. Am. Chem. Soc.* **95**, 7497.
39. O'Leary, M. H. (1978). *In* "Transition States in Biochemical Processes" (R. D. Gandour and R. L. Schowen, eds.), p. 285. Plenum, New York.
40. Rosenberg, S., and Kirsch, J. F. (1979). *Anal. Chem.* **51**, 1375.
41. Sawyer, C. B., and Kirsch, J. F. (1973). *J. Am. Chem. Soc.* **95**, 7375.
42. Rosenberg, S., and Kirsch, J. F. (1981). *Biochemistry* **20**, 3196.
43. Chipman, D. M., and Sharon, N. (1969) *Science* **165**, 454.
44. Quiocho, F. A., and Vyas, N. K. (1984). *Nature (London)* **310**, 381.
45. Secemski, I. I., and Lienhard, G. E. (1971). *J. Am. Chem. Soc.* **93**, 3549.
46. Legler, G. (1973). *Mol. Cell. Biochem.* **2**, 31.
47. Truscheit, E., Frommer, W., Junge, B., Muller, L., Schmidt, D. D., and Wingender, W. (1981), *Angew. Chem., Int. Ed. Engl.* **20**, 744.
48. Withers, S. G., Rupitz, K., and Street, I. P. (1988). *J. Biol. Chem.* **15**, 7929.
49. Jencks, W. P. (1975). *Adv. Enzymol. Relat. Areas Mol. Biol.* **43**, 219.
50. Jones, C. C., Sinnott, M. L., and Souchard, I. J. L. (1977). *J. Chem. Soc. Perkin Trans. 2* p. 1191.
51. Reese, E. T., Parrish, F. W., and Ettlinger, M. (1971). *Carbohydr. Res.* **18**, 381.
52. Levvy, G. A., and Snaith, S. M. (1972). *Adv. Enzymol. Relat. Areas Mol. Biol.* **36**, 151.
53. Leaback, D. H. (1968). *Biochem. Biophys. Res. Commun.* **32**, 1025.
54. Huber, R. E., and Brockbank, R. L. (1987). *Biochemistry* **26**, 1526.
55. Conchie, J., Hay, A. J., Strachan, I., and Levvy, G. A. (1967). *Biochem. J.* **102**, 929.
56. Hackert, M. L., and Jacobson, R. A. (1969). *J. Chem. Soc. D.* p. 1179.
57. Shallenberger, R. S. (1982). "Advanced Sugar Chemistry." AVI Publ., Westport, Connecticut.
58. Lee, Y. C. (1969). *Biochem. Biophys. Res. Commun.* **35**, 161.
59. Lehmann, J., and Schroter, E. (1972). *Carbohydr. Res.* **23**, 359.
60. Wentworth, D. F., and Wolfenden, R. (1974). *Biochemistry* **13**, 4715.
61. Viratelle, O. M., and Yon, J. M. (1980). *Biochemistry* **19**, 4143.
62. Legler, G., Roeser, K.-R., and Illig, H.-K. (1979). *Eur. J. Biochem.* **101**, 85.
63. Bause, E., and Legler, G. (1974). *Z. Phys. Chem. (Leipzig)* **255**, 438.
64. Bause, E., and Legler, G. (1980). *Biochim. Biophys. Acta* **626**, 459.
65. Kurz, G., Lehmann, J., and Vorberg, E. (1981). *Carbohydr. Res.* **93**, C14.
66. Legler, G., and Herrchen, M. (1981). *FEBS Lett.* **135**, 139.
67. Herrchen, M., and Legler, G. (1984). *Eur. J. Biochem.* **138**, 527.
68. Inouye, S., Tsuruoka, T., Ito, T., and Niida, T. (1968). *Tetrahedron* **23**, 2125.
69. Hanozet, G., Pircher, H.-P., Vanni, P., Oesch, B., and Semenza, G. (1981). *J. Biol. Chem.* **256**, 3703.
70. Legler, G., and Julich, E. (1984). *Carbohydr. Res.* **128**, 61.
71. Legler, G., and Pohl, S. (1986). *Carbohydr. Res.* **155**, 119.
72. Fleet, G. W. J., Shaw, A. N., Evans, S. V., and Fellows, L. E. (1985). *J. Chem Soc., Chem. Commun.* p. 841.
73. Fleet, G. W. J. (1985). *Tetrahedron Lett.* **26**, 5073.
74. Grover, A. K., and Cushley, R. J. (1977). *Biochim. Biophys. Acta* **482**, 109.
75. Semenza, G., and Balthazar, A. L. (1974). *Eur. J. Biochem.* **41**, 149.
76. Lai, H.-Y. L., and Axelrod, B. (1973). *Biochem. Biophys. Res. Commun.* **54**, 463.
77. DeBruyne, C. K., and Yde, M. (1977). *Carbohydr. Res.* **56**, 153.
78. Legler, G., and Herrchen, M. (1983). *Carbohydr. Res.* **116**, 95.
79. Loeffler, R. S. T., Sinnott, M. L., Sykes, B. D., and Withers, S. G. (1979). *Biochem. J.* **177**, 145.
80. Sinnott, M. L., Withers, S. G., and Viratelle, O. M. (1978). *Biochem. J.* **175**, 539.
81. Tenu, J.-P., Viratelle, O. M., Garner, J., and Yon, J. (1971). *Eur. J. Biochem.* **20**, 363.

82. Legler, G., Sinnott, M. L., and Withers, S. G. (1980). *J. Chem. Soc., Perkin Trans.* 2 p. 1376.
83. Goldsmith, E. J., Fletterick, R. J., and Withers, S. G. (1987). *J. Biol. Chem.* **262**, 1455.
84. Clissold, S. P., and Edwards, C. (1988). *Drugs* **35**, 214.
85. Lehmann, J., and Zieger, B. (1977). *Carbohydr. Res.* **58**, 73.
86. Hehre, E. J., Genghof, D. Z., Sternlicht, H., and Brewer, C. F. (1977). *Biochemistry* **16**, 1780.
87. Kasumi, T., Tsumuraya, Y., Brewer, C. F., Kersters-Hilderson, H., Claeyssens, M., and Hehre, E. J. (1987). *Biochemistry* **26**, 3010.
88. Hehre, E. J., Brewer, C. F., and Genghof, D. S. (1979). *J. Biol. Chem.* **254**, 5942.
89. Kitahata, S., Brewer, C. F., Genghof, D. S., Sawai, T., and Hehre, E. J. (1981). *J. Biol. Chem.* **256**, 6017.
90. Hehre, E. J., Sawai, T., Brewer, C. F., Nakano, M., and Kanda, T. (1982). *Biochemistry* **21**, 3090.
91. Kasumi, T., Brewer, C. F., Reese, E. T., and Hehre, E. J. (1986). *Carbohydr. Res.* **146**, 39.
92. Marshall, P. J., and Sinnott, M. L. (1983). *Biochem. J.* **215**, 67.
93. Lehmann, J., and Schlesselmann, P. (1983). *Carbohydr. Res.* **113**, 93.
94. Chiba, S., Brewer, C. F., Okada, G., Matsui, H., and Hehre, E. J. (1988). *Biochemistry* **27**, 1564.
95. Hehre, E. J., Brewer, C. F., Uchiyama, T., Schlesselmann, P., and Lehmann, J. (1980). *Biochemistry* **19**, 3557.
96. Schlesselmann, P., Fritz, H., Lehmann, J., Uchiyama, T., Brewer, C. F., and Hehre, E. J. (1982). *Biochemistry* **21**, 6606.
97. Nakano, M., Brewer, C. F., Kasami, T., and Hehre, E. J. (1989). *Carbohydr. Res.* **194**, 139.
98. Weiser, W., Lehmann, J., Chiba, S., Matsui, H., Brewer, C. F., and Hehre, E. J. (1988). *Biochemistry* **27**, 2294.
99. Kanda, T., Brewer, C. F., Okada, G., and Hehre, E. J. (1986). *Biochemistry* **25**, 1159.
100. Hehre, E. J., Kitahata, S., and Brewer, C. F. (1986). *J. Biol. Chem.* **261**, 2147.
101. Fowler, A. V., and Zabin, I. (1977). *Proc. Natl. Acad. Sci. U.S.A.* **74**, 1500.
102. Kalnins, A., Otto, K., Ruther, U., and Miller-Hill, B. (1983). *EMBO J.* **2**, 593.
103. Case, G. S., Sinnott, M. L., and Tenu, J.-P. (1973). *Biochem. J.* **133**, 99–104.
104. Wallenfels, K., and Weil, R. (1972). "The Enzymes," 3rd Ed., Vol. 7, p. 617.
105. Tenu, J.-P., Viratelle, O. M., and Yon, J. M. (1972). *Eur. J. Biochem.* **26**, 112.
106. Withers, S. G., Jullien, M., Sinnott, M. L., Viratelle, O. M., and Yon, J. M. (1978). *Eur. J. Biochem.* **87**, 249.
107. Sinnott, M. L., Viratelle, O. M., and Withers, S. G. (1975). *Biochem. Soc. Trans.* **20**, 1006.
108. Rickenberg, H. V. (1969). *Biochim. Biophys. Acta* **250**, 530.
109. Huber, R. E., Parfett, C., Woulfe-Flanagan, H., and Thompson, D. J. (1979). *Biochemistry* **18**, 4090.
110. Woulfe-Flanagan, H., and Huber, R. E. (1978). *Biochem. Biophys. Res. Commun.* **82**, 1079.
111. Sinnott, M. L. (1971). *Biochem. J.* **125**, 717.
112. Wallenfels, K., and Mahortra, O. M. (1961). *Adv. Carbohydr. Chem.* **16**, 240.
113. Wallenfels, K., and Kurz, G. (1962). *Biochem. Z.* **335**, 559.
114. Deschavanne, P. J., Viratelle, O. M., and Yon, J. M. (1978). *J. Biol. Chem.* **253**, 833.
115. Viratelle, O. M., Yon, J. M., and Yariv, J. (1977). *FEBS Lett.* **79**, 109.
116. Huber, R. E., Gaunt, M. T., and Hurlburt, K. L. (1984). *Arch. Biochem. Biophys.* **234**, 151.
117. Huber, R. E., and Gaunt, M. T. (1983). *Arch. Biochem. Biophys.* **220**, 263.
118. Huber, R. E., Kurz, G., and Wallenfels, K. (1976). *Biochemistry* **15**, 1994.
119. Burstein, C., Cohn, M., Kepes, A., and Monod, J. (1965). *Biochim. Biophys. Acta* **95**, 634.
120. Jobe, A., and Bourgeois, S. (1972). *J. Mol. Biol.* **69**, 397.
121. Silverstein, R., Voet, J., Reed, D., and Abeles, R. H. (1967). *J. Biol. Chem.* **242**, 1338.
122. Chambert, R., Treboul, G., and Dedonder, R. (1974). *Eur. J. Biochem.* **41**, 285.

123. Mooser, G., and Iwaoka, K. (1989). *Biochemistry* **28**, 443.
124. Shifrin, S., and Hunn, G. (1969). *Arch. Biochem. Biophys.* **130**, 530.
125. van der Groen, G., Wouters-Leysen, J., Yde, M., and De Bruyne, C. K. (1973). *Eur. J. Biochem.* **38**, 122.
126. Viratelle, O. M., Tenu, J.-P., Garner, J., and Yon, J. (1969). *Biochem. Biophys. Res. Commun.* **37**, 1036.
127. Sinnott, M. L., and Viratelle, O. M. (1973). *Biochem. J.* **133**, 81.
128. Sinnott, M. L., and Withers, O. M. (1974). *Biochem. J.* **143**, 751.
129. Stokes, T. M., and Wilson, I. B. (1972). *Biochemistry* **11**, 1061.
130. Fink, A. L., and Angelides, K. J. (1975). *Biochem. Biophys. Res. Commun.* **64**, 701.
131. Deschavanne, P. J., Viratelle, O. M., and Yon, J. M. (1978). *Proc. Natl. Acad. Sci. U.S.A.* **75**, 1892.
132. Viratelle, O. M., and Yon, J. (1973). *Eur. J. Biochem.* **33**, 110.
133. Sinnott, M. L., and Souchard, I. J. L. (1973). *Biochem. J.* **133**, 89.
134. Richards, J. H. (1970). "The Enzymes," 2nd Ed., Vol. 2, p. 330.
135. Berg, U., Gallo, R., and Metzger, J. (1976). *J. Org. Chem.* **41**, 2621.
136. Cupples, C. G., Miller, J. H., and Huber, R. E. (1990). *J. Biol. Chem.* **265**, 5512.
137. Ring, M., Bader, D. E., and Huber, R. E. (1988). *Biochem. Biophys. Res. Commun.* **152**, 1050.
138. Edwards, R. A., Cupples, C. G., and Huber, R. E. (1990). *Biochem. Biophys. Res. Commun.* **171**, 33.
139. Bader, D. E., Ring, M., and Huber, R. E. (1988). *Biochem. Biophys. Res. Commun.* **153**, 301.
140. Cupples, C. G., and Miller, J. H. (1988). *Genetics* **120**, 637.
141. Malcolm, B. A., Rosenberg, S., Corey, M. J., Allen, J. S., de Baetselier, A., and Kirsch, J. F. (1990). *Proc. Natl. Acad. Sci. U.S.A.* **86**, 133.
142. Reddy, V. A., and Maley, F. (1990). *J. Biol. Chem.* **265**, 10817.
143. Naider, F., Bohak, Z., and Yariv, J. (1972). *Biochemistry* **11**, 3202.
144. Fowler, A. V., Zabin, I., Sinnott, M. L., and Smith, P. J. (1978). *J. Biol. Chem.* **253**, 5283.
145. Ring, M., Armitage, I. M., and Huber, R. E. (1985). *Biochem. Biophys. Res. Commun.* **131**, 675.
146. Hassid, W. Z. (1967). *In* "Metabolic Pathways," (D. M. Greenberg, ed.), 3rd Ed., Vol. 1, p. 307. Academic Press, New York.
147. Voet, J., and Abeles, R. H. (1970). *J. Biol. Chem.* **245**, 1020.
148. Cohn, M. (1961). "The Enzymes," 2nd Ed., Vol. 5, p. 179.
149. Doudoroff, M. (1961). "The Enzymes," 2nd Ed., Vol. 5, p. 292.
150. Glaser, L. (1964). *In* "Comprehensive Biochemistry" (M. Florkin and E. H. Stotz, eds.), Vol. 15, p. 93. Elsevier, New York.
151. Gold, A. M., and Osber, M. P. (1971). *Biochem. Biophys. Res. Commun.* **42**, 469.
152. Doudoroff, M., Barker, H. A., and Hassid, W. Z. (1947). *J. Biol. Chem.* **168**, 725.
153. Wolochow, H., Putman, E. W., Doudoroff, M., Hassid, W. Z., and Barker, H. A. (1949). *J. Biol. Chem.* **180**, 1237.
154. Dedonder, R. (1966). *In* "Methods in Enzymology" (E. F. Neufeld and V. Ginsburg, eds.), Vol. 8, p. 500. Academic Press, New York.
155. Tanaka, T., Oi, S., and Yamamoto, T. (1979). *J. Biochem. (Tokyo)* **85**, 287.
156. Dedonder, R. (1972). *In* "Biochemistry of the Glycosidic Linkage" (R. Piras and H. G. Pontis, eds.), Vol. 2, p. 21. Academic Press, New York.
157. Hestrin, S., Feingold, D. S., and Avigad, G. (1955). *J. Am. Chem. Soc.* **77**, 6710.
158. Tanaka, T., Yamamoto, S., Oi, S., and Yamamoto, T. (1981). *J. Biochem. (Tokyo)* **90**, 521.
159. Defour, A. (1981). "Approche Tactique pour une Analyse par Spectroscopie de Masse de la Structure Primaire des Proteines: Application à la Levanesaccharase de *B. subtilis*. " Thèse de Doctorat d'Etat. Université Paris VII, Paris.

160. Steinmetz, M., Le Coq, D., Aymmerich, A., Gonzy-Treboul, G., and Gay, P. (1985). *Mol. Gen. Genet.* **200**, 220.
161. LeBrun, E., and van Rapenbusch, R. (1980). *J. Biol. Chem.* **255**, 12034.
162. Chambert, R., and Gonzy-Treboul, G. (1976). *Eur. J. Biochem.* **62**, 55.
163. Chambert, R., and Gonzy-Treboul, G. (1976). *Eur. J. Biochem.* **71**, 493.
164. Hehre, E. J. (1941). *Science* **93**, 237.
165. Ebert, K. H., and Schenk, G. (1968). *Adv. Enzymol.* **30**, 179.
166. Hehre, E. J. (1951). *Adv. Enzymol.* **11**, 297.
167. Carlsson, J., and Elander, B. (1973). *Caries Res.* **7**, 89.
168. Walker, G. J., Morrey-Jones, J. G., Svensson, S., and Taylor, C. (1983). *In* "Glucosyltransferases, Glucans, Sucrose and Dental Caries" (R. J. Doyle and J. E. Ciardi, eds.), p. 179. IRL Press, Washington, D.C.
169. Walker, G. J., Brown, R. A., and Taylor, C. (1984). *J. Dent. Res.* **63**, 397.
170. Hamada, A., and Slade, H. D. (1980). *Microbiol. Rev.* **44**, 331.
171. Hamada, S., Koga, T., and Ooshima, T. (1984). *J. Dent. Res.* **63**, 401.
172. Loesche, W. J. (1986). *Microbiol. Rev.* **50**, 353.
173. Curtiss III, R. (1986). *J. Dent. Res.* **65**, 1034.
174. Kuramitsu, H. K. (1987). *Eur. J. Epidemiol.* **3**, 257.
175. Linzer, R., Reddy, M., and Levine, M. (1986). *In* "Molecular Microbiology and Immunobiology of *Streptococcus mutans*" (S. Hamada, S. M. Michalek, H. Kiyono, L. Menaker, and J. R. McGhee, eds.), p. 29. Elsevier, New York).
176. Mukasa, H. (1986). *In* "Molecular Microbiology and Immunobiology of *Streptococcus mutans*" (S. Hamada, S. M. Michalek, H. Kiyono, L. Menaker, and J. R. McGee, eds.), p. 121. Elsevier, New York.
177. Walker, G. J., and Jacques, N. A. (1987). *In* "Sugar Transport and Metabolism in Gram-Positive Bacteria" (J. Reizer and A. Peterkofsky, eds.), p. 39. Wiley, New York.
178. Gibbons, R. J., and Banghart, S. B. (1967). *Arch. Oral Biol.* **12**, 11.
179. Guggenheim, B., and Schroeder, H. E. (1967). *Helv. Odontol. Acta.* **11**, 131.
180. Gibbons, R. J. (1968). *Caries Res.* **2**, 164.
181. Montville, T. J., Cooney, C. L., and Sinskey, A. J. (1978). *Adv. Appl. Microbiol.* **24**, 55.
182. Parker, R. B., and Creamer, H. R. (1971). *Arch. Oral Biol.* **16**, 855.
183. Dewar, M. D., and Walker, G. J. (1975). *Caries Res.* **9**, 21.
184. Walker, G. J., Pulkownik, A., and Morrey-Jones, J. G. (1981). *J. Bacteriol.* **127**, 201.
185. Ciardi, J. E. (1983). *In* "Glucosyltransferases, Glucans, Sucrose and Dental Caries" (R. J. Doyle and J. E. Ciardi, eds.), p. 51. IRL Press, Washington, D.C.
186. Shimamura, A., Tsumori, H., and Mukasa, H. (1983). *FEBS Lett.* **157**, 79.
187. Takehara, T., Hanada, N., and Saeki, E. (1984). *Microbios Lett.* **27**, 113.
188. Niven, C. F., Jr., Smiley, K. L., and Sherman, J. M. (1941). *J. Biol. Chem.* **140**, 105.
189. Garszczynski, S. M., and Edwards, J. R. (1973). *Arch. Oral Biol.* **18**, 239.
190. Rosell, K.-G., and Birkhed, D. (1974). *Acta Chem. Scand. Ser. A* **28**, 589.
191. Ebisu, S., Kato, K., Kotani, S., and Misake, A. (1975). *J. Biochem. (Tokyo)* **78**, 879.
192. Birkhed, D., Rosell, K.-G., and Granath, K. (1979). *Arch. Oral Biol.* **24**, 53.
193. Shimamura, A., Tsuboi, K., Nagase, T., Ito, M., Tsumori, H., and Mukasa, H. (1987). *Carbohydr. Res.* **165**, 150.
194. Hehre, E. J., and Suzuki, H. (1966). *Arch. Biochem. Biophys.* **113**, 675.
195. Mazza, J. C., Akgerman, A., and Edwards, J. R. (1975). *Carbohydr. Res.* **40**, 402.
196. Genghof, D. S., and Hehre, E. J. (1972). *Proc. Soc. Exp. Biol.* **140**, 1298.
197. Figures, W. R., and Edwards, J. R. (1976). *Carbohydr. Res.* **48**, 245.
198. Binder, T. P., and Robyt, J. F. (1983). *Carbohydr. Res.* **124**, 287.
199. Walker, G. J. (1978). *Int. Rev. Biochem.* **16**, 75.

200. Robyt, J. F., and Eklund, S. H. (1982). *Bioorg. Chem.* **11,** 115.
201. Robyt, J. F., and Eklund, S. H. (1983). *Carbohydr. Res.* **121,** 279.
202. Kuramitsu, H. K. (1975). *Infect. Immun.* **12,** 738.
203. Scales, W. R., Long, L. W., and Edwards, J. R. (1975). *Carbohydr. Res.* **42,** 325.
204. Schachtele, C. F., Harlander, S. K., and Germaine, G. R. (1976). *Infect. Immun.* **13,** 1976.
205. Luzio, G. A., Grahame, D. A., and Mayer, R. M. (1983). *Arch. Biochem. Biophys.* **216,** 751.
206. Smith, D. J., Taubman, M. A., and Ebersole, J. L. (1979). *Infect. Immun.* **23,** 446.
207. Hestrin, S. (1961). *In* "Biological Structure and Function" (T. W. Goodwin and O. Lindberg, eds.), Vol. 1, p. 315. Academic Press, New York.
208. Ditson, S. L., Sung, S. M., and Mayer, R. M. (1986). *Arch. Biochem. Biophys.* **249,** 53.
209. Binder, T. P., Cote, G. L., and Robyt, J. F. (1983). *Carbohydr. Res.* **124,** 275.
210. Ferretti, J. J., Gilpin, M. L., and Russell, R. R. B. (1987). *J. Bacteriol.* **169,** 4271.
211. Mooser, G., and Wong, C. (1988). *Infect. Immun.* **56,** 880.
212. Wong, C., Hefta, S. A., Paxton, R. J., Shively, J. E., and Mooser, G. (1990). *Infect. Immun.* **58,** 2165.
213. Mooser, G., Shur, D., Lyou, M., and Watanabe, C. (1985). *J. Biol. Chem.* **260,** 6907.
214. Janin, J., and Wodak, S. J. (1983). *Prog. Biochem. Biophys. Mol. Biol.* **42,** 21.
215. Koshland, D. E., Jr. (1967). *Annu. Rev. Biochem.* **37,** 359.
216. Jeanes, A. (1977). *In* "Extracellular Microbial Polysaccharides" (P. A. Sandford and A. Laskin, eds.), p. 284. American Chemical Society, Washington, D. C.
217. Tsumori, H., Shimamura, A., and Mukasa, H. (1985). *J. Gen. Microbiol.* **131,** 3347.
218. McCabe, M. M. (1985). *Infect. Immun.* **50,** 771.
219. Janson, D. A., and Cleland, W. W. (1974). *J. Biol. Chem.* **249,** 2567.
220. Webb, B. C., Todhunter, J. A., and Purich, D. L. (1976). *Arch. Biochem. Biophys.* **173,** 282.
221. Smith, E., and Morrison, J. F. (1969). *J. Biol. Chem.* **244,** 4224.
222. Jomain-Baum, M., and Schramm, V. L. (1978). *J. Biol. Chem.* **253,** 3648.
223. Fu, D., and Robyt, J. F. (1988). *Carbohydr. Res.* **183,** 97.
224. Fink, A. L. (1977). *Acc. Chem. Res.* **10,** 233.
225. Parnaik, V. K., Luzio, G. A., Grahame, D. A., Ditson, S. L., and Mayer, R. M. (1983). *Carbohydr. Res.* **121,** 257.
226. Mayer, R. M. (1987). *In* "Methods in Enzymology" (V. Ginsburg, ed.), Vol. 138, p. 649. Academic Press, New York.
227. Luzio, G. A., Parnaik, V. K., and Mayer, R. M. (1983). *Carbohydr. Res.* **121,** 269.
228. DeToma, F., and Abeles, R. H. (1970). *Fed. Proc., Fed. Am. Soc. Exp. Biol.* **29,** 461.
229. Brown, A., and Bruce, T. C. (1973). *J. Am. Chem. Soc.* **95,** 1593.
230. Fife, T. H., and De, N. C. (1974). *J. Am. Chem. Soc.* **96,** 6158.
231. Mooser, G., Hefta, S. A., and Paxton, R. J., Shively, J. E., and Lee, T. D. (1991). *J. Biol. Chem.* **266,** 8916.
232. Koga, T., and Inoue, M., (1981). *Carbohydr. Res.* **93,** 125.
233. Reddy, V. A., and Maley, F. (1990). *J. Biol. Chem.* **265,** 10817.
234. Taussig, R., and Carlson, M. (1983). *Nucleic Acids Res.* **11,** 1943.
235. Sinnott, M. L., and Smith, P. J. (1978). *Biochem. J.* **175,** 525.
236. Marshall, P. J., Sinnott, M. L., Smith, P. J., and Widdows, D. (1981). *J. Chem. Soc., Perkin Trans. 1* p. 366.
237. Halazy, S., Danzin, C., Ehrhard, A., and Gerhart, F. (1989). *J. Am. Chem. Soc.* **111,** 3484.
238. Caron, G., and Withers, S. G. (1989). *Biochem. Biophys. Res. Commun.* **163,** 495.
239. Tong, M. K., and Ganem, B. (1988). *J. Am. Chem. Soc.* **110,** 312.
240. Withers, S. G., Street, I. P., Bird, P., and Dolphin, D. H. (1987). *J. Am. Chem. Soc.* **109,** 7530.
241. Parsons, S. M., and Raftery, M. A. (1969). *Biochemistry* **8,** 4199.
242. Yamada, H., Imoto, T., and Noshita, S. (1982). *Biochemistry* **21,** 2187.

243. Eshdat, Y., McKelvy, J. F., and Sharon, N. (1973). *J. Biol. Chem.* **248**, 5892.
244. Eshdat, Y., Dunn, A., and Sharon, N. (1974). *Proc. Natl. Acad. Sci. U.S.A.* **71**, 1658.
245. Kuroki, R., Yamada, H., Moriyama, T., and Imoto, T. (1986). *J. Biol. Chem.* **261**, 13571.
246. Quaroni, A., and Semenza, G. (1976). *J. Biol. Chem.* **251**, 3250.
247. Legler, G., and Harder, A. (1978). *Biochim. Biophys. Acta* **524**, 102.
248. Legler, G., and Bieberich, E. (1988). *Arch. Biochem. Biophys.* **260**, 437.
249. Nitta, Y., Isoda, Y., Toda, H., and Sakiyama, F. (1989). *J. Biochem. (Tokyo)* **105**, 573.
250. Svensson, B. (1988). *FEBS Lett.* **230**, 72.
251. Svensson, B., Clarke, A. J., Svendsen, I., and Møller, H. (1990). *Eur. J. Biochem* **188**, 29.
252. Kuramitsu, H. K., and Mooser, G. (1990). Unpublished.
253. Wakarchuk, W. W., Greenberg, N. M., Kilburn, D. G., Miller, Jr., R. C., and Warren, R. A. J. (1988). *J. Bacteriol.* **170**, 301.
254. Yaguchi, M., Roy, C., Rollin, C. F., Paice, M. G., and Jurasek, L. (1983). *Biochem. Biophys. Res. Commun.* **116**, 408.
255. Shiroza, T., and Kuramitsu, H. K. (1988). *J. Bacteriol.* **170**, 810.
256. Canfield, R. (1963). *J. Biol. Chem.* **238**, 2598.
257. Hunziker, W., Spiess, M., Semenza, G., and Lodish, H. F. (1986). *Cell (Cambridge, Mass.)* **46**, 227.
258. Shiroza, T., Ueda, S., and Kuramitsu, H. K., (1987). *J. Bacteriol.* **169**, 4263.
259. Hondo, A., Kato, C., and Kuramitsu, H. K. (1990). *J. Gen. Microbiol.* **136**, 2099.
260. Gilmore, K. S., Russell, R. R. B., and Ferretti, J. J. (1990). *Infect. Immun.* **58**, 2452.
261. Toda, H., Hondo, K., and Narita, K. (1982). *Proc. Jpn. Acad.* **55B**, 208.
262. Matsuura, Y., Kusunoki, M., Harada, W., and Kakudo, M. (1984). *J. Biochem. (Tokyo)* **95**, 697.
263. Kluh, I. (1981). *FEBS Lett.* **136**, 231.
264. Buisson, G., Duée, E., Haser, R., and Payan, F. (1987). *EMBO J.* **6**, 3909.
265. Rogers, J. C. (1985). *J. Biol. Chem.* **260**, 3731.
266. Melasniemi, H., Paloheimo, M., and Hemiö, L. (1990). *J. Gen. Microbiol.* **136**, 447.
267. Amemura, A., Chakraborty, R., Fujita, M., Noumi, T., and Futai, M. (1988). *J. Biol. Chem.* **263**, 9271.
268. Hoefsloot, L. H., Hoogeveen-Westerveld, M., Kroos, M. A., van Beeumen, J., Reuser, A. J., and Oostra, B. A. (1988). *EMBO J.* **7**, 1697.
269. Hong, S. H., and Marmur, J. (1986). *Gene* **41**, 75.
270. Binder, F., Huber, O., and Böck, A. (1986). *Gene* **47**, 269.
271. Baecker, P. A., Greenberg, E., and Preiss, J. (1986). *J. Biol. Chem.* **261**, 8738.
272. Boel, E., Hjort, I., Svensson, B., Norris, F., Norris, K. E., and Fiil, N. P. (1984). *EMBO J.* **3**, 1097.
273. Mikami, B., Morita, Y., and Fukazawa, C. (1988). *Seikagaku* **60**, 211.
274. Toda, H., Nitta, Y., Isoda, Y., Kim, J. P., and Sakiyama, F. (1988), see Ref. 249.
275. Kreis, M., Williamson, M., Buston, B., Pywell, J., Hejgaard, J., and Svendsen, I. (1987). *Eur. J. Biochem.* **169**, 517.
276. Kawazu, T., Nakanishi, Y., Uozumi, N., Sasaki, T., Yamagata, H., Tsukagoshi, N., and Udaka, S. (1987). *J. Bacteriol.* **196**, 1564.
277. Raynal, A., Gerbaud, C., Francingues, M. C., and Guerineau, M. (1987). *Curr. Genet.* **12**, 175.
278. Kohchi, C., and Toh-e, A. (1986). *Mol. Gen. Genet.* **203**, 89.
279. Martin I., Débargouillé, M., Ferrari, E., Klier, A., and Rapoport, G. (1987). *Mol. Gen. Genet* **208**, 177.
280. Fouet, A., Klier, A., and Rapoport, G. (1986). *Gene* **45**, 221.

281. Gunasekaran, P., Karunakaran, T., Cami, B., Mukundan, A. G., Preziosi, L., and Baratti, J. (1990). *J. Bacteriol.* **172,** 6727.

282. Sato, Y., and Kuramitsu, H. K. (1988). *Infect. Immun.* **56,** 1956.

283. Scholle, R. R., Robb, S. M., Robb, F. T., and Woods, D. R. (1989). *Gene* **80,** 49.

284. Quiocho F. A. (1986). *Annu. Rev. Biochem.* **55,** 287.

285. Fukada, H., Sturtevant, J. M., and Quiocho, F. A. (1983). *J. Biol. Chem.* **258,** 13193.

286. Svensson, B., Gibson, R. M., Haser, R., and Astier, J. P. (1987). *J. Biol. Chem.* **262,** 13682.

287. Matsuura, Y., Kusunoki, M., Date, W., Harada, S., Bando, S., Tanaka, N., and Kakudo, M. (1979). *J. Biochem.* (*Tokyo*) **86,** 1773.

288. Matsuura, Y., Kusunoki, M., Harada, W., Tanaka, N., Iga, Y., Yasuoka, N., Todo, H., Narita, K., and Kakudo, M. (1980). *J. Biochem.* (*Tokyo*) **87,** 1555.

289. Payan, F., Haser, R., Pierrot, M., Frey, M., Astier, J. P., Abadie, B., Duée, E., and Buison, G. (1980). *Acta Crystallogr. Sect. B: Struct. Crystallogr. Cryst. Chem.* **B36,** 416.

290. Hofmann, B. E., Bender, H., and Schulz, G. E. (1989). *J. Mol. Biol.* **209,** 793.

291. Cothia, C. (1988). *Nature* (*London*) **333,** 598.

292. Phillips, D. C., Sternberg, M. J. E., Thornton, J. M., and Wilson, I. A. (1978) *J. Mol. Biol.* **119,** 329.

293. Achari, A., Marshall, S. E., Muirhead, H., Palmieri, R. H., and Noltmann, E. A. (1981). *Philos. Trans. R. Soc. London Ser. B* **293,** 145.

294. Stuart, D. I., Levine, M., Muirhead, H., and Stammers, D. K. (1979). *J. Mol. Biol.* **134,** 109.

6

Catalytic Strategies in Enzymic Carboxylation and Decarboxylation

MARION H. O'LEARY

Department of Biochemistry
University of Nebraska—Lincoln
Lincoln, Nebraska 68583

THE ENZYMES, Vol. XX

I. Introduction

Carbon dioxide stands between the world of organic carbon and the world of inorganic carbon. Living things absorb CO_2 from the atmosphere by photosynthesis and return it to the atmosphere by respiration. The biochemical reactions associated with these processes are the subject of this chapter.

Carboxylic acids are both the first product of CO_2 absorption and the ultimate substrate for respiration; thus, the key reactions can be written as

$$R—H + CO_2 \rightleftharpoons R—CO_2H \qquad (1)$$

representing carboxylation in the forward direction and decarboxylation in the reverse direction. In essence, these reactions represent the interchange of a hydrogen and a carboxyl group in an organic compound.

Although carboxylations and decarboxylations are often discussed separately (in part because different groups of enzymologists work on the two types of reactions), this need not be so. Carboxylation is the microscopic reverse of decarboxylation, and effects that stabilize the transition state for one reaction (and thus accelerate the reaction) will, of necessity, stabilize the transition state for the other. Thus, mechanistic information derived from studies of carboxylation may be useful in understanding decarboxylation, and vice versa.

Our purpose here is to understand how enzymes catalyze carboxylations and decarboxylations. Our first concern is the "chemical mechanism" of the reaction; that is, a description of the intermediates that occur along the reaction path and the chemical transformations that connect these intermediates. Second is the "mechanistic strategy" of the reaction; that is, a description of the underlying catalytic forces and effects used by the enzyme to achieve the rate acceleration and specificity observed. For most carboxylases and decarboxylases, the chemical mechanisms are relatively well understood. On the other hand, mechanistic strategies are much more elusive, depending as they do on a detailed knowledge of enzyme structure and action.

This is not a comprehensive review of all types of carboxylases and decarboxylases; instead, we hope to show that there is a basic unity of mechanistic strategies for carboxylations and decarboxylations that can equally well be applied to cases we do not discuss.

II. Some Background Chemistry

We begin with a discussion of some basic chemical, kinetic, and thermodynamic issues that are common to all the reactions we will consider.

A. ANION INTERMEDIATES

Carboxylation and decarboxylation represent the interchange of a C–H and a C–C bond. This interchange is, so far as is known, never concerted; instead, the reaction [Eq. (2)] involves a negatively charged intermediate.

$$R\text{—}H + CO_2 \rightleftharpoons R^- + H^+ + CO_2 \rightleftharpoons R\text{—}CO_2^- + H^+ \qquad (2)$$

Although such intermediates are seldom observed, their existence is part of the central dogma of enzyme mechanisms, and they serve as a starting point in many discussions of mechanism.

Thus, when we consider catalytic strategies for carboxylations and decarboxylations, we will see some factors that primarily affect the proton transfer step, some that affect the stability of the anion, and some that affect the carboxylation step.

B. CARBON DIOXIDE

The physical and chemical properties of carbon dioxide (1–3) are fundamental to understanding carboxylations and decarboxylations. The contemporary biosphere is in a dynamic steady state in which the concentration of CO_2 in the atmosphere is about 350 ppm and is slowly increasing. At 25°C, water equilibrated with the atmosphere contains 10 μM CO_2 (2). Biological carboxylations must either develop a strategy for dealing with this low concentration, or else they must use HCO_3^- (which is often more abundant) instead.

1. CO_2 and Bicarbonate

Dissolved CO_2 is in equilibrium with carbonic acid, bicarbonate ion, and carbonate ion. Of these, the most important *in vivo* is bicarbonate ion. The effective pK_a for the interconversion of CO_2 and bicarbonate, defined by the equilibrium [Eq. (3)],

$$CO_2 + H_2O \rightleftharpoons HCO_3^- + H^+ \tag{3}$$

is about 6.3 (2). Thus, above pH 6.3, the concentration of HCO_3^- exceeds that of CO_2. Systems operating near pH 8 (e.g., the chloroplasts of green plants) are confronted with a HCO_3^- concentration that exceeds the CO_2 concentration by about a factor of 50. This represents a serious disadvantage in C_3 plants, where carbon uptake by ribulose-bisphosphate carboxylase requires CO_2, rather than HCO_3^-. C_4 plants, which are more efficient, use HCO_3^- instead, and this provides a significant advantage in the initial carbon uptake process.

Carbon dioxide and bicarbonate have very different chemical reactivities. CO_2 is quite reactive toward nucleophiles (3), reacting, for example, with hydroxide to form bicarbonate, with ammonia and primary and secondary amines to form carbamates, with enolates to form carboxylic acids, and with a variety of organometallic compounds (e.g., Grignard reagents and organolithium reagents) to form carboxylic acid salts (3). Bicarbonate is much less reactive.

2. CO_2 and Bicarbonate as Enzyme Substrates

In the absence of carbonic anhydrase (carbonate dehydratase), the interconversion of CO_2 and HCO_3^- is slow, requiring more than a minute at 15°C. Because of this slow interconversion, carboxylations show slightly different kinetics if CO_2 is used than if HCO_3^- is used. Some enzymes take up CO_2 and some take up HCO_3^-; the only enzyme capable of using both CO_2 and HCO_3^- is carbonic anhydrase. Methods for identifying the one-carbon substrate for carboxylations are well established (4), and the answer is known for a variety of enzymes (5). In general, all enzymes use CO_2 except biotin-dependent enzymes and phosphoenolpyruvate carboxylase, which use HCO_3^-.

C. STEREOCHEMISTRY AND CONFORMATION IN CARBOXYLATION AND DECARBOXYLATION

Carboxylation and decarboxylation usually involve interchange of a hydrogen and a carboxyl group. This process can occur with retention of configuration, with inversion of configuration, or with racemization (6–8). All three stereochemical outcomes have important mechanistic implications. Retention may indicate that the same enzyme catalytic group is responsible for proton removal and for carboxyl binding (9). Inversion precludes this possibility. Racemization may indicate that the proton transfer step is not enzyme mediated.

The stereochemical courses of a variety of carboxylations and decarboxylations have been determined (6–8). Similar enzymes usually, though not invariably, show the same stereochemistry, and a classification scheme for decarboxylases based on stereochemistry has been suggested (10).

The geometry of the transition state for the carboxylation/decarboxylation step is an important aspect of mechanism. The anion intermediate [Eq. (2)] is usually a conjugated anion (often an enolate) in which the negative charge lies above and below a planar atomic framework. Attacking or departing CO_2 will approach from above or below the plane of the conjugated system (Scheme I), rather than from within the plane. The distinction between the two faces of the planar system can usually be made based on the stereochemistry of the carboxylated substrate or product.

SCHEME I. Conformation for decarboxylation.

D. THERMODYNAMICS OF CARBOXYLATIONS AND DECARBOXYLATIONS

Like all catalysts, enzymes increase the rate of attainment of equilibria, but they do not change equilibrium constants. The first limitation on carboxylations and decarboxylations is the thermodynamic limitation: certain reactions are favorable and others are unfavorable.

One of the common interconversions in biochemistry is the decarboxylation of oxaloacetate to form pyruvate and CO_2 [Eq. (4)].

$$^-O_2C\text{—}CH_2\text{—}CO\text{—}CO_2^- + H^+ \rightleftharpoons CO_2 + CH_3\text{—}CO\text{—}CO_2^- \qquad (4)$$

The equilibrium lies on the right: $K = 34\ M$ at pH 8 (11, 12). Thus, the decarboxylation of oxaloacetate is quite favorable, whereas the direct carboxylation of pyruvate is unfavorable.

On the other hand, the enolate of pyruvate is often an intermediate in these reactions, and if the equilibrium is considered from the point of view of the enolate [Eq. (5)],

$$^-O_2C\text{—}CH_2\text{—}CO\text{—}CO_2^- \rightleftharpoons CO_2 + {}^-CH_2\text{—}CO\text{—}CO_2^- \qquad (5)$$

the equilibrium constant is about $10^{-7}\ M$, thus favoring carboxylation (13). We will see below how various enzymes take advantage of these equilibria.

E. ENOLATES AND OTHER ANIONS

As we have noted, most carboxylations and decarboxylations occur by way of anionic intermediates, and the stabilities of these anions are important in determining rates of carboxylations and decarboxylations. The simplest conceivable case is formation of acetic acid from methane and CO_2. Methane has a pK_a of approximately 48 (14); consequently, removal of a proton from methane is precluded in nature and acetic acid is not formed in nature by the carboxylation of methane, nor is acetic acid decarboxylated. Instead, other substrates are used wherein the pK_a is more favorable.

The best means of lowering the pK_a is by delocalization of the negative charge by use of an enolate or its functional equivalent. Enolates are the key intermediates in a wide variety of carboxylations and decarboxylations. However, even simple carbonyl compounds have high pK_a values that preclude existence of a significant proportion of enolate near neutral pH. The loss of a hydrogen from the methyl group of pyruvate [Eq. (6)], for example,

$$CH_3—CO—CO_2^- \rightleftharpoons H^+ + {}^-CH_2—CO—CO_2^- \tag{6}$$

has a pK_a near 17 (15), and thus will occur to a negligible degree in aqueous solution except at extremely high pH. Still, this enolate is known to be an intermediate in a variety of enzymic reactions, and it is clear that enzymes must provide additional stabilization for such enolates.

Metal Stabilization of Enolates

In most enzyme systems, enolate intermediates are stabilized by metal ion complexation. Although few good numbers are available, it appears that metal ion complexation of the oxygen of the keto and enol forms can increase the acidity of an adjacent carbon–hydrogen bond by four to six orders of magnitude. For example, complexation with Mg^{2+} lowers the pK_a at C-3 of oxaloacetate from 13 to 9 (16), and a similar shift is seen with pyruvate (15). Enolates of α-keto acids can be effectively stabilized by metal complexation. For example, in the cases of pyruvic acid and oxaloacetic acid, both the keto oxygen and one of the carboxyl oxygens coordinate to the metal (Scheme II).

SCHEME II. Metal complexes of pyruvate and oxaloacetate.

F. CATALYTIC STRATEGIES

Before we begin a detailed discussion of particular enzymes, it is useful to outline the catalytic strategies that we will consider. It is hoped that this represents a relatively complete list of the major catalytic forces in enzymes.

1. *Thermodynamic Effects*

Nature circumvents thermodynamics by coupling unfavorable equilibria to favorable equilibria. This occurs, for example, in the carboxylation of pyruvic acid by malic enzyme, where the oxaloacetate that is formed initially is immediately reduced to malate. The same strategy is used in isocitrate dehydrogenase. In the carboxylation of ribulose bisphosphate, a thermodynamically unfavorable carboxylation is made irreversible by cleavage of the six-carbon product to form two three-carbon products (vide infra).

The unfavorable enolate-forming equilibrium that initiates many carboxylations [Eq. (2)] is sometimes circumvented by generating the enolate from a high-energy compound rather than by direct proton abstraction. For example, in the case of phosphoenolpyruvate carboxylase, the enolate of pyruvate is generated by phosphate transfer from phosphoenolpyruvate to HCO_3^-, forming the enolate anion in an environment where it usually reacts to form product rather than reacting with solvent to form pyruvate. Phosphoenolpyruvate carboxykinase uses a similar strategy.

2. *Stereochemical and Geometric Control*

As we noted above, the transition state for a carboxylation/decarboxylation step ordinarily has a geometry in which the one-carbon fragment is above the plane of the conjugated system (cf. Scheme I). Enzymic control of this conformation can give rise to a modest increase in rate [one to two orders of magnitude (*17*)], and this factor probably operates for most decarboxylases. Equally important, enzymic control can avoid conformations of the substrate that would give rise to undesirable reactions. This is a particularly important factor in decarboxylations that require pyridoxal 5′-phosphate, where transamination would otherwise compete with decarboxylation.

3. *Binding of* CO_2 *to Enzymes*

Little is known about the binding of CO_2 to enzymes. Although carboxylases generally show saturation kinetics with respect to CO_2, this does not require the existence of an enzyme–CO_2 complex of appreciable stability. In fact, it is likely tht CO_2 binds poorly or not at all to most enzymes.

Enzymes that catalyze carboxylations are faced with a dilemma: of the two possible carbon substrates, CO_2 is the more reactive, but its concentration is never high and its ability to bind to enzymes is probably limited. HCO_3^- has the potential for better binding to enzymes, but it is not very reactive. Different enzymes have evolved different strategies for dealing with this problem. The most creative solution is perhaps the enzymes that bind HCO_3^- but then convert it into CO_2 prior to the carboxylation step; this category includes phosphoenolpyruvate carboxylase and the biotin-dependent carboxylases. (*Editor's note:* See Chapter 7 by Kluger for additional discussion of this point.)

4. *Stabilization of Anions*

Most carboxylations and decarboxylations involve anionic intermediates [cf. Eq. (2)]. The stability of the anion is an important factor in determining reaction rates. An unstable anion will be very reactive toward CO_2, but it will be difficult to form. A stable anion will be less reactive toward CO_2, but it will be easy to form. Many carboxylations involve removal of a proton from an acid of pK_a 15–18. Enzymic stabilization of such anions is very important.

In the case of decarboxylations, the negative charge produced in the decarboxylation step may be neutralized by some distal positive charge. This occurs in the case of Schiff base-dependent decarboxylations, thiamin pyrophosphate-dependent decarboxylations, and in a few other cases. Such systems are often referred to as "electron sinks."

It should be noted that we have focused our attention thus far on the anionic intermediates that occur in these reactions, whereas reaction rates depend on structures and energetics of transition states. The connection comes about because factors that stabilize anionic intermediates invariably stabilize adjacent transition states by similar amounts and by similar mechanisms.

5. *Medium Effects on Reaction Rates*

In many decarboxylations, particularly those involving pyridoxal 5'-phosphate and thiamin pyrophosphate, the starting state for the decarboxylation is a zwitterion in which the positive and negative charges are widely separated. In the decarboxylation step, this zwitterion is converted into a neutral intermediate. Such reactions are faster in less polar environments. Some enzymes appear to take advantage of this fact.

6. *Solvation of CO_2 and Carboxyl Group*

The interconversion of carbon dioxide, which is nonpolar and uncharged, and a carboxyl group, which is polar and charged, is very sensitive to solvation. A polar environment will favor carboxylation whereas a nonpolar environment will favor decarboxylation. As a result, solvation effects may change both equilibrium constants and rate constants.

In a decarboxylation, energy is required to desolvate the carboxyl group prior to or concomitant with the decarboxylation step. This desolvation may be accomplished when the substrate binds to the enzyme. In such a case, binding energy is used to overcome some of the kinetic barrier to the reaction (*17*). Evidence for such an effect can be seen in the X-ray structure of histidine decarboxylase and in models for decarboxylations catalyzed by pyridoxal 5'-phosphate and thiamin pyrophosphate-dependent decarboxylases.

III. Interconversion of Pyruvate and Oxaloacetate

It is likely that nature has found more ways to catalyze the carboxylation of pyruvate and its reverse, the decarboxylation of oxaloacetate, than any other reaction. In spite of the fact that the overall equilibrium strongly favors decarboxylation, a variety of enzymes catalyze the synthesis of oxaloacetate from pyruvate. We will examine a number of these enzymes in the following sections.

Interestingly, although mechanistic details vary, it is likely that a metal-chelated enolate of pyruvate is an intermediate in all these reactions and that all transition states for the carboxylation/decarboxylation step are similar.

A.   OXALOACETATE DECARBOXYLASE

The decarboxylation of oxaloacetate is catalyzed by oxaloacetate decarboxylase [Eq. (7)] from a variety of species.

$$^-O_2C—CH_2—CO—CO_2^- + H^+ \rightarrow CO_2 + CH_3—CO—CO_2^- \tag{7}$$

The enzyme requires a divalent metal ion (*18*) but has no other cofactors. The decarboxylation is invariably stereospecific, but some oxaloacetate decarboxylases operate with retention of configuration and some operate with inversion (*19*).

1. *Models*

Oxaloacetate spontaneously decarboxylates in aqueous solution at neutral pH (*20*). The reaction is catalyzed by metal ions (*16, 21*). The immediate product of the decarboxylation is an enol, enolate, or metal-chelated enolate (Scheme III).

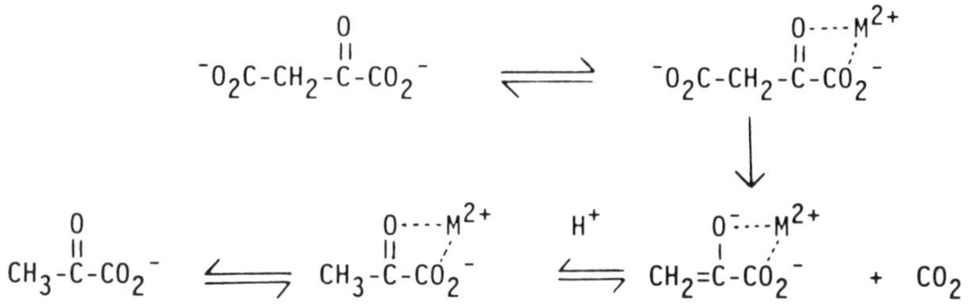

SCHEME III. Oxaloacetate decarboxylation mechanism.

2. Mechanism

The mechanism of the enzymic decarboxylation of oxaloacetate presumably resembles that of the metal ion-catalyzed reaction (Scheme III), in which the enzyme-bound metal chelates to the α-carboxyl and the keto carbonyl of the substrate prior to decarboxylation (cf. Scheme II). Interestingly, an enzyme identified as "oxaloacetate decarboxylase" was later identified as pyruvate kinase (22). The fact that pyruvate kinase complexes pyruvate in a manner similar to that shown in Scheme II (23) lends credence to this mechanism.

The fact that the reaction is stereospecific suggests that a specific enzyme group is responsible for protonation of the enolate product (if the enolate were released and protonated in solution, this would give rise to racemization). The proton-donating group may be lysine, based on the structure of pyruvate kinase.

Heavy-atom isotope effects for the spontaneous and enzymic decarboxylations are large and indicate that in both cases, the decarboxylation step is fully rate determining. The near equality of isotope effects in the two cases suggests that transition states for the two reactions are similar (24, 25).

3. Catalytic Strategy

The principal catalytic factor at work here is the use of the metal ion to stabilize the forming enolate. Enzymic control of the conformation of the bound substrate may also be a factor. The enzymic reaction is faster than the metal ion-catalyzed decarboxylation by only about 10^3 (16), and most of this additional factor is probably due to conformational and environmental control.

B. MALIC ENZYME

Malic enzyme catalyzes the oxidative decarboxylation of malic acid or, in the reverse direction, the reductive carboxylation of pyruvic acid (26) [Eq. (8)].

$$^-O_2C\!-\!CH_2\!-\!CHOH\!-\!CO_2^- + NADP^+ \rightleftharpoons CH_3\!-\!CO\!-\!CO_2^- + CO_2 + NADPH \quad (8)$$

The reaction is readily reversible: $K_{eq} = 0.03\ M$ (12) and different forms of the enzyme use either NAD^+ or $NADP^+$. The enzyme uses CO_2 rather than HCO_3^- (27), and requires a divalent metal ion.

1. Mechanism

Available evidence indicates that malic enzyme operates in two steps: oxidation of malate to form enzyme-bound oxaloacetate, and decarboxylation of this material (27, 28). Malic enzyme will also decarboxylate oxaloacetate (28). Isotope effects indicate that oxidation and decarboxylation are separate steps and that both steps are partially rate determining (29). When the enzyme is supplied with oxaloacetate in the presence of NADPH, the substrate partitions in part toward malate formation and in part toward pyruvate formation (30).

2. *Catalytic Strategy*

The oxidation part of the mechanism seems to be like that catalyzed by malate dehydrogenase. The decarboxylation is, so far as can be told, like the decarboxylation of oxaloacetate by oxaloacetate decarboxylase. Isotope effects indicate that the decarboxylation transition states are similar (*29*).

The important difference between the two enzymes is that malic enzyme can synthesize oxaloacetate from pyruvate and CO_2. The oxaloacetate thus formed is immediately reduced, thus shifting the equilibrium in the required direction.

C. Phosphoenolpyruvate Carboxylase

The direct carboxylation of pyruvate to form oxaloacetate is thermodynamically unfavorable. The enzyme phosphoenolpyruvate (PEP) carboxylase solves this problem by starting instead with phosphoenolpyruvate.

$$CH_2\!=\!\overset{\displaystyle O-PO_3^{2-}}{\underset{\displaystyle |}{C}}-CO_2^- + HCO_3^- \rightarrow {}^-O_2C-CH_2-CO-CO_2^- + P_i^{2-} \tag{9}$$

Reaction (9) is thermodynamically favorable by about 8 kcal/mol (*31*). The enzyme that catalyzes this reaction is found in plants and in a variety of microorganisms, but it is absent from mammalian tissues (*31–35*). It contains a metal ion but has no other cofactors.

1. *Mechanism*

The substrate for the carboxylation is HCO_3^- rather than CO_2 (*4, 36*). When the carboxylation is conducted with $HC^{18}O_3^-$, the P_i produced contains a single atom of ^{18}O (*4, 37*), apparently because the reaction occurs by way of a carboxy phosphate intermediate (Scheme IV). Carboxy phosphate is a known compound, but it is unstable, hydrolyzing in aqueous solution within a few seconds (*38*). Presumably at the active site of the enzyme, carboxy phosphate is protected from hydrolysis. Studies of phosphorus stereochemistry (*39*) and carbon, hydrogen, and oxygen isotope effects (*40, 41*) are consistent with this mechanism. Isotope effect studies indicate that the phosphate transfer step is rate determining (*40*).

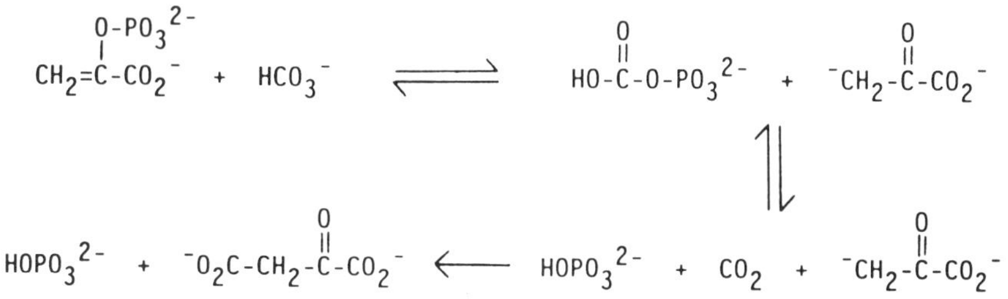

Scheme IV. PEP carboxylase mechanism.

Reaction of carboxy phosphate with the enolate to form oxaloacetate and P_i might occur by either of two mechanisms. The enolate might directly attack the carbonyl carbon of carboxy phosphate (particularly if the carboxyl group remained protonated) and displace the phosphate. Alternatively, carboxy phosphate might decompose spontaneously in the active site of the enzyme to form CO_2 and P_i, and CO_2 could then react with the enolate. Although the former mechanism was accepted for a number of years (33), more recent evidence suggests that the second mechanism is actually correct: when the reaction is conducted with methyl-PEP and $HC^{18}O_3{}^-$, more than a single atom of ^{18}O is incorporated into the P_i produced (42, 43) and substrate recovered after partial reaction has ^{18}O in the nonbridging positions of the phosphate group (43). This isotope scrambling presumably occurs because of the reversible formation of CO_2 and P_i at the active site of the enzyme. Studies with alternate substrates are also consistent with this pathway. With methyl-PEP (44), bromo-PEP (45), fluoro-PEP (45), and chloro-PEP (46), it appears that following the formation of bound CO_2 and P_i, CO_2 dissociates from the enzyme, followed by P_i and the enolate. Protonation of the enolate to form a pyruvate derivative is stereochemically random (43, 44). PEP itself gives about 4% of the product of this pathway (47).

2. *Catalytic Strategy*

PEP carboxylase avoids the thermodynamic problem of enolate formation by beginning with PEP instead of pyruvate. Provided that the enolate is shielded from solvent, this offers a thermodynamically favorable approach to the carboxylation step. PEP carboxylase achieves a high concentration of CO_2 at the active site of the enzyme by starting with $HCO_3{}^-$ and using the phosphate bond energy to dehydrate it. Protection of the active site from solvent is also important in order to ensure that CO_2 is available for reaction with the enolate, rather than dissociating.

The role of the metal ion in the reaction is not known. Analogy with pyruvate kinase (23) suggests that the enolate is likely to be metal coordinated (Scheme II).

D. PHOSPHOENOLPYRUVATE CARBOXYKINASE

PEP carboxykinase also catalyzes the formation of oxaloacetate from PEP [Eq. (10)] (32, 48).

$$
\overset{\displaystyle O-PO_3{}^{2-}}{\underset{\displaystyle CH_2=C-CO_2{}^-}{|}} + NDP + CO_2 \rightleftharpoons {}^-O_2C-CH_2-CO-CO_2{}^- + NTP \tag{10}
$$

In this case, the phosphate is transferred to a nucleotide diphosphate (NDP), forming a nucleotide triphosphate (NTP). Unlike the case of PEP carboxylase, this reaction is readily reversible [K_{eq} = 0.32 M^{-1} (49)]. The one-carbon substrate is CO_2 rather than HCO_3^- (49). The enzyme requires metal ions both to coordinate the nucleotide and (presumably) to bind the enolate (48).

1. *Mechanism*

Like PEP carboxylase, this enzyme begins by transferring a phosphate (in this case to a nucleotide diphosphate) to form an enolate. The enolate then reacts with CO_2.

This is one of the few enzymes for which there is evidence for the existence of an enzyme–CO_2 complex. Carbon isotope effects vary with substrate concentration and suggest that the enzyme reacts by a random kinetic mechanism in which CO_2 may bind to the enzyme prior to the binding of PEP (50). Magnetic resonance studies are also consistent with the existence of an enzyme–metal–CO_2 complex (48, 51).

2. *Catalytic Strategy*

Like PEP carboxylase, PEP carboxykinase forms an enolate by phosphate transfer from PEP, thus eliminating the thermodynamically unfavorable proton removal step. Like PEP carboxylase, the enolate then reacts with CO_2. The enzymes differ in the means that are used for delivering CO_2 to the active site; PEP carboxylase has an active mechanism for delivery, whereas PEP carboxykinase uses the second substrate to trap CO_2 on the enzyme. Both enzymes may use similar metal ion complexation for stabilizing the enolate intermediate.

E. BIOTIN-DEPENDENT ENZYMES

A number of carboxylases function by means of covalently bound biotin (52–57). These enzymes invariably use HCO_3^- rather than CO_2 as the substrate, and a metal ion is also involved.

The prototypical biotin enzyme is pyruvate carboxylase, which catalyzes reaction (11).

$$CH_3-CO-CO_2^- + ATP + HCO_3^- \rightleftharpoons {}^-O_2C-CH_2-CO-CO_2^- + P_i + ADP \quad (11)$$

1. *Models*

It has long been clear that a carboxylated form of biotin is an intermediate in the reaction, and the key to understanding the chemistry of biotin is the structure of this carboxylated intermediate (57). Although a variety of structures have been considered, it is now clear that the N-carboxylated compound is the actual intermediate, and biotin phosphates are not important in the catalytic mechanism.

2. Mechanism

Mechanisms of biotin-dependent reactions invariably involve two separate sequences and two separate enzyme sites, with covalently bound biotin or carboxybiotin moving on a "swinging arm" from one site to the other.

The first sequence begins with the reaction of ATP with HCO_3^- to form carboxy phosphate [Eq. (12)] in a manner reminiscent of PEP carboxylase.

$$ATP + HCO_3^- \rightleftharpoons ADP + {}^-O_2C—O—PO_3^{2-} \qquad (12)$$

As in the case of PEP carboxylase, one of the important pieces of evidence favoring this mechanism is the transfer of ^{18}O from HCO_3^- to P_i (58).

In the second step, carboxy phosphate reacts with biotin to form N-carboxybiotin, which is the final product of the first sequence [Eq. (13)].

$$^-O_2C—O—PO_3^{2-} + E–biotin \rightleftharpoons P_i + E–biotin—CO_2^- \qquad (13)$$

The detailed mechanism of this step is controversial (*Editor's note:* See Chapter 7 by Kluger for additional discussion of this point). The pK_a of the nitrogen atom of biotin is about 17 (59), and the nitrogen is not particularly reactive unless it is deprotonated. Although the simplest mechanism might involve attack of biotin anion on carboxy phosphate, this appears not to be the case. Knowles (57) argues convincingly that the problem is actually solved in a very parsimonious way reminiscent of that described above for PEP carboxylase, with no intervention of external acids, bases, or nucleophiles. Although compelling evidence is still lacking, a large body of circumstantial evidence suggests that within the enzyme–biotin–carboxy phosphate complex, carboxy phosphate decomposes to CO_2 and P_i, after which the inorganic phosphate trianion acts as a base, removing a proton from the –NH of biotin, simultaneously increasing the reactivity of the nitrogen and making possible N-carboxylation.

After completion of the first sequence, carboxybiotin moves to the second site, where the carboxyl group is transferred from biotin to pyruvate, forming oxaloacetate. In essence, a proton and a carboxyl group trade places in this step. Isotope effects indicate that proton removal from pyruvate is not concerted with carboxylation (60, 61). The lack of positional isotope exchange (62) during this process presumably is because the active complex is isolated from solvent, rather than because of a lack of exchangeable sites.

The simplest mechanism for this step is analogous to that given above for formation of carboxybiotin: carboxybiotin initially decarboxylates to give CO_2 and a nitrogen anion; this anion then removes a proton from the methyl group of pyruvate and the resulting enolate reacts with CO_2.

The role of the metal ion in this reaction is unknown. By analogy with other pyruvate-dependent enzymes, the metal may be used to stabilize the enolate. Alternatively, the role of the metal may be to stabilize the biotin anion by complexation to oxygen.

3. *Catalytic Strategy*

Biotin-dependent carboxylases use HCO_3^-, which enables them to operate more efficiently with a one-carbon substrate than would be the case if they operated with CO_2. Apparently these enzymes use the energy derived from ATP hydrolysis to dehydrate HCO_3^- at a site well insulated from solution. The decarboxylation of carboxy phosphate at a protected site on the enzyme is used to generate a strong base that then removes a proton from biotin, forming a highly reactive biotin anion. This combination enables the enzyme to deliver a high concentration of CO_2 at the active site, as originally suggested by Jencks (*17*), and to generate the enolate under conditions where it is needed.

IV. *β*-Keto Acid Decarboxylations Not Involving a Metal Ion

The enzymes described above that convert oxaloacetate to pyruvate and CO_2 appear to use metal chelation to stabilize the enolate formed by decarboxylation. Many other β-ketoacid decarboxylases use a similar mechanism. However, there are a few decarboxylations of β-keto acids or their functional equivalents in which no metal ion is involved. One is the case of acetoacetate decarboxylase, which functions by means of a Schiff base mechanism. A few additional examples are described below. All these cases involve particularly stable enolates.

A. PREPHENATE DEHYDROGENASE

A variety of enzymes catalyze the oxidative decarboxylation of β-hydroxy acids. Isotope effect studies of malic enzyme (*29*), isocitrate dehydrogenase (*63*), and 6-phosphogluconate dehydrogenase (*64*) indicate that all three of these oxidative decarboxylations occur by stepwise mechanisms in which hydride transfer occurs first, forming a β-keto acid that then undergoes decarboxylation. Hydride transfer and decarboxylation are both partially rate determining.

The oxidative decarboxylation of prephenic acid to form *p*-hydroxyphenyl pyruvate is similar, in that it is coupled to an NAD^+-linked oxidation (Scheme V). However, there are several important differences. In the first place, this enzyme

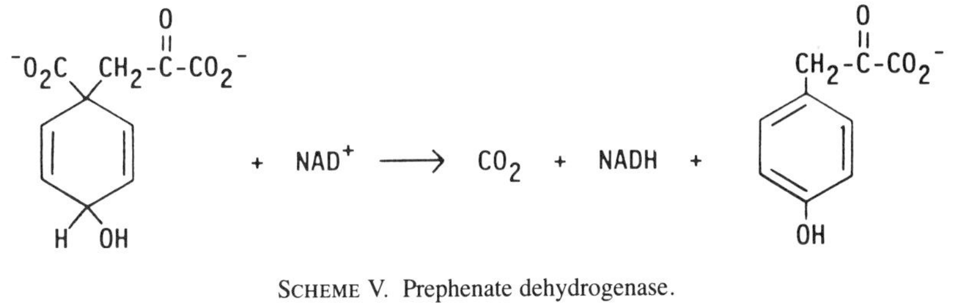

SCHEME V. Prephenate dehydrogenase.

does not require a metal ion (*65*). Second, the reaction catalyzed by prephenate dehydrogenase is totally irreversible, unlike other oxidative decarboxylations we have considered. Third, isotope effect studies indicate that hydride transfer and decarboxylation occur in the same step, with no keto acid intermediate (*65*).

From the point of view of catalytic strategy, all three of these facts are probably connected. The product of the reaction is a phenol, rather than an enolate, and metal ion stabilization of the product is apparently not needed. The driving force associated with formation of an aromatic compound is evidently sufficient that decarboxylation can be concerted with hydride transfer. The carbon isotope effect on this reaction is surprisingly small, perhaps because the transition state is quite early (*65*). The isotope effects also indicate that substrate binding is associated with a conformation change, which may seat the substrate in a reactive conformation in the active site.

B. 6-Phosphogluconate Dehydrogenase

The oxidative decarboxylation of 6-phosphogluconic acid occurs in a stepwise mechanism involving a β-keto acid intermediate (*64*). However, it appears that no metal ion is required in the reaction (*66*). 2-Deoxy-6-phosphogluconate is also a substrate for this enzyme, but the reaction is about two orders of magnitude slower than reaction of the natural substrate. The keto acid produced in the initial oxidation is released into solution and subsequently undergoes slow enzyme-catalyzed decarboxylation (*67*).

C. UDPglucuronate Decarboxylase

This enzyme also catalyzes a decarboxylation via a β-keto acid intermediate, and it appears that no metal is involved (*68, 69*).

D. Acetolactate Decarboxylase

Like 6-phosphogluconate dehydrogenase and UDPglucuronate decarboxylase, this enzyme decarboxylates a β-keto acid with an α-oxygen function without the involvement of a metal ion or other cofactor (*70*). Apparently the electronic properties of this oxygen function are sufficient to stabilize the enolate anion intermediate, so that no metal ion is needed.

V. Ribulose-1,5-bisphosphate Carboxylase

Ribulose-1,5-bisphosphate carboxylase/oxygenase (RuBP carboxylase) catalyzes the carboxylation of ribulose bisphosphate to give 3-phosphoglyceric acid [Eq. (14)] (*71, 72*).

$$
\begin{array}{l}
\text{CH}_2\text{—OPO}_3{}^{2-} \\
|\\
\text{C}=\!\text{O} \\
|\\
\text{HC—OH} \\
|\\
\text{HC—OH} \\
|\\
\text{CH}_2\text{—OPO}_3{}^{2-}
\end{array}
\quad + \text{CO}_2 + \text{H}_2\text{O} \rightarrow 2
\begin{array}{l}
\text{CO}_2{}^{-} \\
|\\
\text{HC—OH} \\
|\\
\text{CH}_2\text{—OPO}_3{}^{2-}
\end{array}
+ \quad 2\text{H}^+
\qquad (14)
$$

RuBP carboxylase is the central CO_2-fixing enzyme in all plants; thus, it is the most abundant enzyme in the biosphere (*71, 72*). The enzyme requires a divalent metal ion but has no other cofactors. CO_2 is the substrate, rather than $HCO_3{}^-$.

Practically every kind of approach in enzymology has been applied to this enzyme, including kinetics, chemical modification, X-ray crystallography, site-directed mutagenesis, and isotope effects (*71, 72*). Recent studies by X-ray crystallography (*73–78*) will particularly occupy our attention, as this is the only carboxylase for which an X-ray crystal structure has been reported.

A. UNIQUE FEATURES

RuBP carboxylase has a number of unusual features. First, the enzyme must be activated by reaction with CO_2 and metal prior to catalysis. Activation involves reaction of CO_2 with an ε-amino group of a lysine residue (Lys-191 in the enzyme from *Rhodospirillum rubrum*, Lys-201 in that from spinach) to form a carbamate that is then coordinated to the metal (*75*). This bound metal then forms part of the substrate-binding site.

Second, the enzyme also catalyzes the oxygenation of ribulose bisphosphate, a reaction in which O_2 serves as an alternate substrate and is competitive against CO_2 [Eq. (15)].

$$
\begin{array}{l}
\text{CH}_2\text{—OPO}_3{}^{2-} \\
|\\
\text{C}=\!\text{O} \\
|\\
\text{HC—OH} \\
|\\
\text{HC—OH} \\
|\\
\text{CH}_2\text{—OPO}_3{}^{2-}
\end{array}
\quad + \text{O}_2 \rightarrow
\begin{array}{l}
\text{CO}_2{}^{-} \\
|\\
\text{HC—OH} \\
|\\
\text{CH}_2\text{—OPO}_3{}^{2-}
\end{array}
+
\begin{array}{l}
\text{CO}_2{}^{-} \\
|\\
\text{CH}_2\text{—OPO}_3{}^{2-}
\end{array}
+ 2\text{H}^+
\qquad (15)
$$

The relative rates of carboxylation and oxygenation vary with enzyme source (*78, 79*) and metal ion.

Third, the higher plant enzyme is composed of eight copies each of a large subunit and a small subunit. The small subunit is absent in the enzyme from certain organisms (e.g., *R. rubrum*), and its function is unknown.

B. MECHANISM

The mechanism originally suggested by Calvin (*80*) (Scheme VI) still provides the best framework for understanding the overall reaction. The first step is removal of a proton at C-3 to form an enol or enolate intermediate. Proton transfer between two oxygens leads to the formation of the enolate at C-2, which reacts with CO_2 (or O_2, in the case of the oxygenation reaction). The "six-carbon intermediate" then reacts with water to give two molecules of 3-phosphoglyceric acid.

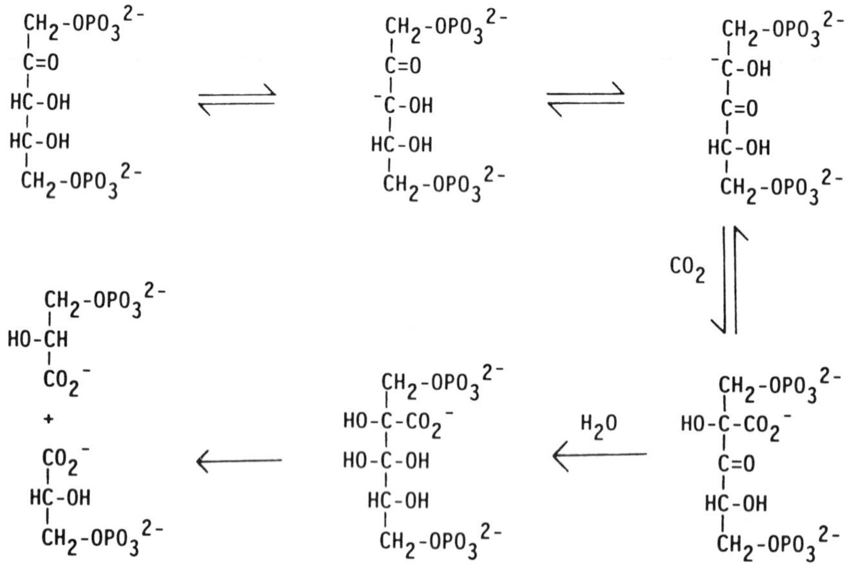

SCHEME VI. RuBP carboxylase mechanism.

C. EVIDENCE FROM X-RAY ANALYSIS

X-Ray crystal structures have been reported both for the enzyme from spinach (*74, 76*) and that from *R. rubrum* (*73, 75, 77*). A structure has also been reported for the enzyme with 2-carboxyarabinitol bisphosphate, an analog of the six-carbon intermediate, bound to the active site (*75*), as well as for the enzyme with 3-phosphoglycerate bound to the active site (*77*).

RuBP binds to the enzyme in an extended conformation, and both phosphate groups have extensive interactions with polar groups of the enzyme (*75*). The metal ion is coordinated to the carbamate and is also bound to the oxygen at C-2 of RuBP (*75*).

The reaction is initiated by hydrogen abstraction from C-3, forming an eno-late, which is presumably stabilized by metal coordination. The enzyme shows a bell-shaped pH–rate profile, with a pK of 7.1 (V) or 7.5 (V/K) on the acid side (*81*), and this group is believed to be the proton-abstracting group. Kinetic (*81*) and site-directed mutagenesis (*82, 83*) studies indicate that this group is probably Lys-166 (in the *R. rubrum* sequence; Lys-175 in the spinach sequence), and X-ray studies (*75*) are consistent with this conclusion.

In the six-carbon intermediate, both the oxygen at C-2 and the newly formed carboxyl group appear to be coordinated to the metal. Cleavage of this interme-diate presumably occurs by attack of external water.

D. KINETICS

Extensive kinetic and isotope effect studies have also provided important in-formation about mechanism. Replacement of the hydrogen at C-3 by deuterium (the site of proton abstraction during reaction) gives rise to a small hydrogen isotope effect under optimum conditions, and this effect is independent of CO_2 concentration (*81, 84*).

During catalytic turnover, label at this same position is washed out of the substrate, provided that the CO_2 concentration is low; the exchange is com-pletely suppressed at high CO_2 concentration (*85*).

For the spinach enzyme, the carboxylation shows a relatively large carbon isotope effect of 1.029 with normal substrate and 1.021 with deuterated substrate (*86*). The isotope effect is smaller for the *R. rubrum* enzyme (*87*). Isotope ef-fects are independent of RuBP concentration and nearly independent of pH (*86*). This indicates that hydrogen transfer and carboxylation are separate steps in the enzymic mechanism (*88*).

These results are most consistent with a mechanism in which CO_2 does not bind until after the enolate is formed and rearranges. In fact, neither kinetics nor crystallographic studies provide any definitive evidence for the existence of a binding site for CO_2. It is possible that the enzyme-bound enolate reacts with dissolved CO_2 in a bimolecular reaction. If this is true, the reaction with oxygen results from a competition between bimolecular reactions of the enolate with CO_2 and with O_2. Variations in the partitioning between the two reactions might result from the different steric requirements of the two transition states (CO_2 must approach side-on to permit carbon–carbon bond formation; O_2 approaches nearly end-on). Consistent with this view is the fact that COS is a substrate for the enzyme with essentially the same V_{max} as CO_2 (*89*). Alternatively, variations in the partitioning might result from varying degrees of reversal of the carboxyl-ation step.

The six-carbon intermediate can be prepared by denaturation of the en-zyme–substrate complex during steady-state turnover. In the absence of enzyme, the intermediate rapidly decarboxylates to form a five-carbon product. In the presence of activated enzyme and Mg^{2+}, the six-carbon intermediate is converted into two molecules of 3-PGA (90). Enzymic control of the fate of this intermediate may be accomplished by control of the conformation of the single bond between C-2 and C-3. EPR spectroscopy (91, 92) and X-ray crystallography (75) of the complex with 2-carboxyarabinitol bisphosphate indicate that the metal is coordinated to the carboxyl group and to the hydroxyl group at C-2 (75), so it is not surprising that the enzyme might control the course of this step.

E. CATALYTIC STRATEGY

Carboxylation of RuBP to form the six-carbon intermediate is significantly endergonic, whereas the overall reaction is exergonic by about 6 kcal/mol. Overall, the reaction is driven by cleavage of the six-carbon intermediate. Clearly the enzyme controls the conformation of the six-carbon intermediate in a way that favors cleavage and disfavors decarboxylation. The enzyme from *R. rubrum* is less successful at this than the more complex higher plant enzyme. Metal complexation is probably a key to control, and the small subunit might also be involved.

The proton abstraction by lysine that initiates the catalytic sequence is thermodynamically unfavorable, but this step is probably assisted by metal complexation of the carbonyl oxygen. It is not known whether the metal complex reorganizes during the interchange of the two enolates. If in fact the conformation of this six-carbon intermediate favors cleavage, rather than loss of CO_2, then the enzyme must undergo a conformation change following the carboxylation step and prior to cleavage. This change might be triggered by binding of the new carboxyl group to the metal.

VI. Schiff Base-Dependent Decarboxylations

A recurrent theme in our discussion of decarboxylases has been the necessity to stabilize a negative charge following the decarboxylation step. The enzymes discussed in this section stabilize the negative charge by use of a nitrogen, rather than an oxygen, and the decarboxylation often results in neutralization of a positive charge on the nitrogen. Reactions in this class are always irreversible.

A. ACETOACETATE DECARBOXYLASE

The decarboxylation of acetoacetic acid [Eq. (16)] has been studied extensively by Westheimer and associates (93, 94).

$$H^+ + CH_3-CO-CH_2-CO_2^- \rightarrow CO_2 + CH_3-CO-CH_3 \qquad (16)$$

The enzyme contains no metals and no cofactors.

1. Models

The decarboxylation of acetoacetate is acid catalyzed (20). Metals do not catalyze the spontaneous decarboxylation of acetoacetate, presumably because the substrate and the product acetone enol are poor ligands for the metal (21). Primary amines catalyze the decarboxylation of acetoacetate by a Schiff base mechanism (Scheme VII), and this provides the best model for acetoacetate decarboxylase (94).

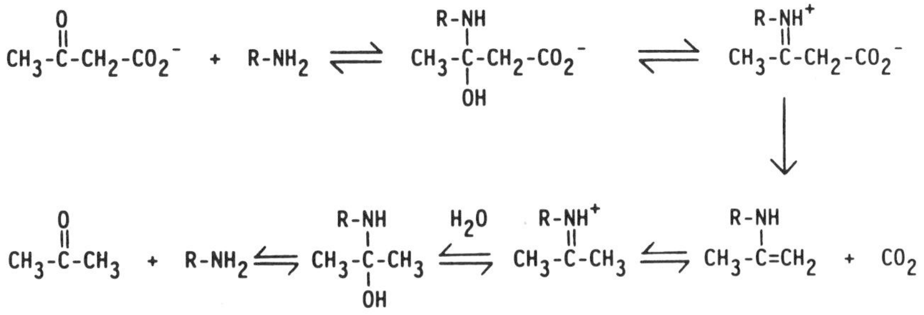

$$CH_3\text{-}\overset{\overset{\displaystyle O}{\|}}{C}\text{-}CH_2\text{-}CO_2^- \quad + \quad R\text{-}NH_2 \quad \rightleftharpoons \quad CH_3\text{-}\overset{\overset{\displaystyle R\text{-}NH}{|}}{\underset{\underset{\displaystyle OH}{|}}{C}}\text{-}CH_2\text{-}CO_2^- \quad \rightleftharpoons \quad CH_3\text{-}\overset{\overset{\displaystyle R\text{-}NH^+}{\|}}{C}\text{-}CH_2\text{-}CO_2^-$$

$$CH_3\text{-}\overset{\overset{\displaystyle O}{\|}}{C}\text{-}CH_3 \quad + \quad R\text{-}NH_2 \overset{}{\rightleftharpoons} CH_3\text{-}\overset{\overset{\displaystyle R\text{-}NH}{|}}{\underset{\underset{\displaystyle OH}{|}}{C}}\text{-}CH_3 \overset{H_2O}{\rightleftharpoons} CH_3\text{-}\overset{\overset{\displaystyle R\text{-}NH^+}{\|}}{C}\text{-}CH_3 \overset{}{\rightleftharpoons} CH_3\text{-}\overset{\overset{\displaystyle R\text{-}NH}{|}}{C}\text{=}CH_2 \quad + \quad CO_2$$

SCHEME VII. Decarboxylation of acetoacetate.

2. Mechanism

Compulsory oxygen-18 exchange from the carbonyl group of the substrate accompanies the enzymic decarboxylation (95). Reaction of the enzyme with acetoacetate in the presence of $NaBH_4$ results in inactivation of the enzyme and formation of a product in which a single lysine amino group has been alkylated (96). This and a variety of other lines of evidence indicate that the reaction occurs by means of a Schiff base mechanism analogous to the amine-catalyzed decarboxylation (Scheme VII). Except for the reactive lysine, the identities of other catalytic groups at the active site are not known.

Replacement of the carboxyl group by —H occurs with racemization (10). This may indicate that protonation of the enamine product does not involve an enzyme catalytic group, but rather involves direct protonation from the solvent. The catalytic amino group has a pK_a near 5.9 (97). Although unusual for a lysine amino group, such a value would be expected if this amino group is adjacent to one or two other amino groups, and in fact the adjacent residue in the amino acid sequence is also lysine (98). The adjacent lysine might be responsible for catalysis of dehydration of the carbinolamine, though evidence for this is lacking.

Carbon isotope effects for both the amine-catalyzed reaction and the enzyme-catalyzed reaction are significantly different from unity but smaller than those expected if decarboxylation is fully rate determining (99). Thus, in both cases Schiff base formation and decarboxylation must be jointly rate limiting and the enzyme must have accelerated both steps by similar amounts.

3. *Catalytic Strategy*

The advantage of this mechanism over preceding mechanisms is that there is no enolate. Instead, decarboxylation results in charge neutralization and formation of an enamine.

The rate of the enzymic decarboxylation of acetoacetic acid exceeds the rate of the spontaneous reaction by perhaps a billionfold. However, simple amines of appropriate pK_a also catalyze the reaction, and the difference in rate between catalysis by the enzyme and catalysis by cyanomethylamine (pK_a 5.5) is only about 100-fold (*100*).

Thus, the majority of the catalytic power of acetoacetate decarboxylase resides in the properties of the catalytic amino group. The remaining rate enhancement may be principally due to conformational effects. In solution, it is likely that an acetoacetate–amine Schiff base will exist in a planar conformation not suitable for decarboxylation. The enzyme can enforce the proper conformation, thus presumably providing the final catalytic factor.

B. DECARBOXYLASES UTILIZING PYRIDOXAL 5'-PHOSPHATE

The decarboxylation of amino acids generally involves pyridoxal 5'-phosphate (PLP) as a cofactor (*101–104*) (Editor's note: For additional features of PLP-dependent decarboxylation, see Chapter 7 by Kluger.). The general reaction is shown in Eq. (17),

$$R\text{—}CHNH_2\text{—}CO_2^- \rightarrow CO_2 + R\text{—}CH_2\text{—}NH_2 \tag{17}$$

and a variety of decarboxylases of relatively high specificity exist. For the most part, these enzymes contain no metal ions. The corresponding PLP-dependent carboxylations are unknown.

1. *Models*

Amino acids react readily with PLP and similar compounds to form Schiff bases. The chemistry of this process is well known (*105*). The most significant catalytic group is the hydroxyl group at the 3-position, which also appears to play an important role in the enzymic reaction (*105*). In both cases, the role of this group is to act as an acid–base catalyst.

The decarboxylation is more enigmatic. Although models for Schiff base formation and transamination are common, models for decarboxylation are rare. If PLP is mixed with an amino acid in aqueous solution, then Schiff base formation will occur, followed by slow transamination. No decarboxylation is observed. If an α-methyl amino acid is used, thus preventing transamination, slow decarboxylation is observed, but only at temperatures above 100°C (*106*). Thus, the problem in enzymic decarboxylation is to catalyze a very slow reaction and avoid a much more favorable reaction.

2. Mechanism

The basic decarboxylation mechanism is analogous to mechanisms of other PLP-dependent processes (Scheme VIII) (*101–104*). PLP is initially bound covalently to the enzyme via a Schiff base linkage to a lysine amino group. Other functional groups of PLP, particularly the phosphate, form strong noncovalent interactions with the enzyme. The substrate binds and reacts to form a bound PLP–substrate Schiff base.

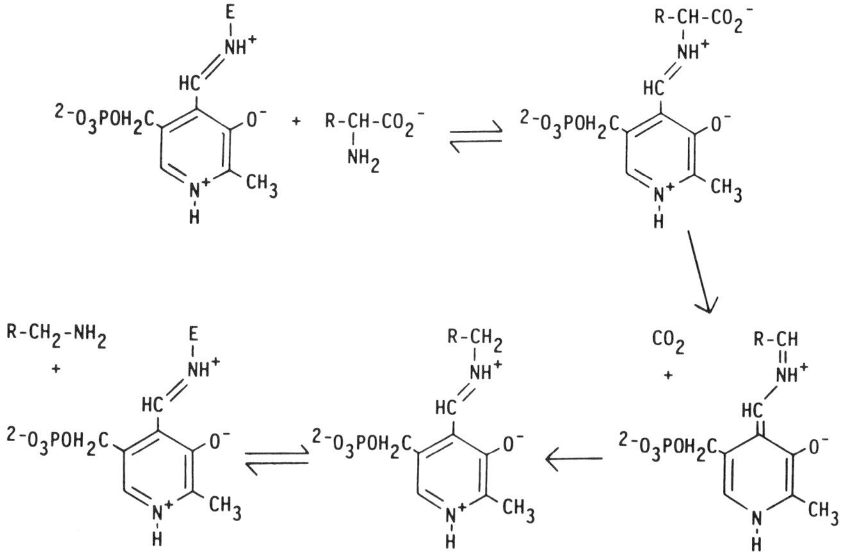

SCHEME VIII. PLP decarboxylation mechanism.

At this point the paths for decarboxylation, transamination, and other reactions diverge. For decarboxylation, the Schiff base loses CO_2 and forms a stabilized quinonoid intermediate. In this step, the negative charge released by cleavage of the carbon–carbon bond is neutralized by the pyridinium nitrogen atom. This intermediate is then reprotonated [ordinarily on the α-carbon of the substrate] with retention of configuration (*6–8*), a Schiff base interchange occurs, and the amine product is released.

The mechanism by which these enzymes facilitate decarboxylation and avoid transamination was originally proposed by Dunathan (*107, 108*), who suggested that reactivity is controlled by the conformation of the single bond between the α-carbon of the amino acid and the Schiff base nitrogen (Scheme IX). Decarboxylation requires the conformation shown in Scheme IX, in which the carboxyl group is out of the plane of the aromatic system and the α-hydrogen and the alkyl group are on the other side of the plane. When decarboxylation occurs,

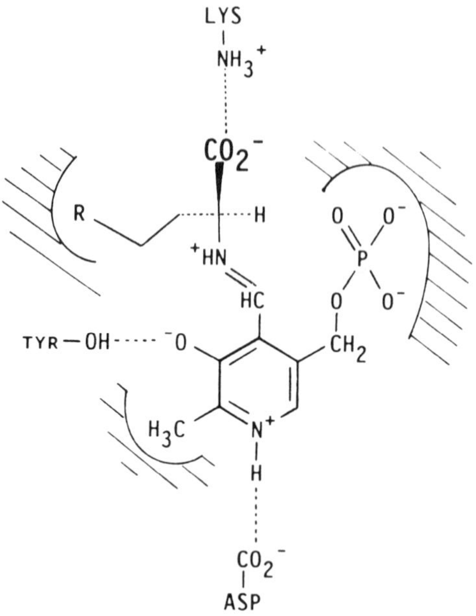

SCHEME IX. Conformation and stereochemistry of decarboxylation and projected active site of a decarboxylase.

the hybridization of the α-carbon changes, and the H and R attached to this carbon move into the plane of the aromatic system. Transamination requires that the groups be rotated differently about the C–N bond, with the scissile hydrogen out of the plane.

Other evidence also indicates that stereochemical control is important. In the case of arginine decarboxylase, increasing or decreasing the length of the side chain by one carbon decreases the decarboxylation rate about 100-fold, and this decrease comes about principally because of a decrease in the rate of the decarboxylation step (109). Presumably when the chain is the wrong length, the enzyme–substrate complex accommodates by rotating the carbon–nitrogen bond, thus destroying the optimum decarboxylation geometry obtained with the natural substrate.

3. Catalytic Strategy

The pyridinium ring in PLP provides a highly optimized electron sink for stabilizing the pair of electrons released by the decarboxylation step. Consequently, the decarboxylation step is irreversible. Beginning the reaction with a Schiff base, rather than with a free aldehyde, also provides an additional catalytic advantage (105), as Schiff base interchange is faster than formation of a Schiff base from an aldehyde and an amine. Precise conformational control is also an integral part of the catalytic process, and this control is also related both to reaction specificity and to stereospecificity.

The decarboxylation step (Scheme IX) can be viewed as a charge neutralization; the negative charge on the carboxylate group is neutralized by the positive charge on the pyridinium ion. Such reactions show very strong medium effects. For example, the decarboxylation of 4-pyridylacetic acid (Scheme X), which occurs by an analogous mechanism, is accelerated about 4000-fold on going from pure water to 75% dioxane (110). This rate change comes about because of greater transition-state stabilization in the less polar solvent. PLP-dependent decarboxylases may use an analogous effect to accelerate the decarboxylation reaction; if the cofactor is buried in a hydrophobic environment, then decarboxylation should be accelerated. Rates and isotope effects for decarboxylations catalyzed by arginine decarboxylase are consistent with the operation of this effect (111). Analogous effects were first noted in the case of thiamin pyrophosphate-catalyzed reactions (112).

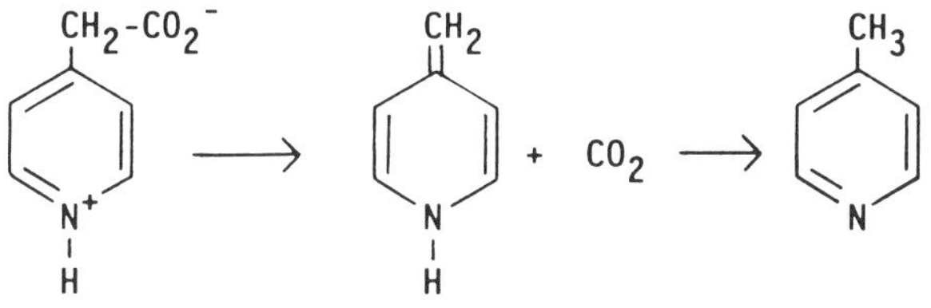

SCHEME X. Decarboxylation of 4-pyridylacetic acid.

A second, related aspect of the medium effect is also likely to be important. Decarboxylation converts a polar, highly solvated $-CO_2^-$ group into a nonpolar, nearly unsolvated CO_2. For the case of benzisoxazole-3-carboxylic acids, carboxyl desolvation has been demonstrated to be an important component of solvent catalysis of decarboxylation (113). Desolvation of the carboxyl group on binding to an enzyme may likewise be an important factor. Carbon isotope effects are consistent with this possibility (111). Oxygen isotope effects on formate dehydrogenase are also consistent with this mechanism (114).

4. *Possible Three-Dimensional Mechanism*

Although no X-ray crystal structure is available for a PLP-dependent decarboxylase, the X-ray structure of mitochondrial aspartate aminotransferase (115, 116) can be used to propose a likely picture of the catalytic mechanism for the analogous glutamate decarboxylase (Scheme IX).

For aspartate aminotransferase, Lys-258 forms a Schiff base with PLP. Formation of the enzyme–substrate Schiff base rotates the coenzyme and places the α-hydrogen to be removed in the immediate vicinity of Lys-258, which is then responsible for proton transfer. Site-directed mutagenesis studies are also consistent with this assignment (*117, 118*). The relationship between the coenzyme and the surrounding solvent can be deduced both from the X-ray structure and from studies of the stereochemistry of reduction of various complexes by NaBH$_4$ (*8, 115*). Corresponding studies with decarboxylases show a corresponding stereochemistry (*119*), and studies of glutamate decarboxylase show corresponding stereochemistry both at the α-carbon and at the aldehyde carbon of the coenzyme (*120, 121*). The orientation of groups shown in Scheme IX is consistent with all this information. It is noteworthy that the carboxyl group to undergo decarboxylation is facing the active site lysine. Since these decarboxylations occur with retention of configuration (*6–8*), this same lysine may be responsible for protonation of the quinonoid intermediate. As in the case of transamination, protonation can occur at either of two sites (*120, 121*), but the proportions of products from protonation at the two sites are different for transamination and for decarboxylation.

C. Pyruvate-Dependent Decarboxylations

Although most amino acid decarboxylases use PLP as a cofactor, a number of decarboxylases use covalently bound pyruvate instead (*122, 123*). The pyruvate-dependent enzyme consists of two types of chains. The pyruvoyl cofactor is formed by the cleavage of a Ser-Ser linkage in a single-chain precursor and is bound as an amide to the N-terminus of one polypeptide chain.

In the case of histidine decarboxylase, both PLP-dependent and pyruvate-dependent enzymes exist. Comparison of properties of these two enzymes provides useful information about the comparative catalytic properties of pyruvate and PLP.

1. *Mechanism*

Like PLP-dependent enzymes, pyruvate-dependent enzymes function by means of a Schiff base mechanism (Scheme XI), with an important difference:

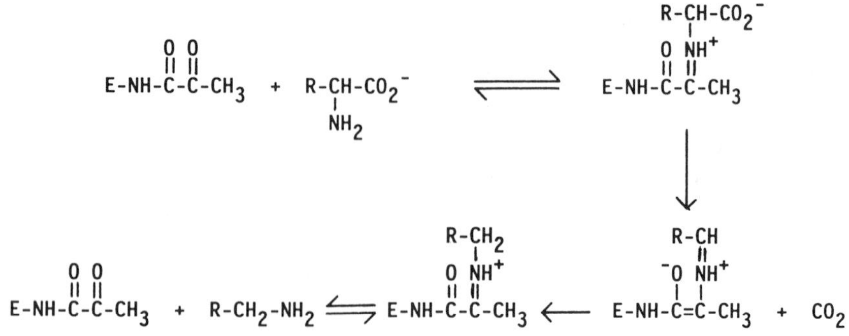

Scheme XI. Pyruvate-dependent decarboxylase mechanism.

the extent of electron delocalization in the decarboxylation step is much less in the pyruvoyl enzymes than in the PLP enzymes. The Schiff base nitrogen is protonated at the time that decarboxylation occurs, and this may contribute to catalysis (124).

2. Crystal Structure

The pyruvate-dependent histidine decarboxylase from *Lactobacillus* 30a has been studied in particular detail by Snell and collaborators (122), and an X-ray structure of the enzyme has recently been published (125, 126). The crystal structure shows a very tight fit between enzyme and substrate, with extensive interactions between the enzyme and the imidazole ring of the substrate. Unlike the case of PLP enzymes, there does not seem to be a significant conformation change on binding the substrate. The conformation of the bound substrate is consistent with the requirement that the carboxyl group be out of the plane of the conjugated system. The carboxyl group of the substrate is bound in a hydrophobic pocket; the only polar group nearby is the carboxyl group of Glu-197.

The pH dependence of the kinetics of histidine decarboxylase (127) demonstrates that the histidine is zwitterionic when it binds to the enzyme. The extra proton on nitrogen must, of course, be removed before the Schiff base is formed. The carboxylate of Glu-197 at the active site may accept this proton. In turn, this same group may then be responsible for proton donation to the Schiff base following decarboxylation. This is consistent with the occurrence of retention of configuration in the overall replacement of $-CO_2^-$ by $-H$ (128) and with studies of enzymes altered at Glu-197 (129). When Glu-197 is replaced by Asp, the protonation that follows decarboxylation occasionally occurs on the pyruvate side, thus giving rise to decarboxylation-dependent transamination (129).

The pyruvoyl amide that acts as an electron sink in the decarboxylation step is hydrogen bonded to an $-NH$ of the polypeptide backbone, but its environment is otherwise unexceptional.

3. Comparison of PLP- and Pyruvate-Dependent Histidine Decarboxylases

Kinetics and isotope effects have been used to compare the pyruvate-dependent histidine decarboxylase from *Lactobacillus* 30a with the corresponding PLP-dependent enzyme from *Morganella morganii* (124, 130). Carbon isotope effects for the two enzymes are similar and indicate that the decarboxylation is principally, though not entirely, rate determining. Enzyme–substrate Schiff base formation is also partially rate determining. Nitrogen isotope effects, though superficially different, are consistent with this conclusion and indicate that in both cases the Schiff base intermediate must be protonated on nitrogen in order for reaction to occur (124, 130). What is most surprising is that both enzymes have the same turnover number (131); neither has a significant catalytic advantage in terms of either V_{max} or V_{max}/K_m. Thus, the two types of enzymes operate at similar rates by similar mechanisms having two rate-determining steps, and at least from the point of view of decarboxylation both catalytic mechanisms are equally functional.

4. Catalytic Strategy

The catalytic strategy is familiar from our discussion of PLP-dependent reactions: reaction via a Schiff base, probable medium control of the decarboxylation, and desolvation of the carboxyl group on binding to the enzyme. What is most surprising is that pyruvate, with its very small electron sink, works as efficiently as PLP, which allows for more extensive electron delocalization. The specialness of PLP in enzymic catalysis must lie in other factors.

VII. Thiamin Pyrophosphate-Dependent Reactions

The decarboxylation of pyruvic acid to yield carbon dioxide and a two-carbon fragment is catalyzed by a class of enzymes that require thiamin pyrophosphate (TPP) and a divalent metal ion. We will restrict our attention to the simplest member of the class, pyruvate decarboxylase (*132, 133*), which catalyzes reaction (*18*).

$$H^+ + CH_3—CO—CO_2^- \rightarrow CO_2 + CH_3—CHO \tag{18}$$

This is the most studied TPP-dependent enzyme, and it is likely that mechanisms of other thiamin pyrophosphate-dependent decarboxylations are similar. (*Editor's note:* Other aspects of thiamin pyrophosphate chemistry are presented in Chapter 7 by Kluger.)

A. MODEL REACTIONS

An understanding of mechanisms of thiamin pyrophosphate-dependent processes must begin with the classic work of Breslow (*105, 134*), who showed that the hydrogen at C-2 of thiamin pyrophosphate can be removed by bases and that the resulting anion is highly reactive. The pK_a at this site is 18 (*135*). In an enzyme active site, this pK_a value may be considerably lower, as the value decreases with decreasing medium polarity. Reaction of the anion with a variety of carbonyl compounds (e.g., acetaldehyde, pyruvate) gives rise to characterizable adducts (*132*).

B. MECHANISM

The mechanism of decarboxylation of pyruvic acid is shown in Scheme XII. The first step is formation of the anion, which then adds to C-2 of pyruvate, forming a covalent adduct. This compound has been prepared chemically and its

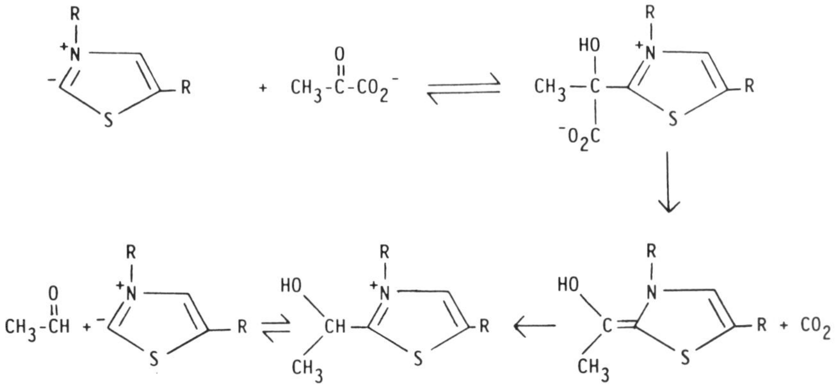

SCHEME XII. Mechanism of decarboxylation of pyruvate.

reactions studied (*132*). In solution, it decarboxylates readily. Isotope effects indicate that formation of this intermediate, rather than decarboxylation, is rate determining in the overall reaction (*136–139*).

Following decarboxylation, the two-carbon–TPP adduct is protonated. This compound has also been synthesized and studied (*132*); interestingly, it does not release acetaldehyde in nonenzymic reactions, whereas it obviously does in the enzymic reaction (*132*). Conformational control on the enzyme may be responsible. Consistent with this, glyoxylic acid is decarboxylated by pyruvate decarboxylase, but the product, hydroxymethyl-TPP, does not release formaldehyde from the enzyme (*132*).

C. CATALYTIC STRATEGY

As in the case of pyridoxal phosphate, the key to reaction in this case is the use of a heterocyclic compound as an electron sink in the decarboxylation step. Conformational control of the TPP–pyruvate adduct may also be important. The enzyme active site is probably nonpolar, and this provides a significant catalytic factor (*112*).

VIII. Conclusion

The strategies used by enzymes in catalyzing carboxylations and decarboxylations are not qualitatively different from the strategies used in catalyzing other types of reactions. However, a number of interesting issues arise from the necessity to deal with CO_2 and from other factors.

A. THERMODYNAMICS

Carboxylation is, in the first place, a thermodynamic problem. Nature solves this problem by associating the carboxylation with some exergonic process. This may be reduction of a product (as in the case of malic enzyme and isocitrate dehydrogenase), hydrolysis of ATP (as in the case of pyruvate carboxylase), hydrolysis of phosphoenolpyruvate (as in the case of phosphoenolpyruvate carboxylase), or cleavage of a carbon–carbon bond (as in the case of ribulose-bisphosphate carboxylase).

By comparison, decarboxylation is largely a kinetic problem. Enzymes have developed a variety of strategies for stabilizing the anionic intermediate that is produced in the decarboxylation step. Metal ion stabilization of enolates is a common theme, particularly for decarboxylation of β-keto acids. The most elegant solutions are perhaps the extensive electron delocalizations seen in pyridoxal phosphate and thiamin pyrophosphate.

Desolvation of the substrate carboxyl group on binding to the enzyme may be a significant catalytic factor in most decarboxylations and carboxylations, but little direct evidence exists.

B. ANION INTERMEDIATES

Many carboxylations are initiated by a proton removal to form an anion. In a number of cases, the proton being removed has a pK_a value greater than 15. In aqueous solution at neutral pH, the concentrations of such anions that might be achieved would always be very low. On the other hand, such anions are clearly important in carboxylations and decarboxylations, and it is clear that enzyme active sites can stabilize these anions.

C. HOW ENZYMES DEAL WITH CO_2

The concentration of CO_2 in natural systems is seldom high, and the concentration of HCO_3^- often exceeds that of CO_2 by more than an order of magnitude. Because it is nonpolar, CO_2 binds only poorly to most enzymes. HCO_3^-, being polar, often binds quite well. On the other hand, CO_2 is reactive toward a variety of reagents, whereas HCO_3^- is unreactive. Natural strategies for dealing with this combination of factors vary. Some enzymes have adapted to the low natural concentration of CO_2. Others bind HCO_3^- but then convert it into CO_2 at the active site.

Although CO_2 is the substrate for the majority of carboxylations, it is not clear how many of these enzymes bind CO_2 prior to reaction. That is, it is not clear how many of these enzymes actually have a Michaelis complex with CO_2 of appreciable stability. Kinetic data are largely silent on this point. The existence of saturation kinetics with respect to CO_2 does not require the formation of a Michaelis complex. Most kinetic studies of carboxylases and decarboxylases indicate that CO_2 is the last substrate to bind or the first product to be released. Accurate kinetics with regard to CO_2 concentration are notoriously difficult to measure because of problems associated with measuring and maintaining low concentrations of CO_2. Most kinds of plastic and rubber are somewhat permeable to CO_2, and simply sealing a flask or spectrophotometer cell may not be adequate to exclude extraneous CO_2. Inhibitors competitive against CO_2 are rare; in the case of ribulose-bisphosphate carboxylase, both O_2 and COS can be used (89). Little use has been made of these substances in other cases.

In the case of ribulose-bisphosphate carboxylase, van Dyk and Schloss (81) argue that no CO_2 Michaelis complex is formed. This may also be true for a variety of other carboxylases. On the other hand, in the case of PEP carboxykinase, both magnetic resonance and isotope effect studies make it likely that a CO_2-containing Michaelis complex is formed.

D. MEDIUM EFFECTS

Studies in solution demonstrate that rates of polar reactions can often be changed by a large factor simply by changing the medium within which the reaction occurs. It is clear that this is an important factor in many carboxylations and decarboxylations. The clearest cases are decarboxylations requiring pyridoxal phosphate and thiamin pyrophosphate.

E. STEREOCHEMICAL AND GEOMETRIC CONTROL

Stereochemical control in carboxylations and decarboxylations is important both because of the rate acceleration that may be achieved and because such control may prevent the occurrence of other, deleterious reactions. Perhaps the most interesting cases are decarboxylations requiring pyridoxal phosphate, where the analogous model reactions are almost unobservable because of competing transamination reactions.

F. FINAL COMMENT

Progress in understanding the mechanisms of carboxylations and decarboxylations has come about through studies of a wide variety of types. Two of the most powerful techniques, X-ray crystallography and site-directed mutagenesis, have been used infrequently in such studies. It is to be hoped that this deficit will be rectified in the near future.

REFERENCES

1. Edsall, J. T., and Wyman, J. (1958). *In* "Biophysical Chemistry," Chap. 10. Academic Press, New York.
2. Butler, J. N. (1982). "Carbon Dioxide Equilibria and Their Applications." Addison-Wesley, Reading, Massachusetts.
3. Inoue, S., and Yamazaki, N. (1982). "Organic and Bio-Organic Chemistry of Carbon Dioxide." Halsted, New York.
4. O'Leary, M. H., and Hermes, J. D. (1987). *Anal. Biochem.* **162**, 358.
5. Inoue, S., and Yamazaki, N. (1982). "Organic and Bio-Organic Chemistry of Carbon Dioxide," Chap. 4. Halsted, New York.
6. Vederas, J. C., and Floss, H. G. (1980). *Acc. Chem. Res.* **13**, 455.
7. Retey, J., and Robinson, J. A. (1982). "Stereospecificity in Organic Chemistry and Enzymology." Verlag Chemie, Deerfield Beach, Florida.
8. Floss, H. G., and Vederas, J. (1982). *In* "Stereochemistry" (C. Tamm, ed.) p. 161. Elsevier, New York.
9. Although this mechanism has often been suggested, to date, there is good evidence for it only in the case of the pyruvate-dependent histidine decarboxylase (*vide infra*).
10. Rozell, J. D., Jr., and Benner, S. A. (1984). *J. Am. Chem. Soc.* **106**, 4937.
11. Calculated from equilibrium constants for malic enzyme and malate dehydrogenase Ref. (*12*).
12. Schimerlik, M. I., Rife, J. E., and Cleland, W. W. (1975). *Biochemistry* **14**, 5347.
13. Leussing, D. L., and Emly, M. (1984). *J. Am. Chem. Soc.* **106**, 443.
14. March, J. (1985). "Advanced Organic Chemistry," 3rd Ed., p. 222. Wiley (Interscience), New York.
15. Miller, B. A., and Leussing, D. L. (1985). *J. Am. Chem. Soc.* **107**, 7146.
16. Leussing, D. L. (1982). *Adv. Inorg. Biochem.* **4**, 171.
17. Jencks, W. P. (1975). *Adv. Enzymol.* **43**, 219.
18. Herbert, D. (1952). *Symp. Soc. Exp. Biol.* **5**, 52.
19. Piccirilli, J. A., Rozell, J. D., Jr., and Benner, S. A. (1987). *J. Am. Chem. Soc.* **109**, 8084.
20. Pollack, R. M. (1978). *In* "Transition States of Biochemical Processes" (R. D. Gandour and R. L. Schowen, eds.), p. 467. Plenum, New York.
21. Hay, R. W. (1976). *In* "Metal Ions in Biological Systems" (H. Sigel, ed.), Vol. 5, p. 127. Dekker, New York.
22. Creighton, D. J., and Rose, I. A. (1976). *J. Biol. Chem.* **251**, 69.
23. Kofron, J. L., Ash, D. E., and Reed, G. H. (1988). *Biochemistry* **27**, 4781.
24. Grissom, C. B., and Cleland, W. W. (1986). *J. Am. Chem. Soc.* **108**, 5582.
25. Kiick, D. M., and Cleland, W. W. (1989). *Arch. Biochem. Biophys.* **270**, 647.
26. Kun, E. (1963). "The Enzymes," 2nd Ed., Vol. 7, p. 157.
27. Hausler, R. E., Holtum, J. A. M., and Latzko, E. (1987). *Eur. J. Biochem.* **163**, 619.
28. Hsu, R. Y., Mildvan, A. S., Chang, G.-G., and Fung, C.-H. (1976). *J. Biol. Chem.* **251**, 6574.
29. Hermes, J. D., Roeske, C. A., O'Leary, M. H., and Cleland, W. W. (1982). *Biochemistry* **21**, 5106.
30. Grissom, C. B., and Cleland, W. W. (1985). *Biochemistry* **24**, 944.
31. Vennesland, B., Tchen, T. T., and Loewus, F. A. (1954). *J. Am. Chem. Soc.* **76**, 3358.
32. Utter, M. F., and Kolenbrander, H. M. (1972). "The Enzymes," 3rd Ed., Vol. 6, p. 117.
33. O'Leary, M. H. (1982). *Annu. Rev. Plant Physiol.* **33**, 297.
34. Andreo, C. S., Gonzalez, D. H., and Iglesias, A. A. (1987). *FEBS Lett.* **213**, 1.
35. Gonzalez, D. H., and Andreo, C. S. (1989). *Trends Biochem. Sci.* **14**, 24.
36. Cooper, T. G., and Wood, H. G. *J. Biol. Chem.* **246**, 5488.

37. Maruyama, H., Easterday, R. L., Chang, H.-C., and Lane, M. D. (1966). *J. Biol. Chem.* **241,** 2405.
38. Wimmer, M. J., Rose, I. A., Powers, S. G., and Meister, A. (1979). *J. Biol. Chem.* **254,** 1854.
39. Hansen, D. E., and Knowles, J. R. (1982). *J. Biol. Chem.* **257,** 14795.
40. O'Leary, M. H., Rife, J. E., and Slater, J. D. (1981). *Biochemistry* **20,** 7308.
41. O'Leary, M. H., and Paneth, P. (1987). *In* "Biophosphates and Their Analogues—Synthesis, Structure, Metabolism and Activity" (K. S. Bruzik and W. J. Stec, eds.), p. 303. Elsevier, Amsterdam.
42. Fujita, N., Izui, K., Nishino, T., and Katsui, H. (1984). *Biochemistry* **23,** 1774.
43. O'Laughlin, J. T., and O'Leary, M. H. (1988). Unpublished.
44. Gonzalez, D. H., and Andreo, C. S. (1988). *Biochemistry* **27,** 177.
45. Diaz, E., O'Laughlin, J. T., and O'Leary, M. H. (1988). *Biochemistry* **27,** 1336; Jane, J. W., Urbauer, J. L., O'Leary, M. H., and Cleland, W. W. (1992). *Biochemistry* in press.
46. Liu, J., Peliska, J. A., and O'Leary, M. H. (1990). *Arch. Biochem. Biophys.* **277,** 143.
47. Ausenhus, S. A., and O'Leary, M. H. (1992). *Biochemistry*.
48. Nowack, T. (1986). *In* "Manganese in Metabolism and Enzyme Function" (V. L. Schramm and F. C. Wedler, eds.), p. 165. Academic Press, New York.
49. Jomain-Baum, M., and Schramm, V. L. (1978). *J. Biol. Chem.* **253,** 3648.
50. Arnelle, D. A., and O'Leary, M. H. *Biochemistry* in press.
51. Hebda, C. A., and Nowack, T. (1982) *J. Biol. Chem.* **257,** 5515.
52. Scrutton, M. C., and Young, M. R. (1972). "The Enzymes," 3rd Ed., Vol. 6, p. 1.
53. Alberts, A. W., and Vagelos, P. R. (1972). "The Enzymes," 3rd Ed., Vol. 6, p. 37.
54. Wood, H. G. (1972). "The Enzymes," 3rd Ed., Vol. 6, p. 83.
55. Wood, H. G., and Zwolinski, G. K. (1976). *Crit. Rev. Biochem.* **4,** 47.
56. Mildvan, A. S., Fry, D. C., and Serpersu, E. H. (1989). *In* "A Study of Enzymes" (S. A. Kirby, ed.), Vol. 2, p. 105. CRC Press, Boca Raton, Florida.
57. Knowles, J. R. (1989). *Annu. Rev. Biochem.* **58,** 195.
58. Kaziro, Y., Hass, L. F., Boyer, P. D., and Ochoa, S. (1962). *J. Biol. Chem.* **237,** 1460.
59. Perrin, C. L., and Dwyer, T. J. (1987). *J. Am. Chem. Soc.* **109,** 5163.
60. O'Keefe, S. J., and Knowles, J. R. (1986). *J. Am. Chem. Soc.* **108,** 328.
61. O'Keefe, S. J., and Knowles, J. R. (1986). *Biochemistry* **25,** 6077.
62. Ogita, T., and Knowles, J. R. (1988). *Biochemistry* **27,** 8028.
63. Grissom, C. B., and Cleland, W. W. (1988). *Biochemistry* **27,** 2934.
64. Rendina, A. R., Hermes, J. D., and Cleland, W. W. (1984). *Biochemistry* **23,** 6257.
65. Hermes, J. D., Tipton, P. A., Fisher, M. A., O'Leary, M. H., Morrison, J. F., and Cleland, W. W. (1984). *Biochemistry* **23,** 6263.
66. Villet, R. H., and Dalziel, K. (1972). *Eur. J. Biochem.* **27,** 251.
67. Rippa, M., Signorini, M., and Dallocchio, F. (1973). *J. Biol. Chem.* **248,** 4920.
68. Ankel, H., and Feingold, D. S. (1966). *Biochemistry* **5,** 182.
69. Ankel, H., and Feingold, D. S. (1965). *Biochemistry* **4,** 2468.
70. Loken, J. P., and Stormer, F. C. (1970). *Eur. J. Biochem.* **14,** 133.
71. Miziorko, H. M., and Lorimer, G. H. (1983). *Annu. Rev. Biochem.* **52,** 507.
72. Andrews, T. J., and Lorimer, G. H. (1987). *In* "The Biochemistry of Plants" (M. D. Hatch and N. K. Boardman, eds.), Vol. 10, p. 131. Academic Press, New York.
73. Schneider, G., Lindqvist, Y., Branden, C.-I., and Lorimer, G. (1986). *EMBO J.* **5,** 3409.
74. Chapman, M. S., Suh, S. W., Curmi, P. M. G., Cascio, D., Smith, W. W., and Eisenberg, D. S. (1988). *Science* **241,** 71.
75. Andersson, I., Knight, S., Schneider, G., Lindqvist, Y., Lundqvist, T., Branden, C.-I., and Lorimer, G. H. (1989). *Nature* (*London*) **337,** 229.

76. Chapman, M. S., Suh, S. W., Cascio, D., Smith, W. W., and Eisenberg, D. (1987). *Nature (London)* **329**, 354.
77. Lundqvist, H., and Schneider, G. (1988). *J. Biol. Chem.* **263**, 3643.
78. Jordan, D. B., and Ogren, W. L. (1981). *Nataure (London)* **291**, 513.
79. Jordan, D. B., and Orgren, W. L. (1983). *Arch. Biochem. Biophys.* **227**, 425.
80. Calvin, M. (1956). *J. Chem. Soc.* p. 1895.
81. Van Dyk, D. E., and Schloss, J. V. (1986). *Biochemistry* **25**, 5145.
82. Hartman, F. C., Soper, T. S., Niyogi, S. K., Mural, R. J., Foote, R. S., Mitra, S., Lee, E. H., Machanoff, R., and Larimer, F. W. (1987). *J. Biol. Chem.* **262**, 3496.
83. Lorimer, G. H., and Hartman, F. C. (1988). *J. Biol. Chem.* **263**, 6468.
84. Sue, J. M., and Knowles, J. R. (1982). *Biochemistry* **21**, 5410.
85. Pierce, J., Lorimer, G. H., and Reddy, G. S. (1986). *Biochemistry* **25**, 1636.
86. Roeske, C. A., and O'Leary, M. H. (1984). *Biochemistry* **23**, 6275.
87. Roeske, C. A., and O'Leary, M. H. (1985). *Biochemistry* **24**, 1603.
88. O'Leary, M. H. (1989). *Annu. Rev. Biochem.* **58**, 377.
89. Lorimer, G. H., and Pierce, J. (1989). *J. Biol. Chem.* **264**, 2764.
90. Pierce, J., Andrews, T. J., and Lorimer, G. H. (1986). *J. Biol. Chem.* **261**, 10248.
91. Miziorko, H. M., and Sealy, R. C. (1984). *Biochemistry* **23**, 479.
92. Pierce, J., and Reddy, G. S. (1986). *Arch. Biochem. Biophys.* **245**, 483.
93. Fridovich, I. (1972). "The Enzymes," 3rd Ed., Vol. 6, p. 255.
94. Westheimer, F. H. (1963). *Proc. Chem. Soc.* p. 253.
95. Hamilton, G. A., and Westheimer, F. H. (1959). *J. Am. Chem. Soc.* **81**, 6332.
96. Warren, S., Zerner, B., and Westheimer, F. H. (1966). *Biochemistry* **5**, 817.
97. Schmidt, D. E., and Westheimer, F. H. (1971). *Biochemistry* **10**, 1249.
98. Laursen, R. A., and Westheimer, F. H. (1966). *J. Am. Chem. Soc.* **88**, 3426.
99. O'Leary, M. H., and Baughn, R. L. (1972). *J. Am. Chem. Soc.* **94**, 626.
100. Westheimer, F. H. (1971). *Proc. Robert A. Welch Found. Conf. Chem. Res.* **15**, 7.
101. Boeker, E. A., and Snell, E. E. (1972). "The Enzymes," 3rd Ed., Vol. 6, p. 217.
102. Braunstein, A. E. (1973). "The Enzymes," 3rd Ed., Vol. 9, p. 379.
103. Metzler, D. E. (1979). *Adv. Enzymol.* **50**, 1.
104. Christen, P., and Metzler, D. E. (1985). "Transaminases." Wiley, New York.
105. Bruice, T. C., and Benkovic, S. J. (1966). "Bioorganic Mechanisms," Vol 2, p. 181. Benjamin, New York.
106. Kalyankar, G. D., and Snell, E. E. (1962). *Biochemistry* **1**, 594.
107. Dunathan, H. C. (1966). *Proc. Natl. Acad. Sci. U.S.A.* **55**, 712.
108. Dunathan, H. C. (1971). *Adv. Enzymol.* **35**, 79.
109. O'Leary, M. H., and Piazza, G. J. (1978). *J. Am. Chem. Soc.* **100**, 632.
110. Marlier, J. F., and O'Leary, M. H. (1986). *J. Am. Chem. Soc.* **108**, 4896.
111. O'Leary, M. H., and Piazza, G. J. (1981). *Biochemistry* **20**, 2743.
112. Crosby, J., Stone, R., and Lienhard, G. E. (1970). *J. Am. Chem. Soc.* **92**, 2891.
113. Kemp, D. S., and Paul, K. G. (1975). *J. Am. Chem. Soc.* **97**, 7305.
114. Hermes, J. D., Morrical, S. W., O'Leary, M. H., and Cleland, W. W. (1984). *Biochemistry* **23**, 5479.
115. Kirsch, J. F., Eichele, G., Ford, G. C., Vincent, M. G., and Jansonius, J. N. (1984). *J. Mol. Biol.* **174**, 497.
116. Ford, G. C., Eichele, G., and Jansonius, J. N. (1980). *Proc. Natl. Acad. Sci. U.S.A.* **77**, 2559.
117. Malcom, B. A., and Kirsch, J. F. (1985). *Biochem. Biophys. Res. Commun.* **132**, 915.
118. Kochhar, S., Finlayson, W. L., Kirsch, J. F., and Christen, P. (1987). *J. Biol. Chem.* **262**, 11446.
119. Vederas, J. C., Reingold, I. D., and Sellers, H. W. (1979). *J. Biol. Chem.* **254**, 5053.

120. Sukhareva, B. S., Dunathan, H. C., and Braunstein, A. E. (1971). *FEBS Lett.* **15**, 241.
121. Yamada, H., and O'Leary, M. H. (1978). *Biochemistry* **17**, 669.
122. Recsei, P., and Snell, E. E. (1984). *Annu. Rev. Biochem.* **53**, 357.
123. Tabor, C. W., and Tabor, H. (1984). *Adv. Enzymol.* **56**, 251.
124. Abell, L. M., and O'Leary, M. H. (1988). *Biochemistry* **27**, 5933.
125. Parks, E. H., Ernst, S. R., Hamlin, R., Xuong, Ng. H., and Hackert, M. L. (1985). *J. Mol. Biol.* **182**, 455.
126. Gallagher, T., Snell, E. E., and Hackert, M. L. (1989). *J. Biol. Chem.* **264**, 12737.
127. Snell, E. E., and Huynh, Q. K. (1986). *In* "Methods in Enzymology" (F. Chytil and D. B. McCormick, eds.), *122*, Vol. 122, p. 135. Academic Press, New York.
128. Battersby, A. R., Nicoletti, M., Staunton, J., and Vleggaar, R. (1980). *J. Chem. Soc., Perkin Trans. 1* p. 43.
129. McElroy, H. E., and Robertus, J. D. (1989). *Protein Eng.* **3**, 43.
130. Abell, L. M., and O'Leary, M. H. (1988). *Biochemistry* **27**, 5927.
131. Tanase, S., Guirard, B. M., and Snell, E. E. (1985). *J. Biol. Chem.* **260**, 6738.
132. Kluger, R. (1987). *Chem. Rev.* **87**, 863.
133. Schellenberger, A., and Schowen, R. L. (1988). "Thiamin Pyrophosphate Biochemistry." CRC Press, Boca Raton, Florida.
134. Breslow, R. (1958). *J. Am. Chem. Soc.* **80**, 3719.
135. Washabaugh, M. W., and Jencks, W. P. (1988). *Biochemistry* **27**, 5044.
136. O'Leary, M. H. (1976) *Biochem. Biophys. Res. Commun.* **73**, 614.
137. DeNiro, M. J., and Epstein, S. (1977). *Science* **197**, 261.
138. Jordan, F., Kuo, D. J., and Monse, E. U. (1978). *J. Am. Chem. Soc.* **100**, 2872.
139. Jordan, F., Kuo, D. J., and Monse, E. U. (1978). *J. Org. Chem.* **43**, 2828.

7

Mechanisms of Enzymic Carbon–Carbon Bond Formation and Cleavage

RONALD KLUGER

Department of Chemistry
University of Toronto
Toronto, Ontario, Canada M5S1A1

I. Introduction

Enzyme-catalyzed reactions that result in the formation of carbon–carbon bonds provide the scaffolding for the structural diversity in the organic and biochemical materials of life. Initially, an observer of charts of metabolic and

THE ENZYMES, Vol. XX

biosynthetic pathways is struck by the large number of reactions contributing to the formation and cleavage of carbon–carbon bonds and the lack of any apparent relationship among them. Closer scrutiny reveals that there are relatively few reaction patterns and analysis of these reveals that there are probably even fewer generally utilized mechanisms. The purpose of this chapter is to identify the mechanistic patterns by which enzymes catalyze the making and breaking of carbon–carbon bonds. The objective will be to find common themes in mechanisms and to identify the basis for the need for each mechanism. The principles of evolution and natural selection can be applied at the mechanistic level and it is logical to expect that we are seeing the outcome in terms of efficiency in reaction mechanisms. I have recently surveyed the basis of these mechanisms in terms of the principles of physical organic chemistry (*1*) and the reader will find that survey useful for background information.

II. Enzymes in Carbon–Carbon Bond Formation

Examination of the Enzyme Commission (EC) divisions of enzymes (*2*) reveals a large number of groups that promote carbon–carbon bond formation or cleavage. These are classified by the types of functional groups involved in the reaction rather than by the reaction pattern or the mechanism of the transformation. A selected list of enzymes involved in carbon–carbon bond formation illustrating this impression includes the following groups:

2. Transferases
    2.1. Transferring one-carbon groups
        2.1.1. Methyltransferases (utilizing *S*-adenosylmethionine)
        2.1.2. Hydroxymethyl-, formyl-, and related transferases (utilizing tetrahydrofolate)
        2.1.3. Carboxyl- and carbamoyltransferases (transcarboxylases, utilizing biotin)
    2.2. Transferring aldehyde or ketone residues
        2.2.1. Transketolases and transaldolases (utilizing thiamin diphosphate)
    2.5. Transferring alkyl groups
        2.5.1.1. Dimethylallyl *trans*transferase (prenyltransferase; dimethylallyl PP:isopentenyl PP)
    4. Lyases: formally classified as cleaving bonds, but in principle catalyzing the reverse as well
    4.1. Carbon–carbon lyases
        4.1.1. Carboxy-lyases
            a. α-Ketoacid decarboxylases (utilizing thiamin diphosphate)
            b. β-Ketoacid decarboxylases
            c. Amino acid decarboxylases (utilizing pyridoxal phosphate)
        4.1.2. Aldehyde-lyases (a few utilize coenzymes, including pyridoxal phosphate)
        4.1.3. Oxo-acid lyases (such as citrate synthase)
    6. Ligases
    6.4. Forming carbon–carbon bonds (biotin-dependent carboxylases)

III. Mechanistic Patterns in Carbon–Carbon Bond Formation without Coenzymes

A. CARBANIONIC TRANSITION STATES

The formation of carbon–carbon bonds by the addition of a nucleophilic carbon center to a carbonyl or other unsaturated electrophilic carbon center is a common enzymic reaction pattern. Examples in which reaction occurs without a catalytic cofactor will be discussed in this section. The mechanisms can be classified by the participants in the carbon–carbon bond-forming process: (1) Carbanions adjacent to imines, better described as enamines, adding to carbonyl centers (Scheme 1). These are found in the aldolase enzymes catalyzing carbon–carbon bond formation and cleavage (3). Bond formation is thermodynamically favored, but coupling to metabolic processes makes the reaction efficient in either direction. They are normally associated with catalysis of carbon–carbon bond cleavage in sugars. (2) Enolate anions typically derived from a ketone or thiol ester, which add to the carbonyl group of the ketone in an α-keto acid (Scheme 2) (4). These are found in the enzymes classed as synthases or (carbon–carbon) lyases. The involvement of coenzyme A as a derivative of the substrate converts the material to one that more readily forms a carbanion, and in this sense coenzyme A serves a role that may be considered to be catalytic.

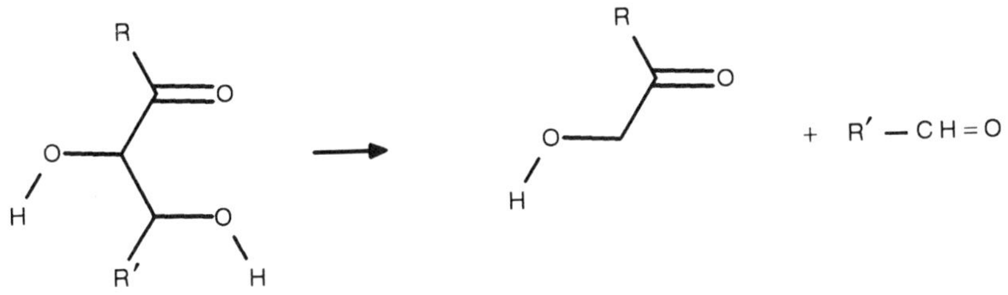

SCHEME 1. Aldolase reaction.

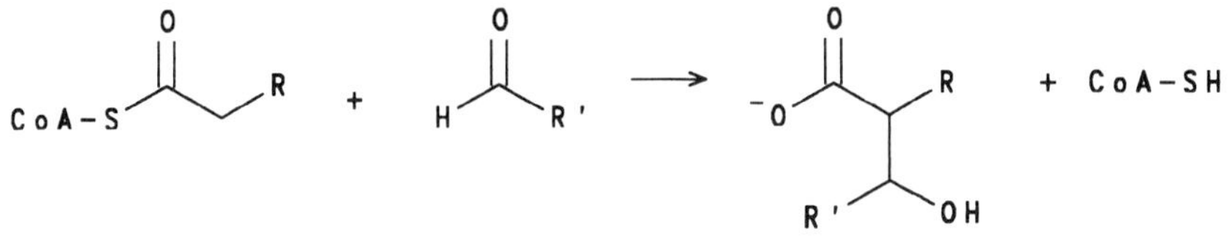

SCHEME 2. Synthase reaction.

It has been proposed that some aldolases function by a similar mechanism (*3*). In these cases metal ion complexes of sugars may act as electrophiles that react with enolate carbanions derived from the other reaction partner (Scheme 3). Alternatively, the metal ion may stabilize the incipient enolate.

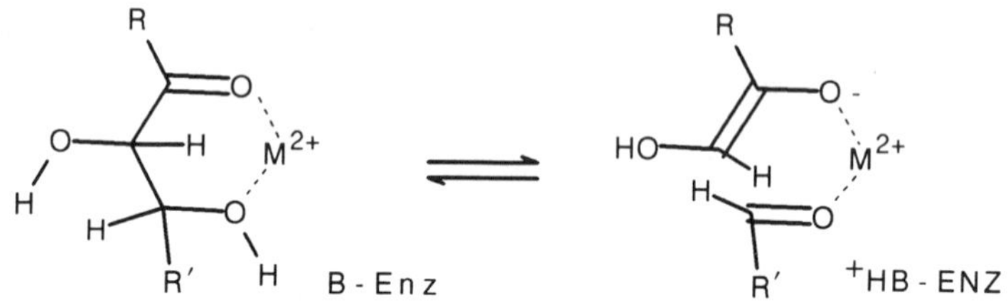

SCHEME 3. Metal ion catalysis in the aldolase reaction.

1. Aldolases Utilizing Enamine Intermediates

Aldolases produce an imine as the initial intermediate from the reaction of a carbonyl group of the substrate with the ε-amino group of a lysine residue of the enzyme (Scheme 4). Reactions of this adduct produce enamines that are tautomers of imines. The imine derivatives of the enzyme can be trapped irreversibly (Scheme 5) by reducing agents, such as sodium borohydride, and by nucleophiles, such as cyanide (*5*). The enamine structure has a zwitterionic resonance contributor with negative charge localized at the carbon α to the carbon attached

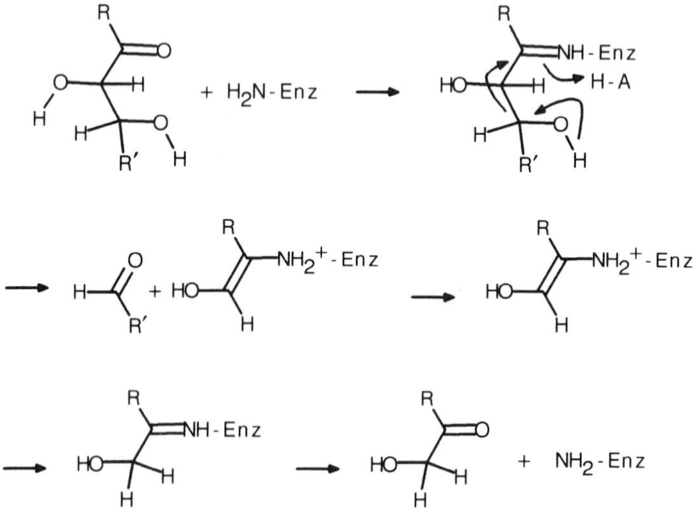

SCHEME 4. Aldolase mechanism.

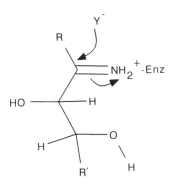

SCHEME 5. Trapping of enamines.

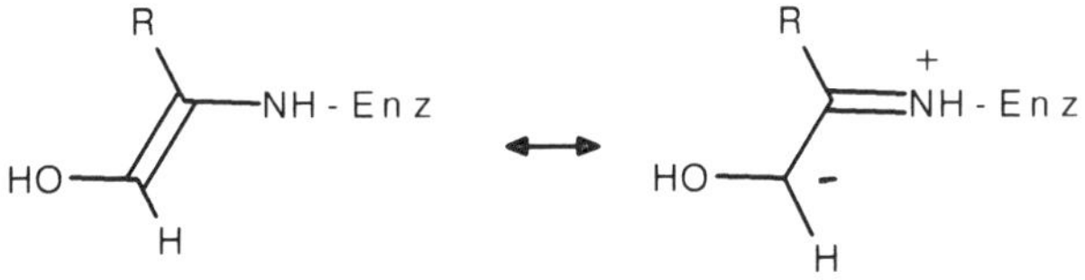

SCHEME 6. Enamine resonance.

to nitrogen (Scheme 6). The polarization of the carbon–carbon bond of the enamine permits the α-carbon to function as a nucleophile and to accept electron density in bond cleavage.

These enzymes catalyze preequilibrium proton exchange at the nucleophilic carbon center at a rate consistent with the intermediate involvement of the conjugate base in the condensation reaction (6). The reaction is formally electrophilic substitution of a carbonyl carbon for a proton at the α-carbon atom of the enamine. Stereochemical studies have shown that the proton and carbonyl bind to the same face of the enamine carbon (carbanionic center) (Scheme 7) (7).

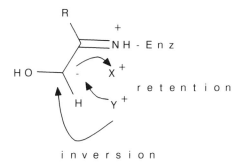

SCHEME 7. Stereochemical considerations.

Substitution with the leaving group and incoming group on the same face of a carbanion is indicative of a reaction in which the carbanion is stabilized by solvent or neighboring electrophile.

a. *Reaction Models for Aldolases.* Good analogies for this mechanism can be found in organic reactions. Amine-catalyzed dealdolization was shown to involve intermediate imine–enamine reactions by early workers. They studied the mechanism of dealdolization of diacetone alcohol (*8, 9*) and speculated that this mechanism would apply to aldolase. Interestingly, the general utility of enamines as carbanion equivalents in organic synthesis was demonstrated decades later.

Other enzymes that utilize Schiff base intermediates for carbon–carbon bond formation have been reviewed by Snell and Dimari in Volume II of "The Enzymes" (*3*).

b. *Demonstration of Aldolase Mechanism.* Horecker and co-workers demonstrated the existence of an intermediate imine–enamine system in fructose-1,6-bisphosphate aldolase from rabbit muscle. They used sodium borohydride to reduce the imine intermediate, which irreversibly inactivated the enzyme. The nature of the adduct of the glycerophosphate adduct with the protein was determined by acid hydrolysis of the protein, which yielded 2 mol of ε-*N*-β-glycerallysine per mole of enzyme (*4*).

2. Direct Carbon–Carbon Bond Formation from Acetyl Coenzyme A

A group of carbon–carbon lyases promote carbon–carbon formation between the α-carbon of a thiolester (typically acetyl-CoA) and the carbonyl carbon of a ketone or aldehyde. These enzymes catalyze a reversible Claisen-type condensation reaction, analogous to an aldol condensation, which is made irreversible by the hydrolysis of the thiolester (see Scheme 2). Unlike the aldolases, these enzymes do not catalyze preequilibrium proton exchange at the carbanionic center and the reactions proceed with backside displacement as indicated by the observation of net inversion of relative configuration. The lack of exchange does not reveal whether an intermediate carbanion forms because there is no way to tell if such a carbanion would be accessible to the medium. The finding that the electrophiles depart and enter from opposite faces of the carbanion is consistent with the carbanion being stabilized as an ionic pair with the entering and leaving groups. The details of such patterns are discussed below.

Walsh refers to the enzymes described in this section (which catalyze carbonyl condensation reactions involving one component that is not an aldehyde or ketone) as "Claisen enzymes" (*6*). I shall use that terminology here, although these reactions are not formally Claisen condensations, but they are distinguished by their mechanisms from the other adolases mentioned in Section III,A,1.

a. *Carbanions in Claisen Enzymes.* The question of whether catalysis by Claisen enzymes involves a free carbanion (enolate) or a transition state in which proton removal to form the enolate is concerted with addition to the carbonyl group of the electrophilic partner has been the subject of a number of studies. The involvement of an enolate in a bimolecular condensation reaction with water has no counterpart in synthetic organic chemistry. However, the high pK_a of the carbon acid (≈ 20) makes it certain that only a very low concentration of free carbanion (enolate) will be present in solution (*10*). If the effective concentration of the other components is high (through enzymic binding) and the rate constant for the reaction between the bound components is large, then the rate of the reaction will also be large.

The α-protons of the substrates in malate and citrate synthases do not undergo catalyzed exchange reactions with the solvent (*4*). These enzymes catalyze the addition of the equivalent of the conjugate carbon base of acetyl-CoA to the keto group of a 2-keto acid and therefore the carbanion is an implied intermediate. The lack of exchange shows that if a carbanion is formed, it is not able to equilibrate with solvent. Of course, if proton transfer and carbon–carbon bond formation occur in a single step, then no proton exchange would be observed. Such a reaction at first seems to provide a reasonable explanation for the results, but the nature of such a single-step reaction deserves extended comment, which follows later.

b. *Trapping of Carbanions as Evidence of Their Existence in Mechanisms.* Oxidizing reagents that react with carbanions have been utilized to provide indirect evidence of the involvement of carbanions in catalytic pathways. Tetranitromethane serves as a source of the equivalent of the powerful electrophile, NO_2^+ (*11*). In aldolase-catalyzed reactions and other reactions where independent evidence exists for reversible carbanion formation, tetranitromethane intercepts the carbanion and generates nitroformate. However, this reagent does not affect Claisen enzymes (*4*). Again, this can be interpreted as indicating that the carbanion is inaccessible, is locally stabilized by a counterion, or that it does not form.

c. *Stereochemical Patterns of Claisen Enzymes.* Stereochemical studies reveal patterns for the substitution process in Claisen enzymes. In these reactions, carbon–carbon bond formation occurs on the face of the nucleophilic carbon center opposite to that from which the proton is abstracted. Srere summarized the stereochemical patterns in enzymes responsible for the formation of citrate and its breakdown (*4*). Citrate is formed by citrate synthase from the reaction of acetyl-CoA and oxaloacetate (citrate synthase) and is cleaved by citrate lyase to acetate and oxaloacetate. Citrate cleavage enzyme converts citrate, coenzyme A, and ATP to acetyl-CoA, oxaloacetate, and ADP.

The condensation of the acetyl methyl carbon of acetyl-CoA with the carbonyl group of oxaloacetate is common to the surmised transition state of all these enzymes. The acetyl methyl carbon becomes deprotonated during the course of the condensation process and the 2-maleyl group, from oxaloacetate, replaces the proton. Normally one would expect that removal of the proton would be an inefficient process, given the high thermodynamic barrier to formation of such a carbanion and given the pK_a of about 20 of carbon acids of this type. However, the only way to avoid the carbanion is to have the proton removal be part of a concerted process with carbon–carbon bond formation. Later in this chapter, I show why we expect the barrier to such a concerted process to be considerably higher than the barriers in reactions that proceed via the carbanion.

Stereochemical studies have shown that the addition to the methyl carbon adjacent to the carbonyl group takes place with inversion of relative configuration (*12–15*). While most enzyme are specific for the *si* face of the carbonyl group of oxaloacetate, some bacterial citrate synthases react at the *re* face. The facial selectivity therefore does not have any mechanistic significance, since it is inconsistent, but the substitution pattern at carbon may indicate an evolved feature based on mechanistic advantage (*16*).

The implications of mechanisms from such data can be derived from the pioneering work of Cram on the stereochemical consequences of carbanionic mechanisms. The stereochemical outcome is the result of several competing factors. Carbanionic reactions in general are not concerted (*17*). Formation of an intimate ion pair between the conjugate base of the substrate and the conjugate acid of a catalytic base in a poor solvent forces attack of the incoming electrophile to the opposite face of the carbanion (*18*). Preassociation of the incoming electrophile could serve to trap the high-energy species so that the reaction would proceed efficiently. The intermediate forms but it does not equilibrate with the solvent and therefore no exchange is observed. Alternatively, retention and racemization result from factors due to the nature of the counterion and solvent.

d. *Reaction Coordinate Diagram for Claisen Enzymes.* The use of a three-coordinate reaction diagram can help in understanding the mechanistic questions and possibility of an alternative pathway (Scheme 8). The third coordinate (not plotted) corresponds to free energy. On one axis we plot the progress of carbon–carbon bond formation and on the other axis we plot the progress of C–H bond cleavage. Since C–H bond cleavage presumably involves transfer of the proton to a basic group on the enzyme, proton transfer to a "B–E" species (for "base–enzyme") is synchronous with C–H bond cleavage. This diagrammatic method was originally used by More O'Ferrall to analyze competing E1, $E1_{cb}$, and E2 reactions (*19*) and was generalized by Jencks (*20*). More recently, Grunwald used this as a basis for a semiquantitative approach to analysis of competing reaction pathways (*21*). Despite the emphasis in Grunwald's paper,

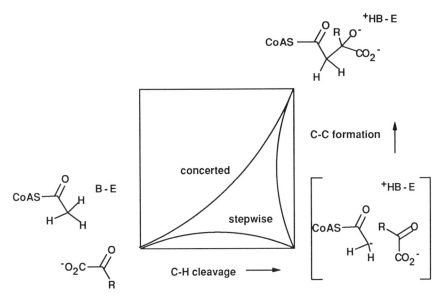

SCHEME 8. Reaction of Claisen enzymes.

as we see later, the methodology is also useful for a qualitative understanding of selection between competing pathways.

   e. *Mechanisms for Claisen Enzymes.* The stepwise mechanism for Claisen enzymes must involve as a first step formation of the conjugate base of the carbon acid by transfer of a proton to a basic group on the enzyme (B–E). The next step is addition of the carbanion (enolate) to the carbonyl carbon of the other reactant. A More-O'Ferrall–Jencks diagram is used to visualize the mechanistic possibilities and probabilities. Progress of the first step is plotted on one axis and progress of the next step is plotted on the perpendicular axis. The complete diagram is a rectangle in which the route described follows two edges in succession. The path of a concerted reaction avoids the corner corresponding to the intermediate carbanion.

   Conceptualizing the structure of any transition state for a concerted reaction requires analysis as to the meaning of departure from the edges of the diagram. Grunwald's procedure enhances this approach (*21*). The corner of the diagram corresponding to the intermediate carbanion is set opposite to a corner representing another intermediate that corresponds to a path in which events take place in the opposite order, even if this is improbable. A diagonal connecting reactants and products is termed the main reaction, and the perpendicular diagonal, between intermediates, is called the disparity reaction. If there is a well in the energy surface of the disparity reaction, then a concerted reaction will be able to occur. Since energy surfaces are continuous, finding such a well anywhere is sufficient to assure that a concerted reaction is possible. Alternatively, if the stepwise path involves an intermediate with no possibility of a lifetime of more than one vibration ($\sim 10^{-13}$ sec), the concerted path is a necessary alternative. An example of such a diagram is given in the following section.

f. *Malate Synthase Intermediates.* As an illustration, for the malate synthase reaction, we plot reaction progress in the general direction of C–H bond cleavage on the x axis and we plot the formation of a new carbon–carbon bond on the y axis. The coordinates are labeled (0,0) for the start of the reaction, (1,1) for the products, and (0,1) and (1,0) for the intermediates. Energy increases in the direction above the x–y plane and decreases below it. Carbon–carbon bond formation is exergonic so that (1,1) is downhill from (0,0). Breaking the C–H bond (by transfer of the proton to a base on the enzyme, B–E) without formation of a carbon–carbon bond leads to the carbanion (enolate) intermediate, which characterizes the stepwise mechanism. This is above the reactants by the free energy corresponding to the formation of the enolate by the base on the enzyme (B–E becomes $^+$HB–E). Since the formation of a carbon–carbon bond without breaking the C–H bond leads to unconventional bonding, this axis label is not directly carbon–carbon bond formation, but indicates that whatever bonding that can occur without departure of the proton is complete at the point marked (0,1).

g. *Use of Grunwald Diagrams for Malate Synthase.* Recognition of the disparity reaction is a useful way to devise a basis from which to interpolate the structure of the transition state in a concerted reaction and to consider the existence of such a path (Scheme 9). The line for the main reaction mode then can be visualized as sliding like a string along the "solid" energy coordinate of the disparity reaction. The disparity coordinate can be analyzed for the possibility of a concave section through which the main reaction can take a lower energy route than permitted by the stepwise processes. The intermediates and products are shown as coenzyme A derivatives (although hydrolysis occurs during the course of the reaction), and proton transfer to oxygen has not been included. The intermediate in the disparity reaction results from the improbable situation of initial carbon–carbon bond formation without proton transfer. Since this necessarily involves an expansion of valence at carbon, the intermediate contains two 5-coordinate carbon atoms analogous to edge-protonated cyclopropane species where the proton is delocalized across the σ-bonds and there are two electrons distributed among three nuclear centers. This should be a considerably higher free energy point than the enolate and its partners at the opposite corner.

To the extent that the concerted path for the main reaction involves a transition state that does not have a full carbanionic center, it must bear some resemblance to the intermediate of the disparity reaction. It is unlikely that the intermediate

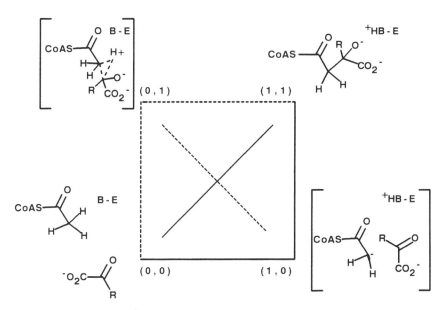

SCHEME 9. Grunwald disparity diagram for malate synthase.

with 5-coordinate carbon atoms would exist in a stepwise process under normal circumstances. The energetic problem with the concerted reaction, besides its higher entropic demands, is that it must bear some component of this high-energy structure. We see no apparent reason for there to be any concave section on the axis of the disparity reaction in this case: partial transfer of the proton from the substrate to a base of the enzyme does not stabilize the situation until the proton is fully transferred. Therefore, the carbanion–enolate in the stepwise main reaction should always lie on a lower energy path. The only alternative reason for a concerted reaction would be if the carbanion were too unstable to exit. In that case, the lifetime of the species is less than a vibration and by the definitions of transition-state theory, this is not distinguishable from a transition state.

h. *Double-Isotope Fractionation and Carbanion Intermediates.* The proof of the existence of a carbanion–enolate intermediate has been attempted by a number of approaches. The most direct evidence is the observation of isotopic exchange of the α-hydrogen for deuterium or tritium (in deuterated or tritated water) at a kinetically competent rate. If the carbanion is accessible to the solvent, instead of reacting with the carbonyl carbon of the second substrate, the carbanion, in principle, should be able to react with the solvent to produce the isotope-exchanged reactant. Dissociation of the reactant from the enzyme leads to the observed exchange. However, in all enzyme reactions for which this method has been attempted, exchange is only observed for materials that are not substrates (4). This does not rule out carbanions as intermediates in these cases but rather suggests that the enzyme operates most effectively by protecting the carbanion or its equivalent from reaction with solvent so that it can react more efficiently with the second substrate. Therefore, other methods have been developed to test for the involvement of an intermediate.

In an important study, Cleland, O'Leary, and their co-workers established a sophisticated use of isotope effects to establish whether a process involving two bond-making or bond-breaking events occurs in one or more steps (22). The method involves the observation of the magnitude of the *change* in the kinetic isotope effect on the second process of the sequence due to isotopic substitution affecting the first process. If the reaction can occur in two steps, isotopes are substituted at sites associated with each step (*Editor's note:* See chapter by Cleland, Vol. XIX), first separately then in tandem. If the intrinsic effect on the second step (that which is observed when there is no other isotopic substitution) is greater than the observed effect (when the second isotope is present), the first step must be partially rate limiting.

In a single-step process that involves both isotopically sensitive processes (as primary isotope effects; the bond to the isotope is broken), the isotope effects should be independent, according to the fundamental assumption of transition-state theory, because the bond to the atom (the isotopic atom in this case) in the transition state has no residual vibrational component (1). Therefore, the isotope effect is due exclusively to differences in ground-state vibrational levels. In a single step in which both isotopic bonds are involved for the two processes of interest, the isotope effects must function independently. In a sequential system, the first isotope effect perturbs the proportioning of the second.

   i.   *Isotope Fractionation in Malate Synthase Reactions.* The malate synthase reaction is analogous to the citrate synthase reaction and the mechanism was the subject of study by Knowles using the double-isotope fractionation method (23). In this Claisen enzyme-catalyzed reaction, malate is produced from the condensation of acetyl-CoA onto glyoxalate. The enzyme does not form an imine intermediate and there is no coenzyme requirement. Eggerer and Klette (24) found that malate synthase does not catalyze exchange of the methyl protons of acetyl-CoA with solvent in the absence of glyoxalate. The stereochemical course of the reaction is inversion at the acetyl carbon, a result that is consistent with either a stepwise or a concerted process (23, 25). Therefore, these results do not establish whether the reaction involves an intermediate.

Knowles and co-workers specifically tested whether cleavage of the carbon–hydrogen bond and formation of the carbon–carbon bond occur in the same step using the double-isotope fractionation method (23). The only reasonable chemical mechanism that would have these occur in separate steps would have carbon–hydrogen bond cleavage precede carbon–carbon bond formation. In a double-isotope fractionation study, it first must be determined that there is a hydrogen isotope effect on V_{max}/K_m. Then the isotope effect for V_{max}/K_m, with carbon-12 replaced with carbon-13 at a position involved in carbon–carbon bond formation, is determined for substrates with hydrogen and deuterium in the bond that is cleaved. Cleland's shorthand nomenclature is convenient to use: $V_{max}/K_m = V/K$. The ratio of V/K for substrates with carbon-12 to those with carbon-13 is written as $^{13}(V/K)$. The ratio for substrates containing hydrogen in the position where carbon–hydrogen bond cleavage occurs is $^{13}(V/K)_H$ and the ratio for substrates in which a C–D bond is broken is $^{13}(V/K)_D$.

The carbon isotope effect on V/K at C-1 of glyoxylate (which is the electrophile) was determined using natural abundance carbon isotopic mixtures of acetyl-CoA and trideuterioacetyl-CoA as the nucleophilic reactant. It is known for this enzyme that the H/D isotope effect on V/K and V is significant (26). The observed carbon isotope effect is based on analysis of product distributions and not on measurement of rates. Therefore, the ratio gives a direct reflection of the competition in the carbon–carbon bond-forming step between the isotopomers. If proton removal occurs in a step that precedes the step in which the carbon isotope is manifested, and both steps are partially rate determining, the observed carbon isotope effect will be reduced. If the reaction is concerted, then deuteration will have no effect on the observed $^{13}(V/K)$, since in the transition state both bonds are broken and the H/D effect will not affect the proportioning of the intermediate and the observed $^{13}(V/K)$ isotope effect. The analysis of the isotopic composition of the product compared to that of the reactant must be done at early stages of the reaction. The need for accurate measurements requires that the carbon source measured for isotopic distribution be done by isotope ratio mass spectrometry of carbon dioxide.

Knowles and co-workers observed $^{13}(V/K)$ at C-1 of glyoxalate by analyzing the isotopic composition of the malate produced in the reaction at early and late stages (23). In order to measure this ratio accurately they developed a procedure to convert C-2 of malate to carbon dioxide. Through a series of enzymic reactions, C-2 was converted to the carbonyl carbon of the acetyl group of acetyl-carnitine (through the reaction sequence malate → pyruvate → acetyl-CoA → acetylcarnitine). The labeled carbon then was carried through another cycle back to malate, but in this case the label was in the carboxylate carbon at C-4 (acetyl-carnitine → acetyl-CoA → 4-[^{13}C]malate). Malic enzyme converts this to pyruvate and labeled carbon dioxide, whose isotopic ratio is determined by mass spectrometry.

The results of the determination of $^{13}(V/K)_H$ for the condensation of acetyl-CoA with glyoxalate catalyzed by malate synthase gave 1.0037 with a standard deviation of 0.00035. When trideuterioacetyl-CoA was used as a substrate in place of acetyl-CoA, the isotope effect, $^{13}(V/K)_D$, was also 1.0037 ($\pm$ 0.0007). The deuterium isotope effect itself, $^{D}(V/K)$, was determined to be 1.3 and the K_m was 1.0, in agreement with earlier reports of Eggerer's group (26). Since there is a $^{13}(V/K)$ effect, one would conclude that a step involving formation of a bond to C-1 of glyoxalate is kinetically significant. Since there also is a $^{D}(V/K)$ effect, one would also conclude that carbon–hydrogen bond cleavage is also kinetically significant. Therefore, it is necessary to find out if these kinetically significant steps are in fact one step or are separate.

Based on the background I have presented on double-isotope fractionation, the observation that for malate synthase $^{13}(V/K)_D$ and $^{13}(V/K)_H$ are identical supports a mechanism in which C–H bond breaking and C–C bond formation occur in the same step: that is, the reaction is truly concerted. Since both isotope effects are seen, one would expect that both centers must be involved in rate-determining processes. In a simplified energy diagram, the presence of a deuterium isotope effect ahead of a carbon isotope effect causes the (apparent) observed carbon isotope effect to become larger. In a two-step steady-state situation, the observed rate constant, k_{obs}, depends on $k_2/(k_{-1} + k_2)$. If isotopic substitution decreases k_1 and k_{-1}, then the observed rate constant becomes less affected by the magnitude of k_2. Therefore, in this case the apparent carbon-13 isotope effect will decrease if deuteration reduces the size of k_1 and k_{-1}. If the carbon-13 isotope effect is constant, then the expression for k_{obs} does not apply. If the two processes occur in the only kinetically significant step, then proton removal will not affect the observed carbon-13 isotope effect. The conclusion that the reaction is concerted is consistent with observations by Knowles but does not agree with chemical reasonableness based on the stereochemical observation of inversion in these systems and the Grunwald diagram analysis, which shows a significant disadvantage for a concerted mechanism versus a stepwise mechanism. However, despite all the preceding information, the results do not *require* that the reaction occurs by a concerted mechanism: they simply rule out a mechanism in which the two steps are each partially rate determining. If the processes occur in different steps and if one or both steps occur after the rate-determining step, then the isotope effects will not interact. Two further pieces of information clarify the picture and establish that (1) a kinetically significant step that is independent of isotope effects precedes both C–H bond breakage and carbon–carbon bond formation, (2) the observed carbon isotope effect is due to the effects of carbon-13 on an uncatalyzed preequilibrium process, and (3) the carbon–carbon bond-forming step is kinetically insignificant (23).

A comparison of the primary hydrogen isotope effect measured in two different ways for the malate synthase reaction provides the clue that permits interpretation of the unusual result in the double-isotope fractionation study. It leads to the conclusion that a kinetically significant step that is isotope insensitive precedes C–H bond cleavage. Therefore, the observed hydrogen isotope effect is not a measure of the properties of the rate-determining step.

In an earlier study, Lenz and Eggerer found that the *intramolecular* $^D k$ effect is 3.8 (*25*). An intramolecular isotope effect measures the results of competition between breaking a C–H versus a C–D bond in the methyl group of partially deuterated acetyl-CoA. The usual kinetic isotope effect is *intermolecular,* where samples of different isotopic composition react under parallel conditions. If the step in which C–H bond-breaking occurs is after the rate-determining step, there should be no observable intermolecular isotope effect. In the case of an intramolecular isotope effect, the measured quantity is the isotopic distribution in the products. Every C–D bond cleavage process competes internally with C–H bond cleavage in the same reacting molecule and the deuterated species is necessarily carried forward by C–H cleavage. Therefore, even if the bond-breaking step occurs after the rate-determining step, the product distribution will reflect a preferential cleavage of the C–H bond. In the case of an intermolecular measurement, the C–D bond is necessarily cleaved in the deuterated substrate and therefore an isotope effect is noted only if C–D cleavage slows the net rate, and this must precede the rate-determining step. Thus, the intramolecular effect is not a measure of the relative rates of reaction of deuterated and undeuterated substrates, but only of the branching in the step that determines whether the product results from the loss of deuterium or hydrogen, a distinction that must occur whether or not the step affects the rate of the reaction.

Under conditions in which an intermolecular isotope effect is measured, if the rate law is a mixture of expressions that are dependent and independent of isotope effects, then the observed rate constants for deuterated and undeuterated substrates will have a smaller ratio than the product distribution measured for the intramolecular effect. In the case of malate synthase, the measured intermolecular isotope effect is only 1.36 whereas the intramolecular effect is 3.8 (*23, 25*). If proton abstraction occurred in the first irreversible step, there would be no difference between the measured intramolecular and intermolecular and isotope effects. The fact that the two isotope effects are different indicates that the C–H bond cleavage step is preceded by an isotopically insensitive, but kinetically significant, step. Therefore, if the C–H bond cleavage step is concerted with carbon–carbon bond formation and that step is partially rate limiting (which it must be, since an isotope effect is observed), that concerted step becomes more rate limiting on deuteration and the magnitude observed for $^{13}(V/K)$ increases.

Because there is an isotopically insensitive, kinetically significant step preceding the C–H bond cleavage step, the observation of an invariant $^{13}(V/K)$ on deuteration is obviously not the result of a concerted mechanism, since that would require the two events to be in one step, and it was shown that the two processes are on either side of the rate-determining process in the reaction sequence. Chemical reasonableness requires that carbon–carbon bond formation cannot precede carbon–hydrogen bond cleavage (see the Grunwald diagram, Scheme 9), yet there is an invariant $^{13}(V/K)$ effect that is greater than unity.

Is any reasonable mechanism consistent with the data? The answer lies in an observation of a probable isotope effect in a coupled nonenzymic phenomenon. The double-isotope fractionation method does not enter into the analysis. The keto group of glyoxalate is actually present as a covalent hydrate to the extent of about 99% of the total glyoxalate concentration (27). However, the ketone is the form that will react in the enzymic process and the concentration of ketone determines the rate of reaction and binding to the enzyme. The equilibrium between ketone and hydrate is not catalyzed by the enzyme and as a result the isotope effect on this equilibrium will appear in the measured kinetic isotope effects. Of course, the extent of this equilibrium will not be affected by deuteration of the methyl group of acetyl-CoA. Therefore, the observed $^{13}(V/K)$ is not an indication of kinetically significant carbon–carbon bond formation but of a preequilibrium hydration, a process that is independent of the enzyme. The value for $^{13}(V/K)$ of 1.0037 is consistent with measured equilibrium isotope effects in related molecules (23). Therefore, the deuteration of acetyl-CoA has no effect on the observed kinetic $^{13}(V/K)$ because that value in fact is due to a preequilibrium and not the rate-determining step. Since proton removal is kinetically significant, if this were concerted with carbon–carbon bond formation, the observed $^{13}(V/K)$ would necessarily have increased because it is in a step after a kinetically significant, isotopically insensitive step. It is concluded that based on the magnitude of the ratio of the intramolecular and intermolecular isotope effects a concerted reaction would have seen $^{13}(V/K)$ increase to 1.011. (The intramolecular effect is about three times the intermolecular effect and the heavy atom effect is predicted to change by the same ratio if the two processes are concerted.) Therefore, the data are consistent with the stepwise mechanism.

The formation of the glyoxalate hydrate is obviously unproductive and is probably an artifact resulting from the assay being done on the isolated system. Bernhard found that many enzymes can bind to other enzymes that occupy adjacent positions in metabolic pathways (28), transferring substrates and products without the intermediate involvement of solvent. In the case of malate synthase, the oxidation of glycolate to glyoxalate might produce material in the unhydrated form, which would be used directly. In that case, the equilibrium isotope effect on hydration would not be observed.

What is the kinetically significant but isotopically insensitive step that precedes C–H bond breaking and carbon–carbon bond formation? The magnitude of V/K (which equates to a second-order rate constant) is small enough to show that the reaction is not diffusion limited. It is thus suggested that the enzyme undergoes a conformational change after binding the two substrates (23).

j. *Concertedness in Condensation Reactions.* The work discussed above indirectly establishes that carbon–carbon bond formation in malate synthase is a stepwise process, a consequence that might have been deduced from inspection of the Grunwald diagram. First, a proton is removed from acetyl-CoA to form the enolate, and then the enolate carbon from which the proton has been removed reacts with the carbonyl carbon of glyoxalate. A concerted reaction would avoid formation of the enolate, but this apparently is avoided because the transition state must be higher in free energy for the concerted process than it is for stepwise reaction. The issue of concerted versus stepwise reactions makes most sense when the two mechanisms are known to be competitive energetically. Is it reasonable to expect that such is the case here?

In nonenzymic reactions, competition between concerted and stepwise reactions is not known for condensation processes. In all cases of intermolecular and intramolecular condensation reactions, there is initial formation of an enolate followed by the formation of the carbon–carbon bond. However, in the case of traditional elimination and substitution reactions, there is a competition between stepwise and concerted processes. The concerted processes (E2 and S_N2) compete with stepwise processes (E1 and $E1_{cb}$, S_N2). In the case of the E2 reaction, a proton is removed and a carbon-based leaving-group bond is broken in a single step, whereas the stepwise mechanisms do these steps in sequence. Depending on the extent of proton transfer and leaving-group departure, the E2 reaction transition state can have E1 or $E1_{cb}$ character. The synthase reactions do not have a good analogy to this since no leaving group is involved. Formally there are three chemical processes involved in the condensation reaction: (1) removal of the proton from the α-carbon, (2) addition of the carbanion to the carbonyl group of the reaction partner, and (3) protonation of the alkoxide product. However, in the consideration of stepwise versus concerted reaction in the study by Knowles, only the synchronization of the first two processes is of significance because protonation on oxygen was not evaluated (*23*).

k. *Rationale for Stereochemical Observations.* The observed stereochemistry of inversion in the enolate reactions is consistent with a mechanism in which a base on the enzyme removes the α-proton and forms a stabilized ion pair. The addition reaction then proceeds from the opposite face of the carbanion.

l. *Enol Mechanism for Claisen Enzymes.* An alternative mechanism for Claisen enzymes involves initial conversion of the nucleophilic substrate into the corresponding enol. In this case, addition of the carbon of the enol to the carbonyl of the electrophilic reactant can be assisted by acid–base catalysis. Application of the mechanism to the malate synthase reaction is shown in the Scheme 10. This mechanism avoids formation of both intermediates considered in the

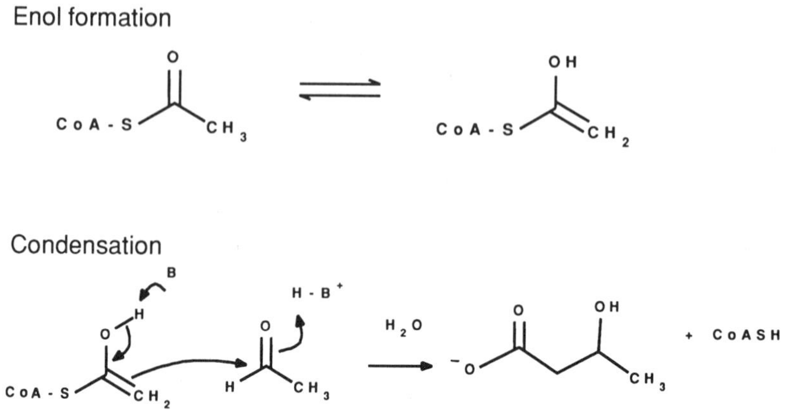

Enol formation

Condensation

Scheme 10. Enol intermediate mechanism.

carbanion mechanism by providing the alternative route through the enol. Although enols are higher in energy than the corresponding tautomeric carbonyl compound, they are lower in energy in neutral solution than the enolate. The involvement of general acid–base catalysis can serve to enhance the rate of bond formation and avoid the formation of protonated carbonyl compounds as intermediates. Formally, this mechanism is a subset of the stepwise mechanisms I have presented as occurring via the corresponding enolate.

3. *Generalization*

Enzyme-catalyzed reactions in which carbon–carbon bonds are formed between a carbon atom adjacent to a carbonyl group in one molecule and the carbonyl carbon of another molecule are stepwise processes. The bond-forming step is not concerted with any other process. Stereochemical patterns are indicative of the means by which the carbanionic species is stabilized. Informative isotope effects are derived from the use of carbon-13 as a substituent at the atoms that undergo bond formation. The use of Grunwald diagrams provides insights into the basis of the mechanistic pattern, showing that the reactions are stepwise because there is no energetic advantage to be gained from a concerted process. Only in the case wherein an intermediate would be too unstable to exist can we expect to find a concerted process, and none has yet been demonstrated.

B. CARBOCATIONIC TRANSITION STATES

The formation of a carbon–carbon bond via intermediates centered about an electron-deficient carbon atom is generally classed as an S_N1 reaction, but the formal description involving a free carbonium ion as an intermediate is an over-simplification and is not applicable to a reaction in the environment provided by an enzyme. An alternative way of describing an $S_N 1$ process is that the displacement at carbon is necessarily stepwise and that departure of the leaving group occurs in a step that precedes the entry of the nucleophile. The leaving group need not dissociate from its complex with the carbon-centered intermediate prior to its departure. On an enzyme, such a mechanism requires a reasonable leaving group and a carbon center that can readily become electron deficient.

Most reactions involving the enzymes listed in Section II include the combination of a carbanion center with an electrophilic center that is part of an unsymmetrical double bond, such as a carbonyl or imino function. The cases in which carbonium ion-type intermediates form involve a completely different structure. This is not a case of competing mechanisms among common functional groups, but rather a specific relationship between functional group structure and mechanism.

1. *Prenyl Transfer*

The formation of terpenes and other isoprenoid lipids involves the condensation of dimethylallyl pyrophosphate with isopentenyl pyrophosphate (*29*). In this case (Scheme 11), the leaving group is the pyrophosphate (PP) of an allylic pyrophosphate, such as dimethylallyl pyrophosphate, generating an allylic carbonium ion derivative (and inorganic pyrophosphate, which forms as an ion pair with the carbonium ion). The electron-rich agent that adds to the electron-deficient center to form the carbon–carbon bond is the π-electron density of the isopentenyl pyrophosphate. The stereochemical course of the biosynthesis of complex natural products has been determined and is consistent with such a mechanism (*30*).

In principle, carbon–carbon bond formation via prenyl transfer might occur via a carbanion generated from addition of an enzymic nucleophile to one of the

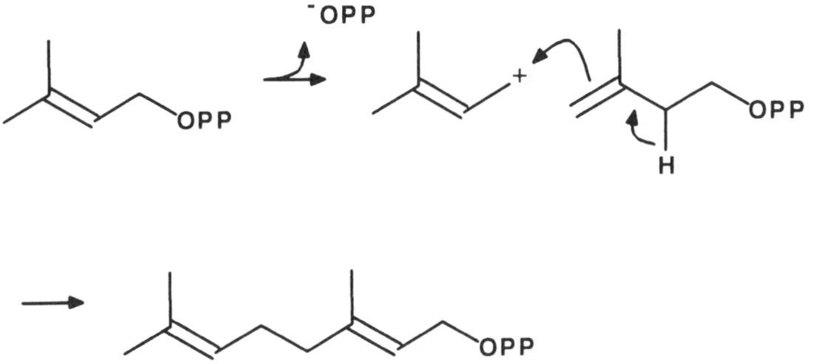

SCHEME 11. General prenyl transfer.

substrates (*30*). This carbanion would displace the pyrophosphate group of the al-lylic substrate in an S_N2 or S_N1 process. A subsequent elimination reaction leads to expulsion of the enzymic nucleophile. The stereochemical course of the reaction in a nonenzymic system would normally be a probe of the nature of the carbonium ion intermediate. The formation of a symmetrically solvated intermediate would give complete epimerization of the reaction center. Ion pairing between the carbonium ion and the leaving group would lead to a degree of inversion while solvent participation would lead to retention of relative configuration. This analysis does not apply in an enzyme-catalyzed reaction because the chirotopic environment [terminology of Mislow and Siegel (*31*)] of the enzyme site controls the stereochemical outcome, whatever the nature of the intermediate. Isotope scrambling experiments (*32*) could also give similar information, but these have not been observed in enzymes catalyzing this class of reaction. An illustration of such a scrambling reaction is shown in Scheme 12.

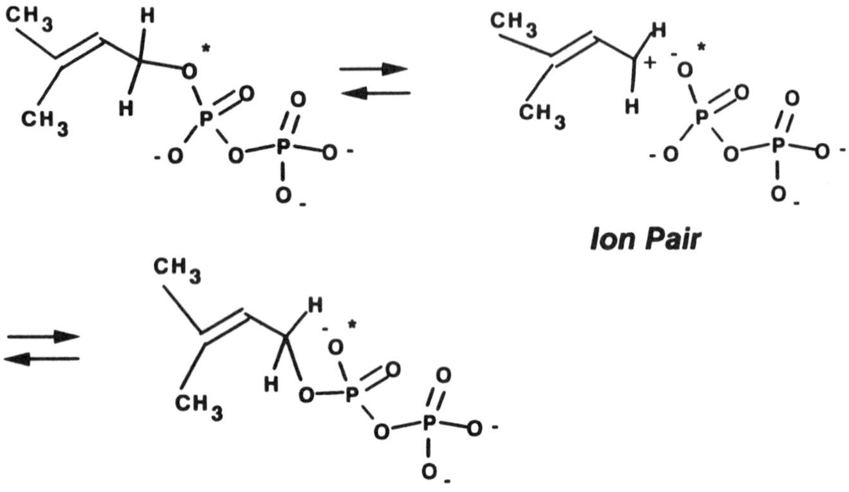

SCHEME 12. Internal return in carbonium ions.

2. Squalene Synthase and Related Processes

The head-to-head union of two farnesyl pyrophosphate molecules to produce squalene is promoted by squalene synthase (*33*). The enzyme utilizes NADPH as a cofactor. If the cofactor is omitted, another species, presqualene pyrophosphate, accumulates (*34*). On addition of NADPH, this is converted to squalene. The structure of presqualene pyrophosphate was elucidated by Epstein and Rilling, who found that the material contains a highly substituted cyclopropane ring as a central feature (*35*). Poulter and associates (*36*), as well as van Tamelen and

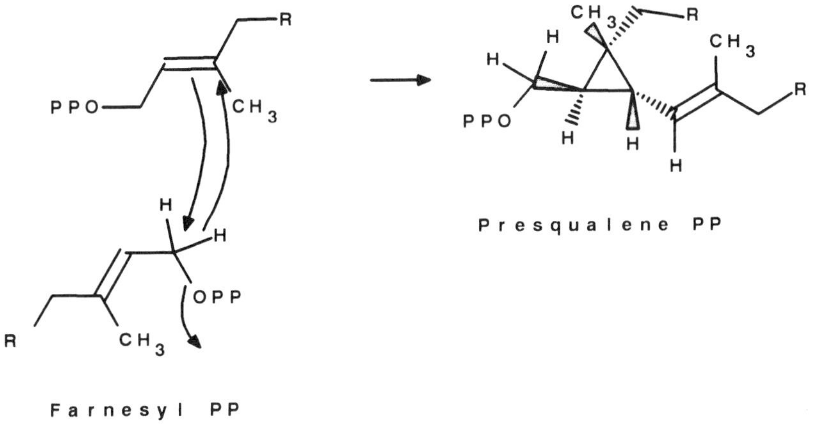

SCHEME 13. Formation of presqualene PP from farnesyl PP.

Schwartz (37), independently proposed the route that leads to presqualene pyrophosphate and to squalene. The head-to-head carbon–carbon bond-forming reaction (Scheme 13) that produces presqualene pyrophosphate formally involves the combination of the elements of two farnesyl moieties, one as the pyrophosphate and the other as the ion from which pyrophosphate has been cleaved. In contrast, the more common head-to-tail carbon–carbon bond-forming process involves two different initial species as reactants.

The conversion of presqualene pyrophosphate to squalene (Scheme 14) is an

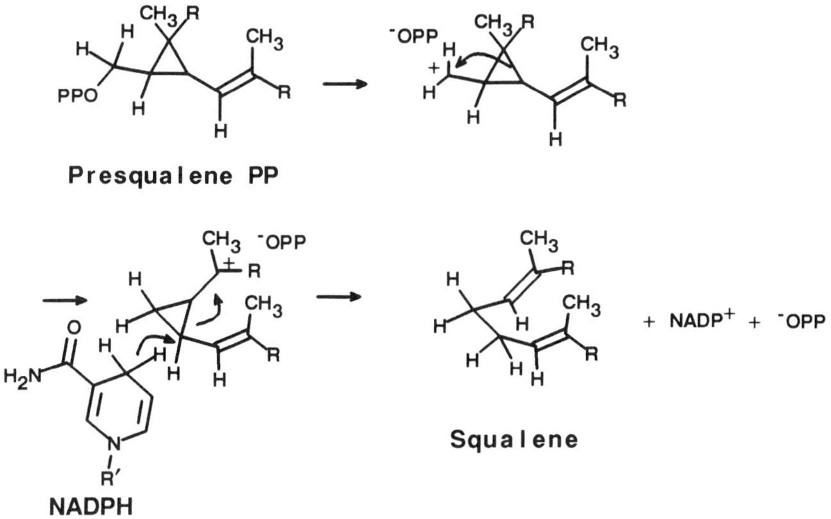

SCHEME 14. Formation of squalene from presqualene PP.

example of the rearrangement of a substituted cyclopropylcarbinyl carbonium ion (*38*). This ion has been studied as an example of a σ-delocalized system and the formation of squalene is a logical possibility of one of the established rearrangements of the ion. However, the enzyme is clearly involved in directing the rearrangement, since Poulter has shown that the spontaneously formed products that would result are derived from structural isomers of the squalene skeleton (*39*).

Although the head-to-head linkage of farnesyl units produces squalene, there are many other terpene-related materials that also appear to result from linkages other than common $1'$–4 (head-to-tail) condensation between an allylic pyrophosphate and a homoallylic pyrophosphate. Huang and Poulter have provided a mechanistically detailed picture of the mechanism of $1'$–3 bond formation in the biosynthesis of the triterpene botryococcene (Scheme 15) (*40*). This C_{30} compound is formed from two equivalents of farnesyl pyrophosphate. Presqualene pyrophosphate, the precursor to squalene, is also a precursor to botryococcene. As noted previously, cyclopropylcarbinyl model systems for presqualene pyrophosphate do not produce the structure characteristic of squalene. In fact, they react almost exclusively to produce the $1,3'$ linkage characteristic of botryococcene.

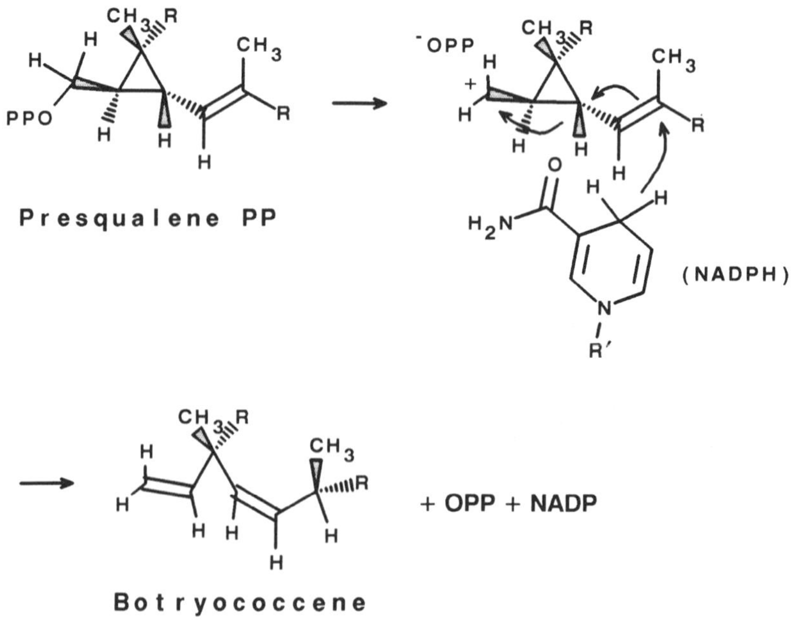

SCHEME 15. Formation of botryococcene from presqualene PP.

Using stereospecifically labeled precursors, Huang and Poulter have provided convincing evidence that squalene and botryococcene arise from a common cyclopropylcarbinyl intermediate, presqualene pyrophosphate (40). Huang and Poulter propose that the relative binding position of NADPH and the cyclopropane intermediate will control the outcome with respect to the final carbon–carbon bond that is produced (40). This work provides not only strong evidence for the existence of carbonium ion intermediates, but also implicates them in a wide variety of enzyme-catalyzed carbon–carbon bond formation processes.

Further information about the carbonium ion intermediates in these systems comes from studies in which ammonium analogs of the carbonium ion are used as inhibitors. Poulter and co-workers (41, 42) have shown that in these reactions the ammonium analog is a very weak inhibitor, but in the presence of pyrophosphate ion, binding becomes dramatically stronger. In a molecule in which the pyrophosphate is covalently attached to form an internal ion pair, an extremely powerful inhibitor is generated.

3. Generalization

Enzymes catalyze the formation of carbon–carbon bonds between allylic and homoallylic pyrophosphate species by mechanisms that are very different from those for carbonyl compounds. Here, carbonium ions, stabilized as ion pairs and generated from allylic pyrophosphates, are likely to be the intermediates that add to the π-electron density of carbon–carbon double bonds to form new carbon–carbon single bonds. Reaction patterns are consistent with model systems and the mechanisms are based on analogies with the models, stereochemical information (which is subject to interpretation), and the structural requirements for inhibitors. Detailed kinetic studies, including isotope effects, which provide probes in the aldolase and Claisen enzymes discussed in Section II, have not yet been performed in these systems. The possibility for surprising discoveries remains and further work is needed to confirm the proposed mechanisms and to generalize them.

IV. Mechanistic Patterns for Carbon–Carbon Bond Formation Involving Selected Coenzymes

REACTION PATTERNS

Carbon-carbon bond formation and cleavage are reactions in which coenzymes play a major role. The coenzymes as a group possess functionality that is not normally found in the side chain of a protein and thus can provide for chemical mechanisms that are not accessible in a protein without such a cofactor. The coenzyme-dependent reactions follow distinct mechanistic patterns and operate on a recognizable group of substrates. The reaction pattern is often identified before the coenzyme when a new process is discovered. Interesting cases arise when the expectation of mechanism and coenzyme turn out to be incorrect and we learn that another mechanistic path is possible. Some examples of such discoveries will be included at the end of this section. However, the purpose in this section is to define the common features of carbon–carbon bond formation and cleavage that are most typical of these coenzyme-dependent reactions. Reactions that are covered in detail in other chapters of "The Enzymes," such as decarboxylation, will not be covered here. However, that material is integral to the objectives of the present chapter, and the reader should include that information as part of the overall pattern being presented.

1. Carboxylation with Biotin as Cofactor

The formation of a bond between the carboxylate group derived from bicarbonate and a carbon adjacent to a carbonyl group is indicative of a reaction catalyzed by an enzyme that utilizes biotin as a cofactor (Scheme 16). The recent review by Knowles covers many recent discoveries relating to the role of enzymes in these reactions (43). (*Editor's note:* For additional aspects of biotin-dependent carboxylation, see Chapter 6 by O'Leary.) Most biotin-dependent enzymes promote a two-step process in which N-carboxybiotin serves as an intermediate in a process involving the exchange of the carboxylate group derived from bicarbonate for a proton at the α-carbon of the carbonyl compound (44).

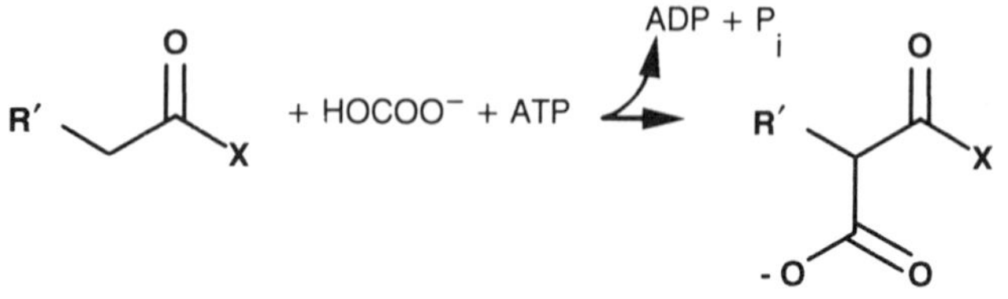

SCHEME 16. Generalized biotin-mediated substitution.

The formation of N-carboxybiotin from bicarbonate and biotin on an enzyme involves the stoichiometric hydrolysis of ATP to produce ADP and inorganic phosphate (Scheme 17). This is particularly significant because the hydrolysis of ATP is "coupled" to formation of the carbon–carbon bond (45, 46) but is not explicitly involved in the apparent stoichiometry of the biosynthetic process. Therefore, the function of ATP is cryptic, as it is in most processes in which ATP hydrolysis accompanies a biosynthetic process. Although biotin-dependent reactions are not a general model for other ATP-dependent processes, the patterns that emerge from the study of such a mechanism guides one in thinking about the other processes. In the biotin-dependent case, the reaction with ATP

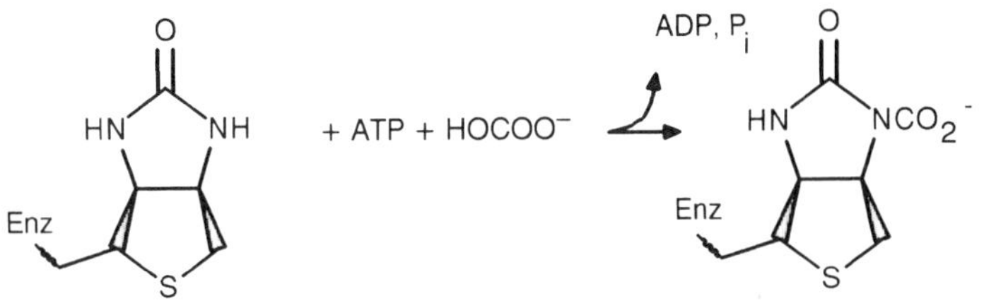

SCHEME 17. Formation of N-carboxybiotin.

changes the process from a two-component system (biotin and bicarbonate) that produces N-carboxybiotin and hydroxide to a three-component system that produces N-carboxybiotin, ADP, and inorganic phosphate. The net conversion of ATP to ADP and inorganic phosphate permits a thermodynamic advantage over the production of hydroxide in a direct process.

The excess oxygen from bicarbonate that is not transferred as the carboxyl group has been shown by isotopic labeling of the oxygen atoms to be transferred to the terminal phosphoanhydride group of ATP, appearing in the inorganic phosphate product (47). The specific mechanistic role of ATP in the process is cryptic because its components are not transferred to any product other than ADP and inorganic phosphate. This has led to a wide range of mechanistic proposals that must be based on a combination of related enzymic data and models.

a. *Mechanism of ATP-Dependent Carboxylations.* No partial exchange reactions are catalyzed by any biotin enzyme (43). The incubation of labeled ADP and unlabeled ATP in the absence of other substrates does not lead to interconversion. Climent and Rubio (48) report that biotin carboxylase promotes the hydrolysis of ATP in the presence of bicarbonate and absence of biotin. They cite this as conclusive evidence in favor of a partial reaction between ATP and bicarbonate occurring in the normal catalytic cycle and propose that this supports the existence of carboxy phosphate as an intermediate. The rate of this reaction is only 0.005 times that of the reaction in the presence of biotin (where ATP hydrolysis is coupled to carboxylation). Why is the reaction slower in the absence of biotin if the normal generation of carboxy phosphate does not depend on biotin? An alternative interpretation is that in the absence of biotin, the site that normally contains biotin is occupied by water or bicarbonate. Since the enzyme is likely to provide a Brönsted base to remove a proton from biotin (49), it will also enhance the nucleophilicity of other bound species. If biotin normally attacks either ATP or bicarbonate (mechanisms that do not involve carboxy phosphate), then one would predict that when water or bicarbonate binds in the biotin site, they would promote the cleavage of ATP, but at a slow rate. Alternatively, the slow rate of the bicarbonate-dependent ATPase reaction can be considered as a kinetically incompetent reaction whose interpretation is ambiguous.

A more subtle exchange, which is indicative of internal return in the formation of an intermediate, can be detected in some enzymes through the use of what Middelfort and Rose call positional isotope exchange (32). However, this class of experiments has also given no evidence of an intermediate in that no positional isotope exchange is observed (43). Together, the exchange results show that all the reaction components, biotin, ATP, bicarbonate, and substrate, are necessary in order for an enzyme to produce N-carboxybiotin.

The oxygen-labeling studies mentioned previously showed that oxygen from [^{18}O]bicarbonate is incorporated into the inorganic phosphate derived from ATP (Scheme 18). This logically implicates a direct interaction of bicarbonate and the terminal phosphorus of ATP. The lack of exchange and the direct interaction can be most readily accommodated by a mechanism involving a rate-determining transition state in the formation of N-carboxybiotin, consisting of the enzyme, bicarbonate, and ATP. In this transition state, an oxygen ligand on bicarbonate attacks ATP, and biotin attacks bicarbonate.

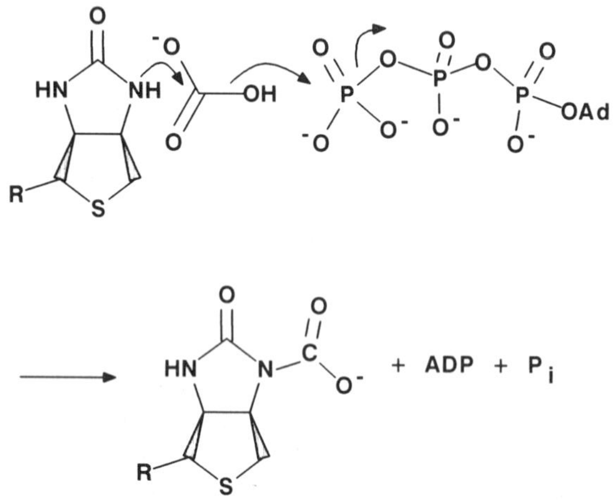

SCHEME 18. Concerted reaction: ATP + bicarbonate + enzyme.

While satisfying the observed criteria, this mechanism is not in accord with expectations from reasonable chemical analogies. A direct displacement on a carboxyl center with hydroxide as a leaving group is unreasonable. Furthermore, there is then no mechanistic function for the cleavage of ATP in promoting the reaction. Knowles has observed that the function of promoting the departure of hydroxide could just as well be done by a proton derived from any Brönsted acid. The choice of this mechanism is made less compelling by the well-known kinetic complications that are possible in enzymic reactions. The lack of exchange does not exclude stepwise processes if the enzyme utilizes a complex mechanism that is the result of favorable evolution. I have considered the mechanistic alternatives in detail and the conclusions will be summarized here (*50*).

The mechanism involving a single transition state is based on the further assumption that no further step is necessary to give all the products. This leads to what is a chemically unreasonable assumption of an S_N2 reaction at bicarbonate in which phosphate from ATP is simultaneously displaced. Such a reaction, in which hydroxide is both a leaving group and a nucleophile is extremely unlikely. Less demanding mechanisms require *ad hoc* assumptions to explain the lack of exchange.

In fact, the data do not require that all the bond-making and bond-breaking processes of the reaction be complete in only one step. Instead of envisioning the expulsion of hydroxide from carbonate, biotin can add to bicarbonate while one of the oxygen ligands attacks ATP without departing. In this case, ATP acts as a Lewis acid catalyst and the oxygen ligand is converted into a better leaving group (*50*).

A general approach to the question of possible mechanisms ignores the lack of observable exchange processes, since this can be rationalized as originating from the demands of ordered binding. The three reactants that are bound to the enzyme (biotin, ATP, and bicarbonate) produce three enzyme-bound products (*N*-carboxybiotin, ADP, and inorganic phosphate). The first mechanism I mentioned, which has all products resulting in a single step, appears to be mechanistically unreasonable. The chemically more reasonable mechanism (see Scheme 19) involves formation of a phosphorylated tetrahedral intermediate in a single step from ATP, bicarbonate, and biotin. This intermediate would decompose to

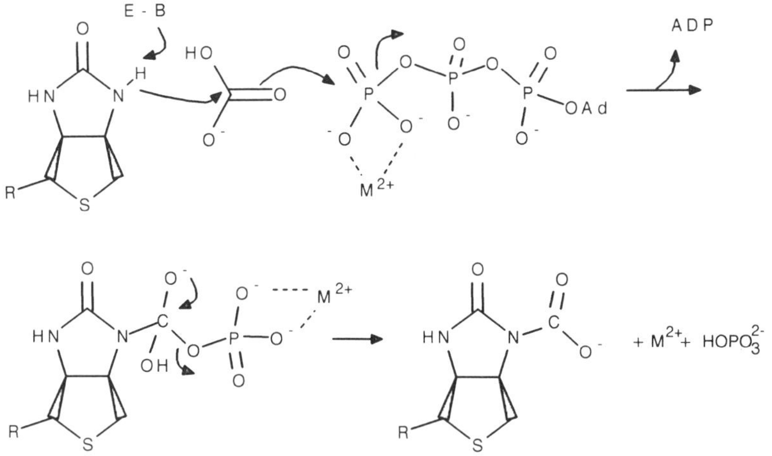

SCHEME 19. A trimolecular transition state leading to the formation of a phosphorylated tetrahedral intermediate.

N-carboxybiotin. The mechanism involves trapping of the addition product between bicarbonate and the conjugate base of biotin by the terminal group of ATP. This mechanism might give positional isotope exchange if the groups were bound loosely.

If the lack of exchange is disregarded, we can consider mechanisms in which two reactants can produce an intermediate that then reacts with the third component. The first possible combination we consider involves the initial reaction of bicarbonate and ATP followed by the reaction of the intermediate with biotin (see Scheme 19). The transfer of oxygen from bicarbonate to phosphate occurs in the first step with the formation of ADP and carboxy phosphate. The latter species might either react directly with biotin or initially decompose to inorganic phosphate and carbon dioxide, which in turn reacts with biotin. The barrier to addition of the conjugate base of biotin to carboxy phosphate should not be significantly lower than the addition to bicarbonate. The advantage of phosphorylation is in the enhancement of the leaving group, not in the electrophilicity of the carbonate, since either a proton or a phosphate is electron withdrawing.

Since carboxy phosphate is expected to have a very fleeting existence (51), what advantage would be gained if the system had evolved to utilize this intermediate? The leaving group (phosphate) is already bound whereas addition to bicarbonate requires transfer of the phosphate from ATP. However, although the oxyanion derived from the tetrahedral intermediate should be a better nucleophile than bicarbonate (based on pK_a), the route via carboxy phosphate is at a disadvantage. On the other hand, the concentration of tetrahedral intermediate will certainly be low, but this can be overcome by enzymic binding of the reacting species so that it will be trapped by ATP.

Sauers *et al.* proposed that carboxy phosphate may decompose to give carbon dioxide and inorganic phosphate prior to reaction with biotin (51). These workers suggest that this provides a "tamed" form of carbon dioxide that is low in entropy in that it is localized in the vicinity of biotin and thus is a much better reactant than bicarbonate.

Another possibility within this class of mechanism involves the initial reaction of biotin with ATP, forming ADP and a phosphorylated biotin species. It has been proposed, based on model studies, that such a species would be O-phosphobiotin (52, 53). This reacts with bicarbonate to produce N-carboxybiotin and inorganic phosphate (Scheme 20). The transfer of oxygen from bicarbonate occurs in the second step in this case. Models for the O-phosphorylation of biotin suggest that such a process can occur readily.

The results of stereochemical studies by Hansen and Knowles have placed restrictions on the possible steps of such a mechanism (54). However, the most reasonable version of this mechanism is consistent with the stereochemical results. The stereochemical studies show that the net effect at phosphorus is overall inversion. That is, the oxygen from bicarbonate is added to the face opposite to that from which ADP leaves at the terminal phosphate of ATP.

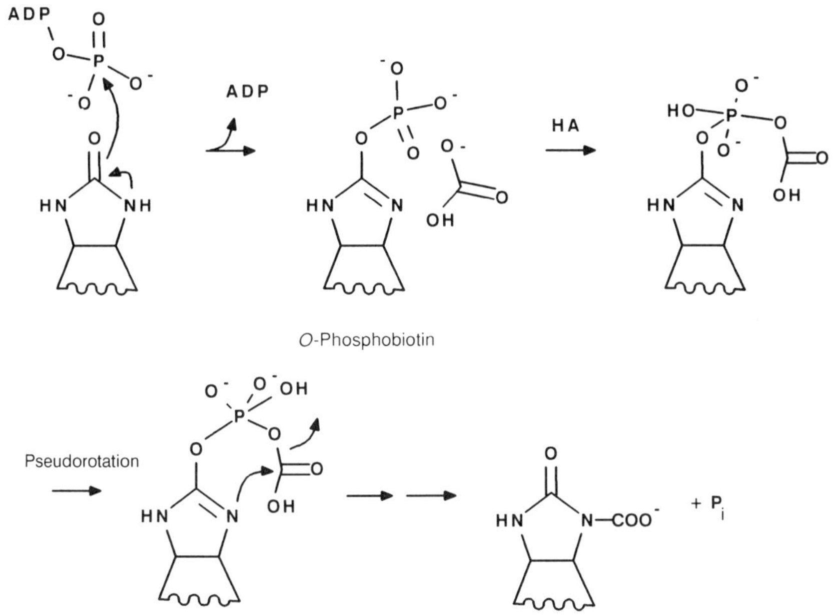

O-Phosphobiotin

SCHEME 20. Phosphorylation of biotin followed by carboxylation.

The attack of biotin on ATP is a direct-displacement process that should lead to inversion at phosphorus (55). Therefore the reaction of O-phosphobiotin with bicarbonate must be a separate process and occur with retention of relative configuration about phosphorus (53). Based on the work of Westheimer (56, 57), it is known that substitution at phosphorus can occur with retention or inversion at phosphorus. The process involving retention of relative configuration involves a step in which the pentacovalent adduct formed from the addition of a nucleophile to a phosphate undergoes an isomerization belonging to the class of molecular rearrangements known as pseudorotations. Addition, pseudorotation, and decomposition lead to net retention of configuration about phosphorus. The possibility of such mechanisms in enzymic reactions was noted over 20 years ago by Usher, who called such a process an "adjacent" mechanism (Scheme 21), while the process leading to inversion is an "in-line" mechanism (Scheme 22) (58). The nomenclature refers to the relative positions of the attacking and leaving groups.

In the decomposition of O-phosphobiotin, it is most reasonable to expect that substitution will occur by an adjacent mechanism rather than by an in-line mechanism (53). The attack of bicarbonate on O-phosphobiotin generates carboxy phosphate, which in turn can decompose to give carbon dioxide, as was discussed previously. If carbon dioxide is to be generated it must be near the N-1' position of biotin with which it must react. In the adjacent mechanism, the car-

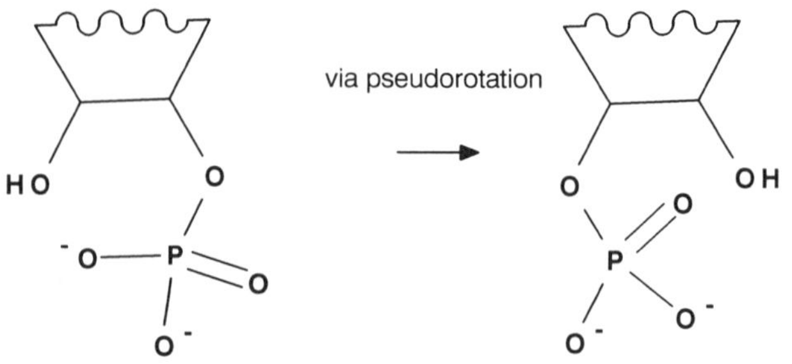

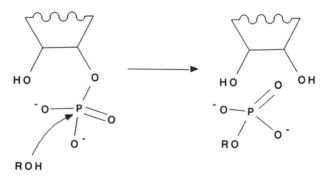

SCHEME 21. Adjacent mechanisms for substitution at phosphorus.

bon atom is considerably closer to the nitrogen than it is in the in-line mechanism. Molecular mechanics calculations in our laboratory indicate that in the adjacent mechanism, the carboxyl carbon is 3.1 Å from the nitrogen, whereas in the in-line mechanism, the carboxyl carbon is separated by 4.5 Å. Furthermore, intervening atoms block direct reaction in the in-line mechanism.

A third member of the "bimolecular then unimolecular" reaction class is a variant of the previous mechanism. In this case, the conjugate base of biotin reacts with bicarbonate to produce an addition intermediate that then reacts with ATP (Scheme 23). It is likely that the phosphorus of the terminal group of ATP would preassociate with an oxygen of bicarbonate. In particular, if the anionic center of bicarbonate associates with a cation, the π-electron density of bicarbonate would align with the phosphorus of the terminal phosphate of ATP. The addition of the conjugate base of a urea to a carboxylate is an appropriate model for this mechanism. The intermediate should be very reactive toward ATP based on the observation that the conjugate base of a carbonyl hydrate reacts rapidly with an internal phosphate ester (59).

Gelb has observed that the transfer of a carboxylate from the substrate to biotin in the reverse of the carboxylation process or in the transcarboxylase reaction does not involve phosphorylation of the carboxylate (Scheme 24). Therefore,

SCHEME 22. In-line mechanism for substitution at phosphorus.

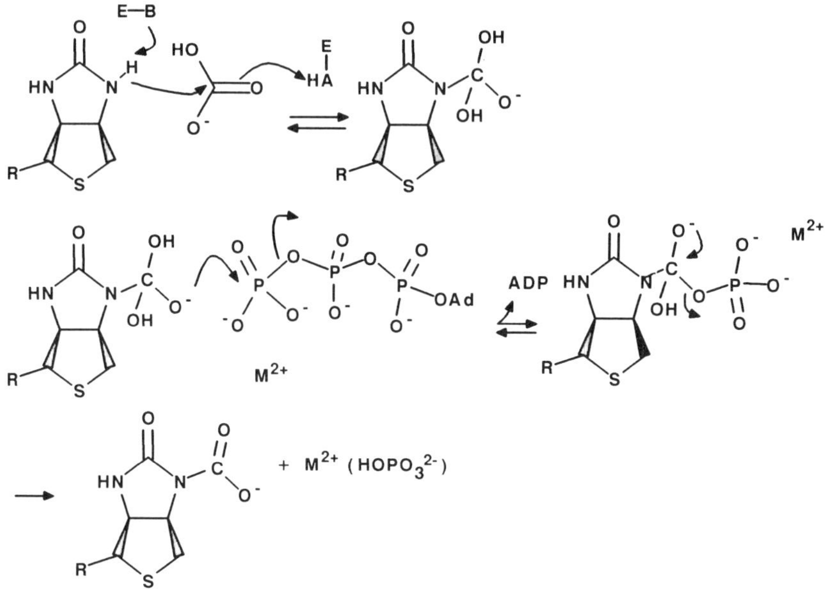

SCHEME 23. Reaction of biotin with bicarbonate followed by reaction with ATP to form a phosphorylated tetrahedral intermediate.

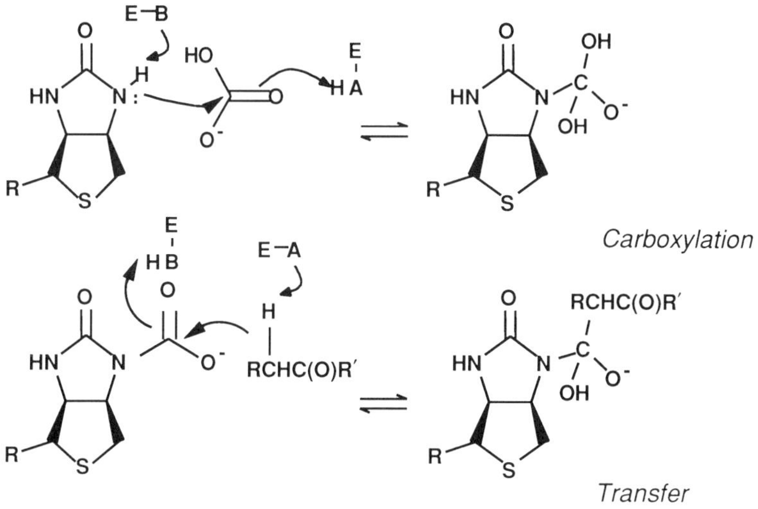

Carboxylation

Transfer

SCHEME 24. Parallel reactions of bicarbonate and a carboxylate substrate with the conjugate base of biotin.

direct attack by the conjugate base of biotin on bicarbonate is energetically similar to other reactions in the sequence (*60*).

b. *Relating Structure and Function.* The unique structure of biotin has been the subject of speculation with regard to its relationship to mechanism. The bicyclic system has no immediately obvious reason for existence. Yet, it is unlikely that the structure is not optimal for its purpose (*61, 62*).

The imidazolidinone ring of biotin is the "business end" and it is well suited for the reactions in which it is involved. Perrin and Dwyer have shown that the exchange of the protons attached to the nitrogen atoms occurs sufficiently rapidly that the conjugate base is a reasonable intermediate for any reaction in which substitution for the proton occurs, as in carboxylation (*49*). The need for acid catalysis, which had been proposed, is not consistent with this observation.

Another interesting feature in which structure and function may be related concerns the conformation of the carboxyl group in *N*-carboxybiotin. Biotin has a dual function: it preserves a carboxyl group after the ATP-dependent carboxylation has occurred and it readily transfers the carboxyl group to an acceptor. Wallace and co-workers showed that the reactivity is "triggered" by binding of *N*-carboxybiotin to the transfer site when substrate or an analog of the substrate is present (*63*). That is, *N*-carboxybiotin must be inherently unreactive in order to preserve the carboxyl group, but it must become reactive in the triggering situation.

The change in reactivity can easily be accomplished by a change in conformation about the bond between the nitrogen of biotin and the carboxyl group (*64*). If the carboxyl group is in the plane of the urea moiety, resonance overlap stabilizes the carbon–oxygen bond (Scheme 25). Rotation of the carboxyl group out of the plane destroys this stabilization and enhances the reactivity of the carboxyl toward nucleophilic attack. Such a change in reactivity can only be accomplished if the ureido group is held in a planar conformation. The fused ring containing sulfur is of the precisely correct geometry to hold the imidazoli-

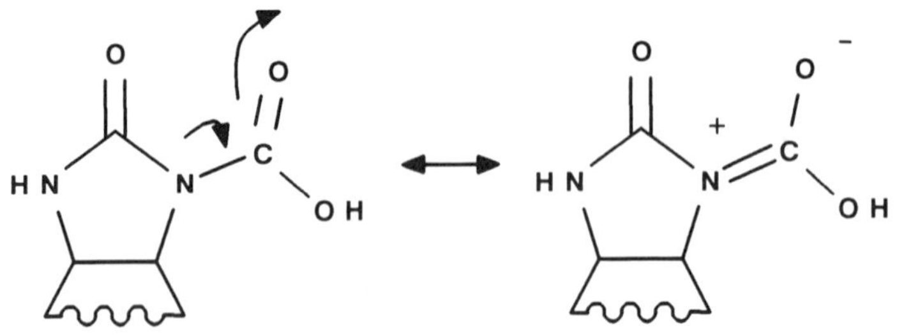

SCHEME 25. Resonance and rotation in carboxybiotin.

dinone ring planar. X-Ray studies of analog in which sulfur is replaced by oxygen or carbon have shown structures in which the imidazolidinone ring is distorted (65). The long carbon–sulfur bond serves to produce the correct bridge length. Shorter bonds cause distortion. Although the sulfur may also provide a mild electronic effect, its major function is structural.

c. *Carboxyl Transfer from Biotin.* Transfer of the carboxyl from *N*-carboxybiotin to an anionic acceptor is analogous to the reactions described for Claisen enzymes earlier in this chapter. The reaction is an electrophilic substitution at the substrate carbon of the carboxyl for a proton. Rétey and Lynen had shown that this reaction occurs with retention of relative configuration (66). In response to this observation, they wrote what is a concerted mechanism (Scheme 26).

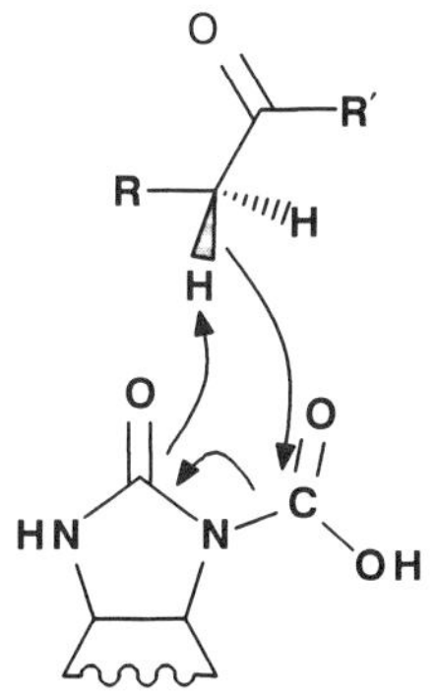

N-Carboxybiotin

SCHEME 26. Carboxyl transfer by a concerted mechanism.

Kuo and Rose showed that the proton that is removed is retained by the enzyme (67). Stubbe and Abeles prepared an alternative substrate in which fluoride elimination competes with carboxylation (68, 69). Neither result defines the mechanism, but they do show that it is likely that the carbanion derived from the substrate is generated as an intermediate and therefore the reaction is not concerted. Definitive results come from double-isotope fraction studies by O'Keefe and Knowles (70) and by Cleland and co-workers (71). As described for Claisen enzymes, this methodology tests whether processes occur in one or two steps. Labeling of the carboxyl to be transferred with carbon-13 and the proton to be transferred as deuterium provided the means to do this test. The results indicate clearly that proton removal from the substrate to generate the carbanion and transfer of the carboxyl occurs in distinct steps. The resulting attack of the carbanion derived from the substrate on the carboxylate generates the conjugate base of biotin.

d. *Carboxylation and Related Reactions.* The formation of carbon–carbon bonds between carbanions and single-carbon units follows patterns that are related to the cofactors. The transfer of the equivalent of carbonic acid is carboxylation, and this is accomplished by reaction with carboxybiotin. As discussed in the section on biotin, *N*-carboxybiotin has the dual function of stabilizing the carboxyl group in the absence of an acceptor and then facilitating the transfer in the presence of an acceptor. Resonance structures and molecular orbital evaluation of structural possibilities provide an insight into the functional significance of the particular groups that are involved in the reaction. We have shown that the planar imidazolidinone ring on biotin is ideal for controlling the reactivity of the carboxyl moiety.

While *N*-carboxybiotin is responsible for transfer of the carboxyl group, biotin derivatives are not involved in the transfer of more reduced one-carbon units. The next lowest oxidation state at carbon involves the transfer of an aldehyde carboxyl —HC=O. This is equivalent to substitution for the hydroxyl group of formic acid. The leaving group in this case is a derivative of tetrahydrofolate, which contains the carbon as a formamide derivative. Amide resonance is a powerful factor in maintaining the C–N bond. Rotation of the carboxyl out of the plane of the amine will weaken the bond by disruption of resonance.

2. *Mechanism of Carbon–Carbon Bond Formation and Cleavage with Pyridoxal Phosphate and Tetrahydrofolate Derivatives*

Many enzymes catalyze reactions in which the amino group of an amino acid is condensed with the aldehyde moiety of enzyme-bound pyridoxal phosphate, producing an enzyme-bound imine. This unsaturated functional group can stabilize an adjacent carbanion derived from the amino acid by a combination of resonance and inductive effects (*72*). The carbanion can be generated by any process that removes a ligand heterolytically so that the carbanion remains. The three ligands are the carboxyl group (which leaves as carbon dioxide), a hydrogen atom (which is transferred as a proton to a base), and the side chain (which is transferred as the equivalent of the carbocation). Processes that are thus promoted include decarboxylation, racemization of the amino acid derivative, and aldol reactions (*73*). The mechanisms of enzymes utilizing pyridoxal phosphate have been reviewed extensively and the function of pyridoxal phosphate is well understood. (*Editor's note:* For additional aspects of pyridoxal phosphate-dependent decarboxylation reactions, see Chapter 6 by O'Leary.) An important mechanistic point is due to the observation by Dunathan that specificity is likely to be controlled enzymically through the torsional angle about the C–N bond originally in the amino acid (*74*). The π-electron system of the imine must overlap the σ-bond to the ligand (that is, the alignment is periplanar or antiperiplanar), which departs so that the transition state leading to the carbanion can be stabilized by $\sigma-\pi$ interactions. The enzyme controls the conformation that determines the reaction pattern.

Although the mode through which conformational specificity is achieved is not known, one possibility is electrostatic interactions between the carboxyl group derived from the amino acid and a cationic site of the enzyme (Scheme 27). If this is the case, then decarboxylases will be oriented correctly regardless of the identity of the other ligands. (However, the interaction of the carboxylate with a cation would probably retard decarboxylation.) That is, a decarboxylase in principle should stabilize a carbanion from either enantiomer of an amino acid if both are capable of binding to the enzyme. If other bonds are activated, then if binding is specified by the interaction of the carboxylate with the enzyme, enantiomers will react differently and enantiotopic ligands will be readily distinguished.

The biosynthesis of δ-aminolevulinate has a key step involving the addition of the equivalent of the carbanion derived from glycine to the thiolester function of

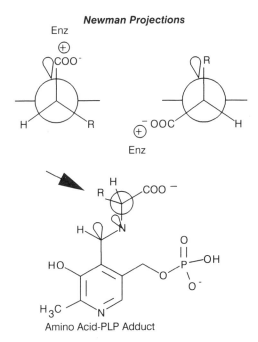

SCHEME 27. Conformations of pyridoxal phosphate (PLP) adducts.

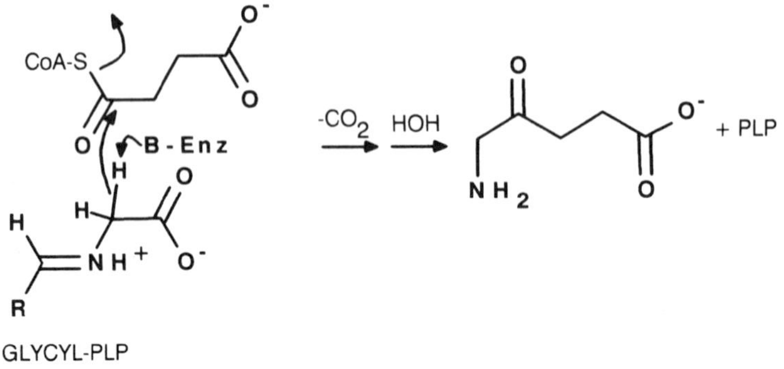

GLYCYL-PLP

SCHEME 28. Formation of δ-aminolevulinate; PLP, pyridoxal phosphate.

succinyl coenzyme A (Scheme 28). The carbanion equivalent is formed by removal of a proton from the α-position by a base on the enzyme.

The most common process involving carbon–carbon formation and pyridoxal phosphate is exemplified by the reaction catalyzed by serine hydroxymethylase, a reaction that utilizes a derivative of tetrahydrofolate as a cofactor. In this system, the adduct of glycine and pyridoxal phosphate is formed and the carbanion is generated as in the previous example. Formally, the production of serine from glycine requires the condensation of the carbanion derived from glycine with formaldehyde. According to the Dunathan hypothesis, the enzyme should distinguish the two enantiotopic hydrogens at the α-position. Akhtar and Jordan observed that such is the case and used this to produce pure enantiomers of mono-deuteroglycine (75).

The addition of formaldehyde to a carbanion does not involve the free formaldehyde molecule, since this material is quite toxic. Instead, an equivalent derivative, N^5,N^{10}-methylene tetrahydrofolate, is utilized (76). The carbon center with two nitrogen substituents is highly polarized by these strongly electron-withdrawing groups and can react with water to produce the carbinolamine derivative of formaldehyde. [Attack of the carbanion equivalent derived from glycine and pyridoxal phosphate would not lead to the product but rather to an aminoalkane derivative (Scheme 29).] The resulting intermediate could then react with the carbanion equivalent by a number of mechanistic possibilities that lead to the addition of the hydroxymethyl group to the carbanionic carbon center.

The transfer of a more highly oxidized one-carbon unit, as in carboxylation reactions, is not accomplished via an analogous derivative of tetrahydrofolate but instead involves N-carboxybiotin. However, transfer of less oxidized one-carbon units involves tetrahydrofolate as a cofactor. These provide for the addition of a hydroxymethyl group or methyl groups. The latter can also be transferred from S-adenosylmethionine.

The basis for the diversity of cofactors for transfer of one-carbon units in biosynthetic pathways is unknown. However, the source of the equivalent of

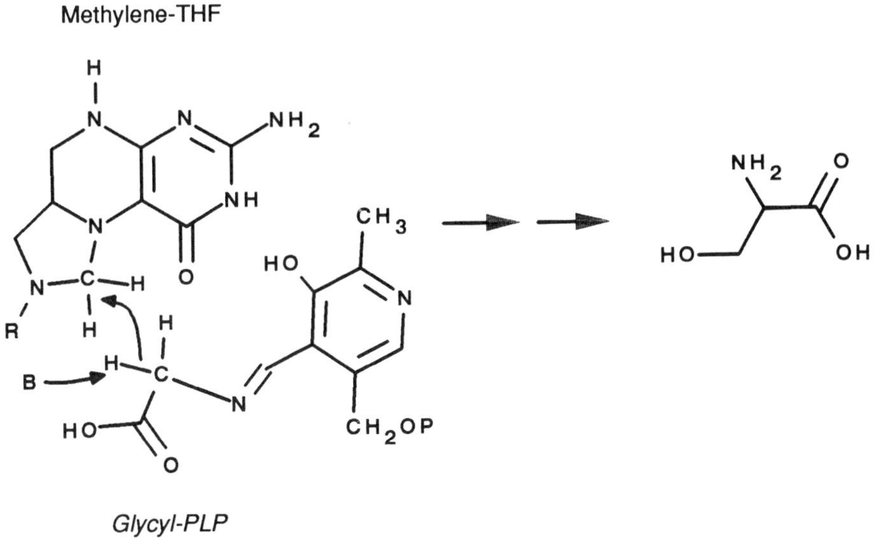

Methylene-THF

Glycyl-PLP

SCHEME 29. Reaction of methylene tetrahydrofolate (THF); PLP, pyridoxal phosphate.

carbon dioxide for carboxylation of biotin is bicarbonate in an ATP-dependent reaction. The reactions involving tetrahydrofolate do not require ATP and utilize a direct reaction. The adducts of formaldehyde and reduced equivalents are more stable than carboxylated derivatives. Therefore, biotin appears to be necessary for controlling the reactivity of the carboxylate group. In addition, tetrahydrofolate derivatives involve double attachments, which might cause a carboxyl group derivative to be too stable.

3. *Thiamin Diphosphate Derivatives in Carbon–Carbon Bond Formation*

Thiamin diphosphate (TDP) functions as a cofactor to overcome a chemically difficult problem in carbon–carbon formation and cleavage (*77*). (*Editor's note:* Thiamin diphosphate-dependent decarboxylation is discussed in Chapter 6 by O'Leary.) The reaction pattern is exemplified by the decarboxylation of pyruvate to give acetaldehyde (or a more oxidized species) and carbon dioxide. In this reaction, the bond that is broken is not inherently activated toward the reaction. The bond that is to be cleaved is between two carbonyl functions. Since these groups are similarly polarized, heterolytic cleavage is not a likely process. Nonenzymically, the direct cleavage of such a bond involves a homolytic (radical) process.

Thiamin diphosphate permits the solution of this problem in a chemically elegant way. It easily forms a carbon–carbon bond to permit the cleavage of another more difficult carbon–carbon bond. The mechanism, which was proposed by

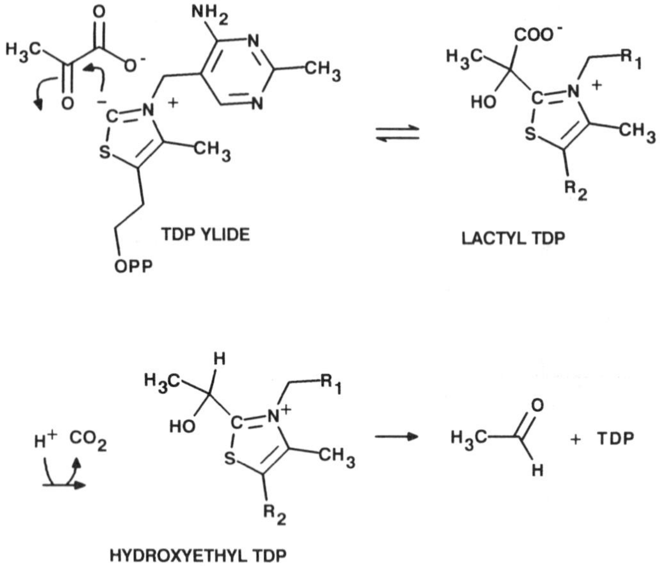

SCHEME 30. Mechanism of thiamin-catalyzed decarboxylation of pyruvate; TDP, thiamin diphosphate.

Breslow in model systems (78, 79) and has been confirmed with the intermediates themselves (80–86), is summarized in Scheme 30.

    a. *Ionization of Thiamin Diphosphate.* The ionization of thiamin diphosphate is the key step in understanding the mechanism. Thiamin diphosphate is a Brönsted acid and the conjugate base, which is an ylide, functions as a nucleophile toward the electrophilic center at the carbon atom of the keto group of pyruvate. The adduct, lactylthiamin diphosphate (84, 85), has a carbon–carbon bond that is now polarized such that heterolysis is promoted. The thiazolium ring derived from thiamin diphosphate serves to delocalize excess electron density generated by the loss of carbon dioxide (which is a Lewis acid). The reaction amounts to an electrophilic substitution reaction (of a proton for carbon dioxide) at a carbanionic center. The reaction cycle is completed by the cleavage of the carbon–carbon bond between thiamin diphosphate and the precursor of the aldehyde.

    Washabaugh and Jencks have studied the ionization of thiamin and have shown that the pK_a is 18.9 (87), contrary to earlier reports of a much lower value (88). This is consistent with the protonation of the thiamin occurring at a diffusion-controlled rate (87, 89, 90). Although the pK_a is high, the rate of proton removal is sufficiently rapid to be competent for an enzyme-catalyzed reaction. The equilibrium constant for the formation of the adduct is about 1 M, indicating that the bond is probably not very strong (*84*). The equilibrium involves the exchange of the C–H bond of thiamin with the C–C bond of lactylthiamin and the conversion of the C=O to an O–H. Since these are occurring together, the C–C bond formation process cannot be isolated thermodynamically. We have shown that the enzyme utilizes the energy of formation of lactylthiamin diphosphate on the enzyme (from pyruvate and thiamin diphosphate) to promote the decarboxylation step (*85*). Chapter 6 by O'Leary in this volume discusses this reaction in the context of other decarboxylation mechanisms.

b. *Transketolase and Transaldolase.* The reactions catalyzed by transketolase and transaldolase also utilize thiamin diphosphate as a cofactor and the mechanism parallels that of pyruvate decarboxylases in the formation of an initial adduct between the ylide derived from thiamin diphosphate and the carbonyl group of the substrate (*91*). A ketol is a 2-hydroxyketone and the enzyme interchanges the hydroxyl alkyl function with an aldehyde as shown in Scheme 31.

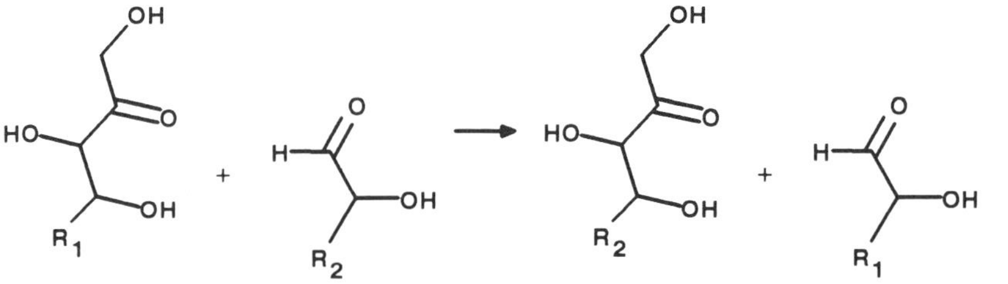

SCHEME 31. The transketolase reaction.

The reaction actually is two processes: α-cleavage and α-condensation via a thiamin diphosphate adduct (Scheme 32). These involve generation of a second carbanion after the ylide has added to the carbonyl of the substrate. The carbanion that is expected to be generated in the reactions is of some interest. Sable demonstrated that such a carbanion can form nonenzymically by showing that 2-(1-hydroxyethyl)thiamin will undergo base-catalyzed exchange of hydrogen for deuterium (in deuterium oxide) at the α-carbon (*92, 93*). The ion is also generated by decarboxylation of lactylthiamin as demonstrated by the fact that reaction occurs with racemization (Scheme 33) (*94*).

c. *Acetolactate Synthase.* The reaction catalyzed by acetolactate synthase combines the mechanisms of pyruvate decarboxylases and transketolases (Scheme

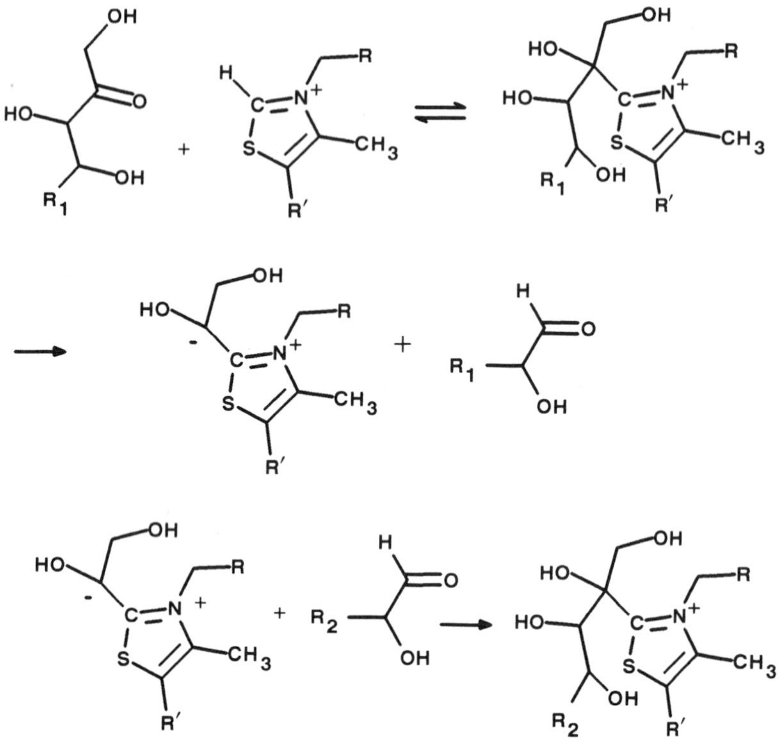

SCHEME 32. Mechanism of the transketolase reaction.

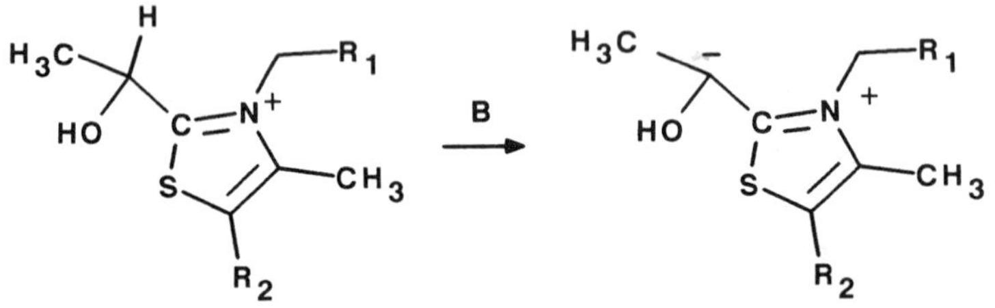

SCHEME 33. Formation of a stabilized carbanion from (hydroxyethyl)thiamin.

34). The product, α-acetolactate, is formally derived from the acyl carbanion generated by the decarboxylation of pyruvate with a second molecule of pyruvate. The formation of lactylthiamin diphosphate on the enzyme generates initially the unprotonated adduct, which contains the enamine equivalent of a carbanion. This can add to the carbonyl group of a second molecule of pyruvate to form acetolactate (Scheme 35).

The carbanion should be a very strong base and therefore would be expected

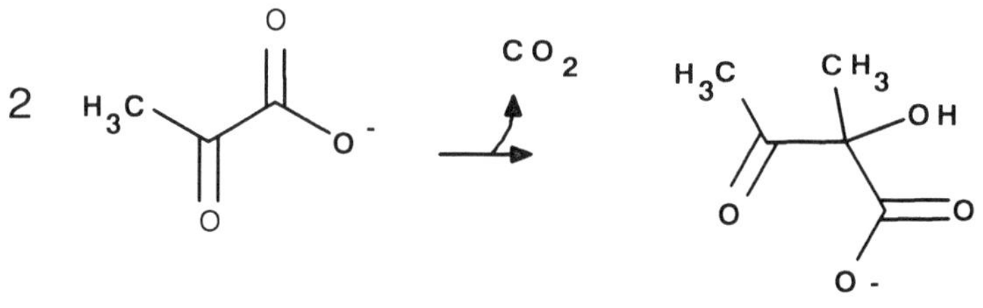

ACETOLACTATE

SCHEME 34. Acetolactate synthase reaction.

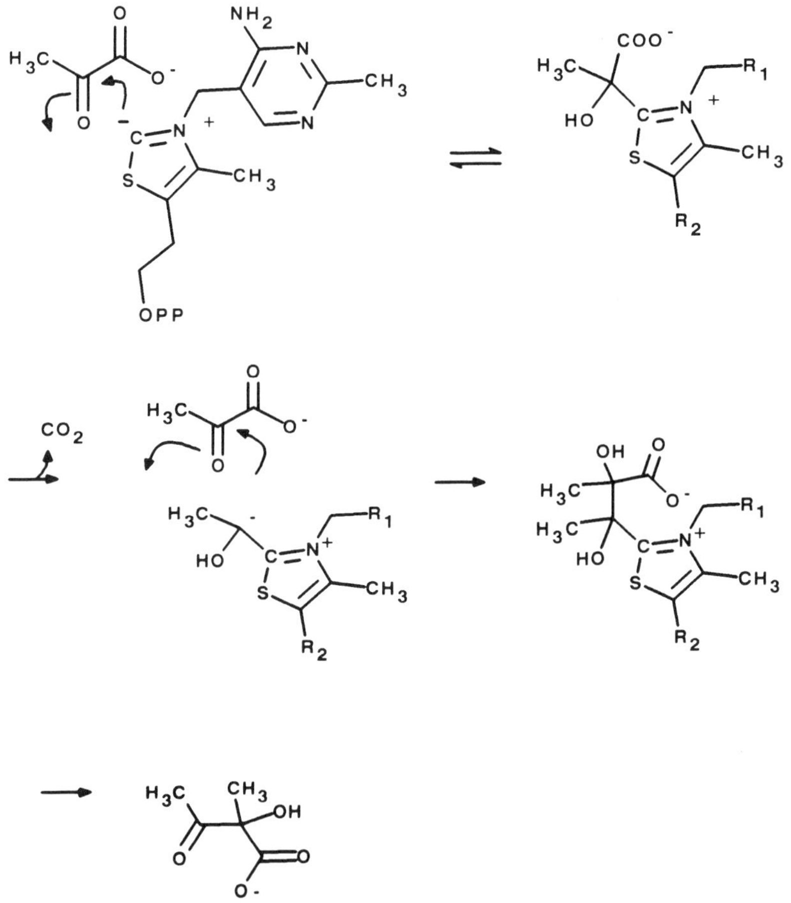

SCHEME 35. Lactylthiamin diphosphate is converted to (hydroxyethyl)thiamin diphosphate.

to add rapidly to any electrophile, including any Brönsted acid. If the carbanion is trapped by the carbonyl group of a bound pyruvate molecule, the alternative quench is avoided as is the thermodynamic problem of deprotonation. The enzyme has been found by LaRossa and Schloss to contain a flavin cofactor in addition to thiamin diphosphate, although there appears to be no function for the flavin in the reaction mechanism (95). While there is no evidence for the function of the flavin, we can speculate that it might serve to protect the carbanion as a reducing equivalent. The carbanion is generated under conditions wherein it is readily lost if a second molecule of pyruvate is not bound. In a plant (which is where the enzyme occurs), a supply of pyruvate is not assured. The alternative is that the carbanion reversibly reduces the bound flavin and is in turn re-reduced in the presence of the second molecule of pyruvate. In this case the carbanion may be stabilized as an adduct of the flavin.

V. Concluding Remarks

In this chapter I have attempted to illustrate some of the variety of mechanistic patterns that can be found among the enzymes that catalyze carbon–carbon bond formation and cleavage. The coverage has not been intended to be exhaustive and many important enzymes and mechanisms have been omitted. For example, rearrangement reactions involving cobalamin involve radical intermediates (96) and the mechanisms are of considerable interest. Carboxylation reactions involving enzymes that utilize vitamin K are in an early stage of mechanistic explication, but appear to follow the patterns of the carbanion systems I have reviewed here (97). Other reactions that appear to be enzyme-catalyzed thermal rearrangements, such as that promoted by chorismate mutase (98), may involve electronic reorganization according to the rules of conservation of orbital symmetry. These and others are best understood in terms of mechanisms derived from roots in physical organic chemistry.

Applications based on the understanding of mechanisms in the design of drugs and specific inhibitors in agriculture are becoming widespread and significant. The line between chemistry and biology is disappearing as we see that mechanisms of reactions apply to understanding processes in both areas. Carbon–carbon bond-forming reactions provide the essential framework for the complex structures we observe. The understanding of mechanisms and a systematic approach to their study should help us to recognize the common patterns in diverse structures and the enzymes that are responsible for their existence.

ACKNOWLEDGMENTS

My work has been supported by ongoing grants from the Natural Sciences and Engineering Research Council of Canada. I thank Professor J. R. Knowles and Professor C. D. Poulter for preprints of their papers, and I thank Dr. Zheng Huang for critical comments.

REFERENCES

1. Kluger, R. (1989). *In* "Enzyme Chemistry" (C. J. Suckling, ed.), 2nd Ed. Chapman & Hall, London.
2. Dixon, M., and Webb, E. C. (1964). "The Enzymes," 2nd Ed.
3. Snell, E. E., and Dimari, S. J. (1970). "The Enzymes," 3rd Ed., Vol. 2, p. 335.
4. Srere, P. A. (1975). *Adv. Enzymol. Relat. Areas Mol. Biol.* **43**, 57.
5. Grazi, E., Cheng, T., and Horecker, B. L. (1962). *Biochem. Biophys. Res. Commun.* **7**, 250.
6. Walsh, C. (1978). "Enzymatic Reaction Mechanisms." Freeman, San Francisco, California.
7. Hanson, K., and Rose, I. A. (1975). *Acc. Chem. Res.* **8**, 1.
8. Westheimer, F. H., and Cohen, H. (1938). *J. Am. Chem. Soc.* **60**, 90.
9. Speck, J. C., and Forist, A. A. (1957). *J. Am. Chem. Soc.* **79**, 4659.
10. Schwarzenbach, C., and Felder, E. (1944). *Helv. Chim. Acta* **27**, 1701.
11. Healy, M. J., and Christen, P. (1973). *Biochemistry* **12**, 35.
12. Rétey, J., Lüthy, J., and Arigoni, D. (1970). *Nature (London)* **226**, 519.
13. Hanson, K. R., and Rose, I. A. (1972). *Crit. Rev. Biochem.* **1**, 33.
14. Klinman, J. P., and Rose, I. A. (1971). *Biochemistry* **10**, 2267.
15. Sprecher, M., Berger, R., and Sprinson, D. B. (1964). *J. Biol. Chem.* **239**, 4268.
16. Rozzell, J. D., Jr., and Benner, S. A. (1984). *J. Am. Chem. Soc.* **106**, 4937.
17. Cram, D. J. (1965). "Fundamentals of Carbanion Chemistry." Academic Press, New York.
18. Cram, D. J., and Haberfield, P. (1961). *J. Am. Chem. Soc.* **83**, 2354.
19. More O'Ferrall, R. A. (1970). *J. Chem. Soc. B* p. 274.
20. Jencks, W. P. (1980). *Acc. Chem. Res.* **13**, 161.
21. Grunwald, E. (1985). *J. Am. Chem. Soc.* **107**, 125.
22. Hermes, J. D., Roeske, C. A., O'Leary, M. H., and Cleland, W. W. (1982). *Biochemistry* **21**, 5106.
23. Clark, J. D., O'Keefe, S. J., and Knowles, J. R. (1988). *Biochemistry* **27**, 5961.
24. Eggerer, H., and Klette, A. (1967). *Eur. J. Biochem.* **1**, 447.
25. Lenz, H., and Eggerer, H. (1976). *Eur. J. Biochem.* **65**, 237.
26. Durchschlag, H., Biedermann, G., and Eggerer, H. (1981). *Eur. J. Biochem.* **114**, 255.
27. Rendina, A. R., Hermes, J. D., and Cleland, W. W. (1984). *Biochemistry* **23**, 5148.
28. Bernhard, S. (1986). *Science* **234**, 1081.
29. Metzler, D. E. (1977). "Biochemistry," pp. 714–727. Academic Press, New York.
30. Popják, G., and Cornforth, J. W. (1966). *Biochem. J.* **101**, 553.
31. Mislow, K., and Siegel, J. (1984). *J. Am. Chem. Soc.* **106**, 3319.
32. Middelfort, C. F., and Rose, I. A. (1976). *J. Biol. Chem.* **251**, 5881.
33. Popják, G., and Agnew, W. S. (1979). *Mol. Cell. Biochem.* **27**, 97.
34. Edmond, J. W., Popják, G., Wong, S. M., and Williams, V. P. (1971). *J. Biol. Chem.* **246**, 6254.
35. Epstein, W. W., and Rilling, H. C. (1970). *J. Biol. Chem.* **245**, 4597.
36. Rilling, H. C., Poulter, C. D., Epstein, W. W., and Larsen, B. (1971). *J. Am. Chem. Soc.* **93**, 1783.
37. van Tamelen, E. E., and Schwartz, M. A. (1971). *J. Am. Chem. Soc.* **93**, 1780.
38. Saunders, M., and Siehl, H. U. (1980). *J. Am. Chem. Soc.* **102**, 6868.
39. Poulter, C. D., Marsh, L. L., Huges, J. M., Argyle, J. C., Satterthwaite, D. M., Goodfellow, R. J., and Moesinger, S. G. (1977). *J. Am. Chem. Soc.* **99**, 3816.
40. Huang, Z., and Poulter, C. D. (1989). *J. Am. Chem. Soc.* **111**, 2713.
41. Poulter, C. D., Capson, T. L., Thompson, M. D., and Bard, R. S. (1989). *J. Am. Chem. Soc.* **111**, 3734.
42. Sandifer, R. M., Thompson, M. D., Gaughan, R. G., and Poulter, C. D. (1982). *J. Am. Chem. Soc.* **104**, 7376.

43. Knowles, J. R. (1989). *Annu. Rev. Biochem.* **58**, 195.
44. Wood, H. G., and Barden, R. E. (1977). *Annu. Rev. Biochem.* **46**, 385.
45. Wimmer, M. J., and Rose, I. A. (1978). *Annu. Rev. Biochem.* **47**, 1031.
46. Wood, H. G. (1976). *Trends Biochem. Sci.* **1**, 4.
47. Kaziro, Y., Hase, L. F., Boyer, P. D., and Ochoa, S. (1962). *J. Biol. Chem.* **237**, 1460.
48. Climent, I., and Rubio, V. (1986). *Arch. Biochem. Biophys.* **251**, 465.
49. Perrin, C. A., and Dwyer, T. J. (1987). *J. Am. Chem. Soc.* **109**, 5163.
50. Kluger, R. (1989). *Bioorg. Chem.* **17**, 287.
51. Sauers, C. K., Jencks, W. P., and Groh, S. (1975). *J. Am. Chem. Soc.* **97**, 5546.
52. Kluger, R., and Adawadkar, P. D. (1976). *J. Am. Chem. Soc.* **98**, 3741.
53. Kluger, R., Davis, P. P., and Adawadkar, P. D. (1979). *J. Am. Chem. Soc.* **101**, 5995.
54. Hansen, D. E., and Knowles, J. R. (1985). *J. Am. Chem. Soc.* **107**, 8304.
55. Thatcher, G. R. J., and Kluger, R. (1989). *Adv. Phys. Org. Chem.* **25**, 99.
56. Westheimer, F. H. (1968). *Acc. Chem. Res.* **1**, 70.
57. Kluger, R., Covitz, F., Dennis, E. A., Williams, L. D., and Westheimer, F. W. (1969). *J. Am. Chem. Soc.* **91**, 6066.
58. Harris, M. R., Usher, D. A., Albrecht, H. P., Jones, G. H., and Moffatt, J. G. (1969). *Proc. Natl. Acad. Sci. U.S.A.* **63**, 246.
59. Blagoeva, I. B., Pojarlieff, I. B., and Kirby, A. J. (1984). *J. Chem. Soc., Perkin Trans. 2* p. 745.
60. Gelb, M. (1989). Personal communication.
61. Stallings, W. C., Monti, C. T., Lane, M. D., and DeTitta, G. T. (1980). *Proc. Natl. Acad. Sci. U.S.A.* **77**, 1260.
62. Berkessel, A., and Breslow, R. (1986). *Bioorg. Chem.* **14**, 249.
63. Goodall, G. J., Prager, R., Wallace, J. C., and Keech, D. B. (1983). *FEBS Lett.* **163**, 6.
64. Thatcher, G. R. J., Poirier, R., and Kluger, R. (1986). *J. Am. Chem. Soc.* **108**, 2699.
65. DeTitta, G. T., Edmonds, J. W., Stallings, W. C., and Donohue, J. (1976). *J. Am. Chem. Soc.* **98**, 1920.
66. Rétey, J., and Lynen, F. (1965). *Biochem. Z.* **342**, 256.
67. Kuo, D. J., and Rose, I. A. (1982). *J. Am. Chem. Soc.* **104**, 3235.
68. Stubbe, J., and Abeles, R. H. (1977). *J. Biol. Chem.* **252**, 8338.
69. Stubbe, J., Fish, S., and Abeles, R. H. (1980). *J. Biol. Chem.* **255**, 236.
70. O'Keefe, S. J., and Knowles, J. R. (1986). *Biochemistry* **25**, 6077.
71. Attwood, P. V., Tipton, P. A., and Cleland, W. W. (1986). *Biochemistry* **25**, 8197.
72. Westheimer, F. H. (1959). "The Enzymes," 2nd Ed., Vol. 1, p. 259.
73. Fersht, A. R. (1985). "Enzyme Structure and Mechanism," 2nd Ed., pp. 72–75. Freeman, New York.
74. Dunathan, H. C. (1966). *Proc. Natl. Acad. Sci. U.S.A.* **55**, 712.
75. Akhtar, M., and Jordan, P. M. (1968). *J. Chem. Soc., Chem. Commun.* p. 1691.
76. Bruice, T. C., and Benkovic, S. J. (1966). "Bioorganic Mechanisms," Chap. 10, Benjamin, New York.
77. Kluger, R. (1987). *Chem. Rev.* **87**, 863.
78. Breslow, R. (1957). *J. Am. Chem. Soc.* **79**, 1762.
79. Breslow, R. (1958). *J. Am. Chem. Soc.* **80**, 3719.
80. Krampitz, L. O., Suzuki, I., and Greull, G. (1961). *Fed. Proc., Fed. Am. Soc. Exp. Biol.* **20**, 971.
81. Krampitz, L. O., Greull, G., Miller, C. S., Bicking, J. B., Skeggs, H. R., and Spragye, J. M. (1961). *J. Am. Chem. Soc.* **80**, 5893.
82. Kramitz, L. O. (1962). *Ann. N.Y. Acad. Sci.* **98**, 466.
83. Holzer, H., and Beaucamp, K. (1959). *Angew. Chem.* **71**, 776.

84. Kluger, R., Chin, J., and Smyth, T. (1981). *J. Am. Chem. Soc.* **103,** 884.
85. Kluger, R., and Smyth, T. (1981). *J. Am. Chem. Soc.* **103,** 214.
86. Kluger, R. (1982). *Ann. N.Y. Acad. Sci.* **378,** 63.
87. Washabaugh, M. W., and Jencks, W. P. (1988). *Biochemistry* **27,** 5044.
88. Hopmann, R. F. W., and Brugnoni, G. P. (1973). *Nature (London) New Biol.* **246,** 157.
89. Washabaugh, M. W., and Jencks, W. P. (1989). *J. Am. Chem. Soc.* **111,** 674.
90. Washabaugh, M. W., and Jencks, W. P. (1989). *J. Am. Chem. Soc.* **111,** 683.
91. Racker, E. (1961). "The Enzymes," 2nd Ed., Vol. 5, p. 397.
92. Mieyal, J. J., Bantle, G., Votaw, R. W., Rosner, I. A., and Sable, H. Z. (1971). *J. Biol. Chem.* **246,** 5213.
93. Mieyal, J. J., Votaw, R. W., Krampitz, L. O., and Sable, H. Z. (1967). *Biochim. Biophys. Acta* **141,** 205.
94. Kluger, R., Karimian, K., and Kitamura, K. (1987). *J. Am. Chem. Soc.* **109,** 6368.
95. LaRossa, R. A., and Schloss, J. V. (1984). *J. Biol. Chem.* **259,** 8753.
96. Halpern, J. (1985). *Science* **227,** 869.
97. Copley, S. D., and Knowles, J. R. (1985). *J. Am. Chem. Soc.* **107,** 5306.
98. Anton, D. L., and Friedman, P. A. (1983). *J. Biol. Chem.* **258,** 14084.

8

Enzymic Free Radical Mechanisms

EDWARD J. BRUSH* • JOHN W. KOZARICH†

*Department of Chemistry
Tufts University
Medford, Massachusetts 02155

†Department of Chemistry and Biochemistry
University of Maryland
College Park, Maryland 20742

THE ENZYMES, Vol. XX

I. Ribonucleotide Reductases

A. BACKGROUND

The ribonucleotide reductases catalyze the reduction of the four common purine and pyrimidine ribonucleoside 5'-di- and triphosphates to the corresponding 2'-deoxyribonucleoside 5'-phosphates (Scheme 1) (*1–4*). Concomitant oxidations of enzyme thiol groups occur, and these must subsequently be reduced for the catalytic cycle to continue (*5*). The reducing equivalents are supplied *in vivo* by NADPH through a transport chain consisting of the proteins thioredoxin and thioredoxin reductase or glutaredoxin, glutathione, and glutathione reductase, whereas dithiothreitol (DTT) may be utilized for *in vitro* experiments (*6–8*).

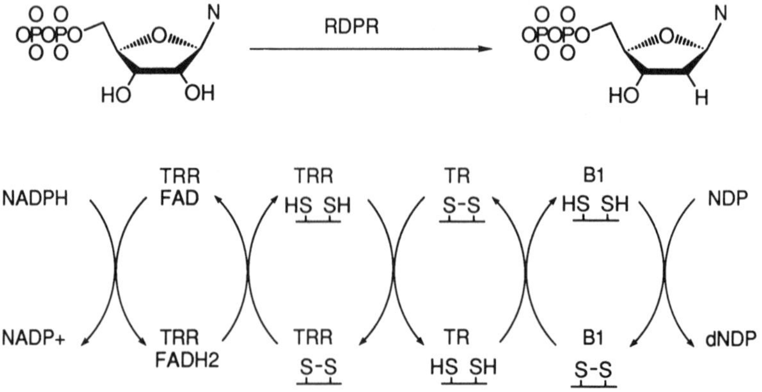

SCHEME 1. Reaction catalyzed by ribonucleotide reductases.

These enzymic reactions are essential to all living cells in that they provide the monomeric precursors for the *de novo* synthesis of DNA. The production of the 2'-deoxyribonucleoside phosphates required for DNA synthesis is carefully regulated through allosteric control of the enzyme by 2'-deoxynucleoside triphosphates and ATP, which regulate both overall activity and substrate specificity.

From their key role in DNA synthesis it is not surprising that the ribonucleotide reductases are ubiquitous in nature. However, three different types of ribonucleotide reductases are now known, each with variations in their cofactor requirements. The most extensively studied and characterized are the ribonucleotide-diphosphate reductases (RDPRs), in particular the enzyme from *Escherichia coli*. This reductase consists of two nonidentical subunits, proteins B1 and B2, which form an active 1:1 complex where the interface between these subunits forms the active site, and each subunit alone is devoid of catalytic activity (9). Protein B2 (M_r 78,000) is a $\beta\beta$ dimer and contains two atoms of tightly bound nonheme iron [Fe(III)] (10). The iron site has been studied by Mössbauer (10) and resonance Raman spectroscopy (11, 12) and is consistent with two μ-oxobridged nonequivalent binuclear high-spin Fe(III) ions in an antiferromagnetically coupled complex (13). What has been the most intriguing characteristic of the *E. coli* enzyme is that the B2 subunit gives rise to an EPR signal characteristic of an organic free radical (14, 15). The EPR spectra of the RDPRs from a variety of sources are strikingly similar (16–19), being composed of a doublet signal centered at $g = 2.0047$. A doublet splitting of 16 G is observed, which collapses into a narrow singlet in D_2O-grown cells (14, 20), suggesting hyperfine interactions with an exchangeable proton. The radical center has been established as being localized on a tyrosyl residue (14, 21) and has been further characterized by an optical absorption maximum at 410 nm. The X-ray crystal structure of the cofactor center in the B2 subunit has recently been reported (22) and is shown in Scheme 2.

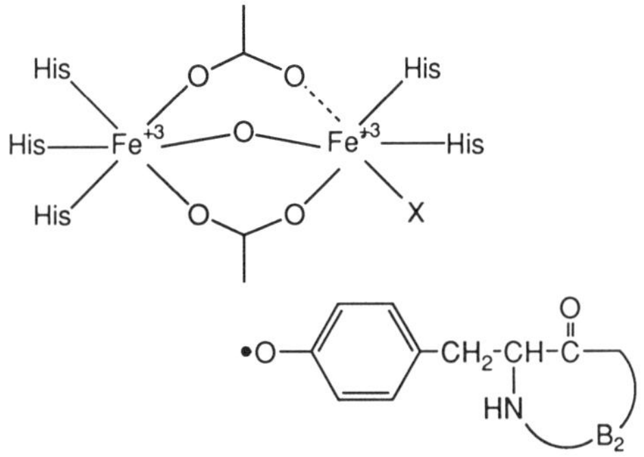

SCHEME 2. Cofactor center of ribonucleotide-diphosphate reductases.

A possible role of the Fe(III) ions may be to stabilize the tyrosyl radical (*13, 23*), which is substantiated by the finding that removal of iron from the enzyme results in loss of activity and EPR signal. Although the enzyme may be reconstituted chemically (*10*), there is evidence that reactivation *in vivo* requires an activating enzyme, Mg^{2+}, Fe(II), oxygen, and the presence of a reductant such as dithiothreitol. The reappearance of both enzyme activity and EPR signal (*10, 24–26*) suggests that the radical is generated by introduction of iron into the apoprotein structure, with the simultaneous oxidation of Fe(II) and tyrosyl residues. The role of oxygen in the reactivation process is curious and has led to the suggestion that oxygen may serve as a regulator of DNA synthesis (*4*). Hydroxyurea, a potent RDPR inactivator and radical scavenger, has no effect on the Mössbauer parameters, indicating that the oxidation state of the iron in B2 is not dependent on the presence of the free radical. Only one tyrosyl radical per B2 subunit has been observed, and a "half-of-the-sites" reactivity model has been proposed (*27*).

Protein B1 (M_r 160,000), a dimer of general structure $\alpha\alpha'$ (*28*), binds the ribonucleotide substrates (*29*) and contains active thiol groups that directly participate in the catalytic process (*30*). The B1 subunit also contains the binding site for the nucleoside diphosphate substrates and the deoxynucleoside triphosphates and ATP, which act as allosteric effectors.

The second class of reductases, the ribonucleoside-5'-triphosphate reductases (RTPRs), are found primarily in prokaryotes and are distinguished from the *E. coli* enzyme in their requirement for adenosylcobalamin (AdoCbl) as cofactor (Scheme 3) (*31*). The enzyme from *Lactobacillus leichmannii* has been extensively studied and consists of a single polypeptide chain of M_r 76,000 (*32*), which contains the binding sites for the 5'-triphosphate as well as the allosteric binding site for the 2'-deoxyribonucleotide triphosphate. Consistent with the reaction mechanisms of AdoCbl enzymes, substrate reduction by RTPR is accompanied by reversible carbon–cobalt bond cleavage (*33, 34*). Similar to the *E. coli* reductase, oxidation of enzyme sulfhydryl groups occurs and their reduction is accomplished *in vivo* using the NADPH–thioredoxin and thioredoxin reductase system and *in vitro* utilizing dihydrolipoate or DTT (*35*). In the presence of substrate a broad EPR resonance near $g = 2.3$ and a narrow but unsymmetric doublet near $g = 2$ are observed (*36*). The doublet signal apparently results from a weak electrostatic exchange interaction between an organic free radical formed by substrate and low-spin Co(II)–B$_{12}$ complex (*20*).

The third class of reductases apparently requires manganese for activity, although few other characterizations have been performed. This enzymic activity is found solely in bacteria, specifically in *Brevibacterium ammoniagenes, Arthrobacter, Micrococcus luteus,* and *Nocardia opaca* (*37, 38*).

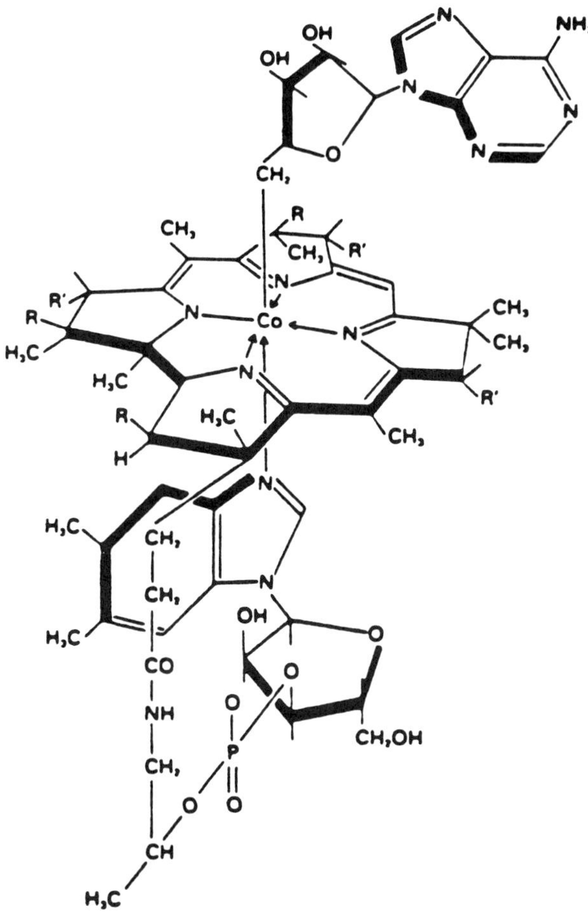

SCHEME 3. Adenosylcobalamin cofactor of ribonucleotide-triphosphate reductase (RTPR).

B. MECHANISTIC PROPOSALS AND CHEMICAL PRECEDENCE

The chemical transformation involves the conversion of a secondary alcohol to a methylene group. This process is chemically difficult, as hydroxide is a poor leaving roup, and no evidence has been found for 2'-phosphorylated or 2'-pyrophosphorylated activated intermediates (39). Furthermore, S_N2 displacement at C-2' is unlikely given the steric crowding by the cis-vicinal base at C-1', whereas cationic intermediates at C-2' formed via an S_N1 mechanism are likewise precluded as a result of the adjacent electron-deficient anomeric center at C-1' (40–42). The discovery of the tyrosyl radical in RDPR has led investigators, in particular Stubbe and co-workers, to propose radical intermediates in the reductive mechanism (for recent reviews see Refs. *37* and *38*). Chemical precedent for the facilitation of elimination reactions via radical intermediates has been observed in the reaction of ethylene glycol and halohydrins (analogs of the nucleoside di- and triphosphates) with Fenton's reagent [Fe(II)/H_2O_2] (*43, 44*) (Scheme 4, compounds **1–3**). Initially, H_2O_2 is reduced by Fe(II), forming hydroxyl radicals that abstract hydrogen from substrate and generate the hydroxyalkyl radical **1** (Scheme 4). Cleavage of the C–X bond occurs readily when X is a halogen, but the poor leaving-group ability of hydroxide requires acid catalysis to produce the radical cation **2**. The driving force for this reaction is tautomerization to the conjugated aldehyde radical, **3**, which undergoes one-electron reduction by Fe(II), forming aldehydes or polymerization products. In the enzymic reaction, the final step is best satisfied by a two-electron reduction of the radical cation, forming a deoxyhydroxyalkyl radical, analogous to the 2'-deoxynucleosides.

Although the tyrosyl radical of RDPR (and the AdoCbl-derived radical of RTPR) is essential for enzyme activity, a specific role in catalysis has not been directly demonstrated. No change in the EPR signal or UV–Vis spectrum of the radical species has been observed in the presence of substrate (*4*). Such studies do not suggest against direct involvement of the radical in the catalytic mechanism, as one must consider that the individual steps involving radical intermediates might not favor accumulation of a new radical species. In the course of

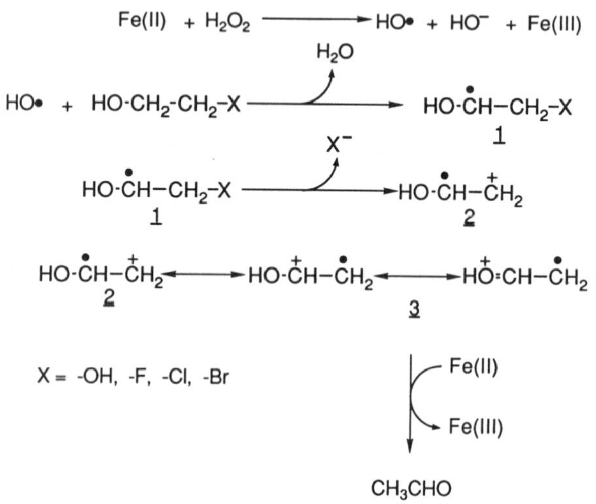

SCHEME 4. Elimination reactions achieved via radical intermediates using Fenton's reagent.

their studies on RDPR and RTPR, Stubbe and co-workers have proposed that the chemical mechanism involves radical intermediates, with a protein dithiol group in the active site serving as the two-electron reductant (*45, 46*). A general mechanistic scheme for RDPR and RTPR catalysis is shown in Scheme 5 (Compounds **4–6**). The hypothesis predicts that a protein-centered radical X■ can abstract a hydrogen atom from the 3'-position of the nucleotide to generate a 3'-nucleotide radical **4**. Generation of this intermediate would facilitate acid-catalyzed cleavage of the 2'-C–OH bond, producing a radical cation intermediate **5**. This intermediate could then be reduced concomitant with oxidation of two thiol groups on the protein. Subsequent reduction of the 3'-deoxynucleotide radical **6** by HX and intermolecular reduction of the enzyme thiol groups regenerate the active form of the enzyme (*46*).

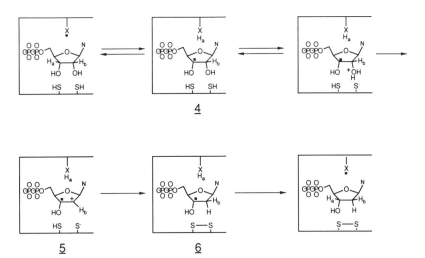

SCHEME 5. Generalized mechanism for RDPR and RTPR catalysis (*56*).

In the subsequent sections, the experimental results for the verification of this mechanism will be discussed. These strategies have involved elegant isotope effect and stereochemical studies with the usual nucleoside di- and triphosphates. Enzymic processing of alternate substrates modified in the 2'-position have led to the discovery of new mechanism-based inhibitors, whose mechanism of inactivation is consistent with the hypothesis for substrate processing described in Scheme 5. Particularly noteworthy is the accumulated evidence that the mechanisms of RDPR and RTPR are quite similar, despite the difference in the cofactors for the respective reactions, i.e., tyrosyl radical and AdoCbl. Finally, the evidence, or the lack of evidence, for the direct involvement of the enzyme radical species, X■, in the catalytic cycle will be discussed.

C. Kinetic Isotope Studies with Normal Substrates

The mechanism proposed in Scheme 5 makes experimentally verifiable predictions concerning the mechanism of the ribonucleotide reductase reaction with normal substrates and substrate analogs. Two crucial predictions are the cleavage of the 3'-carbon–hydrogen bond and the return of (possibly the same) hydrogen to the 3'-position on the same face from which the original atom was abstracted.

The first point was addressed by Stubbe and co-workers (*45–47*) in studies on the tritium isotope effects at the 3'-position by single- and double-label techniques utilizing pyrimidine [3'-^{3}H]UDP/[U-^{14}C]UDP and purine [3'-^{3}H]ADP/[U-^{14}C]ADP. In both cases, conversion to the corresponding 2'-deoxynucleoside 5'-diphosphates by RDPR was subject to isotope selection effect against ^{3}H. The V/K effects ranged from 2 to 4.7 for UDP, and vary with both pH and the allosteric effector used, with the general trend that the lower the pH, the larger the isotope effect observed. A small amount of tritium is released to the solvent, the amount decreasing with decreasing pH and comprising 0.5–0.9% of the total radioactivity at 50% conversion. Similar results are found with ADP, with k_H/k_T values from 1.9 to 1.4. These results are consistent with a mechanism requiring cleavage of the 3'-carbon–hydrogen bond of the substrate. The magnitudes of these primary kinetic isotope effects are not consistent with β-^{3}H secondary isotope effects (*48*), and the observation of ^{3}H$_2$O release substantiates cleavage of the 3'-carbon–tritium bond and suggests that X in Scheme 5 may be an amino acid residue that is subject to a slow exchange during the enzymic process.

Tritium substitution in the 3'-position of ATP and UTP also yielded isotope effects following their reduction by the B$_{12}$-dependent RTPR, suggesting that the initial step of the reactions catalyzed by RTPR is a rate-limiting abstraction of the 3'-hydrogen (*47, 49*). The isotope effect with RTPR was rather insensitive to the pH of the media, although the formation of small quantities of ^{3}H$_2$O was still observed.

The fate of the 3'-H of the substrate after its abstraction by RDPR or RTPR has been studied with [3'-^{2}H]UDP (*46*) and [3'-^{2}H]UTP (*47*), respectively. Examination of the 2'-deoxy products via NMR spectroscopy indicated the presence of deuterium in the 3'-position. Taken together with the tritium isotope effect studies, these results indicate that the hydrogen originally removed from the 3'-position of the substrate is returned to the same position in the products.

D. Alternate Substrates Modified at 2'-Positions

1. *Ribonucleotide-diphosphate Reductase*

The earliest mechanistic studies utilizing alternate substrates were first reported in 1976 by Thelander and co-workers on the inactivation of RDPR by 2'-chloro-2'-deoxyuridine 5'-diphosphate (ClUDP) and the cytidine analog, ClCDP (*9*).

TABLE I

SUBSTRATE ANALOGS OF RIBONUCLEOTIDE REDUCTASES

2'-Substituted 2'-deoxynucleotide	n	B	X	Y	Ref.
2'-Chloro-2'-deoxyuridine	2, 3	U	Cl	H	9, 51, 52, 32
					58, 71, 53, 56
2'-Fluoro-2'-deoxycytidine	2, 3	C	F	H	50, 58
2'-Fluoro-2'-deoxyadenosine	2	A	F	H	50
2'-Azido-2'-deoxycytidine	2	C	N$_3$	H	60
2'-Azido-2'-deoxyuridine	2	U	N$_3$	H	62, 63
2'-Azido-2'-deoxyadenosine	2	A	N$_3$	H	63
2'-Chloro-2'-deoxyadenosine	3	A	Cl	H	58
2'-Bromo-2'-deoxyadenosine	3	A	Br	H	58
2'-Iodo-2'-deoxyadenosine	3	A	I	H	58
2'-Fluoro-2'-deoxyuridine	3	U	F	H	58
9-(2-Chloro-2-deoxy-β-D-arabino-furanosyl)adenine	3	A	H	Cl	58
9-(2-Bromo-2-deoxy-β-D-arabino-furanosyl)adenine	3	A	H	Br	58

They observed specific inactivation of the B1 subunit of the *E. coli* enzyme, the production of free base and chloride, the irreversible modification of protein sulfhydryl groups, and the formation of a new chromophoric absorbance at 320 nm. The tyrosyl radical was not quenched following inactivation, and in this original report a new phosphorus-containing sugar was observed and tentatively identified as 2-deoxyribose 5-diphosphate. Furthermore, inactivation with [^{36}Cl]-, [α-^{32}P]-, or [5-^{3}H]ClCDP did not lead to radiolabeled protein.

Since 1976 a number of 2'-substituted 2'-deoxynucleoside 5'-diphosphates have been shown to be mechanism-based inactivators of the ribonucleotide reductases from a variety of sources (Table I). This work was extended by Stubbe and Kozarich (*50, 51*), who studied the reaction of several 2'-halo-substituted 2'-deoxynucleoside 5'-diphosphates with RDPR. Incubation of RDPR with ClUDP, 2'-deoxy-2'-fluoroadenosine 5'-diphosphate (FADP), or 2'-deoxy-2'-fluorocytidine 5'-diphosphate (FCDP) resulted in time-dependent enzyme inactivation, concomitant and stoichiometric loss of all substituents from the ribose moiety, increase in the UV-Vis absorbance of the protein near 320 nm, and formation of the new phosphorus compound, identified as inorganic pyrophosphate. On the basis of these observations, a preliminary proposal for the reaction of these substrate analogs with RDPR was postulated, based on their working hypothesis of a free radical reaction mechanism (Scheme 6, compounds **7–10**) (*50, 51*). This enzymic reaction mechanism involves the tyrosyl radical of B2 abstracting the 3'-H atom of the substituted nucleoside to generate the 3'-nucleotide radical **7**. This initial abstraction facilitates cleavage of the 2'-carbon–halogen bond (a process not requiring acid catalysis) to generate a radical cation, **8**, which may tautomerize between the 2'- and 3'-positions. The breakage of the carbon–halogen bond at C-2' is analogous to the specific acid-catalyzed C–OH bond cleavage postulated in the normal reaction. The radical cation may then collapse to the 3'-keto-2'-deoxynucleotide radical **9**. This intermediate then reabstracts the hydrogen atom from the protein to regenerate tyrosyl radical and form the 2'-deoxy-3'-ketonucleotide **10**. This step satisfies the observation by Thelander *et al.* (*9*) that during inactivation of RDPR by ClUDP no loss of the protein tyrosyl radical was observed. The 3'-ketone would now dissociate from the active site of the enzyme and undergo buffer-catalyzed abstraction of the 2'- and 4'-α-keto protons of the sugar moiety, permitting elimination of base, and the 5'-substituent, respectively, and the formation of a highly reactive 2-methylene-3(2H)-furanone sugar derivative, **11**. Inactivation presumably results from protein alkylation of the furanone via Michael addition by an amino acid side chain, explaining the loss of protein thiol groups and the formation of the 320-nm chromophore.

The hypothesis in Scheme 6 makes several experimentally verifiable predictions: (1) abstraction of the 3'-hydrogen, (2) suprafacial transfer of the 3'-hydrogen to the 2'-position with the possibility of exchange with solvent, (3) production of 3'-keto-2'-deoxy nucleotide, (4) nonenzymic elimination of base and the 5'-substituent, requiring proton abstraction from the 2'- and 4'-positions, respectively, and (5) covalent labeling of the protein via 2-methylene-3(2H)-furanone. This hypothesis has been thoroughly tested by Stubbe and co-workers; the outcome of their work in support of the mechanism in Scheme 6 is quite persuasive and will be briefly discussed below.

Treatment of either RDPR or RTPR with [3'-³H]ClUDP or [3'-³H]ClUTP, respectively, leads to the formation of ³H₂O and radiolabeling of the inactive protein (*32*). Although only a small isotope effect was observed, the formation of tritiated water is conclusive evidence of 3'-carbon–hydrogen bond cleavage by an exchangeable group on the enzyme. The radiolabeling of the inactive protein is also consistent with Scheme 6 as it is suggestive of the 3'-hydrogen being returned to the substrate analog at the 2'-position, prior to exchange with solvent protons.

When RDPR was incubated with ClUDP in the presence of NaBH₄ or ethanethiol, the enzyme was protected from inactivation (*32, 52*). Furthermore, a

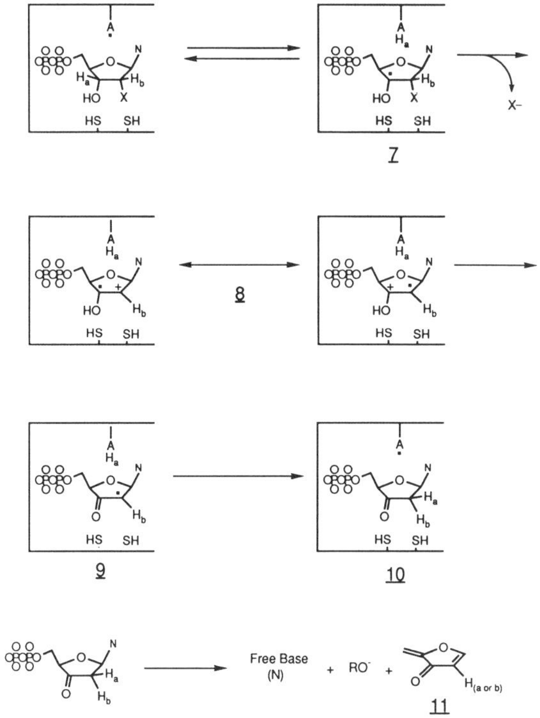

SCHEME 6. Mechanism of inhibition of RDPR by 2'-halo-substituted deoxynucleoside 5'-diphosphates.

radiolabeled metabolite was isolated that, after treatment with alkaline phosphatase, was identified by NMR spectroscopy as a mixture of 2'-deoxyuridine, **12**, and 2'-deoxy-*xylo*-uridine, **13** (Scheme 7). Formation of these compounds is best explained by enzymic conversion of the 2'-chloro nucleotide to the 3'-keto-2'-deoxynucleotide as is indicated in Scheme 6. The keto nucleotide is released to solution and is trapped by the borohydride prior to conversion to the alkylating agent. Furthermore, enzymic analysis and chemical degradation of the isolated

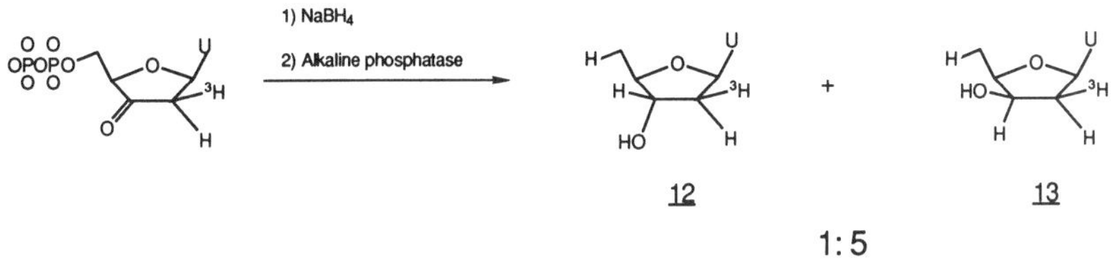

SCHEME 7. Protection of RDPR from ClUDP inactivation by NaBH$_4$.

[2′-³H]deoxyuridine indicated that the hydrogen atom originally present in the 3′-position of ClUDP is transferred across the β face of the nucleotide to the 2′-position of 2′-deoxy-3′-ketoUDP. Hence the overall reaction catalyzed by RDPR on the 2′-substituted substrate analogs is a net 1,2-hydrogen shift. More importantly, this study fixes the protein radical position on the top face of the substrate, with access to both the 2′- and 3′-positions. Furthermore, since ³H from [3′-³H]ClUDP is known to be present in the 2-methylene-3(2H)-furanone alkylating agent (32), the unlabeled hydrogen on the α face of the ketoUDP must be lost, fixing uracil elimination as anti.

The ability of borohydride to trap the 2′-deoxy-3′-keto nucleotide is good evidence that the alkylating agent is formed nonenzymically in solution. As a result, nonspecific alkylation of protein via the 2-methylene-3(2H)-furanone occurs (32). Incubation of RDPR with [5′-³H]ClUDP results in 4.2 equivalents of ³H per equivalent of B1 inactivated, with both subunits of RDPR radiolabeled in the reaction, 2.5 equivalents of ³H on B1, and 1.7 equivalents of ³H on B2 (52). The structure of the alkylating agent was verified in an experiment where RDPR and RTPR were incubated with ClUDP in the presence of DTT or ethanethiol, and protection from inactivation was observed. Metabolites were isolated; their structures, elucidated by IR and NMR, are consistent with the nonenzymic reaction of thiols with the alkylating agent, producing alkylthiofuranones (Scheme 8). Model studies with synthetic 2-methylene-3(2H)-furanone suggest that the new protein chromophore results from a two-step process involving initial attack of a protein nucleophile, presumably a thiolate, to the exocyclic double bond. This initial protein-bound intermediate is suspected to constitute the inactivation step. The observed chromophore is formed by attack of an amino group at the endocyclic double bond of the protein-bound furanone, and subsequent ring opening and tautomerization to form a β-hetero-substituted enone system, 14 (Scheme 9, compounds 14 and 15) (53). The properties of this protein adduct (λ_{max} 320 nm, $\varepsilon = 20,000$ M^{-1} cm^{-1}, resistance to borohydride reduction) are

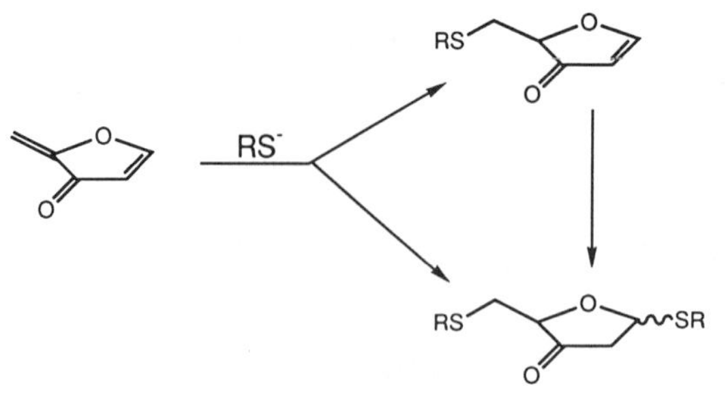

SCHEME 8. Alkylthiofuranone formed by protective thiols and 2-methylene-3(2H)-furanone.

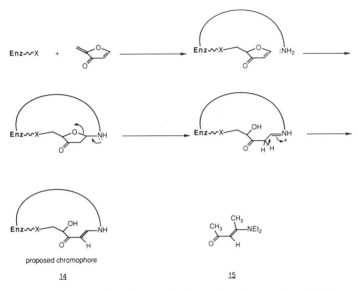

SCHEME 9. Chromophore formed by reaction of protein and 2-methylene-3(2H)-furanone.

favorably similar to a model compound, **15,** prepared from 2,4-pentanedione (λ_{max} 316 nm, $\varepsilon = 24,800$ M^{-1} cm^{-1}, borohydride resistant) (*53*), thereby-consistent with **14** being the structure of the 320-nm chromophore.

Chemical precedent for the nonenzymic elimination of base and inorganic pyrophosphate may be found in the attempted synthesis of 3'-ketothymidine by oxidation of TMP (*54*), which led instead to the formation of P_i, thymine, and unidentified sugars. Furthermore, Hershfield reported that 2'-deoxyadenosine inactivated *S*-adenosylhomocysteine hydrolase and that this inactivation was accompanied by adenine release (*55*).

2. Ribonucleotide-triphosphate Reductase

Strikingly similar results are observed in the reaction of RTPR with 2'-halogenated ribo- and arabinonucleoside triphosphates (Table I). Reaction of RTPR with all analogs led to mechanism-based inactivation of the enzyme and concomitant formation of ^3H_2O (in the case of 3'-^3H analogs), base, triphosphate, and halide. In all cases the inactive enzyme possessed a new UV–Vis absorbance maxima near 320 nm, characteristic of enzyme inactivation by 2-methylene-3(2H)-furanone. Such results are consistent with a common reaction path for all 2'-halonucleotides. With the 2'-chloro analog, destruction of the AdoCbl cofactor to form 5'-deoxyadenosine and cob(II)alamin is observed, although this was not observed to occur with any of the other analogs and was not believed to be intimately involved in the inactivation mechanism. No radioactivity was found in the reisolated cofactor, and evidence was presented indicating the cofactor was not kinetically competent to abstract a hydrogen from the substrate and was rapidly exchanging with the solvent. Also, DTT was observed to protect the enzyme from inactivation, suggesting an alkylation mechanism similar to the RDPR reactions (56).

A mechanism to explain these transformations would be similar to that shown in Scheme 6, the key step again being the initial abstraction of the 3'-hydrogen, followed by a non-enzyme-catalyzed loss of halide to form the resonance-stabilized radical cation. Again, the major difference from the normal catalytic sequence (Scheme 5) is that protonation of the halide leaving group is not necessary, as seen by the ability of the ara-ClATP and ara-BrATP (Table I) to inactivate RTPR comparably with the normal ribo isomers. Normal substrate processing requires enzyme-mediated general acid catalysis to assist elimination of the 2'-hydroxyl group. Thus, ara-ATP has been observed to bind to RTPR and cause reversible cofactor homolysis (33, 57), yet does not cause inactivation, suggesting the absence of an appropriate acid catalyst on the top face of the ribosyl ring.

E.   ROLE OF REDOX-ACTIVE THIOLS

The mechanistic features that lead to partitioning between inactivation and formation of the 2'-deoxynucleotide products are a particularly interesting aspect of the reductase reaction with normal substrates and the 2'-substituted substrate analogs. This partitioning is best explained by considering the oxidation or protonation state of the active site protein thiol groups in these respective reactions. In the reduction of normal substrates, elimination of the 2'-substituent (hydroxide ion) requires specific acid catalysis, presumably via the active site thiols as postulated in Scheme 5. The protonation state of the thiols at this stage then facilitates reduction of the substrate-derived radical cation (46, 58). However, in the mechanism of RDPR and RTPR inactivation by ClUDP, there is no catalytic role for the active site thiols in the mechanism of inactivation, and the subtle differences in protein conformation due to the protonation state of the thiols are capable of allowing a rearrangement reaction to occur with the alternate substrates (32, 52). Elimination of the halide leaving group from the 2'-substituted substrate analogs does not require catalysis, leaving the thiol groups in the incorrect protonation state to catalyze a normal reduction reaction. Disulfide formation from the two protonated thiols in the active site is apparently slow, such that the radical cation in Scheme 6 tautomerizes to the 3'-ketone intermediate and eventually to the 2-methylene-3(2H)-furanone alkylating agent. Consistent with this hypothesis are the observations by Atore and Stubbe (52) that treatment of RDPR with [3'-³H]ClUDP leads to multiple turnovers, inactivation, and release of ³H₂O whether the enzyme thiols are in the fully reduced form in the absence of a reducing system, or if the active site thiols are oxidized. Hence, the redox-active thiols of subunit B1 in RDPR are not required for conversion of ClUDP to the ketone responsible for inactivation.

Recently, the redox-active thiols of RDPR were located and the amino acid sequence in this vicinity were determined (59). Two pairs of redox-active dithiols are suggested to be involved in substrate reduction by RDPR, Cys-752/Cys-757 and Cys-222/Cys-227. The proposed interaction of these thiols in substrate reduction of CDP to dCDP is illustrated in Scheme 10. The carboxylate-terminal dithiols, Cys-752/Cys-757, are intimately involved in substrate reduction. The Cys-222/Cys-227 dithiol couple then transfers the reducing equivalents back into the active site dithiols by disulfide interchange, and may be subsequently reduced by thioredoxin or DTT. The redox-active thiols of RTPR have also been located. Curiously, the amino acid sequence of the RDPR containing the Cys-752/Cys-757 couple is strikingly similar to the sequence of a dithiol-containing peptide isolated from RTPR. It is not yet clear if this sequence homology reflects divergent evolution of RDPR and RTPR from a common ancestor, or rather if convergent evolution from different ancestors is indicated, a result of the similar chemistry performed by these enzymic reactions.

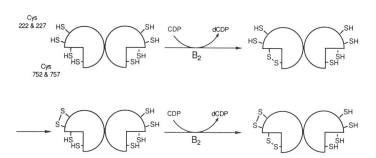

SCHEME 10. Possible interaction of cysteines during RDPR catalysis (59).

F. SEARCH FOR RADICAL INTERMEDIATES: 2'-AZIDO SUBSTRATE ANALOGS

The proposed mechanisms for the utilization of substrates and substrate analogs by the ribonucleotide reductases, Schemes 5 and 6, predict the formation of various radical intermediates. To search for these intermediates and their direct participation in these reactions, EPR spectroscopy has been employed (60). However, these attempts have not revealed the generation of new radical signals.

Work along this line has led investigators to study the reactions of 2'-azido substrate analogs with RDPR. Thelander and co-workers first reported on the inactivation of RDPR by 2'-azido-2'-deoxycytidine 5'-diphosphate (N₃CDP) and 2'-azido-2'-deoxyuridine 5'-diphosphate (N₃UDP) (9). Sjoberg and collaborators subsequently reported an intriguing observation that reaction of RDPR with N₃CDP and N₃UDP led to tyrosyl radical loss and the concomitant, transient formation of a new radical signal (60, 61). The EPR characteristics indicated a hyperfine structure consistent with couplings to a one-hydrogen and one-nitrogen nucleus, a major triplet of about 25 G splitting, with a smaller doublet splitting of about 6 G. The magnitude of the hyperfine coupling for the nitrogen interaction (25 G) is the same order of magnitude if the unpaired spin density resided on the nitrogen atom. Isotopic labeling of the protein with ^{15}N and ^{2}H did not alter the new transient EPR spectrum, so it was concluded that the hyperfine couplings arose from nuclei of the substrate analog rather than the protein. Furthermore, the reaction with N₃CDP is suicidal; the new radical species forms with concomitant loss of the B2 radical signal and B2 radical UV–Vis absorption maximum at 410 nm. The nucleotide was destroyed in the process, forming free nucleic acid base and radical-quenched B2; no CDP or dCDP was observed in the N₃CDP reaction. This is the first unequivocal demonstration of radical chemistry being catalyzed by ribonucleotide reductase.

The first conclusive evidence for the localization of spin density on the nucleotide was provided by the work of Stubbe and co-workers (62). Treatment of RDPR with [2'-^{15}N]N₃UDP caused the triplet splitting to collapse to a doublet, consistent with the new radical species being localized at the nitrogen originally at the 2'-position of substrate. Furthermore, incubation of the enzyme with [1'-, 2'-, or 3'-^{2}H]N₃UDP had no effect on the radical signal, suggesting that the hydrogen coupling to this species is not due to the hydrogen atom on the 1'-, 2'-, or 3'-carbons of the substrate analog.

Detailed studies on the mechanism of RDPR inactivation by these substrate analogs (63) have revealed that rather complex chemical steps are occurring. The potency of the 2'-azido inhibitors is emphasized by the work of Stubbe, where complete inactivation of RDPR occurs with one equivalent of N₃UDP. The tyrosyl radical of the B2 subunit is quenched by N₃UDP, whereas the specific activity of the B1 subunit is reduced by 50%. Since B1 has two active sites, and B2 has only one tyrosyl radical, these data are consistent with stoichiometric inactivation indicating one turnover per enzyme inactivation event and one equivalent of radiolabeling. Hence, modification of the B2 subunit only occurs to the extent of tyrosyl radical destruction. Consistent with this idea is the finding that inactivation by [5'-^{3}H]N₃UDP leads to covalent radiolabeling specific for the B1 subunit of the protein. Finally, isolation of inactive B2 subunit and reactivation treatment with Fe(II) (10) leads to recovery of B2 activity, suggesting regeneration of the B2 protein radical.

Incubation of the enzyme with [3'-^{3}H]N₃UDP resulted in 0.2 mol of ^{3}H released to solvent per mole of enzyme inactivated. Hence, the initial step in the mechanism of RDPR reaction with N_3UDP is consistent with the normal substrates and the 2′-halo analogs in that the nucleotide 3′-carbon–hydrogen bond is cleaved. Furthermore, essentially no ^{3}H was found to be covalently bound to the inactivated protein, consistent with the reactive amino acid residue quantitatively exchanging the abstracted isotope with solvent, and suggesting an isotope effect of ~5 on the 3′-abstraction event. Additional studies indicate the stoichiometric formation of inorganic pyrophosphate and uracil from N_3UDP, although short incubations with $[\beta\text{-}^{32}P]N_3$UDP initially gave a transiently ^{32}P-labeled protein. No azide release was detected; however, with $^{15}N_3$UDP a stoichiometric amount of nitrogen is produced, with the remaining nitrogen atom of the azido group appearing as the radical center in the transient EPR signal seen during inactivation. In contrast to the proposed inactivation mechanism for ClUDP, inactivation by N_3UDP or N_3ADP does not occur if the active site thiols of B1 are oxidized, and there is no protection observed from ethanethiol.

The key feature so far observed in RDPR inactivation mechanisms is that destruction of the tyrosyl radical (one-electron chemistry) may be coupled to cleavage of the 3′-carbon–hydrogen bond of the nucleotide. The 3′-hydrogen is abstracted as a hydrogen atom to produce a nucleotide radical intermediate with substrates and 2′-halogenated analogs (32, 46). However, in these cases, the nucleotide radical exists only transiently, regenerating the tyrosyl radical with each turnover. Apparently, with the azido nucleotides a different mechanism is operative, wherein the azido group functions as a radical trap of the initially formed nucleotide radical, to form a new and relatively stable radical species. Without regeneration of the tyrosyl radical, inactivation occurs after one turnover.

The appearance of a new UV–Vis absorbance at 320 nm in N_3UDP inactivation of RDPR suggests that a species similar to 2-methylene-3($2H$)-furanone is formed. The observation that ^{32}P is initially tightly, but not covalently, bound to the protein following inactivation suggests that the nucleotide phosphates are not released to solution prior to the generation of the new radical species. Hence, the elimination of PP$_i$ might occur only when the active site opens, allowing the nucleotide radical to be quenched. At this time it remains to be determined if the PP$_i$ is still bound to the sugar moiety in the new radical, or if it has already been eliminated but remains held in the closed active site.

A working hypothesis for this inactivation mechanism has been proposed (63) and is presented in Scheme 11 (compounds 16–20). Consistent with the mechanisms of substrate and inhibitor utilization by RDPR, the initial step is proposed to be abstraction of the 3′-hydrogen by the tyrosyl radical X to generate a 3′-nucleotide radical, 16. There are no known good chemical precedents for decomposition of β-azidoalkyl radicals such as 16. The best analogy is an intermolecular reaction of 1-hydroxy-1-methylene radical with primary alkyl azides

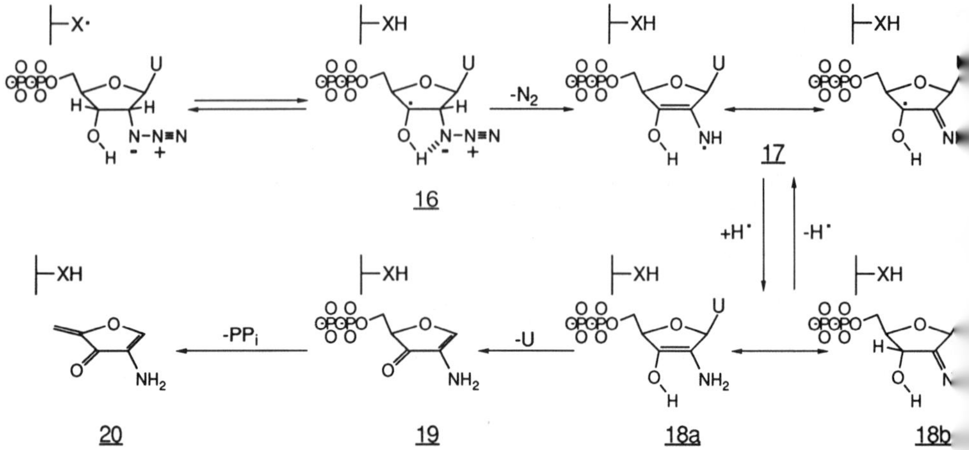

SCHEME 11. Postulated mechanism for the inhibition of RDPR by 2-azidoUDP (63).

to form dialkylaminyl radicals (Scheme 12) (64). Hence, the enzymic mechanism in Scheme 11 may represent the intramolecular variation of that reaction. The succeeding steps in the enzymic mechanism may involve **16** decomposing to liberate N_2 and to yield the EPR-observable delocalized radical, **17**, with arbitrary protonation states of the 3'-oxygen and 2'-nitrogen. Eventually **17** would abstract a hydrogen atom to form **18a** or **18b**. Tautomerization of **18a** with β-elimination of uracil would then form **19**, which in turn would release PP_i to form the unsaturated aminoketone (**20**), which would subsequently alkylate the protein in a Michael reaction. An alternate mechanism involving the redox thiols in the reduction of the azido group has also been suggested (63); this mechanism is based on the known reaction of organic azides with DTT at high pH (65). After abstraction of the 3'-hydrogen, generation of a nitrogen-centered radical would require homolytic carbon–nitrogen bond cleavage in a step having chemical precedent for a trialkylamine (66). However, for both of the chemical precedents suggested, the EPR radical parameters for the iminyl or aminyl radicals make them poor candidates for the observed nucleotide radical.

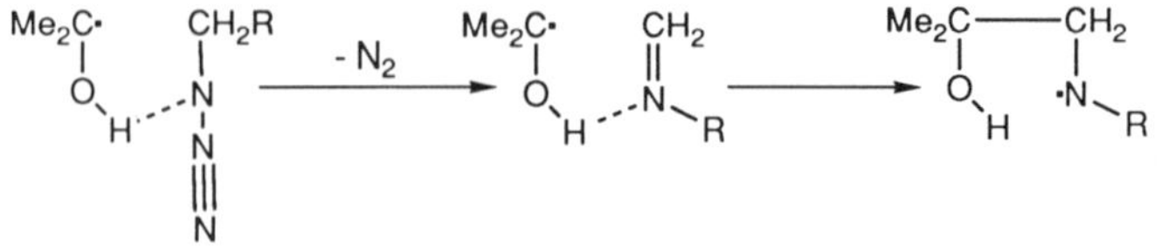

SCHEME 12. Intermolecular reaction of 1-hydroxy-1-methylene radical with primary alkyl azides.

G. ROLE OF ADENOSYLCOBALAMIN IN RIBONUCLEOSIDE-TRIPHOSPHATE REDUCTASE REACTIONS

Given the unique and essential role the ribonucleotide reductases play in DNA synthesis, it is surprising that these enzymes require distinct and elaborate cofactors to catalyze related reactions with similar chemistry. As noted above, the RTPR of *L. leichmannii* differs from the *E. coli* diphosphate reductase in that the former requires AdoCbl as a cofactor. Despite these differences in cofactor structure, Stubbe and co-workers have found remarkable similarities in the mechanisms of the reactions catalyzed by these two reductases (*47, 49*). This point is substantiated by the irreversible inactivation of both classes of enzymes by 2′-chloro-2′-deoxynucleotides, producing a similar product and radiolabel distribution in the reductase-catalyzed breakdown of [3′-³H]- or [5′-³H]ClUDP/ClUTP (*32, 53, 56, 58*). Furthermore, a distinct sequence homology between RDPR and RTPR has been observed in peptide fragments containing the redox-active thiol residues from each protein (*59*).

Evidence has accumulated suggesting that the RTPR is an unusual AdoCbl-dependent enzyme based on the catalyzed reaction mechanism and the role of the cofactor in the reduction process. As with all AdoCbl-requiring enzymes, homolysis of the carbon–cobalt bond has been found to be involved in catalysis (*33, 34*). However, some of the more unusual characteristics may be summarized as follows: (1) RTPR catalyzes an intermolecular redox reaction rather than a rearrangement, where the nucleoside triphosphate is reduced and sulfhydryl groups on the enzyme are oxidized; (2) AdoCbl is not tightly bound to ribonucleotide reductase but rather is freely dissociable from the enzyme; (3) tritium in the 5′-position of the 5′-deoxyadenosine moiety of AdoCbl is not transferred to product, but slowly exchanges with solvent (*67, 68*). In contrast, incubation of substrate and RTPR in ³H₂O leads to stereospecific synthesis of [2′-³H]deoxynucleoside triphosphate (*69*). So, unlike other AdoCbl-dependent enzyme systems, there is no definitive evidence supporting the role of adenosylcobalamin as an intermediate hydrogen-transferring agent between substrate and product. The work done by Stubbe and co-workers indicates that AdoCbl may not function in direct hydrogen abstraction from substrate, and, furthermore, the hydrogen abstracted from normal substrates does not migrate to a new position in the product. In almost all known AdoCbl-mediated reactions, the adenosylcobalamin acts as an intermediate hydrogen carrier abstracting a specific hydrogen from substrate and returning it to a different position in the product (*70*). Hence, with [5′-³H]AdoCbl all the tritium is transferred to product, and none is exchanged with the solvent.

The pertinent data in elucidating the role of AdoCbl as a hydrogen abstractor in RTPR reactions involve studies on the fate of the 3′-hydrogen of substrates and substrate analogs and of the 5′-hydrogen of AdoCbl; these are summarized as follows. The initial step of the reactions catalyzed by RTPR involves abstraction of the substrate 3′-hydrogen. Isotope effects of approximately 1.8 and 2.1 are observed using [3′-^{3}H]UTP and [3′-^{3}H]ATP (*47, 49*) as substrates, respectively, similar in magnitude to the corresponding isotope effects for RDPR (*45, 46, 49*). Small amounts of tritium from 3′-^{3}H-labeled substrate appear as ^{3}H$_2$O; however, no radioactivity is found in the reisolated cofactor under single- or multiple-turnover conditions (*49*). The fate of the 3′-hydrogen has been probed utilizing [3′-^{2}H]UTP, where the isolated product was identified as 2′-deoxy[3′-^{2}H]UTP. This result suggests that the abstracted 3′-hydrogen is returned to that position in the product (*47*). This is evidence suggesting that the cofactor does not serve as a direct H abstracter from substrate, given that with [5′-^{3}H]AdoCbl the ^{3}H appears released quantitatively to ^{3}H$_2$O (*67, 68*).

Incubation of [3′-^{3}H]ClUTP with RTPR and excess AdoCbl leads to the formation of 3′-keto-2′-deoxy[2′-^{3}H]UTP and small amounts of ^{3}H$_2$O (*32, 58, 71*), directly analogous to the inactivation sequence of 2′-substituted substrate analogs with RDPR (*32, 52*). Along with inactivation of RTPR enzymic activity, the AdoCbl cofactor is destroyed, forming 5′-deoxyadenosine and cob(II)alamin, with no tritium detected in these fragments or in the reisolated intact cofactor. In AdoCbl-dependent rearrangements, where the cofactor is destroyed by formation of 5′-deoxyadenosine, the usual source of the third hydrogen is the substrate, but that is apparently not the case here. Cofactor destruction is unique to ClUTP, as no other 2′-analogs surveyed result in such products (*58*). These results with alternate substrates are consistent with those above and suggest that the AdoCbl cofactor does not play a role as a direct hydrogen abstractor in RTPR reactions, and furthermore does not catalyze a rearrangement reaction.

A mechanism consistent with these observations is presented in Scheme 13 (*53, 70*). The first step of the mechanism involves homolytic cleavage of the carbon–cobalt bond of AdoCbl, producing a 5′-deoxyadenosyl radical [cob(II)-alamin], a step that is known to be feasible both chemically (*72, 73*) and kinetically (*68, 69*). The 5′-deoxyadenosyl radical then performs a novel function by acting as a radical chain initiator, abstracting a hydrogen atom from a protein residue XH on the enzyme, generating X·. The identity of X· is not yet clear, although a thiyl or tyrosyl radical has been postulated (*56*). The new protein-based radical then abstracts a hydrogen atom from the 3′-position of the nucleotide triphosphate substrate, producing a 3′-nucleotide radical. The protein residue XH can exchange with solvent, consistent with the observation of enzyme-dependent release of small amounts of 3′-^{3}H as ^{3}H$_2$O (*49*). The nucleotide radical would undergo loss of H$_2$O, probably catalyzed by one of the redox-active thiols, to produce the radical cation intermediate. Reduction of this intermediate by the active site thiol and thiolate would yield the 3′-radical of the 2′-deoxynucleotide triphosphate product, which would reabstract hydrogen atom from XH to generate the 2′-deoxy product and the protein radical. Thus, by return of the abstracted hydrogen to the original 3′-position in the normal sub-

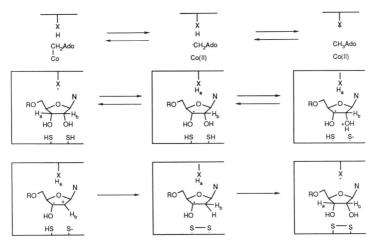

SCHEME 13. Mechanism of RTPR-mediated catalysis using cobalamin; R, triphosphate (49).

strates, RTPR becomes the only known AdoCbl-dependent enzyme that does not catalyze a rearrangement reaction with its normal substrates.

These studies with RTPR illustrate the unusual role of the AdoCbl cofactor in the reduction of nucleotide triphosphates. In the usual rearrangement reactions, the reaction mechanisms involve hydrogen abstraction by a 5′-deoxyadenosyl radical from substrate (73). With RTPR, the abstraction of hydrogen from substrate is indirect, via an AdoCbl-derived protein radical. A further distinction between RTPR and the AdoCbl-requiring enzymes diol dehydratase and ethanolamine ammonia-lyase is that no transfer of 3′-hydrogen from the nucleotide triphosphate substrates to the cofactor was observed. However, the RTPR-catalyzed reaction with ClUTP is formally analogous with the diol dehydratase reaction in that the conversion to 3′-keto-2′-deoxyUTP is an internal redox reaction involving a 3′- to 2′-hydrogen shift. In this regard the AdoCbl cofactor of RTPR serves as an appropriate model for B_{12}-dependent rearrangement reactions. This work has led Stubbe and co-workers to postulate (56) that the enzymes catalyzing AdoCbl-dependent rearrangements and RTPR may have more mechanistic features in common than previously thought, and that the protein radical generated via homolytic cleavage of the carbon–cobalt bond is a common feature of all AdoCbl-dependent enzymes.

H. CONCLUSIONS

A significant observation in studying the mechanism of these enzymic reduction reactions is that even though the structures of the cofactors are unique, and the protein structures are dissimilar, the mechanisms are apparently quite similar. The results of experiments utilizing substrate analogs support this hypothesis and substantiate the remarkable mechanistic similarities between these two enzymes that utilize totally different cofactors to perform the same type of chemistry. Both RDPR and RTPR initiate their reactions by abstraction of the 3'-hydrogen from substrate or substrate analog, with similar kinetic isotope effects. Furthermore, both reductases are inactivated by 2'-halo-2'-deoxynucleotides by similar mechanisms, ultimately involving the formation of alkylated protein. This point is emphasized in the recent studies of Stubbe and co-workers (59), which show that there is a remarkable amino acid sequence similarity in the region around the active site dithiols of the *E. coli* and *L. leichmannii* enzymes.

II. Monoamine Oxidase

A. BACKGROUND

Mitochondrial monoamine oxidase (EC 1.4.3.4; MAO) is a membrane-bound flavoenzyme that catalyzes the oxidative deamination of biogenic monoamines to the corresponding aldehydes (Scheme 14). In general terms, the enzymic reaction involves oxidation of the free base form of the amine to the iminium ion (*74*), with concomitant reduction of the flavin adenine dinucleotide (FAD) cofactor (*75*). The iminium ion is nonenzymically hydrolyzed to the aldehyde, and the reduced flavin is reoxidized by O_2, forming H_2O_2 (*76, 77*). The oxidase is widely distributed in the tissues of mammals and the physiological role of this enzymic reaction is believed to be involved in controlling the levels of diverse biogenic and exogenous amines (*78*). Consistent with this idea is the recent interest in inhibitors of the MAO reaction, as these compounds have been exploited clinically to control depression and hypertension (*79, 80*).

On the basis of studies with selective inhibitors and on the differences in rates at which different substrates are oxidized, two forms of the oxidase have been identified (81, 82). Each isozyme catalyzes the same reaction and all are localized in the tissues, predominantly in the liver and brain of mammals. Furthermore, the amino acid sequences at the site of the covalently bound flavin are the

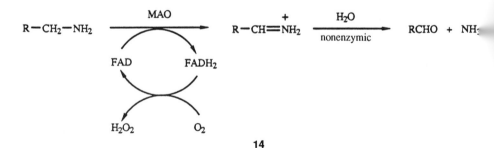

14

SCHEME 14. General reaction catalyzed by mitochondrial monoamine oxidase.

same (83). However, the substrate specificities are different. Type A MAO selectively catalyzes the oxidative deamination of the transmitter amines serotonin and norepinephrine and is inhibited by clorgyline (Table II). Type B MAO preferentially oxidizes benzylamine and β-phenylethylamine and is selectively inhibited by deprenyl (Table II). Tyramine and dopamine are good substrates for both forms of MAO (82, 84–86).

TABLE II

IRREVERSIBLE INHIBITORS OF MONOAMINE OXIDASE: UNSATURATED ALKYLAMINES

Compound	Structure	Reference
Pargyline		92
Clorgyline		122
Deprenyl		122
3-Dimethylamino-1-propyne		93
N-But-2-ynyl-N-benzylmethylamine		123
N-2, 3-Butadienyl-N-benzylmethylamine		123
Propargylamine		91, 124

In 1952, inhibition of the MAO reaction was first correlated with antidepressant effects (*87*). Since that time hundreds of compounds have been identified as MAO inhibitors (*88*), many of which are used *in vivo* to study the role of neurohormones, while others are used clinically as either antidepressive or antihypertensive agents. Although it is generally accepted that MAO inhibitors exert their beneficial effects by inhibiting the brain enzyme (*89*), little is known concerning the mechanism of action of MAO inhibitors or of the chemical mechanism of the MAO reaction.

Studies on the inactivation mechanism of MAO by active site-directed reagents are potentially useful in defining the chemical mechanism of the normal reaction. In this review, only compounds known to be mechanism-based inhibitors will be studied, particularly in their relationship to elucidating the chemical mechanism of this enzymic redox reaction.

B. MECHANISM-BASED INHIBITION OF MONOAMINE OXIDASE: UNSATURATED ALKYLAMINES

A mechanism-based inhibitor may be defined as a chemically unreactive compound that is treated by the target enzyme as a substrate, but instead of forming the usual product, it is converted into a highly reactive species via the normal catalytic mechanism. Prior to release from the active site, the reactive intermediate may alkylate amino acid functional groups, forming a new covalent bond and inactivating the enzyme (*90*). Irreversible, mechanism-based inactivation is typified by first-order, time-dependent loss of enzyme activity; saturation kinetics; inactivation protection by substrates and reversible inhibitors; failure to recover activity following dialysis; and usually a chemical stoichiometry of one covalent adduct formed per enzyme active site.

One class of mechanism-based MAO inhibitors includes the unsaturated alkylamines (propargylamine analogs) (Table II). Although the kinetics of enzyme inactivation for these compounds are consistent with a mechanism-based inhibitor, in only a few cases has the chemical mechanism and site of protein modification been determined. Pargyline (*N*-benzyl-*N*-methyl-2-propynylamine) is a classic example. Pargyline reacts stoichiometrically and irreversibly with the MAO of bovine kidney, with protection from inactivation afforded by substrate benzylamine (*91*). Furthermore, the reaction involves bleaching of the FAD cofactor at 455 nm and the formation of a new absorbing species at 410 nm and a covalent adduct of inactivator with flavin cofactor (*92*).

Abeles and co-workers have studied the inactivation reaction of 3-dimethylamino-1-propyne with bovine liver MAO-A (*93*). Again, the inactivation reaction was found to be stoichiometric and irreversible, with covalent modification of the flavin cofactor suggested by irreversible bleaching of the flavin spectrum. Based on the chemical and spectral properties and pK_a of the enzymically de-

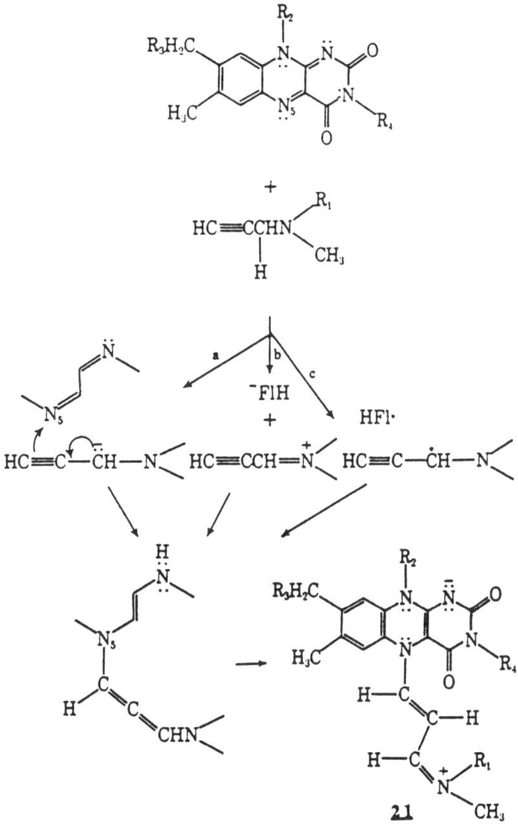

SCHEME 15. Possible reaction mechanisms for inactivation of bovine liver MAO by 3-dimethyl-amino-1-propyne (*123*). R_1, CH_3; R_2, adenine dinucleotide; R_3, apoenzyme; R_4, H.

rived adduct as compared to a variety of model compounds, the structure of the inhibitor–flavin adduct was proposed to be an N-5-substituted dihydroflavin, the flavocyanine **21** (Scheme 15) (*93*). From a knowledge of the reactions of flavins (*94*), three possible mechanisms through which the proposed adduct could be formed were suggested (Scheme 15). In pathway (a), enzyme-catalyzed removal of an α-proton from the inactivator would result in an intermediate carbanion, which could add to N-5 of the oxidized flavin and, following protonation, give the observed adduct. A second possibility, pathway (b), requires complete oxidation of the inhibitor to form reduced flavin and a highly reactive alkyneimine. Nucleophilic addition of N-5 of the reduced flavin across the triple bond of the intermediate in a Michael addition reaction would produce **21**. Pathway (c) involves formation of a radical pair complex implicating flavin and inactivator.

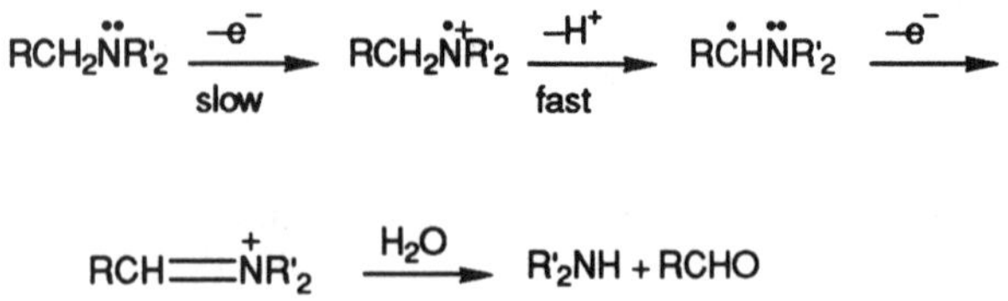

$$RCH_2\overset{..}{N}R'_2 \xrightarrow[\text{slow}]{-e^-} RCH_2\overset{.+}{N}R'_2 \xrightarrow[\text{fast}]{-H^+} R\overset{.}{C}H\overset{..}{N}R'_2 \xrightarrow{-e^-}$$

$$RCH\!=\!\!=\!\!\overset{+}{N}R'_2 \xrightarrow{H_2O} R'_2NH + RCHO$$

22

SCHEME 16. Electrochemical oxidation of amines (122–124).

This complex could be formed either by hydrogen atom transfer from inactivator to flavin, or by an electron transfer from an initially formed inhibitor carbanion to flavin. Collapse of this radical pair followed by protonation could explain formation of the adduct via a partially reduced flavin reacting with a partially oxidized inactivator.

Based on a series of elegant studies on the enzymic and nonenzymic oxidation of amines and substrate analogs, Silverman and co-workers proposed that the mechanism of irreversible inactivation and substrate utilization by MAO is mediated through radical intermediates (95). Precedence for this mechanism is based on the electrochemical oxidation of amines, which is believed to proceed through the radical cation intermediate **22** (Scheme 16) (96–98). Thus, the corresponding mechanism for monoamine oxidation by MAO requires two one-electron transfers from the substrate to flavin (Scheme 17, compounds **23** and **24**). Enzymic reaction is initiated by slow electron transfer of an amine non-bonded electron to the flavin cofactor, producing the amine radical cation **23** and the flavin semiquinone radical. Formation of the amine radical cation facilitates loss of the α-proton, thereby avoiding the removal of nonacidic protons that would be necessary in a carbanionic mechanism. Subsequent electron

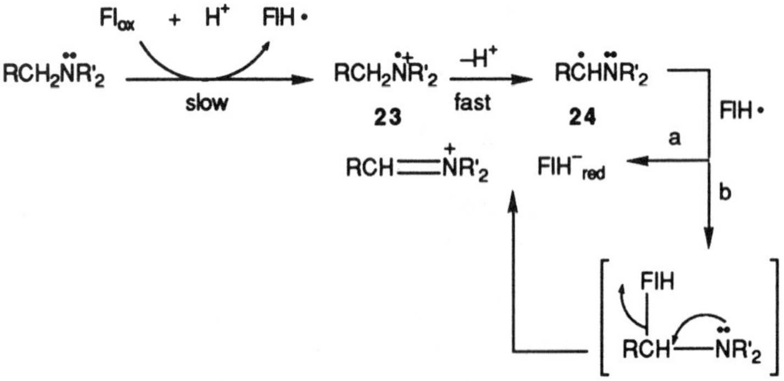

SCHEME 17. Mechanism for MAO-catalyzed monoamine oxidation proposed by Silverman (95).

transfer would produce a carbon-centered radical **24** that may decompose via pathway (a), the transfer of a second electron to flavin. Alternatively, it may decompose via pathway (b), which involves radical combination between flavin and substrate followed by two-electron transfer to give the fully reduced cofactor and iminium ion. The imine would be nonenzymically hydrolyzed to the aldehyde, and the reduced flavin is enzymically reoxidized by O_2 to complete the catalytic cycle.

C. N-CYCLOPROPYL-N-ARYLALKYLAMINES

Information in support of this hypothesis has been obtained in detailed studies on the inactivation kinetics and mechanism of pig and beef liver MAO-A by N-cyclopropyl-N-arylalkylamines. The structures and some kinetic properties of this second group of MAO inhibitors are summarized in Table III. The inactivation characteristics of the N-cyclopropylamines are generally similar to those of the propargylamines: (1) time-dependent, first-order loss of enzyme activity, saturation kinetics, and protection from inactivation by substrate or product; (2) pH-dependent rate of inactivation corresponding to the pH dependence of enzyme activity; (3) little activity recovery after exhaustive dialysis; (4) partitioning between normal product formation and inactivation; and (5) time-dependent conversion of the covalently bound FAD cofactor from the oxidized to a reduced form, which is fairly resistant to reoxidation. An important differ-

TABLE III

MECHANISM-BASED INHIBITORS OF MONOAMINE OXIDASE: N-CYCLOPROPYL-N-ARYLALKYLAMINES

Compound	R_1	R_2	R_3	R_4	Ref.
N-Cyclopropyltryptamine	H	Indoylethyl	H	H	99
N-Cyclopropylbenzylamine	H	Benzyl	H	H	99, 95, 102, 103
N-(1-Methyl)cyclopropylbenzylamine	H	Benzyl	CH$_3$	H	104, 105
Trans-2-Phenylcyclopropylamine	H	H	H	Phenyl	109
(±)-N-cyclopropyl-α-methylbenzylamine	H	α-methylbenzyl	H	H	83, 101
N-Cyclopropyl-α,α-dimethylbenzylamine	H	α,α-dimethylbenzyl	H	H	101
1-Phenylcyclopropylamine	H	H	Phenyl	H	83, 100, 110
N-Cyclopropyl-N-methylbenzylamine	CH$_3$	Benzyl	H	H	103
1-Benzylcyclopropylamine	H	H	Benzyl	H	83, 112

ence between the inhibitors in Tables II and III is that with the propargylamines the bond between the inactivator compound and protein is stable, whereas with the N-cyclopropylamines the protein–inactivator adduct may be bonded reversibly, irreversibly, or both.

The original mechanistic proposal of Silverman and co-workers was that inactivation originates via enzyme-catalyzed oxidation of the N-cyclopropyl carbon to form a reactive cyclopropanoneimine, **25** (Scheme 18, compounds **25–28**) (*95, 99*). This mechanism was based on the observation of time-dependent release of tritium from N-[1-^{3}H]cyclopropylbenzylamine, and of an isotope effect of 1.5 at low concentrations of the corresponding deuterated com-

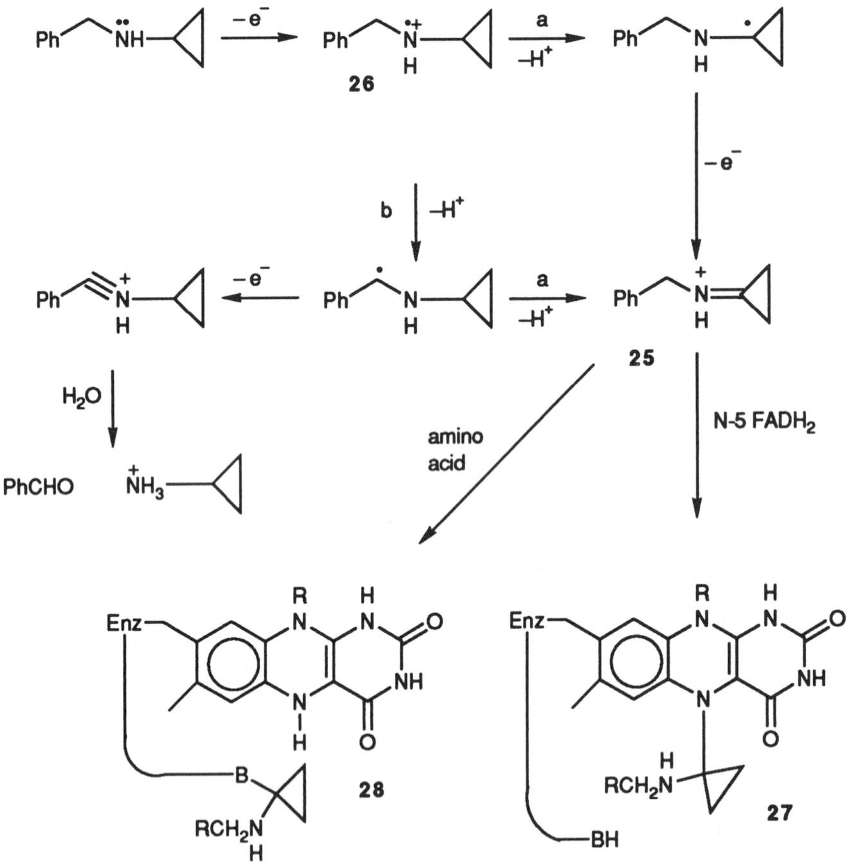

SCHEME 18. Original mechanistic proposal for MAO inactivation by N-cyclopropyl-N-arylalkyl-amines (*95, 99*).

pound (*99*). Furthermore, the optical spectrum of the flavin cofactor was observed to change to the reduced form concomitant with inactivation, consistent with a mechanism-based oxidative process to activate the inhibitor molecule. Hence, it was postulated that **25** inactivates MAO by alkylating N-5 of the reduced flavin to give the mixed diamino ketal of cyclopropanone **27,** locking the cofactor in the reduced form (*99*). Alternatively, the inactivator may become bound to an active site amino acid residue, forming **28,** such that reaction of O_2 with the reduced cofactor is not possible unless denaturing conditions are used. Inactivation of MAO by these compounds requires the cyclopropyl ring, because the corresponding *N*-isopropyl derivatives inhibit reversibly or not at all (*99, 100*).

The oxidation of primary amine substrates by MAO occurs at the α-methylene carbon. This suggests that the mechanism of MAO inactivation by *N*-cyclopropylbenzylamine (*N*-CBA) may involve oxidation at either the benzylmethylene carbon or the *N*-cyclopropyl position (*93, 95*). Oxidation of the benzylmethylene carbon was expected to lead to the formation of a benzylic imine by two one-electron transfer processes [pathway (b), Scheme 18]. Subsequent imine hydrolysis would form benzaldehyde (or benzoic acid via air oxidation) and cyclopropylamine. Using [*phenyl*-^{14}C]*N*-CBA, the ratio of nonamine radioactive metabolites (formed via benzylmethylene oxidation) to ^{14}C-labeled MAO (resulting from inactivation via cyclopropyl carbon oxidation) was 0.6. Partitioning between product formation and inactivation pathways arising from the independent oxidation of both carbons α to the nitrogen was proposed to be consistent with a common intermediate, the nitrogen radical cation **26** (Scheme 18) (*95*).

Electrochemical oxidation of *N*-CBA supports this hypothesis (*95*). These data indicate that the oxidation potential of *N*-CBA is within the capabilities of a flavoenzyme, and a similar partition ratio is observed in the oxidation of the benzylmethylene carbon versus the *N*-cyclopropyl carbon. This preliminary mechanism is thus consistent with the two one-electron transfer process and the nitrogen-centered radical cation proposed for the oxidation of normal substrates by MAO (Scheme 17).

D. REACTIVATION OF *N*-CBA-INACTIVATED MONOAMINE OXIDASE: REVISED INACTIVATION MECHANISM FOR *N*-CYCLOPROPYLARYLALKYLAMINES

Studies on the structure of the enzyme-bound adduct revealed additional information as to the mechanism of inactivation and strengthened the proposals put forth concerning the oxidation of substrates and inhibitors through a nitrogen-centered radical intermediate. The stability of the covalent adduct formed following MAO inactivation by *N*-cyclopropylbenzylamine, *N*-cyclopropyl-α-methylamine, and α,α-dimethylbenzylamine was found to depend on the presence of substrates and nonsubstrate amines (*99, 101, 102*). Incubation of the inactive enzyme with a reactivator leads to release of the covalent adduct with concomitant recovery of enzyme activity.

This reactivation phenomenon was investigated in detail by Yamasaki and Silverman (*102*). Reactivation is a pseudo-first-order, time-dependent process that exhibits saturation kinetics, suggesting reversible binding to the active site. Primary and secondary aromatic amines are the most effective activators, whereas alcohols and thiols have no reactivating effect. In almost all cases, a compound is bound better as a substrate (K_m) than as an activator (K_r), with K_r/K_m = 2.7–690 for the various amines. Furthermore, rates for amine oxidation are much greater than the rates for amine reactivation of N-CBA-inactivated MAO (k_{cat}/k_{react} = 10^4–10^5), with most of the k_{react} values being quite similar (excluding benzylamine). In summary, no correlation was found between effective MAO substrates and effective reactivators.

Three mechanisms have been proposed for the reactivation of N-CBA-inactivated MAO; these are illustrated in Scheme 19 (*102*). In pathway (A), the aldehyde group of the active site adduct forms a Schiff base with the activator amine, which is followed by base-catalyzed β-elimination of the imine of the covalent adduct and subsequent nonenzymic hydrolysis. The product has been identified as acrolein (*103*). Air oxidation of the unblocked reduced flavin cofactor then leads to active enzyme. The base may be an amino acid residue or a second molecule of the activator. Pathway (B) is similar to the first, without prior formation of the Schiff base. An S_N2 displacement reaction is suggested by path-

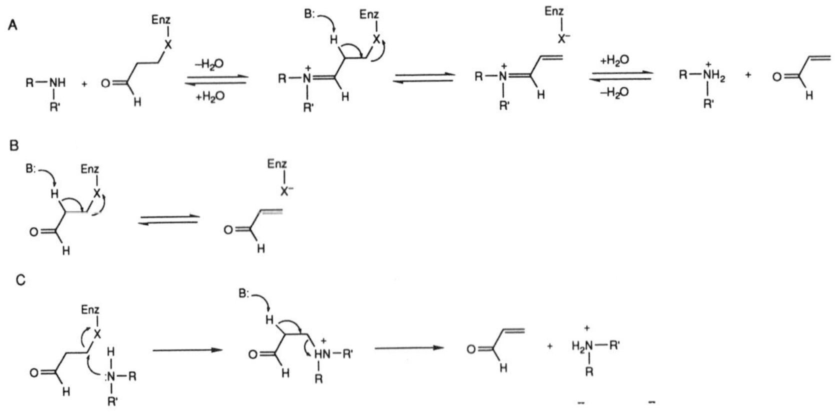

SCHEME 19. Proposed mechanisms for reactivation of N-cyclopropylamine-inactivated MAO (*102*). X represents an active site amino acid residue; B represents a base. R, Aryl, arylalkyl, or alkyl; R', H or alkyl.

way (C) via displacement of the active site nucleophile from the adduct by the amine activator. Pathway (C) is unlikely in this case, since thiols, which are more powerful nucleophiles than amines at pH 7, have no reactivation effect. Furthermore, prior treatment of the inactive enzyme with sodium borohydride prevents reactivation. Reduction of the carbonyl function should not have an effect on an S_N2 displacement reaction, but would be consistent with an elimination reaction, which requires an acidic proton β to the leaving group. The Schiff base mechanism appears to be the most reasonable, especially where tertiary amines do not reactivate the enzyme and where the postulated Schiff base formed between the covalent adduct and reactivator ($\pm$)-α-methyl[α-^{3}H]benzylamine could be trapped with sodium cyanoborohydride (*102*).

The identification of acrolein as the compound released on reactivation of N-CBA-inactivated MAO led to a reevaluation of the mechanism originally proposed in Scheme 18. In that mechanism, the formation of cyclopropanone-imine (**25**) was based primarily on the observation that MAO inactivation by N-[1-^{3}H]CBA led to the release of 0.7 equivalents of nonamine tritium per active site (*99*). Hence, the nonamine tritium was believed to be ^3H_2O originating from imine formation, as shown in Scheme 18. Subsequent, more detailed studies were consistent with the earlier work, indicating that inactivation of MAO by N-[1-^{3}H]cyclopropylbenzylamine results in the incorporation of ~3 equivalents of tritium into the enzyme and release of nonamine tritium, identified as [^{3}H]acrolein (*103*). Reactivation with benzylamine released only 1 equivalent of bound tritium as [^{3}H]acrolein with concomitant recovery of enzyme activity. The adducts formed by N-cyclopropylbenzylamine, N-cyclopropyl-α-methyl-benzylamine, and N-cyclopropyl-N-methylbenzylamine appear to be identical, as acrolein is formed as a metabolite and reactivation of MAO by benzylamine occurs at the same rate for each compound. The reactivated enzyme still contains covalently bound inactivator, suggesting that the released metabolite originates from the active site, whereas the adducts that remain bound result from random labeling on the protein surface. The site of inactivation for each compound appears to be at protein amino acid residues, since the flavin cofactor, reduced during inactivation reaction, is fully and rapidly reoxidized on protein denaturation (*103*).

It is curious to note that N-CBA-inactivated MAO continues to produce acrolein at pH 7, but at pH 9 incorporates up to 40 equivalents of acrolein into the protein in a time-dependent manner. The increased incorporation at higher pH is consistent with the ionization of nucleophilic groups that partake in Michael addition reactions with the liberated acrolein. The continued production of acrolein after complete inactivation results from reactivation–inactivation cycling of the enzyme by the N-CBA inhibitor (*103*). However, with N-(1-methyl)cyclopropylbenzylamine, the inactivated and *reduced* enzyme does not recover full activity on benzylamine treatment, and remains in the reduced form even after denaturation with 6 M urea, indicating that the inactivator is irreversibly bound to the reduced form of the flavin cofactor (104).

Based on these data, a modified mechanism for the inactivation of MAO by N-cyclopropylbenzylamine and its analogs is presented in Scheme 20 (compounds 29–31) (103). This mechanism is consistent with many of the known details of MAO inactivation by N-cyclopropyl-N-arylalkylamines. Oxidation of N-cyclopropylbenzylamine partitions between two sites on the molecule, at the benzylmethylene leading to normal product formation or at the cyclopropyl group facilitating ring opening and enzyme inactivation. Consistent with the originally proposed mechanism for the oxidation of normal substrates, a one-electron transfer to flavin produces the amine radical cation 29, pathway (a). One-electron reduction of the flavin to a semiquinone may be followed by hydro-

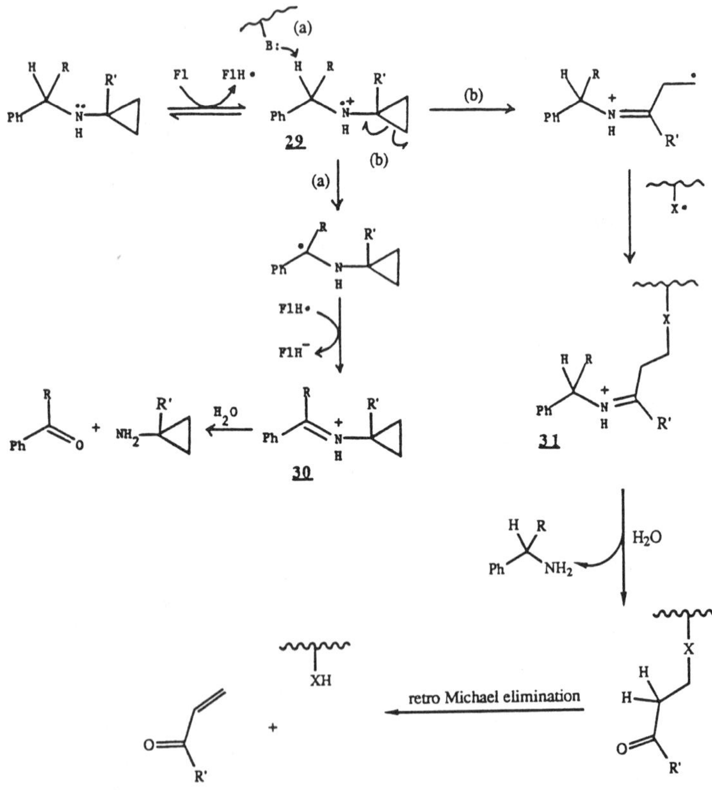

SCHEME 20. Modified mechanism for the inactivation of MAO by N-cyclopropylbenzylamine (101).

gen atom abstraction of an active site amino acid. This is proposed to result in an equilibrium mixture of flavin radical and amino acid radical, possibly a tyrosine or cysteine (*100, 103*). The next step is the abstraction of an activated benzylic proton, followed by electron transfer to the flavin or amino acid radical, producing the imine (**30**), and subsequent hydrolysis to produce an α-phenyl carbonyl and amino cyclopropyl derivative.

In contrast to the original mechanism proposed by Silverman in Scheme 18, this revised inactivation mechanism does not require hydrogen or proton abstraction from C-1 of the cyclopropyl ring (*95, 104*). As illustrated in pathway (b) in Scheme 20, inactivation results from homolytic opening of the cyclopropyl ring, forming a primary alkyl radical. Radical combination may then occur with the flavin semiquinone or the amino acid-centered radical. The protein-bound imine adduct **31** may hydrolyze, releasing a substituted benzylamine, and the resulting 3-amino acid-substituted carbonyl would undergo a retro-Michael elimination of the protein on reactivation treatment, giving HX and an α,β-unsaturated carbonyl compound, as in Scheme 19. Elimination reaction may also occur directly from the initial imine **31,** producing the same products. In the case where the inhibitor is *N*-CBA (R′ = H), acrolein would be formed.

By removing the requirement for proton or hydrogen abstraction from C-1 of the cyclopropyl ring, this mechanism becomes consistent with the potent time-dependent inactivation of porcine and bovine MAO by *N*-(1-methyl)cyclo-propylbenzylamine (R = H, R′ = CH_3 in Scheme 20) (*104, 105*). Furthermore, partitioning between benzylmethylene oxidation and cyclopropyl ring opening is observed, although the ratio of normal product formation to inactivation (0.10–0.22) favors the latter. This result is best explained again by initial formation of the amine radical cation and suggests that cyclopropyl ring opening is 5–10 times faster than benzylmethylene proton abstraction (*104*).

Methylation of the benzylmethylene carbon has also been found to dramatically alter the partition ratio, favoring cyclopropyl ring opening and inactivation 99:1 over normal product formation. These results are consistent with the reactivity of the α-proton of secondary amine radical cations formed in electrochemical oxidations (*96, 101*). It was found that the ratio of α-proton removal from the two N-bound substituents reflects the kinetic acidity of these protons adjacent to the amine radical cation. These results are also a reflection of the expected decrease in the acidity of the α-protons on increasing the number of substituents at the benzylic carbon (*106*). In the enzymic reaction, steric effects may be involved as well. This ability to manipulate the partition ratio is important and has implications in the design of MAO inhibitors as potential antidepressant agents. Hence, α-substitution of MAO inhibitors may lead to an alteration in the partition ratio, favoring enzyme inactivation over metabolite formation (*101*). This would decrease the formation of by-products that may be potential cytotoxic agents.

E. RELATIONSHIP BETWEEN ACTIVE SITE GEOMETRY AND INHIBITOR
    STRUCTURE IN DEFINING SITE OF COVALENT BOND FORMATION IN
    MONOAMINE OXIDASE INACTIVATION

The inactivation of MAO by the *N*-cyclopropyl-*N*-arylalkylamines primarily results in covalent modification of the reduced flavin cofactor. Subsequent studies with structurally modified cyclopropylamine derivatives indicate that modification may partition between different sites on the protein, depending on the inhibitor structure. In all cases, the structures of the reactive inactivator species and of the resulting covalent adduct are consistent with a mechanism involving a nitrogen-centered radical intermediate.

2-Phenylcyclopropylamine is a known mechanism-based inactivator of mitochondrial MAO (*107*) and has been used clinically as an antidepressant agent (*108*). Silverman has reported that the treatment of the inactivated and covalently labeled enzyme with 2,4-dinitrophenylhydrazine led to the release of a product identified as the 2,4-dinitrophenylhydrazone of cinnamaldehyde (*109*). Furthermore, Paech and co-workers have shown that the covalent adduct is not formed with the flavin cofactor, but with an active site amino acid, presumably the sulfhydryl group of cysteine (*107*). Given this information, the knowledge that thiols reversibly add to α,β-unsaturated aldehydes and the proposed formation of the amino radical cation, the mechanism of 2-phenylcyclopropylamine inactivation and the structure of the covalent intermediate are shown in Scheme 21 (*109*). Homolytic opening of the cyclopropyl ring leads to the secondary carbon-centered radical of cinnamaldehydeimine **32** (Scheme 21), which undergoes radical recombination with the protein-bound radical X·. Since X is proposed to be

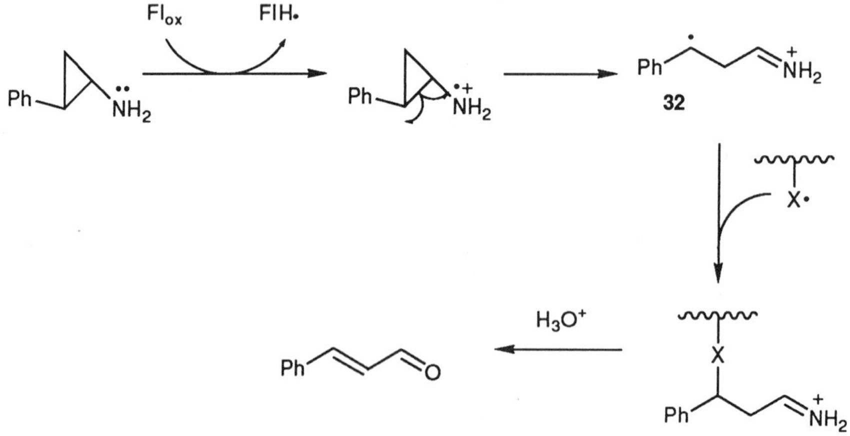

SCHEME 21. Proposed mechanism of MAO inactivation by 2-phenylcyclopropylamine (*109*).

the sulfur of cysteine, the thiol-centered radical could be formed by hydrogen abstraction of the sulfhydryl group by the flavin semiquinone radical. The structure of the covalent intermediate is consistent with the observation that treatment of the inactive enzyme with borohydride increases the stability of the adduct to denaturation and dialysis. Reduction of the aldehyde (or imine) would be expected to decrease the acidity of the α-protons and prevent elimination and formation of the resonance-stabilized cinnamaldehyde product (*109*).

In contrast to the formation of covalent adducts at flavin or at an amino acid residue by *N*-cyclopropyl-*N*-arylalkylamines and 2-phenylcyclopropylamine, respectively, inactivation of MAO by 1-phenylcyclopropylamine leads to *both* types of adducts (*100*). Both inactivation pathways are proposed to originate from an initial one-electron oxidation by flavin to produce a common intermediate, the amine radical cation **33** in Scheme 22 (compounds **33–39**). Homolytic cyclopropyl ring opening would lead to the reactive primary alkyl radical **34**, which could be captured by the active site radical, either flavin semiquinone or amino acid centered. Subsequent hydrolysis of the imine, pathway (a), forms a 3-substituted propiophenone–protein adduct. The structure of the adduct was verified in the isolation of acrylophenone via base-catalyzed retro-Michael reaction of the β-substituted adduct.

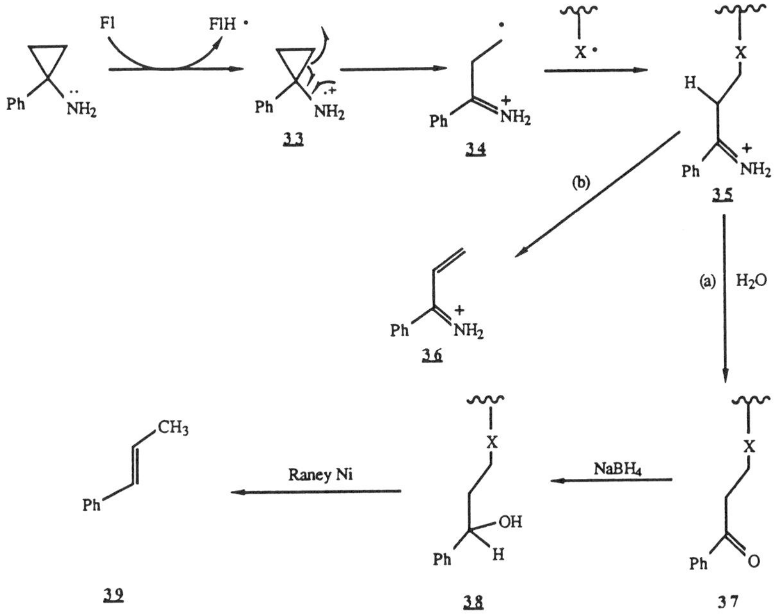

SCHEME 22. Proposed mechanism of MAO inactivation by 1-phenylcyclopropylamine (*100*).

Depending on the length of incubation and concentration of the inactivator, varying amounts of activity could be recovered following dialysis of the inactivated protein. Inactivation reaction occurs concomitantly with flavin reduction, and activity recovery occurs only with the reoxidation of the cofactor. Complete, irreversible inactivation occurs when the flavin cofactor remains reduced even after denaturation; benzylamine has no effect on reactivation of 1-phenylcyclopropylamine-inactivated enzyme. The protein nucleophile is suggested to be the reduced flavin cofactor, with the specific site of labeling probably at the N-5 position (100). This idea is consistent with the isolation of peptide fragments containing flavin and radiolabeled adduct following inactivation by ^{14}C-labeled inactivator and pronase treatment.

Irreversible inactivation of MAO by 1-phenylcyclopropylamine requires eight molecules of inactivator. Prior to the inactivation event, seven of these eight molecules of 1-[phenyl-^{14}C]phenylcyclopropylamine form a reversible covalent bond with an amino acid residue in the active site, which is slowly released on standing or dialysis. The reversible adduct decomposes with a half-life of 65 min, which is seven times faster than the rate of irreversible inactivation and is consistent with the observation that eight molecules of inactivator are required for each irreversible inactivation. The isolation of a nearly quantitative amount (7 equivalents) of acrylophenone suggests that pathway (b) in Scheme 22 is operative, involving a base-catalyzed retro-Michael reaction to eliminate HX from 35 and produce the imine of acrylophenone 36, which will rapidly hydrolyze to the ketone. As this pathway is essentially the same as that for the irreversible inactivation mechanism, the nucleophilic group on the protein must be something other than the N-5 position of flavin, but a good leaving group such as cysteine or the C-4a position of flavin (100). On denaturation, reoxidation of the flavin occurred at a significantly faster rate than release of radiolabeled adduct, suggesting that the nucleophile is not derived from flavin but is an amino acid functional group.

Several approaches have been taken to identify the site of the reversible adduct formation (110). Treatment of MAO inactivated with 1-[phenyl-^{14}C]phenylcyclopropylamine with sodium borohydride prevents the release of the unstable adduct, presumably due to reduction of the ketone 37 or imine 35 to the alcohol 38. Treatment of the reduced adduct with pronase and subsequent reduction with Raney nickel led to the release and isolation of trans-β-methylstrene 39. Raney nickel is known to cleave carbon–sulfur bonds specifically (111). Furthermore, the expected product from cleavage of the carbon–sulfur bond of the reduced adduct is 1-phenyl-1-propanol; however, under the conditions of the Raney nickel reduction, 1-phenyl-1-propanol is dehydrated to trans-β-methylstyrene (110). This result suggests that the amino acid residue is cysteine. The finding that the total number of cysteine residues detected with 5,5′-dithiobis(2-nitrobenzoic acid) decreases by one in the denatured, 1-phenylcyclopropylamine-inactivated enzyme corroborates this idea (*110*). The inactivation of MAO by 1-benzylcyclopropylamine appears to follow a similar mechanism (*83, 112*).

Silverman and Zieske have rationalized how a protein nucleophile other than flavin is involved in MAO inactivation reactions, and why different inactivator compounds specifically react with flavin, protein amino acids, or both (*100*). Hydrogen atom donation from a cysteine residue to the flavin semiquinone radical would produce a thiyl radical, which could then capture the primary or secondary alkyl radical generated on cyclopropyl ring opening from the amine radical cation of the inactivator. The hydrogen atom abstraction reaction between the flavin and active site amino acid may be an equilibrium process such that either species could be present at any turnover. Hence, a combination of steric constraints and proximity to either the flavin semiquinone radical or the thiol radical will determine the site of adduct formation for a particular inactivator structure. A two-dimensional representation is shown in Scheme 23 (compounds **40–42**), which illustrates the proposed equilibrium between the flavin semiquinone radical and amino acid as well as the proposed intermediates for the inactivation of MAO by *N*-(1-methylcyclopropyl)benzylamine **40** (*104*), *trans*-2-phenylcyclopropylamine **41** (*109*), and 1-phenylcyclopropylamine **42** (*100*). The position of the iminium ion is fixed for all three intermediates. The proximity of the inactivator radical center relative to the particular protein radical is consistent with proposed site of attachment of inactivator to protein: **40** is near the flavin radical, such that exclusive flavin attachment occurs, **41** is positioned closer to the amino

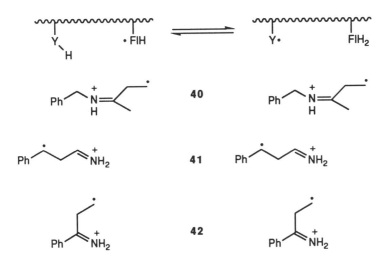

SCHEME 23. Proposed role of flavin semiquinone radical, amino acid residue, and possible intermediates in the inactivation of MAO by *N*-(1-methylcyclopropyl)benzylamine (**40**) (*104*), *trans*-2-phenylcyclopropylamine (**41**) (*109*), and 1-phenylcyclopropylamine (**42**) (*100*).

acid radical, where an unstable (reversible) adduct forms, and **42** appears to be equidistant from each radical site and attachment at both protein positions is observed (*100*).

F. New Classes of Monoamine Oxidase Mechanism-Based Inactivators: Direct Observation of Radical Intermediates

The radical mechanism proposed for substrate utilization and inactivation of MAO is compatible with MAO inactivation by some rather novel compounds (Table IV). 1-Phenylcyclobutylamine has been shown to be both a substrate and an inactivator of MAO (*113*). Inactivation results in covalent and irreversible labeling of the flavin cofactor and requires 325 turnovers to product before inactivation reaction. In the course of MAO processing of 1-phenylcyclo-butylamine, some unusual metabolites were observed: 2-phenyl-1-pyrroline **46**, 3-benzoylpropanal **48**, and 3-benzoylpropionic acid **49**. These results are consistent with the mechanism shown in Scheme 24 (compounds **43**–**49**). The initial step is one-electron oxidation of 1-phenylcyclobutylamine to produce the amine radical cation **43**. Subsequent homolytic cyclobutyl ring opening gives the primary alkyl radical **44**, which is the central intermediate in the partitioning be-

TABLE IV

Novel Irreversible Inhibitors of Monoamine Oxidase

Compound	Structure	Reference
Allylamine	$\diagdown\diagup\diagdown_{NH_2}$	*125*
1-Phenylcyclobutylamine	(cyclobutyl with phenyl and NH_2)	*113*
(Aminoalkyl)trimethylsilanes	$(CH_3)_3Si(CH)_nNH_2$	*114*
1-Amino-1-benzoylcyclobutane	(phenyl–C(=O)–cyclobutyl with NH_2)	*120*

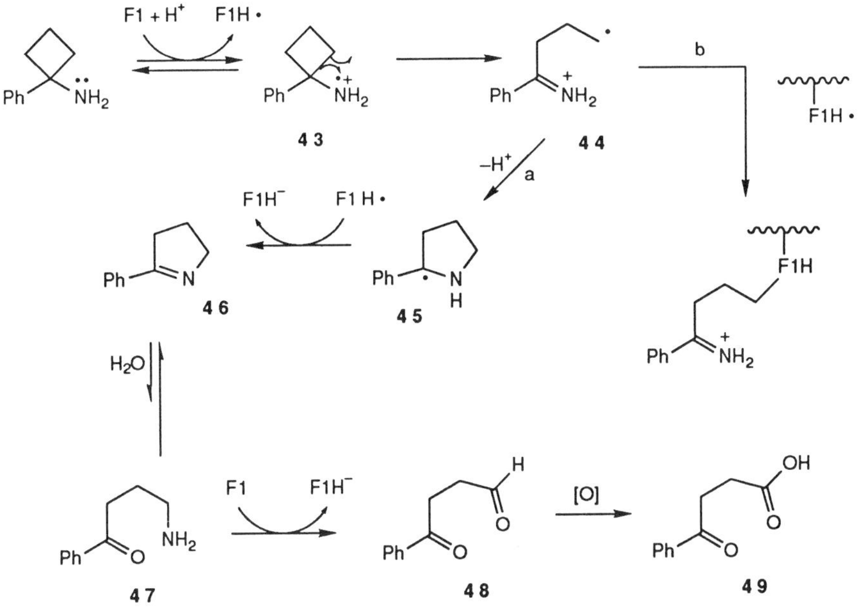

SCHEME 24. Inactivation mechanism of MAO by 1-phenylcyclobutylamine (*113*).

tween metabolite formation [pathway (a)] and inactivation reaction [pathway (b)]. In this respect, **44** is an intramolecular radical trap, since cyclization of **44** to **45** and a second one-electron oxidation would produce metabolite **46**, 2-phenyl-1-pyrroline. The equilibrium between **46** and **47** favors **46**, but **47** is apparently a very good substrate of MAO, which explains why it is not observed as a metabolite. Oxidation of **47** by MAO produces 3-benzoylpropanal **48**, which is then nonenzymically oxidized by O_2 to 3-benzoylpropionic acid **49**. Hence, metabolism and inactivation of MAO by 1-phenylcyclobutylamine are quite consistent with the postulated radical mechanism for MAO-catalyzed amine oxidation (*113*), as there is no known nonradical pathway that may lead to such products as well as to enzyme inactivation.

A second new class of MAO mechanism-based inactivators, (aminoalkyl)trimethylsilanes, have been reported by Silverman and Banik (*114*). The idea for this class of MAO inactivators is based on the known activation of the carbon–silicon bond toward homolytic cleavage reaction when the silicon atom is β to a radical cation (*115, 116*). The aminomethyl-, aminoethyl-, and (aminopropyl)trimethylsilanes are all pseudo-first-order time-dependent inactivators of beef liver MAO that reduce the flavin cofactor during the inactivation reaction. Since denaturation of the inactivated enzyme allows flavin reoxidation, covalent bond formation might be to an amino acid residue (*114*). The stabilities of the enzyme adducts from the (aminoalkyl)trimethylsilanes were found to be differ-

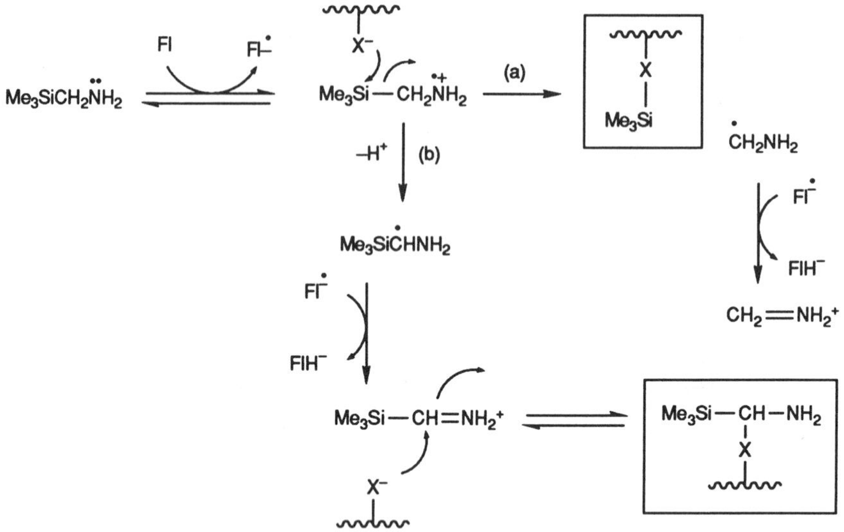

SCHEME 25. Proposed mechanism for MAO inactivation by (aminomethyl)trimethylsilane (*114*).

ent, with enzyme activity returning from the methyl and ethyl derivatives with a half-life of 5.5 days, whereas the propyl compound became reactivated with a half-life of 13 hr. The reason for this difference is suggested to indicate that two different adducts are formed (*114*). Proposed inactivation mechanisms are shown in Schemes 25–27. In Scheme 25 one-electron oxidation produces the amine radical cation, which may partition through two pathways. In pathway (a), nucleophilic attack by an amino acid functional group at electrophilic silicon would lead to trimethylsilation of the enzyme. Alternatively, pathway (b) predicts proton and electron transfer to give the immonium ion of formyltrimethylsilane, which is subject to nucleophilic attack at carbon. In Scheme 26, the reactive species is proposed to be the α-silylimine. Silicon of α-silylcarbonyl compounds

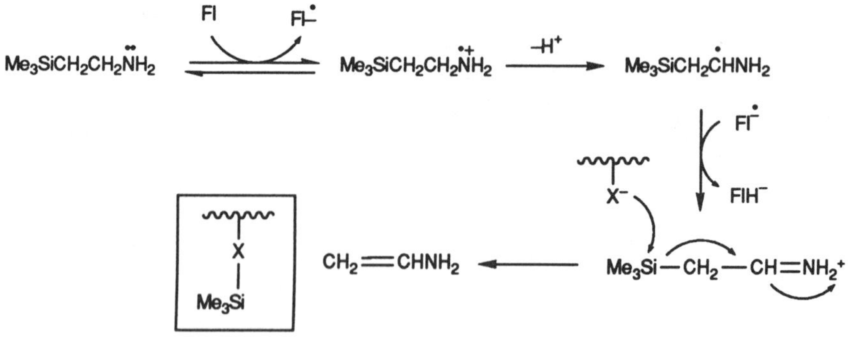

SCHEME 26. Proposed mechanism for MAO inactivation by (aminoethyl)trimethylsilane (*114*).

are highly electrophilic (*117*), so trimethylsilation of MAO as shown in Scheme 26 is reasonable. For the propylsilyl analog, cyclopropanation is suggested, as shown in Scheme 27. The cyclopropanation of trimethylsilylpropyl cationic species also has precedence in the literature (*118, 119*). The mechanisms of these inactivation reactions and the structures of the enzyme-bound adducts are currently under study.

Evidence for radical intermediates in the MAO-catalyzed oxidation of amine substrates and inhibitors is supported by the inactivation of MAO by cyclopropyl-

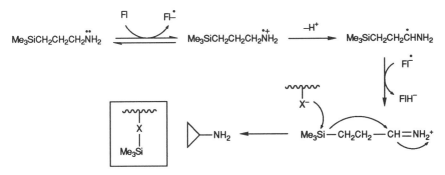

SCHEME 27. Proposed mechanism for MAO inactivation by (aminopropyl)trimethylsilane (*114*).

and cyclobutyl-containing mechanism-based inactivators. In these studies, the covalent protein-bound adducts and metabolites are best rationalized as originating via homolytic ring-opening reactions. However, no ESR evidence has been reported to support the single-electron transfer mechanisms proposed. The first evidence for a radical intermediate in the MAO-catalyzed oxidation of an amine substrate has recently been reported by Silverman and co-workers (*120*). A new substrate for MAO was prepared, the 1-amino-1-benzoylcyclobutane. The basis for the design of 1-amino-1-benzoylcyclobutane first takes into consideration the ability of MAO to catalyze the oxidation of 1-phenylcyclobutylamine, yielding metabolites and enzyme inactivation (*113*). Second, the initial primary alkyl radical is expected to be trapped intramolecularly, forming a new radical species stabilized by the combined effect of donor and acceptor groups, known as the captodative effect (*121*). The failure to detect radical intermediates in previous studies of MAO-catalyzed reactions is thought to result from (1) the kinetic efficiency of the enzyme in processing radical intermediates before release from the active site or (2) the short lifetime of the radical intermediates (because of their thermodynamic instability). Hence, a combination of captodative stabilization of the intermediates and their scavenging using spin traps appeared to be a reasonable strategy for the detection and identification of such intermediates.

1-Amino-1-benzoylcyclobutane, **50**, was found to be both a substrate ($K_m = 330$ mM, $k_{cat} = 0.2$ min^{-1}) and an inactivator ($K_i = 16.7$ mM, $k_{inact} = 0.016$

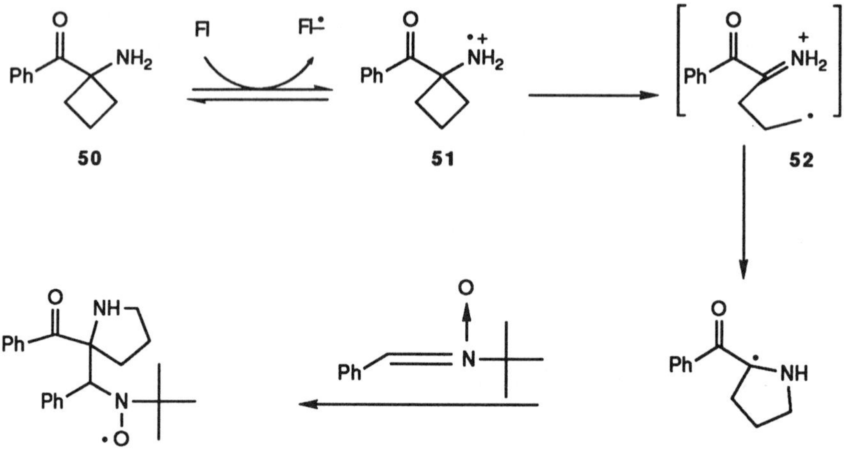

SCHEME 28. Proposed mechanism for MAO processing of 1-amino-1-benzoylcyclobutane and trapping of radical intermediates by α-phenyl-N-tert-butylnitrone (PBN) (*120*).

min^{-1}) of MAO (Table IV). In a manner analogous to the mechanism of oxidation of 1-phenylcyclobutylamine by MAO (Scheme 24), homolytic cyclobutyl ring opening of the initially formed amino radical cation **51** (Scheme 28, compounds **50**–**55**) would produce the unstable primary carbon-centered radical intermediate **52**. Intramolecular trapping of **52** produces the captodative radical **53**, analogous to the formation of **45** in the processing of 1-phenylcyclobutylamine by MAO (Scheme 24). In the presence of the spin trap, α-phenyl-N-tert-butylnitrone (PBN) **54**, a new ESR spectrum relative to the baseline was observed. The spectrum consisted of a triplet of doublets (a_N = 15.9 G; a_b^H = 2.9 G) centered at a g value of ~2.006, which is consistent for the structure of the adduct **55** formed between PBN and **53**. Similar results were obtained when 1-phenylcyclobutylamine was the substrate. Control experiments gave results consistent with the trapped radical product originating from **53**, and not hydrogen peroxide or products of normal substrate turnover. This preliminary study thus provides the first direct evidence for the formation of free radical intermediates in the reaction mechanism of MAO. Conclusive identification of the structure of spin-trapped adduct **55** is in progress.

III. DNA Photolyase

A. BACKGROUND

Irradiation of DNA by ultraviolet light (200–300 nm) results in damage to the biopolymer, with the most prevalent covalent change being the formation of

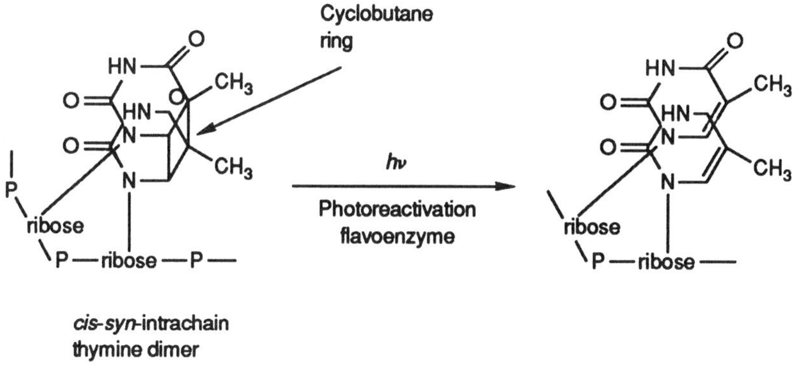

SCHEME 29. Photoactivated scission of pyrimidine dimers (*160*).

intrachain cyclobutane rings. These result from the [2 + 2] cycloaddition of adjacent pyrimidine bases forming intrastrand *cis-syn*-pyrimidine dimers (*126*). If left unrepaired, these lesions would eventually result in mutations, and possibly cancer and death. In order to maintain the crucial genetic information necessary for an organism's survival, nature has developed at least two enzymic processes to repair UV-damaged DNA. The first system requires a dimer-specific endonuclease followed by DNA polymerase (*127*). This reaction, which operates in the dark, has been well characterized and is not limited to the repair of thymidine adducts. The second reactivation system is particularly intriguing because visible light (360–440 nm) is utilized to catalyze photochemical cleavage of the dimer back to the adjacent pyrimidine bases (Scheme 29). These photoreactivation enzymes, the DNA photolyases (DNA-PLs; EC 4.1.99.3; deoxyribodipyrimidine photo-lyase, photoreactivating enzyme), have been known for nearly 40 years (*128*) and are found in a variety of bacteria and yeast (*129–132*) as well as in mammalian sources (*133*). Flavin cofactors have been identified in the DNA-PL of *E. coli* and yeast (*134–136*), and the enzyme from *Streptomyces griseus* contains a derivative of 8-hydroxy-5-deazaflavin, Coenzyme F_{420} (*137, 138*). However, the reaction mechanism whereby DNA-PLs can convert light energy into the chemical energy necessary for the cleavage of a cyclobutane ring is not clearly understood. Furthermore, an incompletely characterized DNA-PL from *E. coli* is reported to utilize RNA as a cofactor (*139, 140*).

B. PHYSICAL CHARACTERIZATION OF *ESCHERICHIA COLI* DNA PHOTOLYASE

The DNA-PL of *E. coli* has been the most extensively studied of the photoreactivating enzymes. The only known substrates for the enzyme are *cis-syn*-cyclobutylpyrimidine dimers, and light is required for the dimer cleavage to occur but not for substrate binding (*141*). The enzyme is specific for UV-irradiated DNA, as untreated DNA does not bind to the lyase (*142*). The pyrimidine dimer structure of the damaged DNA is primarily responsible for this specificity, with electrostatic contributions to binding being limited to one or two phosphates.

The *E. coli* protein contains noncovalently bound flavin (FAD) (*143*) and a second cofactor, which has recently been characterized as a 7,8-dihydropterin substituted at the 5- and 6-positions (*144*), possibly a 5,10-methenyl tetrahydrofolate (*145*). The protein-free pterin is unstable and undergoes a reversible oxygen-dependent bleaching reaction (*146*), although extended exposure of the bleached form to alkaline conditions gives rise to irreversible modification (*144*). Some of the chemical reactions of the *E. coli* pterin cofactor are summarized in Scheme 30. The overproducing plasmid for the *E. coli* enzyme has been constructed, resulting in a 15,000-fold amplification of the lyase (*147*). The overproduced *E. coli* enzyme has been purified to homogeneity and is a monomer of M_r 49,000 (*148*). The assay for photolyase activity is a modified transformation method that measures the increase in biological activity of UV-irradiated DNA observed on treatment with the photoreactivating enzyme (*148, 149*). In the presence of visible light, *E. coli* DNA-PL transforms UV-irradiated DNA dimers into monomers at a rate of 2.4 pyrimidine dimers/min (*148*). The turnover rate *in vivo* has been estimated at 4.5 pyrimidine dimers/min (*150*). The nucleotide sequence of the gene and the amino acid sequence of the *E. coli* DNA-PL apoenzyme have been reported (*151*).

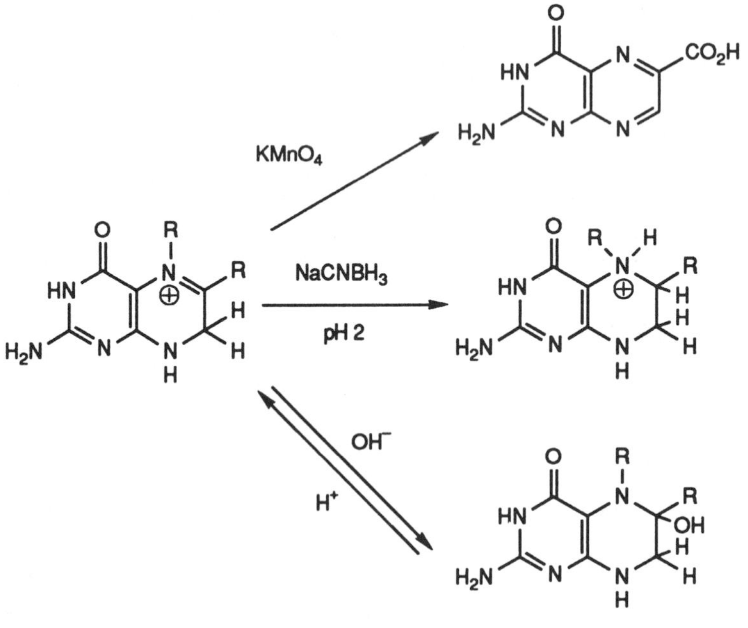

SCHEME 30. Chemical transformation of the *E. coli* pterin cofactor (*144*).

Work by Jorns *et al.* detailing the physical characteristics of *E. coli* DNA-PL has provided valuable insight as to the photoreaction mechanism. The blue-colored solutions of the concentrated protein show absorption maxima at 580, 475, and 384 nm (*146*). Based on the 580-nm absorption maxima, as well as fluorescence and ESR spectral data, the flavin structure in the isolated enzyme has been identified as the blue neutral FAD radical (*146*). The ESR spectrum of the blue enzyme is consistent with that of other neutral flavoprotein radicals, a single broad signal with a bandwidth of 19 G (*146*). The radical appears to be reasonably stable, as neither O_2 nor $K_3Fe(CN)_6$ has any effect on the absorption spectrum of the native enzyme, although oxidized FAD is formed if the protein is aerobically denatured. Dithionite treatment produces reduced flavin, which is rapidly air oxidized back to the radical species (*146*). Long-term storage at $-20°C$ or extended exposure at $0-5°C$ will also lead to varying extents of co-factor oxidation, as evidenced by the isolation of green (partially oxidized) or yellow (fully oxidized) enzyme and a corresponding loss of enzyme activity (*155*). The observation that anaerobic denaturation of the protein results in dis-proportionation of the radical species to oxidized and reduced flavin suggests that the cofactor radical is an N-5-unsubstituted flavin semiquinone (*146*).

Treatment of the blue form of the lyase with dithionite or irradiation at wave-lengths greater than 520 nm in the presence of DTT produces fully reduced flavin cofactor ($FADH_2$) and results in a dramatic increase in activity and quantum yield (*146, 153, 154*). This suggests that the catalytically active oxidation state of the cofactor may be the reduced flavin, and that it is this form of the cofactor that functions as an electron donor in catalysis (*155*). Subsequent studies by Sancar and co-workers indicate that *E. coli* DNA-PL does not contain the flavin semiquinone radical *in vivo*, and that the blue, radical enzyme does not catalyze dimer cleavage (*156*).

The absorption band at 384 nm is composed of contributions of the radical species and the second chromophore, whereas the fluorescence spectra with ex-citation maxima at 398 nm and emission maxima at 470–480 nm are attributed to the pterin alone (*146, 155*). The 7,8-dihydropterin cofactor, $\lambda_{max} = 360$ nm when free in solution and 390 nm when protein bound, is labile at neutral pH, readily decomposing upon denaturation to form products without significant vis-ible absorption maxima. The photoreduction described above also reduces the second cofactor but in an irreversible manner with complete loss of its fluores-cence and visible absorption characteristics (*157*). Reduction of the blue semi-quinone FAD cofactor to the fully reduced form has no effect on the absorption spectrum of the pterin, suggesting that the absorption spectrum of the second cofactor must be independent of the oxidation state of the flavin and that the two cofactors are electronically isolated from each other (*157*). However, reduction of the flavin radical results in an increase in the fluorescence of the second co-factor, possibly indicating that the flavin radical acts as a potent quencher of fluorescence of the 7,8-dihydropterin.

C. Photoreactivating Model Systems

A variety of pyrimidine dimer-cleaving photochemical model systems have been developed to aid in the study and elucidation of the DNA-PL reaction mechanism (*152, 158, 159*). Direct excitation of pyrimidine dimers does not occur on ultraviolet or visible irradiation, suggesting that enzyme cofactors might be involved in a photosensitization process (*146*). Model systems utilizing free flavin derivatives have recently been described (*137, 160*). Rokita and Walsh have demonstrated that lumiflavin, 5-deazariboflavin, and 8-methoxy-7,8-didemethyl-N^{10}-ethyl-5-deazaflavin (Scheme 31) are effective photosensitizers for the thymine dimer cleavage reaction (*160*). These reactions utilized *cis-syn*-thymine dimer as substrate with irradiation at the λ_{max} of the flavin derivative, under strict anaerobic conditions and high pH. Flavin derivatives that contain electron-rich substituents at the 8-position, such as 8-hydroxy-5-deazariboflavin, 8-hydroxyriboflavin, 8-(dimethylamino)riboflavin, and 8-(methylamino)riboflavin, are inactive in the photolysis reaction. However, if electron donation into the aromatic ring system is blocked, as in the 8-methoxy derivative, photosensitization then occurs. These data are consistent with the excited state of lumiflavin, 5-deazariboflavin, and 8-methoxy-7,8-didemethyl-N^{10}-ethyl-5-deazaflavin being involved in the mechanism of thymine dimer cleavage (*160*), and suggest that the enzyme-mediated reaction may also occur by classical photosensitization.

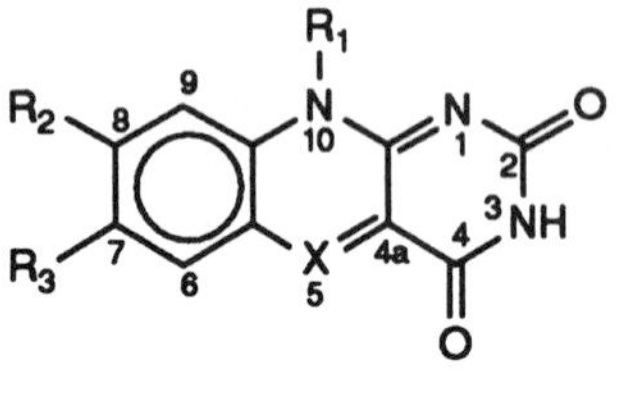

Scheme 31. Structures of flavin derivatives. 5-Deazariboflavin; X, C; R_1, ribityl; R_2, R_3, CH_3. 8-Hydroxy-7,8-didemethyl-5-deazariboflavin: X, C; R_1, ribityl; R_2, OH; R_3, H. 8-Methoxy-7,8-didemethyl-N^{10}-ethyl-5-deazaflavin: X, C; R_1, CH_2CH_3; R_2, OCH_3; R_3, H. Lumiflavin: X, N; $R_1 = R_2 = R_3 = CH$. Riboflavin: X, N; R_1, ribityl; $R_2 = R_3 = CH_3$. 8-Hydoxyriboflavin: X, N; R_1, ribityl; R_2, OH; R_3, CH_3 (*160*).

The reactions catalyzed by these nonenzymic photochemical model systems mimic the enzyme-mediated conversion of light energy directly into chemical energy. The flavin sensitizers act as a catalyst and are not rapidly degraded in the reaction. In the early stages of the reaction, product formation is linear with time and the rate of photolytic cleavage is saturable by the thymine dimer substrate. A general mechanism for flavin- or 5-deazaflavin-photosensitized monomerization of thymine dimers has been suggested and is shown in Scheme 32. The mechanism requires the preliminary formation of a reversible flavin–dimer com-

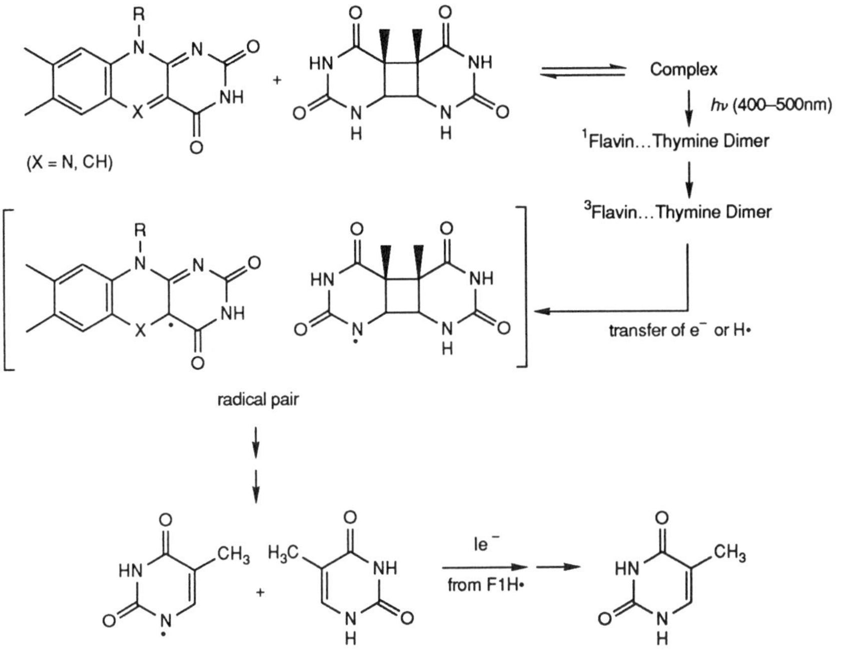

SCHEME 32. General mechanism of flavin- or 5-deazaflavin-photosensitized monomerization of thymine dimers (*161*).

plex, followed by irradiation (400–500 nm). Triplet flavin appears to be the key intermediate in the photochemical mechanism, since two triplet quenchers, oxygen and Dabco (diazabicyclo[2.2.2]octane), inhibit thymine formation. The photosensitized flavin may transfer light energy to the substrate either in a direct manner or by electron or hydrogen shift (*160*). A one-electron (or hydrogen) transfer between triplet flavin and thymine dimer will yield a complex of the thymine radical cation and flavin semiquinone. Apparently, a closely associated flavin–dimer complex is formed because free radical scavengers do not inhibit the photochemical reaction. Monomerization is completed following radical fragmentation and a second one-electron transfer (*161*).

To elucidate further the possible role of flavin as a sensitizer in photolyase catalysis, Jorns conducted a study on the light-catalyzed monomerization of thymine dimers by reduced flavins and flavin analogs as models for the DNA-PL reaction of *E. coli* and yeast (*162*). The reduced forms of 1-deazariboflavin, $N(3)$-methyllumiflavin, 7,8-dimethyl-1,10-ethyleneisoalloxazinium perchlorate, and 5-deazariboflavin were generated anaerobically with excess dithionite. Of these four flavins, cleavage of *cis-syn*-[*methyl-*^{3}H]thymine dimer to [*methyl-*^{3}H]-thymine was observed with reduced 1-deazariboflavin, and to a lesser extent with reduced $N(3)$-methyllumiflavin. Detailed studies on the cleavage of thymidine dimers by reduced 1-deazariboflavin indicate strikingly similar characteristics to the photoreactivation reactions catalyzed by *E. coli* DNA-PL (*162*). The model reaction with 1-deazariboflavin is most effective near neutral pH, where there is a linear dependence on the length of illumination, concentration of the flavin, and saturation kinetics observed on variation of the dimer concentration. The rate is linear in the early stages of the reaction and is relatively constant over the pH range of 8.5–11.5. The oxidized form of the flavin is inactive, and reduced 1-deazariboflavin was shown to remain unchanged after extended irradiation, indicating that this flavin sensitizer does indeed act as a true catalyst in the photolysis reaction. Hence, the role of flavin in the enzymic and model reaction is proposed to be that of a photosensitizer. Furthermore, this model study suggests that the second chromophore, 5,10-methenyl tetrahydrofolate (*145*), is not absolutely required for photoreactivating reaction, since only the reduced flavin catalyzes dimer cleavage in the model reaction. The quantum yield, f, of the model reaction with 1-deazariboflavin is a measure of the efficiency of the photoreactivation process, and was determined to be 9.2×10^{-3}. This is less efficient than the *in vivo* enzymic photoreactivating reaction where $f = 1.0$, indicating that a dimer cleavage occurs for every photon of light absorbed (*163*).

Based on the known chemistry of flavin photolysis reactions, it appears unlikely that thymine dimer cleavage occurs via a direct energy transfer mechanism (*160*). One proposal suggests that in the model reaction with 1-deazariboflavin, the thymine dimer radical anion is formed via electron donation from the excited sensitizer (*164*). Alternatively, electron abstraction by the excited flavin could occur, resulting in the thymine dimer radical cation (*159, 160*), although it is unlikely that reduced flavin would act as an electron acceptor. A schematic for this mechanism is illustrated in Scheme 33, where the initial formation of a sensitizer–dimer complex is consistent with the observed saturation kinetics. The complex is activated by excitation of the ionized sensitizer (pH $>$ 7), and electron donation to the dimer forms the dimer radical anion and the zwitterionic, neutral 1-deazariboflavin radical (*162*). Thymine dimer radical would spontane-

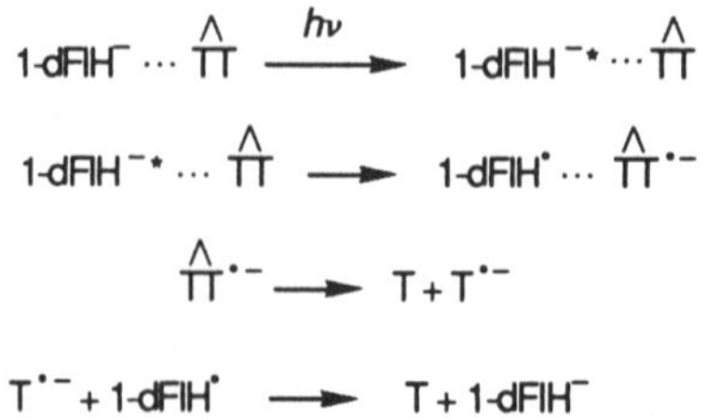

SCHEME 33. Thymine dimer cleavage by electron transfer (*162*).

ously monomerize, forming thymine and monomer radical anion. The latter species donates an electron to the 1-deazaflavin radical, thus completing the catalytic cycle.

D. ROLE OF ENZYME COFACTORS IN DNA PHOTOLYASE CATALYSIS

Based on the model studies discussed above, an analogous mechanism for the DNA-PL-catalyzed cleavage of pyrimidine dimers would occur via initial photoactivation of one or both of the enzyme cofactors, followed by electron transfer between the activated photosensitizer and enzyme-bound dimer. The studies by Jorns indicated that only reduced flavin is required to initiate thymine dimer cleavage in the model reaction (*162*). This is consistent with the observation that the quantum yield for the blue semiquinone radical form of *E. coli* DNA-PL *in vitro* was found to be much lower than the *in vivo* values. Treatment of the isolated enzyme with dithionite to give fully reduced flavin cofactor then results in a 12- to 15-fold increase in the quantum yield, suggesting that the flavin cofactor is in the reduced oxidation state *in vivo* (*163*).

The role of the second chromophore in catalysis is not clear. Selective removal of the pterin by photochemical decomposition (*153*), or irreversible reduction with sodium borohydride (*157*), does not significantly decrease enzyme activity. These studies suggest that enzymic photoreaction can occur by a mechanism that is independent of the second cofactor. The absolute action spectrum of the photolyzed enzyme was studied to help define the possible roles of the enzyme cofactors in catalysis. By comparing the wavelength dependence of the absolute action spectrum for the blue semiquinone radical form of the enzyme to its absorption spectrum, the particular chromophore acting as the sensitizer for the photoreactivation reaction can be identified. The absolute action spectrum of DNA-PL is determined under conditions where the enzyme is saturated with substrate, and this complex is then exposed to limiting light at a specific wavelength. Under these conditions, the rate of photoreactivation reaction compared to a control under full irradiation becomes a measure of the efficiency by which the enzyme utilizes a photon of light (*163*). Strong similarities were observed at 384 nm, consistent with the pterin chromophore acting as a photosensitizer. However, the absolute action spectrum indicates that enzymic photolysis occurs at wavelengths greater than 450 nm, where the pterin cofactor does not absorb. This would be consistent with the flavin cofactor functioning as a photosensitizer alone (*163*). The efficiency of the *in vivo* enzymic photolysis reaction is independent of the amount of light exposure, indicating 100% efficiency for each photon of light absorbed.

Although the second cofactor is is not directly required for enzyme-mediated dimer cleavage, it is possible that both the second cofactor and reduced FAD cofactors might be capable of acting as the photosensitizers. This suggests that reduced FAD or the pterin cofactor act individually or together as sensitizer in the enzyme-catalyzed reaction (*157*). Since the relative action spectrum for the photoreactivation reaction and the absorption spectrum of the second chromophore both display peaks near 390 nm, the second chromophore may be acting as a sensitizer *in vivo* (reduced photolyase), as well as with the isolated blue enzyme (*152*).

E. PROPOSED MECHANISM OF PYRIMIDINE DIMER CLEAVAGE BY *ESCHERICHIA COLI* DNA PHOTOLYASE

A reasonable mechanism for pyrimidine dimer cleavage by blue photolyase and fully reduced enzyme has been proposed by Sancar and Jorns and is illustrated in Scheme 34 (*163*). Photoreactivation reaction is initiated by conversion

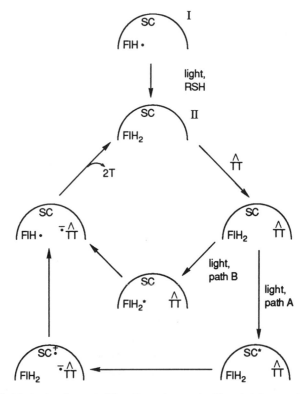

SCHEME 34. Mechanism for pyrimidine dimer cleavage by blue photolyase and fully reduced enzyme (*163*). FlH·, FAD neutral blue radical; FlH$_2$, FADH$_2$; SC, second chromophore; *, excited state due to absorption of light; +, radical cation; ⁻, radical anion; T̂T, pyrimidine dimer; 2T, repaired pyrimidine monomers.

of the neutral blue flavin radical (**I**) to FADH$_2$ (**II**), in a reaction requiring light and thiols. Binding of pyrimidine dimer may occur before or after the preliminary photoreduction step. Absorption of a second photon may lead to excitation of either cofactor, with dimer cleavage initiated by donation of an electron to the pyrimidine dimer. When the second cofactor is the sensitizer [pathway (A)], the radical cation of the pterin and radical anion of the dimer are produced. Subsequent electron transfer from reduced FAD reduces the second cofactor radical cation, while the unstable pyrimidine dimer radical spontaneously monomerizes (*164*). Pyrimidine, including a pyrimidine radical anion, result; the latter species subsequently reduces the FAD radical cation, completing the catalytic cycle (*163*).

An alternate mechanism involving reduced FAD as sensitizer [pathway B] indicates that an electron is directly donated to the pyrimidine dimer, forming FADH· and dimer radical anion. Pyrimidine dimer radicals are unstable (*157*) and will spontaneously decompose to the monomer and monomer radical anion. The latter species then regenerates reduced FAD in a manner similar to that described for pathway (A).

The observation by Jordan and Jorns that the fluorescence of reduced flavin is selectively quenched on the formation of UV-irradiated enzyme–oligothymidylate complexes may suggest the possibility of singlet intermediates in catalysis (*154*). The cleavage of thymine dimers by a singlet-state sensitizer has been postulated in a photoreactivating model reaction utilizing indole derivatives as the photochemical catalysts (*165*). Hence, the *E. coli* photoreactivating reaction would be initiated by electron donation from the singlet excited state of FADH$_2$ to the pyrimidine dimer, forming FADH· and dimer radical anion (Scheme 35). In a reaction analogous to that in Scheme 34, the unstable dimer radical anion would decompose to monomer and monomer radical anion, with the latter species regenerating reduced FAD by a second electron transfer step.

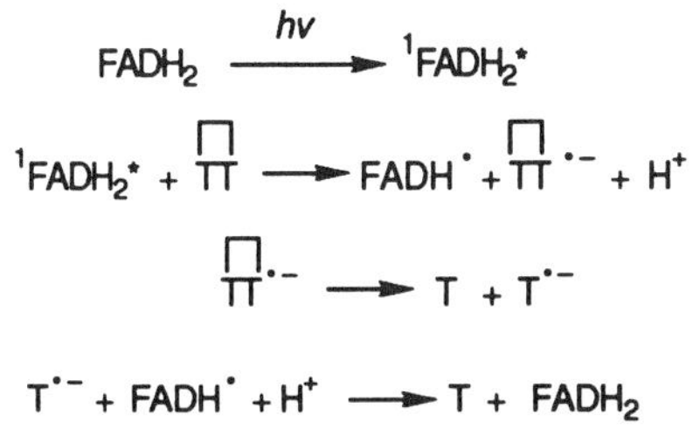

$$FADH_2 \xrightarrow{\ h\nu\ } {}^1FADH_2^*$$

$$^1FADH_2^* + \overset{\sqcap}{TT} \longrightarrow FADH^{\cdot} + \overset{\sqcap}{TT}{}^{\cdot -} + H^+$$

$$\overset{\sqcap}{TT}{}^{\cdot -} \longrightarrow T + T^{\cdot -}$$

$$T^{\cdot -} + FADH^{\cdot} + H^+ \longrightarrow T + FADH_2$$

SCHEME 35. Scission of pyrimidine dimers by a singlet photosensitization mechanism (*154*).

A similar mechanistic pathway for a singlet pterin intermediate has also been proposed, although substrate binding has no observable effect on the fluores-

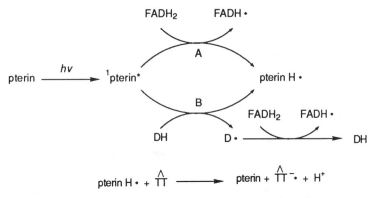

SCHEME 36. Pterin-catalyzed pyrimidine dimer scission (*154*).

cence of pterin (*154*). Such a mechanism is feasible if the singlet pterin initially reacts with the protein or flavin cofactor prior to electron transfer with substrate dimer. Such a mechanism is illustrated in Scheme 36. The singlet pterin cofactor abstracts an electron from reduced FAD [pathway (A)] or from an amino acid residue in the active site [pathway (B)]. The pterin radical may now reduce the dimer, forming monomer and monomer radical anion, with the latter species reducing FAD radical to $FADH_2$. In pathway (B), the flavin radical is formed via abstraction by the amino acid radical, in a step that prevents collapse of the [D·–pterin·] complex in the reverse direction. The presence of an amino acid donor in the active site of *E. coli* DNA-PL has been proposed by Sancar and co-workers (*153*).

The DNA-PL from yeast shares many similarities with the reduced *E. coli* enzyme, including $FADH_2$ and 5,10-methenyl tetrahydrofolate cofactors, and high quantum yield (*143, 145, 146, 155*). These similarities suggest that the mechanism of the yeast enzyme may be similar to that proposed for the *E. coli* photolyase. A second class of photorepair enzymes includes the $FADH_2$- and 8-hydroxy-5-deazaflavin-dependent photolyases (*166, 167*). The common characteristic between these photoreactivating enzymes is the presence of reduced flavin, and it has been suggested that the pterin and deazaflavin cofactors each play a similar role in the photoreactivating reaction as sensitizers in catalysis (*154*).

Recently, Schultz and co-workers describe the generation of antibodies that catalyze the light-dependent cleavage of thymine dimers (*168*), analogous to the photoreactivating *E. coli* DNA-PL reactions. The K_m for the carboxymethylthymine dimer is 6.5 mM, and a nonoptimal k_{cat} of 1.2 min^{-1} is observed, which compares favorably with the turnover number for *E. coli* DNA-PL cleavage of UV-irradiated DNA of 2.4 pyrimidine dimers/min (*148*). From the wavelength dependence of the quantum yield, it was suggested that the "combining site" of the antibody contains a tryptophan that acts as the sensitizer for the antibody-catalyzed reaction. Consistent with this suggestion is the report that the peptide Lys-Trp-Lys and a DNA-binding protein will catalyze the cleavage of thymidine dimer by the reversible donation of an electron from the excited indole of tryptophan to the dimer (164). The photomechanism of the antibody-mediated photoreactivation reaction is currently being studied.

IV. Pyruvate Formate-Lyase

A. BACKGROUND

Pyruvate formate-lyase (EC 2.3.1.54; formate acetyltransferase; PFL) catalyzes the key reaction in anaerobic glucose metabolism in bacteria, the coenzyme A-dependent dismutation of pyruvate into acetyl-CoA and formate. The reaction, first reported by Werkman and co-workers in the early 1940s (169, 170), was described as the "phosphoroclastic" cleavage of pyruvate because acetyl phosphate was detected as a product. Following the discovery of CoA and the elucidation of its role in acetyl transfer reactions (171, 172), the intermediacy of acetyl-CoA in pyruvate dismutation was realized; the overall reaction catalyzed by PFL is generally described by two half-reactions (Scheme 37).

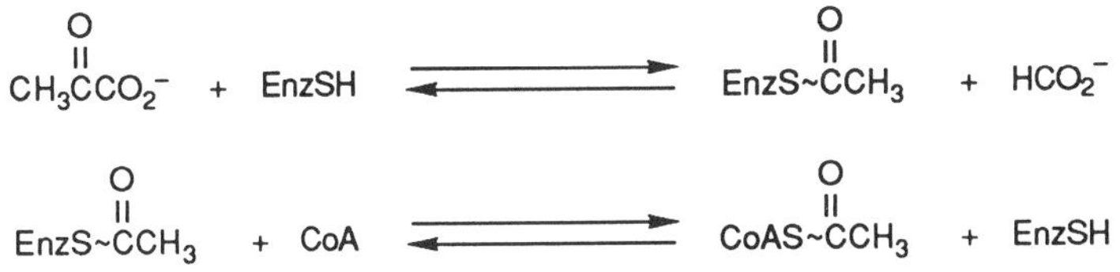

$$CH_3\overset{\overset{\displaystyle O}{\|}}{C}CO_2^- \; + \; EnzSH \; \rightleftharpoons \; EnzS\sim\overset{\overset{\displaystyle O}{\|}}{C}CH_3 \; + \; HCO_2^-$$

$$EnzS\sim\overset{\overset{\displaystyle O}{\|}}{C}CH_3 \; + \; CoA \; \rightleftharpoons \; CoAS\sim\overset{\overset{\displaystyle O}{\|}}{C}CH_3 \; + \; EnzSH$$

SCHEME 37. Half-reactions of pyruvate formate-lyase (187).

The enzyme is a homodimeric protein of M_r 170,000 and contains no known organic or metal ion cofactors. The enzyme is readily inactivated by oxygen and interconverts between active and inactive forms in vivo (173, 174). The activation process occurs under conditions of anaerobiosis and is catalyzed by an Fe(II)-dependent activating enzyme (M_r 30,000) (175). Elegant studies on the in vitro activation of PFL by Knappe and co-workers (176, 177) have revealed that a complex "activation" cocktail is required, which includes the activating enzyme, pyruvate, or oxamate as allosteric effectors, S-adenosylmethionine (SAM), and flavodoxin (175) or photoreduced 5-deazariboflavin (178). A possible role for a B_{12} derivative in the activation or catalytic reaction for PFL is not likely in light of the observation that E. coli 113-3, a methionine/B_{12} auxotroph, possesses PFL activity (*177*). The chemical mechanism of the activation reaction is not clear, although it is known to result in the stoichiometric cleavage of SAM to 5'-deoxyadenosine and methionine (*179, 180*). A transient Fe–adenosyl complex may also be involved and acts as a hydrogen atom abstractor from inactive PFL (*181*). The activating enzyme contains an unidentified organic cofactor whose UV–Vis absorption spectrum suggests a modified flavin (*178*). The role of this cofactor in the activation reaction is not yet known.

B. MECHANISTIC STUDIES

Attempts at elucidating the chemical mechanism of the PFL-catalyzed dismutation reaction have been hampered by the air sensitivity of the active enzyme. However, discovery of the activation reaction and isolation of a plasmid (p29) coding for the structural gene of PFL (*182*) has led to the isolation of large amounts of the inactive form of the enzyme, which may be readily converted to active lyase. The known mechanistic details for the PFL reaction indicate the formation of an acetylthioester–enzyme complex (*174*). This is consistent with the observation of Ping-Pong kinetics and the CoA-independent pyruvate–formate carboxylate exchange reaction (Scheme 37) (*183*). The turnover number for the forward reaction (pyruvate to formate) is 1100 sec^{-1} (pyruvate $K_m = 2 \text{ m}M$, CoASH $K_m = 0.007 \text{ m}M$), and in the reverse reaction 380 sec^{-1} (formate $K_m = 24.5 \text{ m}M$ acetyl-Co-A $K_m = 0.05 \text{ m}M$) (*174*). The equilibrium constant for the first half-reaction is ~50, and the overall equilibrium constant of 750 (30°C, pH 8.1) is in agreement with the chemically determined values.

The primary structure of *E. coli* PFL and its activating enzyme have been determined from the DNA nucleotide sequences (*184*). A particular region of interest is that containing the adjacent cysteine residues at positions 418–419. Knappe and co-workers have suggested that the cysteine residue at 419 is the site of the acetylthioester formation in the catalytic cycle of the enzymic reaction (*185*).

The most striking physical characteristic about PFL is the recent discovery by Knappe and co-workers that activation of the lyase generates an enzyme-associated EPR resonance (*180*). The EPR signal is centered at $g = 2.003$ with a doublet splitting of 1.5 mT, and this is consistent with an organic free radical located on an amino acid residue of the protein. The nature of the amino acid residue supporting the radical species is not yet known, although EPR and ENDOR studies indicate that the unpaired electron is coupled with three sets of hydrogen nuclei, with the major splitting attributed to an acidic proton that exchanges with the solvent (*181*). The signal in the resting form of the free or acetyl enzyme is apparently not directly due to a thiyl radical, as no change in the protein EPR signal is observed in ^{33}S-enriched enzyme samples (*185*), although this does not exclude the potential involvement of a short-lived thiyl radical intermediate during catalysis. The radical signal is present in both the resting and acetyl forms of the active enzyme, but is quenched by the addition of the known PFL inhibitors, oxygen or hypophosphite (*186*), with concomitant enzyme inactivation.

The reaction catalyzed by PFL functions as the anaerobic counterpart of the pyruvate dehydrogenase reaction; however, the chemical mechanisms of these related biological reactions are probably quite different. Heterolytic mechanisms for the PFL-mediated cleavage of pyruvate are illustrated in Scheme 38. In pathway (A), decarboxylation of the hemithioketal results in the formation of a high-energy thiohemiacetal carbanion, which must be followed by electron transfer chemistry to CO_2 to give formate and the acetyl–enzyme. In pathway (B), heterolytic C–C bond cleavage in the opposite direction requires the unlikely formation of a formate dianion intermediate. Since both of these pathways result in chemically unreasonable steps, and are without precedent in physical organic chemistry, alternate mechanistic schemes have been explored.

The discovery that the active form of PFL contains a paramagnetic species suggests the intriguing possibility that this unusual enzymic reaction proceeds via a homolytic chemical mechanism. Testing of this hypothesis by Kozarich and co-workers has led to studies on the inactivation of PFL by a variety of compounds

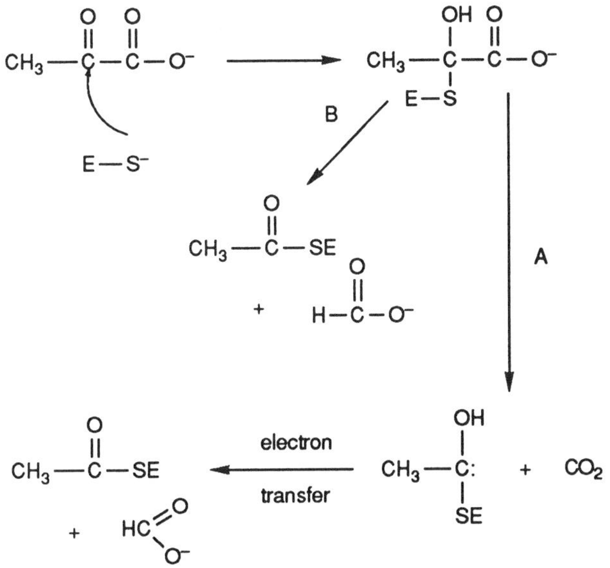

SCHEME 38. Possible heterolytic mechanism for the PFL-mediated cleavage of pyruvate by pyruvate formate-lyase.

TABLE V

Compounds Tested as Inactivators of Pyruvate Formate-Lyase

Compound	Concentration (mM)	Half-life (inact; min)[a]	Substrate protection[b]
O_2	0.25	<1	—
Hypophosphite	0.25	7	Yes; formate
Phosphite	5.0	n.i.	—
Methylphosphinate	5.0	n.i.	—
Acetylphospinate	0.025	2	Yes; pyruvate
Phosphate	100	n.i.	—
Propargylic acid	0.5	1	Yes; pyruvate
Propargyl alcohol	5.0	n.i.	—
2-Butynoic acid	5.0	n.i.	—
Acrylic acid	5.0	3.5	Yes; pyruvate
3-Fluoropyruvate	0.5	8	Yes; pyruvate

[a] n.i., No inactivation.
[b] 10 mM incubated with pyruvate formate-lyase prior to inactivation.

(187, 203). A summary of some recently discovered inactivators of E. coli PFL is presented in Table V. Consistent with these compounds acting as mechanism-based or active site-directed inhibitors is the observation of pseudo-first-order inactivation kinetics, substrate protection (by pyruvate or formate for inhibitors that are pyruvate or formate analogs, respectively), and isotope effects on the rates of inactivation by the deuterated analogs. The details of some of these studies, the proposed inactivation mechanisms, and the implications to the normal enzymic reaction are discussed below.

C. Inactivation of Pyruvate Formate-Lyase by Hypophosphite

The inhibition of pyruvate formate-lyase by hypophosphite was first observed by Novelli in work on the CoA-independent carboxylate exchange reaction between pyruvate and formate (186). In a more detailed study by Knappe et al., time-dependent inactivation is observed to occur with concomitant loss of the enzyme free radical EPR signal (180). The inactivation kinetics are first order and the rate of inactivation is accelerated when the enzyme is in the acetylated form. Furthermore, inactivation by [³H]hypophosphite leads to the stoichiometric release of tritium to H_2O, and treatment of PFL with [³²P]hypophosphite produces an alkali-labile radiolabel that is covalently bound to the inactive enzyme (180).

Recently, Kozarich and co-workers (187) reported an investigation on several critical aspects of PFL inactivation by hypophosphite and their relationship to the normal enzymic reaction. These results, along with those of Knappe *et al.* (*180*), strongly suggest that hypophosphite is recognized by PFL as a formate analog. This is consistent with the 3.6-fold acceleration in the rate of inactivation observed when pyruvate, PFL, and hypophosphite are incubated in the absence of CoASH (*187*). This result is most easily explained by a facilitated binding of hypophosphite to the formate binding site in the acetyl–enzyme. In the absence of CoA, only the first half-reaction is operative (Scheme 37, upper reaction), resulting in a buildup of the acetyl form of the enzyme (*174*). Furthermore, the observation that formate (at 20 mM) affords partial protection to PFL from inactivation only when the enzyme is acetylated is consistent with this proposal. The high concentration of formate required for protection is consistent with the K_m of formate (24.5 mM) in the reverse reaction (*174*).

Remarkably, the inactivation of PFL by hypophosphite was found to be reversible, as treatment of inactive enzyme with the PFL "activating" system results in partial recovery of enzymic activity (*187*). Studies on the stoichiometry and reactivity of the ^{32}P-labeled enzyme by Kozarich and co-workers indicate (*203*) that [^{32}P]hypophosphite-inactivated PFL that has been freed of activating enzyme by gel exclusion chromatography (*178*) can be reactivated to 44% of the original activity. Since reactivation occurs without loss of the original ^{32}P radiolabel, the observed enzyme activity may result from a modified enzyme. If this is the case, the protein may exhibit altered kinetic behavior toward substrates or inhibitors. However, an initial investigation on the substrate kinetics has revealed that the K_m values for pyruvate and formate are nearly identical to a control sample of native PFL inactivated by O$_2$ (*203*). A decrease in the V_{max} was observed for the inhibitor-treated enzyme as well as the controls, and may result from denaturation of the enzyme samples. The progress curve for reactivation of hypophosphite-inactivated enzyme is superimposible on that for an untreated sample, and this reactivated (and modified?) active enzyme is still susceptible to hypophosphite inactivation. These preliminary results may suggest that the modified amino acid residue(s) has little direct involvement in the catalytic mechanism. An alternate possibility is that the protein-bound phosphorus label is capable of exchanging to other sites on the protein during reactivation, or possibly can be exchanged to the activating enzyme, resulting in its inactivation and incomplete PFL activation. Additional work is in progress to substantiate these results on the hypophosphite-modified enzyme.

The key experiment in relating the mechanism of hypophosphite inactivation to the normal PFL reaction is the observation of a primary kinetic isotope effect on the inactivation process with [^{2}H$_2$]hypophosphite (*187*). An isotope effect of 2.6 is reported with the free enzyme (0.25 mM hypophosphite), and a smaller but still primary effect of 1.6 for the acetyl–enzyme. This suggests that phosphorus–hydrogen bond cleavage is at least partially rate limiting during inactivation. These studies have led Kozarich and co-workers to propose a preliminary mech-

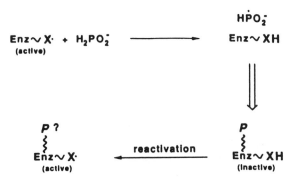

SCHEME 39. Mechanism of hypophosphite inactivation of pyruvate formate-lyase (*187*).

anism for hypophosphite inactivation (Scheme 39) (*187*). Enzyme-catalyzed homolytic phosphorus–hydrogen bond cleavage would quench the enzyme-bound radical and generate a hypophosphite radical anion that could react covalently with the enzyme, and could ultimately be quenched by solvent hydrogens or DTT. This mechanism is consistent with the observation that hypophosphite inactivation occurs with quenching of the EPR-observable radical signal and covalent radiolabeling of the inactive enzyme with ^{32}P (*180, 203*). The EPR spectrum of the hypophosphite radical anion has been characterized (g = 2.0028) (*189*). Irradiation of hypophosphite crystals has resulted in the detection of stabilized radicals with a half-life of 24 hr (*190*), and the excellent reducing properties of hypophosphite are well known in organic chemistry and are consistent with the relatively low energy of the P–H bond (79 kcal/mol) (*191, 192*).

The proposed inactivation mechanism in Scheme 39 predicts that the normal reverse reaction, i.e., hydrogen abstraction from formate, may lead to a formate radical anion intermediate. Consistent with this idea, a V_{max}/K_m isotope effect of 3.6 ± 0.7 has been reported for the formation of pyruvate from [^{2}H]formate and acetyl-CoA (*187*). Thus, rate-limiting carbon–hydrogen bond cleavage occurs in the normal reverse reaction. The generation of the formate radical anion via Fenton chemistry is a facile process; its EPR spectrum has been characterized (g = 2.000) in aqueous solution and its chemical reactivity studied (*193–195*).

This persuasive evidence for homolytic chemistry occurring in the PFL reaction has led Kozarich and co-workers to propose a working hypothesis for the overall enzymic reaction (*187*). A chemical precedent exists in the H_2O_2/Fe(II)-mediated homolytic cleavage of ethyl pyruvate to ethyl formate radical and acetate. This is the Minisci reaction, which has been employed in the selective carboxylation of heteroaromatic bases (Scheme 40) (*196–198*). The initial step in this process is the formation of a hydroperoxy hemiketal with ethyl pyruvate and H_2O_2, followed by Fe(II) reduction of the hemiketal to the alkoxy radical.

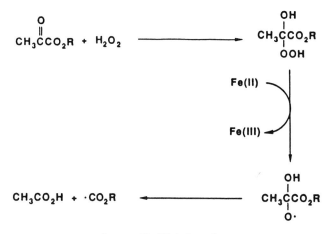

SCHEME 40. Minisci reaction.

This is then followed by homolytic β-scission to form acetate and the ethyl formate radical synthon.

The mechanism for the PFL reaction is proposed to operate in an analogous manner (Scheme 41) (*187*). The enzyme-based radical (structure unknown) initiates reaction by the generation of a thiyl radical at an active site cysteine. Addition of this radical to the carbonyl of pyruvate forms a tetrahedral alkoxy radical intermediate reminiscent of the related intermediate in the Minisci reac-

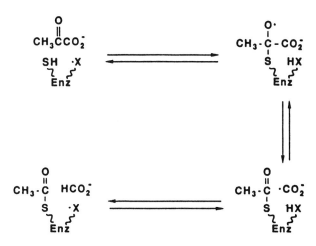

SCHEME 41. Postulated mechanism of pyruvate formate-lyase based on the Minisci reaction (*187*).

tion. Alternatively, this adduct could be generated by the addition of a cysteine thiolate to the carbonyl, forming a thiohemiketal, followed by a one-electron reduction by the enzyme radical species. Subsequent collapse of the alkoxy radical via β-scission yields the acyl–enzyme and the formate radical anion, which is reduced to formate by hydrogen atom abstraction from the quenched enzyme radical, thereby regenerating this species. CoA-dependent thioester exchange would complete the catalytic cycle.

Although thiolate addition would be preferred on thermodynamic grounds, the radical process is favored because the production of thiyl radicals is a facile process. Furthermore, the resulting transiently quenched enzyme radical is immediately positioned to deliver a hydrogen atom to the formate radical anion. There has been growing awareness of the role thiyl radical intermediates may play in biological reaction mechanisms. Thiyl radicals have been proposed as intermediates in the mechanism of styrene–glutathione conjugate formation catalyzed by prostaglandin H synthase and horseradish peroxidase (199), in the pyruvate: ferredoxin oxidoreductase reaction (200), and inferred in the B_{12}-dependent ribonucleotide reductase reaction previously discussed.

This hypothesis does not appear to be consistent with the observation by Knappe and co-workers that the inactivation of PFL by [^{3}H]hypophosphite results in the stoichiometric release of ^{3}H to the solvent (180). The stoichiometric release of tritium on inactivation implies no isotope effect, although the use of the tritium as a tracer complicates the analysis. An intramolecular discrimination by the enzyme between the tritium and the hydrogen on a labeled hypophosphite molecule would necessarily dampen the intermolecular selection effect. Because the deuterium isotope effect data were determined under nonsaturating conditions with nearly complete dideuterated hypophosphite, any intramolecular discrimination is supressed. However, the stoichiometric release of tritium would also necessarily require that both hydrogens of hypophosphite be labilized during inactivation reaction. Three possibilities that would account for the tritium release are (1) an enhanced susceptibility to heterolytic exchange with solvent from an enzyme-bound species, (2) a mechanistic scenario for inactivation, which requires the sequential homolytic abstraction of both hypophosphite hydrogens, and (3) a less likely requirement for multiple turnovers of hypophosphite per inactivation event, which, when the tritium selection effect is factored in, gives the appearance of stoichiometric tritium release. The ultimate distinction among these possibilities should be possible by reexamining the inactivation of PFL by [^{3}H]hypophosphite and by a determination of the oxidation state of the phosphorus covalently bound to the inactivated enzyme.

The structure of the protein-bound phosphoryl intermediate is not yet clear. Preliminary studies by Kozarich and co-workers have indicated that alkaline treatment of [^{32}P]hypophosphite-inactivated PFL releases a radioactive phosphorus compound that is not at the oxidation state of hypophosphite and is indistinguishable from phosphate or phosphonate (*203*).

Recently, Knappe and co-workers have reported the first direct evidence for radical intermediates in the PFL reaction (*185, 201*). Inactivation of acetyl–PFL by hypophosphite at 0°C gives rise to the transient formation of a new, multiplet-structured EPR signal centered at $g = 2.0032$. The hyperfine couplings are reported to be consistent with interactions of one phosphorus and three hydrogens with the paramagnetic center, which is suggested to arise from the reaction of hypophosphite with the acetyl–enzyme intermediate to produce an α-phosphonyl carbon-centered radical of 1-hydroxyethyl phosphonate (Scheme 42). Remarkably, Knappe has indicated that the phosphonate is covalently bound to the lyase by a thiophosphoryl bond to Cys-418, the amino acid adjacent to the postulated acetyl carrier in the normal PFL reaction at Cys-419 (*185*). This characterization is based on ^{31}P NMR studies of the phosphonate-containing peptide from chymotryptic peptide digestion of the inactive protein. A resonance at 44 ppm is suggested to be consistent with the CH–P–S structure. Furthermore, it was reported that 1-hydroxyethyl phosphonate was isolated following cleavage of the peptide-bound phosphonate with phosphodiesterase or bromine.

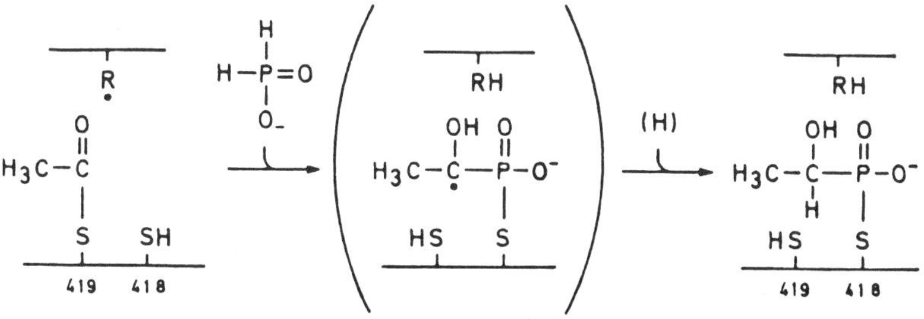

SCHEME 42. Postulated mechanism of inactivation of acetyl-PFL by hypophosphite by Knappe and colleagues (*185*).

A mechanism is proposed where phosphorus–carbon bond formation is mediated by the enzyme-bound free radical, and this reaction scheme is suggested to resemble the normal carbon–carbon bond-forming reaction in the direction of pyruvate synthesis from acetyl-CoA and formate (*185*). The equation proposed by Knappe (Scheme 42) suggests formation of a 1-hydroxyethylphosphonate thioester, presumably through the hypophosphite radical anion (*187*). The proposed mechanism involves carbon–Cys-419 bond cleavage, reduction of the acetylcarbonyl, and a second phosphorus–hydrogen bond cleavage leading to the formation of a (Cys-418) thiophosphonyl ester. The relationship of this mechanistic scheme to the normal enzymic reaction is not clear, as it suggests that in the reverse reaction between formate and acetyl–enzyme, a lactyl–enzyme thioester intermediate would be formed. Subsequent hydrolysis would, however, give lactate instead of pyruvate as a product. It is evident that additional study and characterization of the proposed thiophosphoryl product are required to substantiate the mechanism of this unique inactivation reaction.

D.   INACTIVATION OF PYRUVATE FORMATE-LYASE BY PROPARGYLIC ACID

Kozarich and collaborators have evaluated a variety of pyruvate and formate analogs as potential inhibitors of PFL. In the course of their studies on pyruvate analogs, the acetylenic carboxylic acid, propargylic acid, was found to be an effective active site-directed inhibitor of PFL (203). At 0.5 mM propargylic acid, the decrease in enzyme activity follows first-order kinetics, with a half-life of ~2.5 min. Pyruvate (5.0 mM) affords substantial protection from inactivation ($t_{1/2}$ = 12 min), which is consistent with the inactivation reaction occurring at the active site. Furthermore, the enzyme is apparently quite specific for propargylic acid, since incubation with 5 mM concentrations of acetylene dicarboxylate, 2-butynoic acid, phenylpropargylic acid, and propargyl alcohol has no effect on PFL activity. Radiolabeling experiments using [2,3-^{14}C]propargylic acid indicate a 1:1 ratio of radiolabel to enzyme active sites, although nonspecific radiolabeling occurs on prolonged incubation. Treatment of the inactivated enzyme with the PFL activation system results in the regeneration of enzyme activity, with apparent stoichiometric release of the radiolabel.

Since thiols and thiyl radicals are known to readily undergo addition reactions at unsaturated carbon centers (199, 202, 204), a possible mechanism for this inactivation reaction is shown in Scheme 43. Addition of the active site nucleophilic or radical species followed by protonation or electron transfer, respectively, would yield the thioacrylate derivative and inactive enzyme. Of course, addition to C-2 of propargylic acid is also possible, forming a 2-substituted acrylate derivative instead.

Remarkably, incubation of [3-^{2}H]propargylic acid with PFL yields a primary kinetic isotope effect of 3–4 on the rate of inactivation (203). A problem in obtaining reproducible results was traced to the unexpected rapid exchange of the C-3 deuterium with solvent protons. By ^{1}H NMR methods this exchange was estimated to have a half-life of less than 1 hr at 25°C. Verification of the observed isotope effect was accomplished by taking advantage of this property and observing an increase in the rate of inactivation on incubation of [^{2}H]propargylic acid

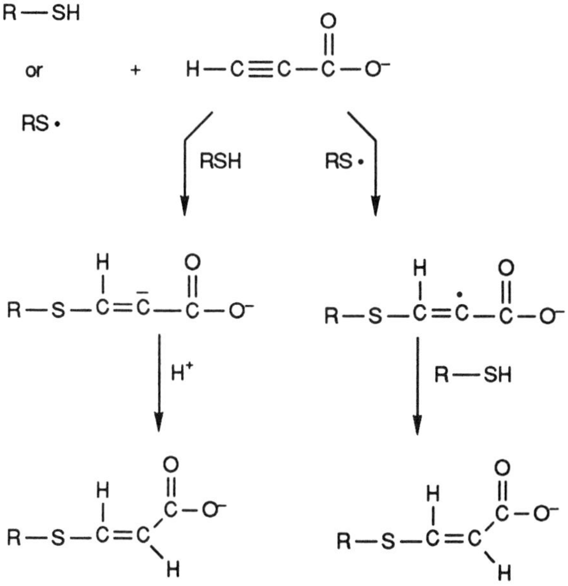

SCHEME 43. Addition of thiols and thiyl radicals to propargylic acid.

in H_2O ($^2H-^1H$ exchange) or a corresponding decrease in the rate of inactivation due to the onset of the isotope effect on incubation of [^1H]propargylic acid stock solutions in D_2O. These experiments verify the exchange of the terminal proton in neutral to weakly basic (pH 8.1) solutions as well as the kinetic isotope effect. From a mechanistic standpoint this result suggests that cleavage of the carbon–hydrogen bond in propargylic acid is at least partially rate limiting in the mechanism of inactivation. One possibility may involve a 1,2-hydrogen radical shift in the radical pathway in Scheme 43, leading to the formation of a C-3 thioacrylate radical which may be stabilized by the sulfur atom. However, 1,2-hydrogen radical shifts are generally not observed in organic chemistry (205). Recently, the occurrence of a 1,2-hydride shift in the oxidation of phenylacetylene to phenylacetic acid by peracids and cytochrome P-450 was reported by Komives and Ortiz de Montellano (206), as indicated by an isotope effect for the bent transition state of 1.7. The proposed mechanisms (Scheme 44) are initiated by oxygen transfer to the triple bond, resulting in the development of positive charge on the carbon adjacent to the phenyl ring, followed by concomitant transfer of the terminal hydrogen to the α-position. Additional studies are in progress to clarify the unusual isotope effect data in the inactivation of PFL by propargylic acid.

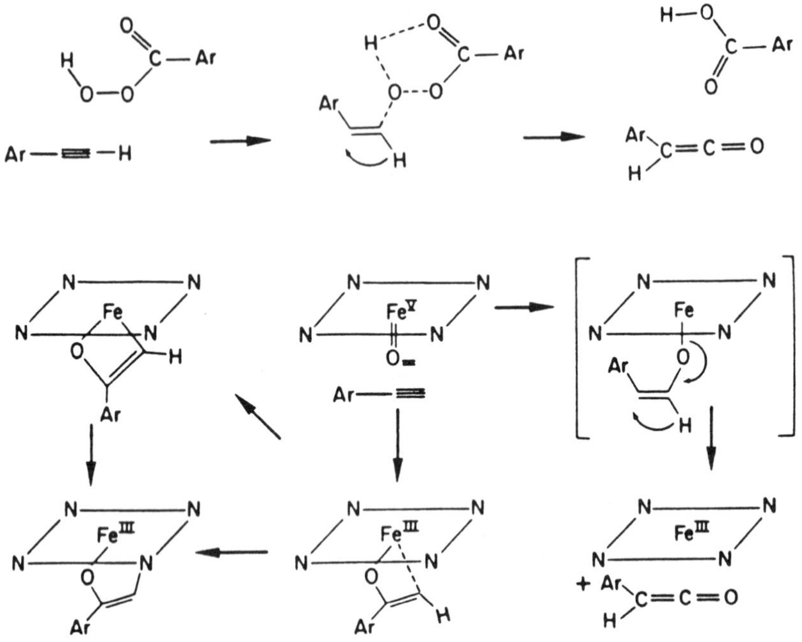

SCHEME 44. 1,2-Hydride shift in the oxidation of phenylacetylene to phenylacetic acid (206).

E.   INACTIVATION OF PYRUVATE FORMATE-LYASE BY ACETYL PHOSPHINATE

The strict specificity of PFL for pyruvate and formate as substrates places some restrictions on the structure of potential active site-directed inhibitors. Therefore, acetyl phosphinate (207) is a natural candidate for a PFL inhibitor, as it is not only a pyruvate analog, but is also a derivative of hypophosphite, a known mechanism-based PFL inactivator. In the absence of CoA, 100 mM acetyl phosphinate is an effective time-dependent inhibitor of PFL, although the inactivation reaction does not appear to be first order. If 55 mM CoA is included in the inactivation reaction mixture, first-order kinetics are now observed, and with 10 mM acetyl phosphinate the half-life of the inactivation reaction is 3 min. In the presence of 5 mM pyruvate (no CoA), PFL is completely protected from inactivation by 100 mM acetyl phosphinate.

These results indicate that acetyl phosphinate is the most potent active site-directed inactivator of PFL now known. Although it is not yet clear what the overall mechanism of inactivation might be, consistent with the mechanism for the normal PFL reaction proposed by Kozarich and co-workers (Scheme 41) (187), there is the intriguing possibility that phosphorus–carbon or phosphorus–hydrogen bond cleavage might occur. These possibilities are currently being probed (*208*) by looking for the formation of pyruvate from acetyl phosphinate and [^{14}C]formate via the acetyl phosphinate-derived acetyl–enzyme, and isotope effects on the rate of inactivation utilizing [^{2}H]acetyl phosphinate.

V. Other Radical Enzymes

A. PYRUVATE: FERREDOXIN OXIDOREDUCTASE

A rather unique α-ketoacid decarboxylase has been discovered, pyruvate: ferredoxin oxidoreductase (EC 1.2.7.1), which catalyzes the CoA-dependent oxidative decarboxylation of 2-oxobutyrate, pyruvate, and 2-oxoglutarate to CO_2 and acetyl-CoA, in a reaction requiring [2Fe–2S] ferredoxin or O_2 as electron acceptor (Scheme 45) (*209–211*). This enzyme activity has been detected in fermentative and photosynthetic anaerobes, but the air-stable enzyme from the aerobic archaebacterium *Halobacterium halobium* has been purified and studied in greatest detail (*212*). The reaction is reminiscent of the CoA-dependent dismutation of pyruvate by *E. coli* pyruvate formate-lyase (*181*), with the obvious exceptions that CO_2 is formed instead of formate and differences in the cofactors involved. The oxidoreductase is an iron–sulfur protein containing two [4Fe–4S] clusters and two molecules of thiamin diphosphate per molecule of enzyme (*212*). The presence of thiamin diphosphate suggests a decarboxylation mechanism similar to the related reactions catalyzed by the pyruvate dehydrogenase enzyme complex, although flavins, lipoic acid, and pantetheine are not present (*212*). Thiol groups are apparently not important in catalysis since thiol-specific reagents such as arsenite and iodoacetamide have no effect on the enzymic reaction (*213*). Preliminary mechanistic studies provided additional evidence that the mechanism of pyruvate dismutation to acetyl-CoA was different than that for the pyruvate dehydrogenase enzyme, as reported by Uyeda and Rabinowitz (*211*). They found that the pyruvate: ferredoxin oxidoreductase from the eubacterium *Clostridium acidiurici* forms acetyl-CoA concomitantly with the reduction of an unknown enzyme-bound chromophore (probably an Fe–S cluster), which can be subsequently reoxidized by electron acceptors such as FAD or ferredoxin. Although this work suggested that the iron–sulfur clusters are acting as electron-transferring groups, the question of the mechanism by which oxidation and subsequent acyl transfer occur in the absence of lipoic acid was still not clear.

$$R-\overset{\displaystyle O}{\overset{\|}{C}}-CO_2^- \ + \ 2\,Fd_{ox} \ + \ CoA \ \longrightarrow \ R-\overset{\displaystyle O}{\overset{\|}{C}}-CoA \ + \ 2\,Fd_{red} \ + \ CO_2$$

SCHEME 45. General reaction catalyzed by pyruvate: ferredoxin oxidoreductase.

In the course of EPR studies on the role of the iron–sulfur clusters, free radical signals were observed and were interpreted as arising from intermediates in the enzyme-catalyzed reaction (*209*). The EPR signals are observed in the isolated enzyme and are enhanced in the presence of substrate, being quite stable at room temperature and in the presence of oxygen. Dithionite quenches the radical signal, revealing the typical EPR spectrum of the iron–sulfur clusters, but studies in oxygen-depleted systems with both pyruvate and CoA present revealed a complex superposition of the iron–sulfur cluster signal with that of the radical, which indicated the possibility of spin–spin coupling interaction between the reduced enzyme and the substrate-induced radical species (*213*). At 92K the signal is centered at $g = 2.006$ and shows hyperfine splitting. Removal of the substrates by gel filtration has no effect on the radical concentration, although subsequent addition of CoASH quenched the observed signals, suggesting that they are intermediates in the catalytic cycle.

The mechanism proposed for the pyruvate:ferredoxin oxidoreductase is drastically different from that of the pyruvate dehydrogenase enzyme complexes, as radical intermediates are proposed, suggesting stepwise one-electron transfers. Furthermore, the preliminary work by Uyeda and Rabinowitz (*211*) on the mechanism of *C. acidiurici* pyruvate:ferredoxin oxidoreductase indicates that there may be subtle differences in the mechanism of the oxidoreductase from different organisms. The key points and contrasts in the mechanism of these enzymes may be summarized as follows. Addition of pyruvate to a stoichiometric amount of enzyme leads to the formation of an equimolar amount of CO_2. In *H. halobium* (*213*) a enzyme-mediated one-electron transfer to an exogenous electron acceptor occurs at this stage, with formation of a stable enzyme-bound radical intermediate, whereas the enzyme from *C. acidiurici* remains in the oxidized state. Addition of CoA leads to the subsequent formation of acetyl-CoA from both enzymes, which reduction (two electrons?) of an unknown chromophore in the *C. acidiurici* enzyme, and transfer of a second electron in the *H. halobium* reaction. The mechanistic cycle is then completed with the reoxidation of both enzymes by suitable electron acceptors.

The mechanism suggested by Kerscher and Oesterhelt is indicated in Scheme 46 for the enzyme from *H. halobium* (*213*). The initial step is identical to that of the 2-oxoacid dehydrogenase complexes and involves binding of pyruvate to thiamin diphosphate and subsequent decarboxylation yielding hydroxyethylthiamin diphosphate. This intermediate undergoes one-electron transfer to the [4Fe–4S] cluster to form the stable free radical. The cluster is then reoxidized by ferredoxin or oxygen to give the enzyme–intermediate complex. Reaction with CoA initiates the second electron transfer to the iron–sulfur cluster, acyl transfer, followed by reoxidation of the enzyme by ferredoxin or O_2 to complete the cycle. Two basic questions are yet unanswered: (1) What is the mechanism of the enzymic reaction between CoASH and hydroxyethyl-TPP in the absence

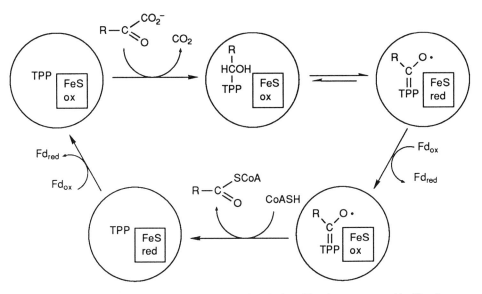

SCHEME 46. Reaction mechanism for pyruvate:ferredoxin oxidoreductase proposed by Kerscher and Oesterhelt (213).

of lipoic acid and given the apparent noninvolvement of protein thiol groups? (2) How does CoASH initiate the electron transfer from the postulated alkoxyethyl-TPP radical to the Fe–S center?

Recently, Wahl and Orme-Johnson reported their studies on the characterization and mechanism of the pyruvate:flavodoxin (ferredoxin) oxidoreductase from *Klebsiella pneumoniae* and *Clostridium thermoacetium* (214). These oxidoreductases appear to be closely related to that from *C. acidiurici* (215) in that the iron is present in two Fe_4S_4 clusters, which act as electron acceptors in the catalytic mechanism. However the *K. pneumoniae* and *C. thermoaceticum* enzymes may be mechanistically distinct from the *H. halobium* oxidoreductase (213) in that free radical intermediates are not detected for the former enzymic reaction. EPR signals in the *Klebsiella* or *C. thermoaceticum* oxidoreductases are only observed in the fully reduced enzyme when the reductants dithionite or pyruvate *and* CoASH are present (214). The suggested mechanism for the pyruvate oxidoreductase from *K. pneumoniae* and *C. thermoaceticum* is initially similar to the mechanism for all TPP enzymes in that decarboxylation of pyruvate leads to the formation of hydroxyethyl-TPP. Two one-electron transfers to each of the two Fe–S clusters occur on the binding of CoASH. However, the mechanism for the formation of acetyl-CoA from the hydroxyethyl-TPP intermediate and of the CoASH-induced electron transfers is not yet clear.

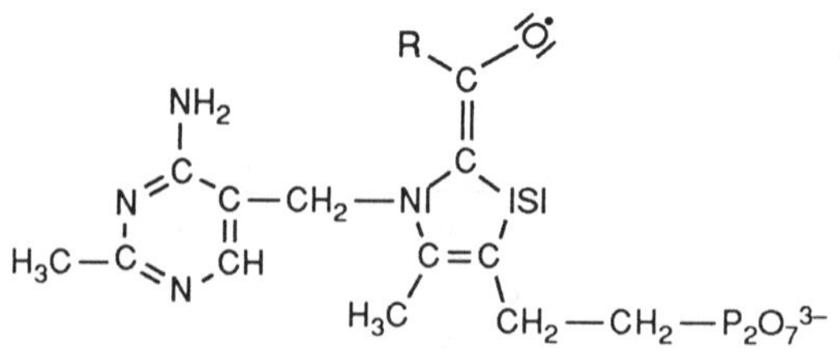

57

Scheme 47. Proposed structure of alkoxyethylthiamin radical (213).

The novel intermediates in the reaction catalyzed by pyruvate:ferredoxin oxidoreductase and related enzymes invoke a rather novel role for thiamin diphosphate in radical enzymic reactions. The pyruvate formate-lyase of *Streptococcus faecalis* has also been observed to contain TPP as cofactor (216) and may involve similar intermediates, although reduction of CO_2 to formate may be required. This enzyme is isolated under strict anaerobic conditions, which has restricted mechanistic studies up to the present time. However, the structure of the radical intermediate has not been definitely established, but the alkoxyethylthiamin radical 57 in Scheme 47 has been suggested (213). The alkoxy radical would be stabilized by delocalization of the lone electron over the heterocyclic ring of thiamin (213). An alternative structure and mechanism is suggested from the work of Docampo *et al.* on the pyruvate:ferredoxin oxidoreductase from *Tritichomonas foetus* (217). Incubation of pyruvate and CoASH with the hydrogenosomal fraction and the spin trap 5,5-dimethyl-1-pyrroline N-oxide (DMPO) gave EPR spectra consistent with radical adducts derived from a pyruvate C-3 radical and thiyl radical of CoASH. Hence, the mechanism of Kerscher and Oesterhelt (213) was modified as indicated in Scheme 48. Following the normal TPP route to binding and decarboxylation of pyruvate, the hydroxyethyl-TPP donates one electron to the Fe–S cluster, forming in this case a carbon-centered radical at C-2 of the enzyme-bound intermediate 58 (Scheme 48), followed by one-electron oxidation of the Fe–S center by ferredoxin or O_2. This stable enzyme–radical complex then binds CoASH, which is proposed to reduce the iron–sulfur cluster, forming the CoA–thiyl radical, which then accepts the acyl group from TPP, generating acetyl-CoA, and followed by reoxidation of the enzyme to complete the catalytic cycle.

Although the chemistry for this mechanism has not been worked out in detail, the idea of a carbon-centered radical at the C-3 position of pyruvate is consistent with EPR studies on the pyruvate:ferredoxin oxidoreductase from *H. halobium* discussed above, where the hyperfine splitting was dependent on the substrate structure and the number of hydrogen nuclei near C-3 (209).

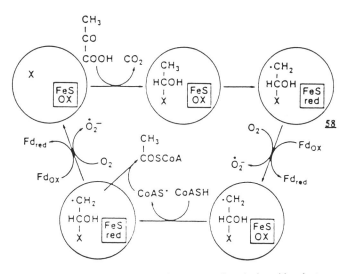

SCHEME 48. Alternate reaction mechanism for pyruvate:ferredoxin oxidoreductase proposed by Docampo *et al.* (*217*).

B. LYSINE 2,3-AMINOMUTASE

The initial step in the microbiological metabolism of lysine to acetate, butyrate, and ammonia is the reversible interconversion of L-lysine and L-β-lysine via exchange of the α-amino group and a β-proton catalyzed by lysine 2,3-aminomutase (Scheme 49). The reaction has been observed in several species of *Clostridium, Nocardia,* and *Streptomyces* (*218, 219*), but the mechanistic studies have concentrated on the enzyme from *Clostridium subterminale* SB4, which has been purified and characterized (*220*). The enzyme is extremely sensitive to reversible oxygen inactivation, with reactivation achieved by anaerobic incubation with a thiol and Fe(II), and a protein-bound Fe(II) is apparently required for activity (*220*). The exchange reaction is also accelerated by *S*-

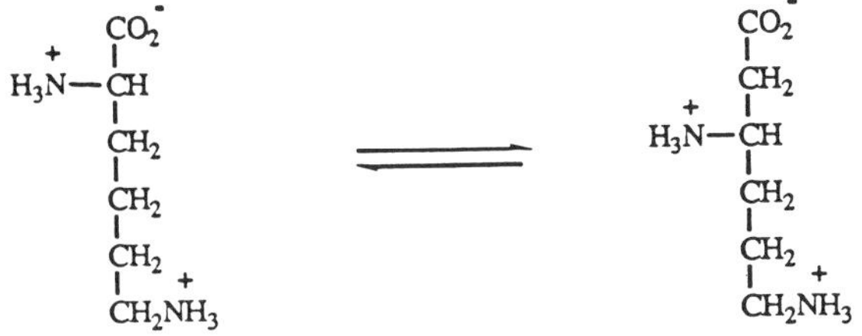

SCHEME 49. General reaction catalyzed by lysine 2,3-aminomutase.

adenosylmethionine, which binds to the enzyme with a K_m of 2.8 nM, whereas its substrate, L-lysine, binds with a K_m of 6.6 mM. The equilibrium constant for the reversible reaction (K = L-β-lysine/L-lysine) is 6.7 at 37°C. Furthermore, the enzyme contains pyridoxal phosphate as an organic cofactor (220), as well as covalently bound lysine (221).

The lysine 2,3-aminomutase reaction is reminiscent of the 1,2 rearrangement reactions catalyzed by B_{12}-dependent enzymes. The B_{12} dependence of several enzymic lysine 2,3-aminomutase reactions has been characterized in a variety of animals and microorganisms (222). However, the enzyme from *C. subterminale* SB4 is apparently not B_{12} requiring (220), suggesting that the enzymes catalyzing α- to β-amino transfers are not all evolutionarily related. Given the unique cofactor requirements of the *C. subterminale* SB4 enzyme, i.e., SAM and Fe(II), several studies aimed at elucidating the reaction mechanism have recently been reported.

Preliminary studies to elucidate the mechanism of this unusual exchange reaction and the role for SAM in catalysis were centered on defining the stereochemistry of the reaction. Early studies by Chirpich *et al.* suggest that the proton in the β-position of L-lysine is conserved, since enzymic exchange reactions with L-lysine run in D_2O or HTO did not produce isotopically labeled L-β-lysine or L-lysine, after correction for the presence of lysine racemase (220). Aberhart *et al.* have utilized ^{2}H NMR and showed that the C-3 hydrogen is indeed conserved (223). Hence, transformation of (2S)-α-lysine to (3S)-β-lysine involves a shielded transfer of the 3-*pro-R* hydrogen of L-lysine to the 2-*pro-R* position of β-lysine, with retention of the 3-*pro-S* hydrogen of L-lysine in C-3 of β-lysine, and the C-2 hydrogen of L-lysine is retained in the 2-*pro-S* position of the β-lysine product. The overall stereochemistry indicates inversion of configuration at C-2 and C-3 (Scheme 50). Furthermore, exchange reaction with L-[2-^{15}N,3-^{13}C]lysine (223) indicates that transfer of the amino group occurs completely intramolecularly, although it is interesting that enzymic reaction with a mixture of L-[3,3-^{2}H]lysine and L-[2-^{13}C]lysine (224) produced mainly monodeuterated β-lysine with an equal distribution ($\pm$ 10%) of deuterium between the C-2-*pro-R* and C-3 positions (223). This result was interpreted as indicating intermolecular exchange of the transferred hydrogen from C-3 to C-2 of a second substrate molecule. Furthermore, the observation that the β-lysine product is monodeuterated (at C-2 or C-3) is indicative of "washout" of the transferred C-3 hydrogen, in contrast to the previously discussed studies of Chirpich *et al.* (220), where little or no exchange of solvent protons into L-lysine or β-lysine was observed. Aberhart *et al.* proposed a mechanism (223, 225) wherein the hydrogen migration was mediated by an unspecified "hydrogen carrier" group on the protein in a manner related to the 1,2 interchange reactions catalyzed by adenosylcobalamin-mediated transformations. Aberhart also speculated that the aziridine intermediate **59** (Scheme 50) might be on the reaction pathway, and

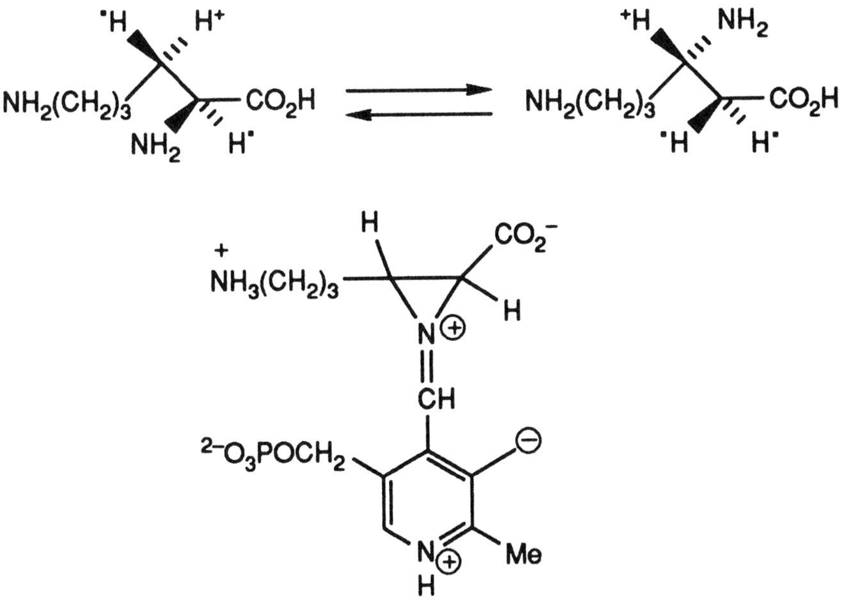

SCHEME 50. Mechanism and stereochemistry of the lysine 2,3-aminomutase reaction proposed by Aberhart *et al.* (*223*).

indicated that such an intermediate requires complete cleavage of the C-3 carbon–hydrogen bond, consistent with his recent determination of an isotope effect of 2.9 ± 0.3 obtained in competition experiments between L-[3,3-^{2}H$_2$]lysine and L-[2-^{2}H]lysine (*226*). There is no isotope effect observed with conversion of the L-[2-^{2}H]lysine, consistent with the idea that lysine C-2 hydrogen bond cleavage does not occur. Similar results and conclusions have been reported by Thiruvengadam *et al.* (*225*) based on their studies on the lysine 2,3-aminomutase from *Streptomyces* L-1689-23.

As alluded to above, the studies of Aberhart and co-workers were further complicated by their observation that the total deuterium content in the products was always less than stoichiometric with the reactant amino acid. Since exchange with the solvent has been ruled out, Moss and Frey approached this problem by investigating the role of SAM in the aminomutase reaction (*221*). Their results suggest that the elusive hydrogen acceptor is SAM. Incubation of L-lysine with S-[2,8,5'-^{3}H]adenosylmethionine catalyzes the interconversion of L-lysine and β-lysine with incorporation of tritium into both isomers. Essentially no tritium is observed in the products of the reaction when S-[2,8-^{3}H]adenosylmethionine or S-adenosyl[*methyl*-^{3}H]methionine is utilized. Furthermore, the distribution of tritium in the lysine isomers is consistent with the reported equilibrium constant for the aminomutase reaction: 84% in the β-lysine and 16% in the L-lysine [K_{eq} = 5.3 for the formation of β-lysine (221)].

The role of the 5'-deoxyadenosyl group of SAM in the hydrogen transfer mechanism of lysine 2,3-aminomutase may be analogous to the related 1,2 exchange reactions mediated by B_{12}-dependent rearrangements involving the 5'-deoxyadenosyl portion of deoxyadenosylcobalamin. Such reactions involve rearrangements of a hydrogen and some other functional group between two adjacent carbons. Furthermore, these B_{12}-mediated reactions require a mechanism consistent with the exchange of a chemically unreactive hydrogen atom. In the usual B_{12} reactions this is accomplished by initial homolytic cleavage of the cobalt–5'-deoxyadenosyl bond to generate the 5'-deoxyadenosyl radical, which then abstracts a hydrogen atom from the substrate, producing a substrate radical intermediate and 5'-deoxyadenosine. Subsequent rearrangement of the substrate radical and reabstraction of hydrogen from 5'-deoxyadenosine yield the products and the 5'-deoxyadenosyl radical.

Based on this precedence and the work of Chirpich et al. (220) as well as that of Aberhart and co-workers (223), Moss and Frey have proposed a related mechanism for lysine 2,3-aminomutase that is based on the generation of a 5'-deoxyadenosyl free radical (Scheme 51, Compounds **60–62**) (221). L-lysine binds to the enzyme pyridoxal phosphate cofactor, forming the aldimine intermediate **60**. The 5'-deoxyadenosyl radical could be generated via homolytic cleavage of the methionine–5'-deoxyadenosyl bond, possibly through the intermediacy of an adenosyl–Fe(II) complex as suggested above for pyruvate formate-lyase activation. The 5'-adenosyl radical abstracts the 3-*pro-R* hydrogen from L-lysine to form a C-3 radical **61,** which cyclizes to the aziridine radical intermediate **62**. This intermediate is particularly attractive for several reasons. The aziridine radical may be stabilized by delocalization through the pyridine ring, thereby substantiating the feasibility of this reaction intermediate. It is also symmetrical, with the exchanging nitrogen bridging C-2 and C-3. The intermediate is therefore well suited for ring opening to form either β-lysine or L-lysine in the reversible enzymic reaction via the respective formation of the C-2 or C-3 radical and the subsequent reabstraction of hydrogen from 5'-deoxyadenosine. This step could also account for the observed "intermolecularity" of hydrogen transfer from C-3 of L-lysine (223). Following H abstraction from substrate by SAM-5'-$CH_2\cdot$, equilibration of the methyl hydrogens would occur via rotation prior to return of hydrogen to C-2 or C-3 of the substrate or product. Rotation of the 5'-deoxyadenosyl methyl group would result at best in a 33% probability of return of the same hydrogen atom, which may be even more biased if deuterium selection effects are considered. Additional verification of this novel mechanism and role of SAM and pyridoxal phosphate awaits studies to confirm the presence of free radical intermediates in the enzymic reaction pathway, the cleavage of

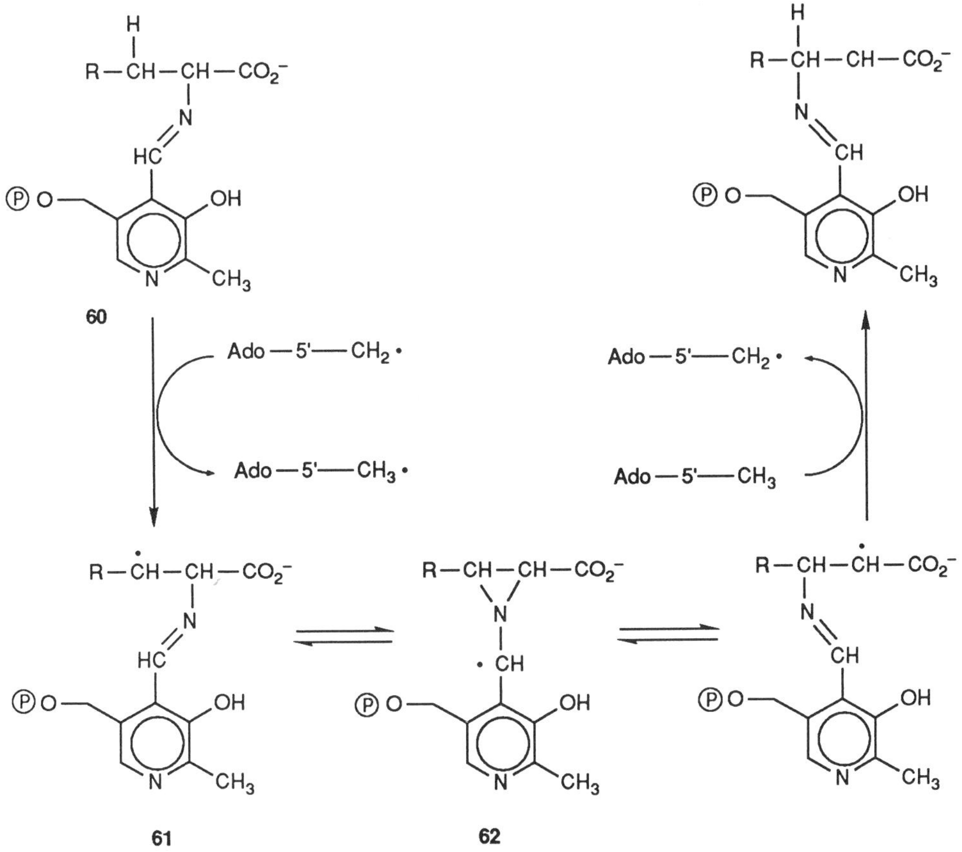

SCHEME 51. Lysine 2,3-aminomutase reaction mechanism proposed by Moss and Frey (*221*).

SAM to the 5'-deoxyadenosyl radical, and the kinetic competence of the intermediates (*221*).

C.   α-HYDROXY-ACID DEHYDRATASES

The reversible dehydration of 2-hydroxy acids to the corresponding alkenes is
a well-known reaction in bacteria (Scheme 52). The interconversion of lactate and
acrylate has been detected in *Clostridium propionicum* (*227*) and *Megasphaera
elsdenii* (*228*); the dehydration of (*R*)-2-hydroxyglutarate to (*E*)-glutaconate by
Acidaminococcus fermentans (*229, 230*) and the conversion of (*R*)-phenyl lactate to (*E*)-cinnamate in *Clostridium sporogenes* (*231*) have also been noted.
This enzymic reaction is of considerable interest to bioorganic chemists regard-

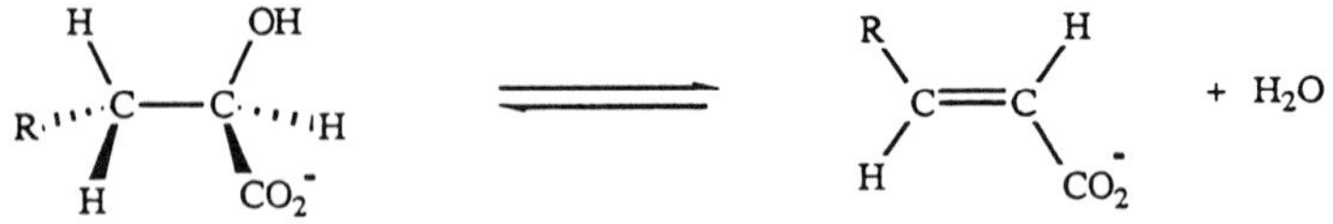

SCHEME 52. General reaction catalyzed by α-hydroxy-acid dehydratase.

ing the mechanism by which an enzyme must activate the nonacidic C-3 hydrogen for removal in the elimination reaction, as an ionic mechanism for the reversible dehydration is unlikely. Furthermore, the hydroxyl group is situated α to a carbonyl and is thus a poor leaving group.

The dehydration of (R)-2-hydroxyglutarate to glutaconate has been studied by Buckel and co-workers ($229, 230$), who have found that the syn elimination of (R)-2-hydroxyglutarate requires a prior activation of the enzyme from cell-free extracts of $A.$ $fermentans$ by ATP, NADH, and $MgCl_2$. The enzymic reaction requires a dithiol, acetyl phosphate, and CoASH, as well as strictly anaerobic conditions, because the active enzyme is irreversibly inactivated on contact with oxygen. Evidence was presented that the adenylation of the protein may be a requisite for activation. Based on preliminary EPR studies and the inactivation of the enzymic reaction by hydroxylamine and uncouplers of oxidative phosphorylation (such as azide, arsenate, and 2,4-dinitrophenol), Schweiger and Buckel postulated a radical mechanism for 2-hydroxyglutarate dehydration via the hydroxyl radical (Scheme 53) (229). Hence, the ATP/NADH-dependent ac-

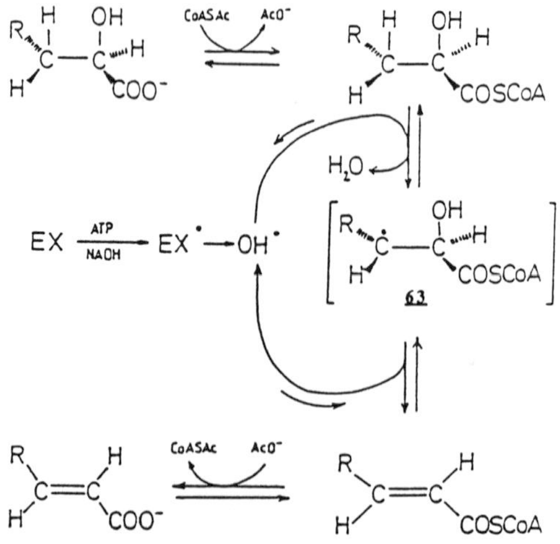

SCHEME 53. Reaction mechanism for 2-hydroxyglutarate dehydration proposed by Schweiger and Buckel (229).

tivation is suggested to give rise to a radical enzyme or a protein-stabilized hydroxyl radical, based on the appearance of a new signal in the EPR spectrum following activation. The substrate, possibly in the form of its CoA thioester, binds with the enzyme and the β-hydrogen is abstracted by an enzyme or hydroxyl radical to form the β-carbon-centered radical of the substrate **63**. Subsequent elimination of hydroxyl radical regenerates the radical form of the enzyme and forms the 2-alkenoic CoA–ester product, which is hydrolyzed by an unspecified means to form the 2-alkenoic acid.

Schweiger and Buckel (*232*) have investigated the conversion of (*R*)-lactate to propionate by cell-free extracts of *C. propionicum*. Very low levels of acrylate were detected [0.5% of the (*R*)-lactate present], which suggests that acrylate is an intermediate in the conversion. This is consistent with their observation of HTO formation utilizing [3-^{2}H]lactate as substrate (*233*). Furthermore, the inhibition of (*R*)-lactate dehydration by hydroxylamine, azide, and 2,4-dinitrophenol is analogous to the inhibition of 2-hydroxyglutarate by these compounds (*229*), and was suggested to be indicative of a radical mechanism (*232*).

Abeles and co-workers extended the work of Buckel on the properties and mechanism of the reductive dehydration of lactate to propionate in *C. propionicum* (*234, 235*). Studies by Abeles concluded that acrylate is indeed an intermediate in lactate dehydration, based on the observation of a primary kinetic isotope effect of 1.8 on the conversion of [3-^{2}H$_3$]lactate to propionate, indicating that β-hydrogen abstraction is at least partially rate limiting. The true substrates for the reaction were determined to be the CoA thioesters of lactate and acrylate, hence the enzyme has been aptly named lactyl-CoA dehydratase (*234*). The reaction was found to be mediated by two proteins, E1 and E2, and like the (*R*)-2-hydroxyglutarate-dehydrating enzyme from *A. fermentans,* both E1 and E2 from *C. propionicum* are oxygen labile. The hydration reaction by the lactyl-CoA dehydratase system requires Mg^{2+} and catalytic amounts ATP or GTP for full activity; ATP could not be replaced by ADP or 5'-adenylylimidodiphosphate, suggesting that dehydratase activation requires ATP hydrolysis. Furthermore, on purification, E2 was found to be a yellow-colored protein containing equal amounts of riboflavin and FMN as organic cofactors (*234*). E1 is metal ion free, but E2 contains 8.2 ± 0.04 mol of Fe/mol E2, 1 mol of which can be removed by 1,10-phenanthroline (*235*). E2 has also been found to contain 7.33 ± 0.68 mol of inorganic sulfur, suggesting the presence of iron–sulfur clusters. The EPR spectrum of E2 indicates the presence of two signals at 4K. Signal 1 is suggested to be an "unusual" [4Fe–4S] cluster ($g_{\perp} = 2.0232$ *and* $g_{\parallel} = 2.0006$) with axial symmetry, and signal 2 is most consistent with a [3Fe–3/4S] cluster ($g_1 = 1.982$, $g_2 = 1.995$, and $g_3 = 2.019$) and is orthorhombic. The observation that signal 2 is changed in the presence of acrylyl- or lactyl-CoA suggests that the substrate binding site is on E2.

Neither E1 nor E2, together or alone, mediated the formation of acrylyl- or lactyl-CoA from the respective acid plus CoA and acetyl phosphate or acetyl-CoA. However, a CoA transferase that catalyzes the formation of the CoA thioesters of lactate, acrylate, and propionate from acetyl-CoA has been isolated from *C. propionicum* (*233*). It is curious that acrylyl-CoA could not be isolated as a product with the crude or purified enzyme system, although acrylyl-CoA is readily converted to lactyl-CoA in the purified system. Abeles suggested that perhaps acrylyl-CoA exists as an enzyme-bound intermediate that requires the presence of a reductase for release as propionate. In this manner the organism is protected from buildup of toxic amounts of any acrylyl intermediates, which are known to undergo Michael addition reactions with biological nucleophiles.

As with the dehydration of 2-hydroxyglutarate, the chemical mechanism of lactyl-CoA dehydration is particularly intriguing, especially with the potential involvement of flavins and Fe–S centers. Abeles has suggested an alternative mechanism. The mechanism of Kuchta and Abeles (Scheme 54) indicates the presence of a protein-bound radical species that may be stabilized by a highly reduced metal ion or flavin. The initial step involves β-hydrogen abstraction, with the C-3 radical combining with the metal ion or flavin, or a direct one-electron transfer to C-3 may occur, leading to the elimination of hydroxide and formation of alkene via an ionic elimination mechanism. The oxidized enzyme is converted back to the reduced form by the initially abstracted hydrogen. Ob-

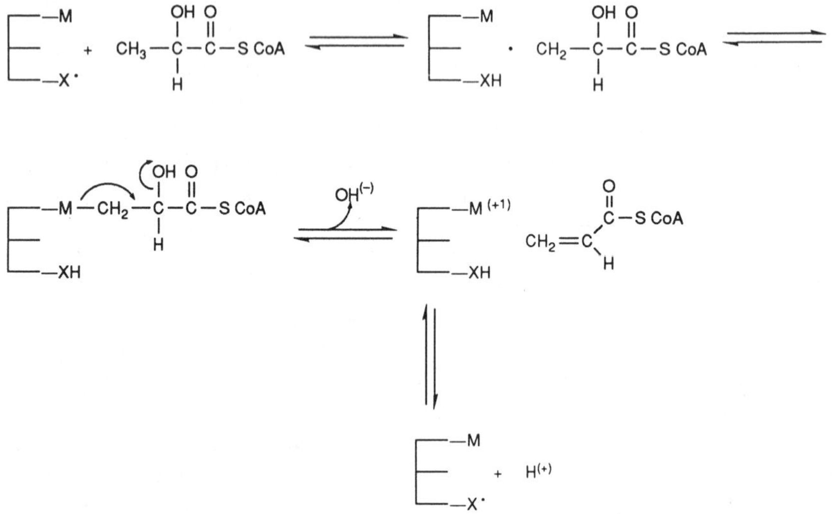

SCHEME 54. Mechanism of lactyl-CoA dehydration proposed by Kuchta and Abeles (*234*).

viously, more work on the details of this mechanism and the redox changes of the protein cofactors is required for further elucidation of this mechanism. Precedence for the mechanism presented in Scheme 54 is found in the metal ion-mediated dehydration of α-(hydroxyl)alkyl radicals to alkene, hydroxide, and one-electron reduced metal ions (236–238).

D. FLAVOPROTEIN PHENOLIC HYDROXYLASES

The mechanism for the hydroxylation of aromatic substrates by flavoprotein monooxygenases has been the subject of significant research interest and controversy over the past decade. These enzymes (p-hydroxybenzoate hydroxylase, phenol hydroxylase, and melilotate hydroxylase) catalyze the initial step in the β-ketoadipic acid pathway, the hydroxylation of substituted phenols into catechols (Scheme 55). Oxygen is required as cosubstrate, which is activated by the reduced FAD cofactor. The complex mechanism for the oxidative half-reaction is thought to consist of at least four steps and three intermediates (239–242) and to involve a controversial 4a,5-ring-opened flavin (242, 249, 250) (Scheme 56). The flavin C4a-hydroperoxy intermediate **64** and flavin C4a-hydroxy intermediate **65** have been assigned the structures shown in Scheme 56 based on the UV absorbance spectra of various model compounds compared with that of the modified enzyme cofactor alkylated at N(5) (243). However, evidence for the intermediacy of various ring-opened flavin species has been tentative at best, as model compounds and model reactions do not support such an intermediate (242).

Recently, Anderson and co-workers have suggested that radical intermediates may be formed in the mechanism of the flavoprotein phenolic hydroxylases (244–246). Their approach has centered on comparing the absorption spectra of inter-

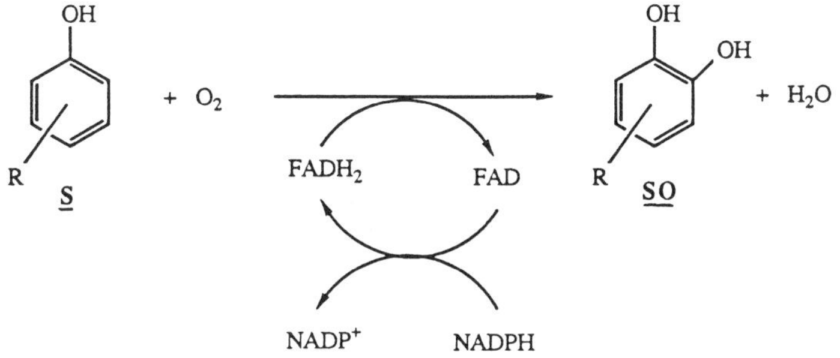

SCHEME 55. General reaction scheme for the hydroxylation of aromatic substrates (S) by flavoprotein monooxygenases to give products (SO).

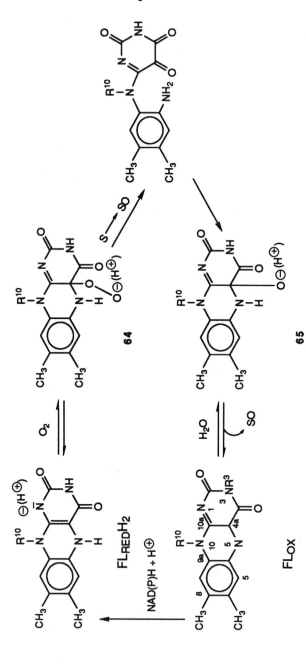

SCHEME 56. Proposed mechanism for hydroxylation of substrate (S) to product (SO) via flavin 4a,5-ring opening (242).

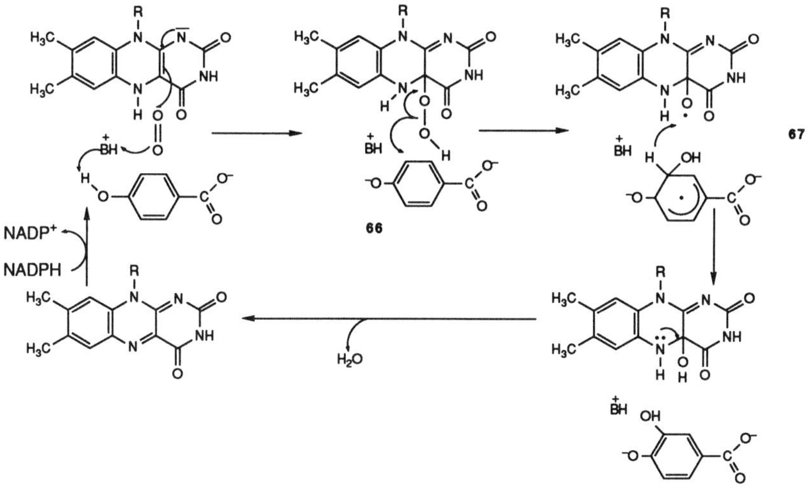

SCHEME 57. Proposed mechanism of Anderson *et al.* involving homolytic scission of the flavin C(4a)-hydroperoxide bond (*244*).

mediate **67** (Scheme 57) formed in the reaction catalyzed by *p*-hydroxybenzoate hydroxylase (*240*) with those of the intermediates produced in the nonenzymic addition of hydroxyl radicals to the 3-position of the enzyme substrates 4-hydroxybenzoic acid, 2,4-dihydroxybenzoic acid, and 4-aminobenzoic acid (*244, 246*). Their results were interpreted as consistent with the transient formation of hydroxyl radical adducts resulting from the interaction of substrates with the flavin C(4a)-hydroperoxy intermediate in the *p*-hydroxybenzoate hydroxylase-catalyzed reaction. Such a reaction requires homolytic scission of the O–O peroxy bond forming the flavin C(4a)-alkoxy radical and the hydroxycyclohexadienyl radical form of the substrate. The proposed radical mechanism is illustrated in Scheme 57 and is similar to the previous mechanistic schemes proposed by Massey and co-workers (*240*). Interaction of oxygen with the reduced flavin yields the flavin C(4a)-hydroperoxy adduct **66,** which undergoes a concerted homolytic O–O bond scission and addition of hydroxyl radical to the C-3 position of the aromatic ring of the substrate, forming the hydroxycyclohexadienyl radical and the flavin C(4a)-alkoxy radical, the suspected elusive intermediate **67** (Scheme 57). The driving force for hydroperoxide homolysis is suggested to be polarization of the O–O bond of the flavin C(4a)-hydroperoxide through dipole moment interactions and hydrogen bonding of the flavin ring with the protein (*244, 245*). Anderson has also indicated that deprotonation of the C-4 hydroxy group of the substrate would further increase the electrophilicity of the C-3 position. The flavin C(4a)-alkoxy radical is suspected to behave chemically as an alkoxyl-type radical instead of a phenoxyl type, given that the isoalloxazine ring of the flavin is in its reduced form. The remaining steps of the mechanism involve reduction of **67** by the substrate hydroxycyclohexadienyl radical and expulsion of water from the flavin, forming product and oxidized flavin **68** (Scheme 57), respectively. Reduction of the enzyme flavin by NADPH completes the catalytic cycle. Suggestions for the pathways of conversion of intermediate **67** to **68** involve (245) (1) abstraction of a hydrogen atom from C-3 of the hydroxycyclohexadienyl radical by the flavin C(4a)-alkoxy radical as shown; (2) electron transfer from the hydroxycyclohexadienyl radical to the flavin alkoxy radical, forming an ionic pair consisting of the cationic form of the product and flavin C(4a)-alkoxide; and (3) β-scission of the tertiary flavin alkoxy radical to form a fragmented flavin ring, which would delocalize the radical prior to electron transfer, and subsequent ring closure via nucleophilic attack. An illustration of these possibilities is shown in Scheme 58.

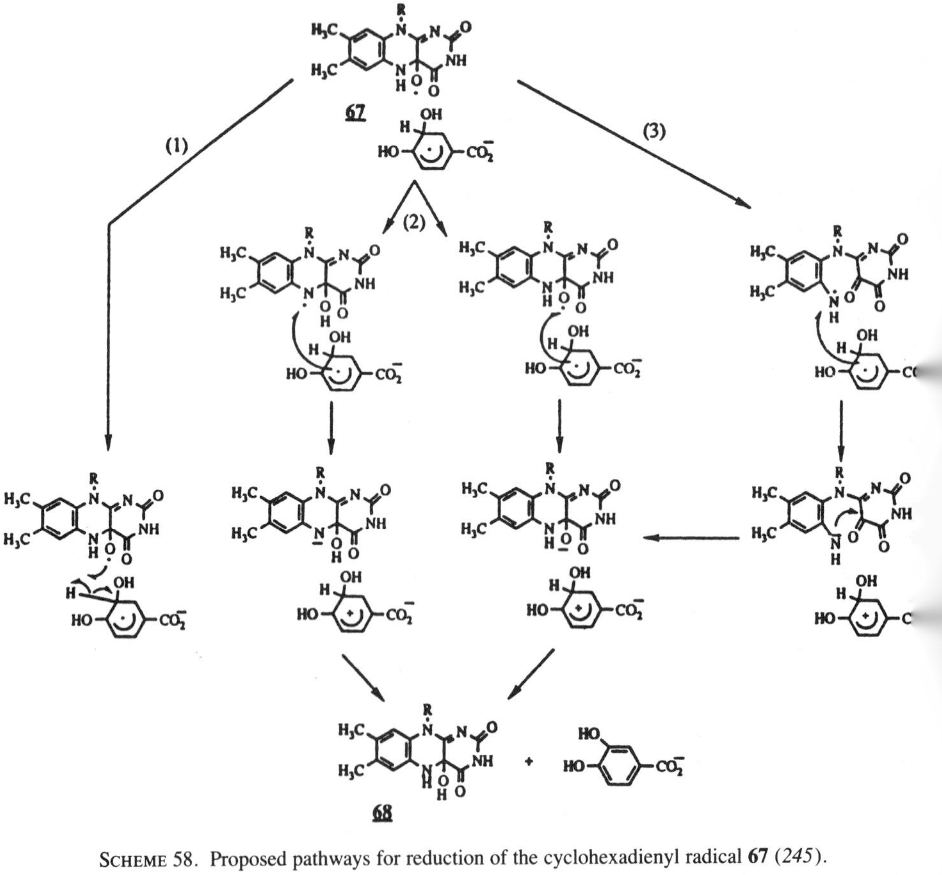

Scheme 58. Proposed pathways for reduction of the cyclohexadienyl radical **67** (245).

Obviously, more work is required to further substantiate the presence of the proposed radical intermediates in the p-hydroxybenzoate hydroxylase reaction, possibly via EPR and spin-trapping studies. Studies by Detmer and Massey (247) on phenol hydroxylase have indicated that the reaction rate constants for the conversion of meta-substituted substrates plotted versus the Hammett s^+ parameters yield a straight line of slope equal to $^-0.5$. This is consistent with the mechanism proposed by Anderson, as the negative slope is expected for an electrophilic aromatic substitution reaction, while the small magnitude of the slope may be indicative of a radical mechanism. Furthermore, recent work by Massey and co-workers on p-hydroxybenzoate hydroxylase utilizing 6-hydroxy-FAD as cofactor and p-aminobenzoate as substrate indicated that the absorption spectrum of intermediate **67** exhibited a satellite band at 440 nm (248). Anderson *et al.* suggest that the satellite band may result from the formation of an aromatic phenoxyl radical at the C-6 position of the isoalloxazine ring of the flavin (244). This species would result from a shift of the initial peroxyl radical center from C(4a) to C-6 via N(5) (245).

REFERENCES

1. Thelander, L., and Reichard, P. (1979). *Annu. Rev. Biochem.* **48**, 133.
2. Stubbe, J. (1989). *Adv. Enzymol.* **63**, 349.
3. Eriksson, S., and Sjoberg, B. M. (1989). *In* "Allosteric Enzymes" (G. Herve, ed.), pp. 189–215. CRC Press, Boca Raton, Florida.
4. Reichard, P., and Ehrenberg, A. (1983). *Science* **221**, 514.
5. Thelander, L. (1974). *J. Biol. Chem.* **249**, 4858.
6. Holmgren, A. (1985). *Annu. Rev. Biochem.* **54**, 237.
7. Blakley, R. L. (1978). *In* "Methods in Enzymology" (P. A. Hoffee and M. E. Jones, eds.), Vol. 51. p. 246. Academic Press, New York.
8. Ashley, G., and Stubbe, J. (1987). *Pharmacol. Ther.* **30**, 301.
9. Thelander, L., Larrson, A., Hobbs, J., and Ecstein, F. (1976). *J. Biol. Chem.* **251**, 1398.
10. Atkin, C. L., Thelander, L., Reichard, P., and Lang, G. (1973). *J. Biol. Chem.* **248**, 7464.
11. Petersson, L., Graslund, A., Ehrenberg, A., Sjoberg, B. M., and Reichard, P. (1980). *J. Biol. Chem.* **255**, 6706.
12. Sjoberg, B. M., Sanders-Loehr, J., and Loehr, T. M. (1987). *Biochemistry* **26**, 4242.
13. Sjoberg, B. M., Loehr, T. M., and Sanders-Loehr, J. (1982). *Biochemistry* **21**, 96.
14. Sjoberg, B. M., Reichard, P., Graslund, A., and Ehrenberg, A. (1977). *J. Biol. Chem.* **252**, 536.
15. Sjoberg, B. M., Reichard, P., Graslund, A., and Ehrenberg, A. (1978). *J. Biol. Chem.* **253**, 6863.
16. Sahlin, M., Graslund, A., Ehrenberg, A., and Sjoberg, B. M. (1982). *J. Biol. Chem.* **257**, 366.
17. Graslund, A., Ehrenberg, A., and Thelander, L. (1982). *J. Biol. Chem.* **257**, 5711.
18. Lankinen, H., Graslund, A., and Thelander, L. (1982). *J. Virol.* **41**, 893.
19. Akerblom, L., Ehrenberg, A., Graslund, A., Lankinen, H., Reichard, P., and Thelander, L. (1981). *Proc. Natl. Acad. Sci. U.S.A.* **78**, 2159.
20. Schepler, K. L., Dunham, W. R., Sands, R. H., Fee, J. A., and Abeles, R. H. (1975). *Biochim. Biophys. Acta* **397**, 510.
21. Box, H. C., Budzinski, E. E., and Freund, H. G. (1974). *J. Chem. Phys.* **61**, 2222.

22. Nordlund, P., Sjoberg, B. M., and Eklund, H. (1990). *Nature (London)* **345**, 593.
23. Sahlin, M., Graslund, A., Petersson, L., Ehrenberg, A., and Sjoberg, B. M. (1989). *Biochemistry* **28**, 2618.
24. Ehrenberg, A., and Reichard, P. (1972). *J. Biol. Chem.* **247**, 3485.
25. Petersson, L., Graslund, A., Ehrenberg, A., Sjoberg, B. M., and Reichard, P. (1980). *J. Biol. Chem.* **255**, 6706.
26. Barlow, T., Eliasson, R., Platz, A., Reichard, P., and Sjoberg, B. M. (1983). *Proc. Natl. Acad. Sci. U.S.A.* **80**, 1492.
27. Sjoberg, B. M. (1986). *In* "Thioredoxin and Glutaredoxin Systems: Structure and Function" (A. Holmgren, C. I. Branden, H. Jornvall, and B. M. Sjoberg, eds.), p. 199. Raven, New York.
28. Eriksson, S., Sjoberg, B. M., Hahne, S., and Karlstrom, O. (1977). *J. Biol. Chem.* **252**, 6132.
29. von Dobeln, U., and Reichard, P. (1976). *J. Biol. Chem.* **251**, 3616.
30. Thelander, L. (1973). *J. Biol. Chem.* **248**, 4591.
31. Blakley, R. L. (1978). *In* "Methods in Enzymology" (P. A. Hoffee and M. E. Jones, eds.), Vol. 51, p. 246. Academic Press, New York.
32. Harris, G., Ator, M., and Stubbe, J. (1984). *Biochemistry* **23**, 5214.
33. Tamao, Y., and Blakley, R. L. (1973). *Biochemistry* **12**, 24.
34. Orme-Johnson, W. H., Beinert, H., and Blakley, R. L. (1974). *J. Biol. Chem.* **249**, 2338.
35. Stubbe, J. (1983). *Mol. Cell. Biochem.* **50**, 25.
36. Hamilton, J. A., Tamao, Y., Blakley, R. L., and Coffman, R. E. (1972). *Biochemistry* **11**, 4696.
37. Ashley, G., and Stubbe, J. (1987). *Pharmacol. Ther.* **30**, 301.
38. Stubbe, J. (1988). *Biochemistry* **27**, 3893.
39. Reichard, P., and Ehrenberg, A. (1983). *Science* **221**, 514.
40. Robbins, M. J., and Hawrelak, S. D. (1978). *Tetrahedron Lett.* p. 3653.
41. Robins, M. J., Sporns, P., and Muhs, W. H. (1979). *Can. J. Chem.* **57**, 274.
42. Robins, M. J., and Wilson, J. S. (1981). *J. Am. Chem. Soc.* **103**, 932.
43. Gilbert, B. C., Larkin, J. P., and Norman, R. E. C. (1972). *J. Chem. Soc., Perkin Trans. 2* p. 794.
44. Walling, C., and Johnson, R. A. (1975). *J. Am. Chem. Soc.* **97**, 2405.
45. Stubbe, J., and Ackles, D. (1980). *J. Biol. Chem.* **255**, 8027.
46. Stubbe, J., Ator, M., and Krenitsky, T. (1983). *J. Biol. Chem.* **258**, 1625.
47. Stubbe, J., Ackles, D., Segal, R., and Blakley, R. L. (1981). *J. Biol. Chem.* **256**, 4843.
48. Melander, L., and Saunders, W. H. (1980). "Reaction Rates of Isotopic Molecules," p. 174. Wiley, New York.
49. Ashley, G. W., Harris, G., and Stubbe, J. (1986). *J. Biol. Chem.* **261**, 3958.
50. Stubbe, J., and Kozarich, J. W. (1980). *J. Biol. Chem.* **255**, 5511.
51. Stubbe, J., and Kozarich, J. W. (1980). *J. Am. Chem. Soc.* **102**, 2505.
52. Ator, M. A., and Stubbe, J. (1985). *Biochemistry* **24**, 7214.
53. Ashley, G. W., Harris, G., and Stubbe, J. (1988). *Biochemistry* **27**, 4305.
54. Pfizner, K. E., and Moffatt, J. G. (1965). *J. Am. Chem. Soc.* **87**, 5661.
55. Hershfield, M. S. (1979). *J. Biol. Chem.* **254**, 22.
56. Ashley, G. W., Harris, G., and Stubbe, J. (1988). *Biochemistry* **27**, 7841.
57. Chen, A. K., Bhan, A., Hopper, S., Abrams, R., and Franzen, J. S. (1974). *Biochemistry* **13**, 654.
58. Harris, G., Ashley, G. W., Robins, M. J., Tolman, R. L., and Stubbe, J. (1987). *Biochemistry* **26**, 1895.
59. Lin, A., Ashley, G. W., and Stubbe, J. (1987). *Biochemistry* **26**, 6905.
60. Sjoberg, B. M., Graslund, A., and Eckstein, F. (1983). *J. Biol. Chem.* **258**, 8060.
61. Sjoberg, B. M., and Graslund, A. (1983). *Adv. Inorg. Biochem.* **5**, 87.

62. Ator, M., Salowe, S. P., Stubbe, J. A., Emptage, M. H., and Robins, M. J. (1984). *J. Am. Chem. Soc.* **106**, 1886.
63. Salowe, S. P., Ator, M. A., and Stubbe, J. (1987). *Biochemistry* **26**, 3408.
64. Roberts, B. P., and Winter, J. N. (1979). *J. Chem. Soc., Perkin Trans.* 2 p. 1353.
65. Staros, J. V., Bayley, H., Standring, D. N., and Knowles, J. R. (1978). *Biochem. Biophys. Res. Commun.* **80**, 568.
66. Gilbert, B. C., Larkin, J. P., and Norman, R. O. C. (1972). *J. Chem. Soc., Perkin Trans.* 2 p. 794.
67. Ables, R. H., and Beck, W. S. (1967). *J. Biol. Chem.* **242**, 3589.
68. Hogenkamp, H. P. C., Ghambeer, R. K., Brownson, C., Blakley, R. L., and Vitols, E. (1968). *J. Biol. Chem.* **243**, 799.
69. Blakley, R. L., Ghambeer, R. K., Batterham, T. J., and Brownson, C. (1966). *Biochem. Biophys. Res. Commun.* **24**, 418.
70. Abeles, R. H., and Dolphin, P. (1976). *Acc. Chem. Res.* **9**, 114.
71. Stubbe, J., Smith, G., and Blakely, R. L. (1983). *J. Biol. Chem.* **258**, 1619.
72. Halpern, J., Kim, S. H., and Leung, T. W. (1984). *J. Am. Chem. Soc.* **106**, 8317.
73. Babior, B. M., and Krouwer, J. S. (1979). *Crit. Rev. Biochem.* **6**, 35.
74. McEwen, C. M., Sasaki, G., and Jones, D. C. (1969) *Biochemistry* **8**, 3952.
75. Oi, S., and Yasunobu, K. T. (1973). *Biochem. Biophys. Res. Commun.* **52**, 631.
76. Yasunobu, K. T., and Oi, S. (1972). *Adv. Biochem. Psychopharmacol.* **5**, 91.
77. Houslay, M. D., and Tipton, K. F. (1973). *Biochem. J.* **135**, 735.
78. Kapeller-Adler, R. (1970). "Amine Oxidases and Methods for Their Study," p. 28. Wiley (Interscience), New York.
79. Squires, R. F. (1972). *Adv. Biochem. Psychopharmacol.* **5**, 355.
80. Neff, N. H., and Goridis, C. (1972). *Adv. Biochem. Psychopharmacol.* **5**, 307.
81. Sandler, M., and Youdin, M. B. H. (1972). *Pharmacol. Rev.* **24**, 331.
82. Johnston, J. P. (1968). *Biochem. Pharmacol.* **17**, 1285.
83. Silverman, R. B., and Hiebert, C. K. (1988). *Biochemistry* **27**, 8448.
84. Neff, N. H., and Yang, H. Y. (1974). *Life Sci.* **14**, 2061.
85. Knoll, J., and Magyar, K. (1972). *Adv. Biochem. Psychopharmacol.* **5**, 393.
86. Zreika, M., McDonald, I. A., Bey, P., and Palfreyman, M. G. (1984). *J. Neurochem.* **43**, 448.
87. Selikoff, I. J., Robitzek, E. H., and Ornstein, G. G. (1952). *Q. Bull. Sea View Hosp.* **13**, 17.
88. Ho, B. T. (1972). *J. Pharm. Sci.* **61**, 821.
89. Murphy, D. L., Garrick, N. A., Aulakh, C. S., and Cohen, R. M. (1984). *J. Clin. Psychiatry* **45**, 37.
90. Walsh, C. (1982). *Tetrahedron* **38**, 871.
91. Hellerman, L., and Erwin, V. G. (1968). *J. Biol. Chem.* **243**, 5234.
92. Chuang, H. Y. K., Patek, D. R., and Hellerman, L. (1974). *J. Biol. Chem.* **249**, 2381.
93. Maycock, A. L., Abeles, R. H., Salach, J. I., and Singer, T. P. (1976). *Biochemistry* **15**, 114.
94. Bruice, T. C. (1975). *Prog. Bioorg. Chem.* **4**, 1.
95. Silverman, R. B., Hoffman, S. J., and Catus, W. B. (1980). *J. Amer. Chem. Soc.* **102**, 7126.
96. Mann, C. K., and Barnes, K. K. (1970). "Electrochemical Reactions in Non-Aqueous Systems," Chap. 9. Dekker, New York.
97. Masui, M., and Sayo, H. (1971). *J. Chem. Soc. B* p. 1593.
98. Lindsay Smith, J. R., and Masheder, D. J. (1976). *J. Chem. Soc., Perkin Trans.* 2 p. 47.
99. Silverman, R. B., and Hoffman, S. J. (1980). *J. Am. Chem. Soc.* **102**, 884.
100. Silverman, R. B., and Zieske, P. A. (1985). *Biochemistry* **24**, 2128.
101. Silverman, R. B. (1984). *Biochemistry* **23**, 5206.
102. Yamasaki, R. B., and Silverman, R. B. (1985). *Biochemistry* **24**, 6543.
103. Vazquez, M. L., and Silverman, R. B. (1985). *Biochemistry* **24**, 6538.

104. Silverman, R. B., and Yamasaki, R. B. (1984). *Biochemistry* **23**, 1322.
105. Silverman, R. B., and Hoffman, S. J. (1981). *Biochem. Biophys. Res. Commun.* **101**, 1396.
106. Jones, J. R. (1973). "The Ionization of Carbon Acids," Academic Press, New York.
107. Paech, C., Salach, J. I., and Singer, T. P. (1980). *J. Biol. Chem.* **255**, 2700.
108. Goodman, L. S., and Gilman, A. (1975). "The Pharmacological Basis of Therapeutics," 5th Ed., P. 180. Macmillan, New York.
109. Silverman, R. B. (1983). *J. Biol. Chem.* **258**, 14766.
110. Silverman, R. B., and Zieske, P. A. (1986). *Biochem. Biophys. Res. Commun.* **135**, 154.
111. Pettit, G. R., and Van Tamelen, E. E. (1962). *Org. React. (N.Y.)* **12**, 356.
112. Silverman, R. B., and Zieske, P. A. (1985). *J. Med. Chem.* **28**, 1953.
113. Silverman, R. B., and Zieske, P. A. (1986). *Biochemistry* **25**, 341.
114. Silverman, R. B., and Banik, G. M. (1987). *J. Am. Chem. Soc.* **109**, 2219.
115. Ohga, K., Yoon, U. C., and Mariano, P. S. (1984). *J. Org. Chem.* **49**, 213.
116. Lan, A. J. Y., Quillen, S. L., Heuckeroth, R. O., and Mariano, P. S. (1984). *J. Am. Chem. Soc.* **106**, 6439.
117. Brooke, A. G. (1974). *Acc. Chem. Res.* **7**, 77.
118. Calas, R., and Valade, J. (1958). *Bull. Soc. Chim. Fr.* p. 919.
119. Sommer, L. H., van Strien, R. E., and Whitmore, F. C. (1949). *J. Am. Chem. Soc.* **71**, 3056.
120. Yelekci, K., Lu, X., and Silverman, R. B. (1989). *J. Am. Chem. Soc.* **111**, 1138.
121. Viehe, H. G., Merenyi, R., Stella, L., and Janousek, Z. (1979). *Angew. Chem., Int. Ed. Engl.* **18**, 917.
122. Egashira, T., Ekstedt, B., and Oreland, L. (1976). *Biochem. Pharmacol.* **25**, 2583.
123. Krantz, A., and Lipkowitz, G. S. (1977). *J. Am. Chem. Soc.* **99**, 4156.
124. McEwen, C. M., Jr., Sasaki, G., and Jones, D. C. (1969). *Biochemistry* **8**, 3963.
125. Silverman, R. B., Hiebert, C. K., and Vazquez, M. L. (1985). *J. Biol. Chem.* **260**, 14648.
126. Hariharan, P. V., and Cerutti, P. A. (1977). *Biochemistry* **16**, 2791.
127. Demple, B., and Linn, S. (1980). *Nature (London)* **287**, 203.
128. Dulbecco, R. (1950). *J. Bacteriol.* **59**, 329.
129. Saito, W., and Werbin, H. (1970). *Biochemistry* **9**, 2610.
130. Rupert, C. S., and To, K. (1976). *Photochem. Photobiol.* **24**, 229.
131. Setlow, J. K. (1966). *Curr. Top. Radiat. Res.* **2**, 195.
132. Jagger, J., Stafford, R. S., and Snow, J. M. (1969). *Photochem. Photobiol.* **10**, 383.
133. Sutherland, B. M., Runge, P., and Sutherland, J. C. (1974). *Biochemistry* **13**, 4710.
134. Iwatsuki, N., Joe, C. O., and Werbin, H. (1980). *Biochemistry* **19**, 1172.
135. Sancar, A., Smith, F., and Sancar, G. (1984). *J. Biol. Chem.* **259**, 6028.
136. Jorns, M., Sancar, G., and Sancar, A. (1984). *Biochemistry* **23**, 2673.
137. Eker, A. P. M., Dekker, R. H., and Berends, W. (1981). *Photochem. Photobiol.* **33**, 65.
138. Eker, A., Pol, A., VanderMyer, P., and Vogels, G. (1980 *FEMS Microbiol. Lett.* **8**, 161.
139. Sutherland, B. M., Chamberlin, M. J., and Sutherland, J. C. (1973). *J. Biol. Chem.* **248**, 4200.
140. Snapka, R. M., and Sutherland, B. M. (1980). *Biochemistry* **19**, 4201.
141. Sancar, G. B., Smith, F. W., Reid, R., Payne, G., Levy, M., and Sancar, A. (1987). *J. Biol. Chem.* **262**, 478.
142. Sancar, G. B., Smith, F. W., and Sancar, A. (1985). *Biochemistry* **24**, 1849.
143. Sancar, A., and Sancar, G. B. (1984). *J. Mol. Biol.* **172**, 223.
144. Wang, B., Jordon, S. P., and Jorns, M. S. (1988). *Biochemistry* **27**, 4222.
145. Johnson, J. L., Hamm-Alvarez, S., Payne, G., Sancar, G. B., Rajagoplan, K. V., and Sancar, A. (1988). *Proc. Natl. Acad. Sci. U.S.A.* **85**, 2046.
146. Jorns, M. S., Sancar, G. B., and Sancar, A. (1984). *Biochemistry* **23**, 2673.
147. Sancar, G. B., Smith, F. W., and Sancar, A. (1983). *Nucleic Acids Res.* **11**, 667.

148. Sancar, A., Smith, F. W., and Sancar, G. B. (1984). *J. Biol. Chem.* **259**, 6028.
149. Rupert, C. S., Goodgal, S. H., and Herriott, R. M. (1958). *J. Gen. Physiol.* **41**, 451.
150. Harm, W. (1970). *Mutat. Res.* **10**, 227.
151. Sancar, G. B., Smith, F. W., Lorence, M. C., Rupert, C. S., and Sancar, A. (1984). *J. Biol. Chem.* **259**, 6033.
152. Pac, C., Kubo, J., Majima, T., and Sukurai, H. (1982). *Photochem. Photobiol.* **36**, 273.
153. Heelis, P. F., Payne, G., and Sancar, A. (1987). *Biochemistry* **26**, 4634.
154. Jordan, S. P., and Jorns, M. S. (1988). *Biochemistry* **27**, 8915.
155. Jorns, M. S., Baldwin, E. T., Sancar, G. B., and Sancar, A. (1987). *J. Biol. Chem.* **262**, 486.
156. Payne, G., Heelis, B. R. R., and Sancar, A. (1987). *Biochemistry* **26**, 7121.
157. Jorns, M. S., Wang, B., and Jordon, S. P. (1987). *Biochemistry* **26**, 6810.
158. Rosenthal, I., Rao, M. M., and Salomon, J. (1975). *Biochim. Biophys. Acta* **378**, 165.
159. Ben-Hur, E., and Rosenthal, I. (1970). *Photochem. Photobiol.* **11**, 163.
160. Rokita, S. E., and Walsh, C. T. (1984). *J. Am. Chem. Soc.* **106**, 4589.
161. Walsh, C. (1986). *Acc. Chem. Res.* **19**, 216.
162. Jorns, M. S. (1987). *J. Am. Chem. Soc.* **109**, 3133.
163. Sancar, G. B., Jorns, M. S., Payne, G., Fluke, D. J., Rupert, C. S., and Sancar, A. (1987). *J. Biol. Chem.* **262**, 492.
164. Helene, C., Charlier, M., Toulme, J. J., and Tolume, F. (1978). "DNA Repair Mechanisms," p. 141. Academic Press, New York.
165. Van Camp, J. R., Young, T., Hartman, R. F., and Rose, S. D. (1987). *Photochem. Photobiol.* **45**, 365.
166. Eker, A. P. M., Hessels, J. K. C., and Dekker, R. H. (1986). *Photochem. Photobiol.* **44**, 197.
167. Eker, A. P. M., Hessels, J. K. C., and van de Velde, J. (1988). *Biochemistry* **27**, 1758.
168. Cochran, A. G., Sugasawara, R., and Schultz, P. G. (1988). *J. Am. Chem. Soc.* **110**, 7888.
169. Kalnitsky, G., and Werkman, C. H. (1943). *Arch. Biochem.* **2**, 113.
170. Utter, M. F., and Werkman, C. H. (1944). *Arch. Biochem.* **5**, 413.
171. Chantrenne, H., and Lipmann, F. (1950). *J. Biol. Chem.* **187**, 757.
172. Stadtman, E. R., Novelli, G. D., and Lipmann, F. (1951). *J. Biol. Chem.* **191**, 365.
173. Knappe, J., Schacht, J., Mockel, W., Hopner, T., Vetter, H., and Edenharder, R. (1969). *Eur. J. Biochem.* **11**, 316.
174. Knappe, J., Blaschkowski, H. P., Grobner, P., and Schmitt, T. (1974). *Eur. J. Biochem.* **50**, 253.
175. Blaschkowski, H. P., Neuer, G., Ludwig-Festl, M., and Knappe, J. (1982). *Eur. J. Biochem.* **123**, 563.
176. Knappe, J., Bohnert, E., and Brummer, W. (1965). *Biochim. Biophys. Acta* **107**, 603.
177. Chase, T., Jr., and Rabinowitz, J. C. (1968). *J. Bacteriol.* **96**, 1065.
178. Conradt, H., Hohmann-Berger, M., Hohmann, H. P., Blaschkowski, H. P., and Knappe, J. (1984). *Arch. Biochem. Biophys.* **228**, 133.
179. Knappe, J., and Schmitt, T. (1976). *Biochem. Biophys. Res. Commun.* **71**, 1110.
180. Knappe, J., Neugebauer, F. A., Blaschkowski, H. P., and Glazler, M. (1984). *Proc. Natl. Acad. Sci. U.S.A.* **81**, 1332.
181. Knappe, J. (1987). *In* "*Escherichia coli* and *Salmonella typhimurium*—Cellular and Molecular Biology" (F. C. Neidhardt, editor-in-chief), Vol. 1, p. 151. American Society for Microbiology, Washington, D. C.
182. Pecher, A., Blaschkowski, H. P., Knappe, J., and Bock, A. (1982). *Arch. Microbiol.* **132**, 365.
183. Utter, M. F., Lipmann, F., and Werkman, C. H. (1945). *J. Biol. Chem.* **158**, 521.
184. Rodel, W., Plaga, W., Frank, R., and Knappe, J. (1988). *Eur. J. Biochem.* **177**, 153.
185. Plaga, W., Frank, R., and Knappe, J. (1988). *Eur. J. Biochem.* **178**, 445.
186. Novelli, G. D. (1955). *Biochim. Biophys. Acta* **18**, 594.
187. Brush, E. J., Lipsett, K. A., and Kozarich, J. W. (1988). *Biochemistry* **27**, 2217.

188. Deleted in proof.
189. Gilbert, B. C., Larkin, J. P., Norman, R. O. C., and Storey, P. M. (1972). *J. Chem. Soc., Perkin Trans. 2* p. 1508.
190. Morton, J. R. (1962). *J. Mol. Phys.* **5,** 217.
191. Kornblum, N., Cooper, G. D., and Taylor, J. E. (1950). *J. Am. Chem. Soc.* **72,** 3013.
192. Dean, J. A. (1987). "Handbook of Organic Chemsitry." McGraw-Hill, New York.
193. Norman, R. O. C., and West, P. R. (1969). *J. Chem. Soc. B* p. 389.
194. Beckwith, A. L. J., and Norman, R. O. C. (1969). *J. Chem. Soc. B* p. 400.
195. Anderson, N. H., Dobbs, A. J., Edge, D. J., Norman, R. O. C., and West, P. R. (1971). *J. Chem. Soc. B* p. 1004.
196. Bernardi, R., Caronna, T., Galli, R., Minisci, F., and Perchinunno, M. (1973). *Tetrahedron Lett.* p. 645.
197. Heinisch, G., and Lotsch, G. (1985). *Angew. Chem., Int. Ed. Engl.* **24,** 692.
198. Heinisch, G., and Lotsch, G. (1985). *Tetrahedron* **41,** 1199.
199. Stock, B. H., Schreiber, J., Guenat, C., Mason, R. P., Bend, J. H., and Eling, T. E. (1986). *J. Biol. Chem.* **261,** 15915.
200. Docampo, R., Moreno, S. N. J., and Mason, R. P. (1987). *J. Biol. Chem.* **262,** 12417.
201. Unkrig, V., Knappe, J., and Neugebauer, F. A. (1988). *In* "Organic Free Radicals" (H. Fischer, and H. Heimgartner, eds.), p. 141. Springer-Verlag, Berlin, Heidelberg.
202. Shine, H. J., Bandlish, B. K., Mani, S. R., and Padilla, A. G. (1979). *J. Org. Chem.* **44,** 915.
203. Lipsett, K. A., Brush, E. J., Ulissi, L. A., and Kozarich, J. W. (1992). Manuscript in preparation.
204. Oae, S. (1975). "Organic Chemistry of Sulfur." Plenum.
205. Freidlina, R. K., and Terent'ev, A. B. (1980). *In* "Advances in Free Radical Chemistry" (G. H. Williams, ed.), Vol. 6, p. 1.
206. Komives, E. A., and Ortiz de Montellano, P. R. (1987). *J. Biol. Chem.* **262,** 9793.
207. Baillie, A. C., Wright, K., Wright, B. J., and Earnshaw, C. G. (1988). *Pestic. Biochem. Physiol.* **30,** 103.
208. Ulissi-DeMario, L. A., Brush, E. J., and Kozarich, J. W. (1991). *J. Am. Chem. Soc.* **113,** 4341.
209. Cammack, R., Kerscher, L., and Oesterhelt, D. (1980). *FEBS Lett.* **118,** 271.
210. Kerscher, L., Nowitzki, S., and Oesterhelt, D. (1982). *Eur. J. Biochem.* **128,** 223.
211. Uyeda, K., and Rabinowitz, J. C. (1971). *J. Biol. Chem.* **246,** 3120.
212. Kerscher, L., and Oesterhelt, D. (1981). *Eur. J. Biochem.* **116,** 587.
213. Kerscher, L., and Oesterhelt, D. (1981). *Eur. J. Biochem.* **116,** 595.
214. Wahl, R. C., and Orme-Johnson, W. H. (1987). *J. Biol. Chem.* **262,** 10489.
215. Uyeda, K., and Rabinowitz, J. C. (1971). *J. Biol. Chem.* **246,** 3111.
216. Lindmark, D. G., Paolella, P., and Wood, N. P. (1969). *J. Biol. Chem.* **244,** 3605.
217. Docampo, R., Moreno, S. N. J., and Mason, R. P. (1987). *J. Biol. Chem.* **262,** 12417.
218. Stadtman, T. C. (1970). *Adv. Enzymol. Relat. Areas Mol. Biol.* **38,** 413.
219. Ohsugi, M., Kahn, J., Hensley, C., Chew, S., and Barker, H. A. (1981). *J. Biol. Chem.* **256,** 7642.
220. Chirpich, T. P., Zappia, V., Costilow, R. N., and Barker, H. A. (1970). *J. Biol. Chem.* **245,** 1778.
221. Moss, M., and Frey, P. A. (1987). *J. Biol. Chem.* **262,** 14859.
222. Poston, J. M. (1980). *J. Biol. Chem.* **255,** 10067.
223. Aberhart, D. J., Gould, S. J., Lin, H. J., Thiruvengadam, T. K., and Weiller, B. H. (1983). *J. Am. Chem. Soc.* **105,** 5461.
224. Aberhart, D. J., and Cotting, J. A. (1988). *J. Chem. Soc., Perkin Trans. 1* p. 2119.

225. Thiruvengadam, T. K., Gould, S. J., Aberhart, D. J., and Lin, H. J. (1983). *J. Am. Chem. Soc.* **105**, 5470.
226. Aberhart, D. J. (1988). *J. Chem. Soc., Perkin Trans. 1* p. 343.
227. Cardon, B. P., and Barker, H. A. (1947). *Arch. Biochem. Biophys.* **12**, 165.
228. Ladd, J. N., and Walker, D. J. (1959). *Biochem. J.* **71**, 364.
229. Schweiger, G., and Buckel, W. (1984). *Arch. Microbiol.* **137**, 302.
230. Buckel, W. (1980). *Eur. J. Biochem.* **106**, 439.
231. Pitsch, C., and Simon, H. (1982). *Hoppe-Seyler's Z. Physiol. Chem.* **363**, 1253.
232. Schweiger, G., and Buckel, W. (1985). *FEBS Lett.* **185**, 253.
233. Schweiger, G., and Buckel, W. (1984). *FEBS Lett.* **171**, 79.
234. Kuchta, R. D., and Abeles, R. H. (1985). *J. Biol. Chem.* **260**, 13181.
235. Kuchta, R. D., Hanson, G. R., Holmquist, B., and Abeles, R. H. (1986). *Biochemistry* **25**, 7301.
236. Cohen, H., Meyerstein, D., Shusterman, A. J., and Weiss, M. (1984). *J. Am. Chem. Soc.* **106**, 1876.
237. Ryan, D. A., and Esperson, J. H. (1982). *Inorg. Chem.* **21**, 527.
238. Sorek, Y., Cohen, H., Mulac, W. A., Schmidt, K. H., and Meyerstein, D. (1983). *Inorg. Chem.* **22**, 3040.
239. Entsch, B., Massey, V., and Ballou, D. P. (1974). *Biochem. Biophys. Res. Commun.* **57**, 1018.
240. Entsch, B., Ballou, D. P., and Massey, V. (1976). *J. Biol. Chem.* **251**, 2550.
241. Masey, V., Claiborne, A., Detmer, K., and Schoper, L. M. (1982). *In* "Oxygenases and Oxygen Metabolism" (M. Nozaki, S. Yamamoto, Y. Ishimura, M. J. Coon, L. Ernster, and R. W. Eastabrook, eds.), p. 185. Academic Press, New York.
242. Wessiak, A., Schopfer, L. M., and Massey, V. (1984). *J. Biol. Chem.* **259**, 12547.
243. Ghisla, S., Entsch, B., Massey, V., and Husain, M. (1977). *Eur. J. Biochem.* **76**, 139.
244. Anderson, R. F., Patel, K. B., and Stratford, M. R. L. (1987). *J. Biol. Chem.* **262**, 17475.
245. Anderson, R. F. (1988). *In* "Oxidases and Related Redox Systems" (T. E. King, H. S. Mason, and M. Morrison, eds.), p. 167. Alan R. Liss, New York.
246. Anderson, R. F., Patel, K. B., and Stratford, M. R. L. (1987). *J. Chem. Soc. Faraday Trans. 1* **83**, 3177.
247. Detmer, K., and Massey, V. (1985). *J. Biol. Chem.* **260**, 5998.
248. Entsch, B., Massey, V., and Claiborne, A. (1987). *J. Biol. Chem.* **262**, 6060.
249. Hamilton, G. A. (1971). *Prog. Bioorg. Chem.* **1**, 83.
250. Wessiak, A., Noar, B., and Bruice, T. C. (1984). *Proc. Natl. Acad. Sci. U.S.A.* **81**, 332.

9

Molecular Mechanism of Oxygen Activation by P-450

YOSHIHITO WATANABE* · JOHN T. GROVES[†]

Division of Molecular Engineering
Graduate School of Engineering
Kyoto University
Kyoto 606, Japan

[†]*Department of Chemistry*
Princeton University
Princeton, New Jersey 08544

THE ENZYMES, Vol. XX

I. Introduction

The P-450 enzyme comprises a unique class of hemoprotein that has proto-porphyrin IX at the active site as a prothetic group (Fig. 1), an unusual cysteine thiolate ligand to the heme iron, and a characteristic absorption band at 450 nm. The reduced form of P-450 readily binds CO to form a CO complex ($1a$–d). In 1962, Omura and Sato showed P-450 to be a hemoprotein based on the ethyl isocyanate difference spectrum of the reduced pigment (2). At the same time, Mason and colleagues also concluded that P-450 was a low-spin ferric hemoprotein by EPR measurements (3).

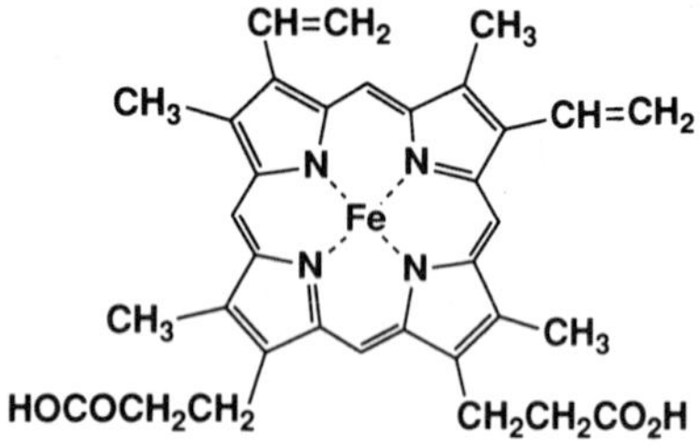

FIG. 1. Protoporphyrin IX–Fe complex.

The physiological function of P-450 is the oxidative metabolism of foreign compounds (such as drugs and cancer reagents) and steroids in liver, lung, and placenta, via an oxygen transfer mechanism. Estabrook and co-workers demonstrated for the first time the participation of P-450 in the C-12 hydroxylation of steroids catalyzed by adrenal cortex microsomes (4). Since then, many other reactions, including hydroxylation and oxidative dealkylations, have been explored. A family of P-450 enzymes has been isolated. Some of these, especially the forms involved in steroid metabolism, mediate regio-, chemo-, and stereo-specific hydroxylation. The stoichiometry of the monooxygenation is shown in Eq. (1) ($1a$–d), where the S represents substrate.

$$S + O_2 + 2e^- + 2H^+ \rightarrow SO + H_2O \tag{1}$$

The mechanism of oxygen activation by P-450 has been extensively studied ($1c$, $5a$, b); however, crucial intermediates in the catalytic cycle are so unstable that it is not possible to examine their structures and reactivities. Accordingly, model systems for P-450 with synthetic metalloporphyrins have illuminated the enzymic reaction mechanism. For instance, the so-called picket fence porphyrin, $meso$-tetra($\alpha,\alpha,\alpha,\alpha$-pivalamidophenyl)porphyrin (TpivPP) Fe(II), prepared by Collman and co-workers, forms a stable complex with O_2 ($6a$, b). An X-ray crystal structure of the oxy form of this picket fence porphyrin has contributed significantly to understanding the detailed structures of oxy forms of *P*-450, hemoglobin, and related enzymes. Model systems of the *P*-450 reactions also have been devised. In 1954, Udenfriend *et al.* found that the reductive activation of molecular oxygen by a combination of Fe(II), EDTA, and ascorbic acid allowed the oxidation of benzene to the corresponding phenol (*7*). The active species responsible for the oxidation is not clear; however, Hamilton and coworkers proposed that oxygen was transferred from an iron-bound dioxygen adduct based on observable differences in reactivity compared to the hydroxy radical (Scheme I) (*8*). Since then, many model systems for *P*-450 utilizing nonheme iron have been reported (*9*). The first example of *P*-450-type oxidation catalyzed by synthetic iron porphyrin complexes was reported by Groves *et al.* in 1979 (*10*). This work demonstrated the facile oxidation of alkenes and alkanes by the model system employing Fe(III)TPP(Cl) (TPP: 5, 10, 15, 20-tetraphenylporphyrin) and iodosylbenzene, PhIO. Since then, there have been a large number of studies related to the *P*-450-type reactions.

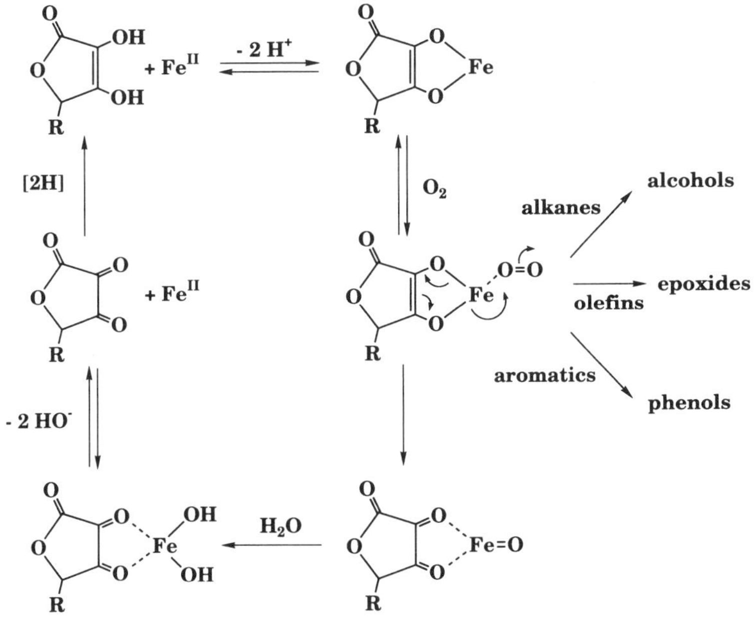

R: -CH(OH)CH₂OH

SCHEME I. Reductive oxygen activation by a simple iron complex.

In this review, we will focus on the mechanistic aspects of the oxygen activation and reactions of the *P*-450 systems involving metalloporphyrins. For this purpose, we will briefly outline some of the chemistry of *P*-450 and examine in detail the mechanisms of oxygen transfer by comparing the enzymic system to the model studies. At this time, applications of these model systems in synthetic chemistry have not been extensively pursued.

II. Structure and Reactivity of P-450

The cloning of various *P*-450 enzymes has permitted the determination of their amino acid sequences (*11*). *P*-450$_{cam}$ of *Pseudomonas putida* (*12*) has recently been crystallized; its structure was determined to 2.4 Å resolution in 1984 (*13*) and to 2.2 Å resolution in 1986 (*14*). Figure 2 illustrates stereoscopic views of the camphor binding site, which clearly shows the hydrogen bonding interaction of the carbonyl oxygen of *d*-camphor with the phenolic OH in Tyr-96 (*14*).

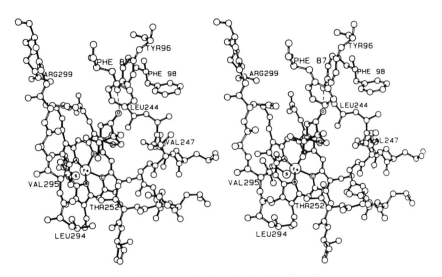

FIG. 2. Stereoscopic view of active site of *P*-450$_{cam}$.

Some typical monooxygenation reactions catalyzed by *P*-450 are listed in Table I (*15*). Actually, the enzyme produces a strong oxidant in these reactions, but the precise reactive species in these oxygenations is still obscure. In the following sections, we will describe, step by step, what is known of the oxygen activation mechanism of *P*-450.

TABLE I

TYPICAL *P*-450-CATALYZED REACTIONS

Type of oxidation	Substrates	Type of oxidation	Substrate
Aromatic hydroxylation	3,4-Benzopyrene	S-Oxygenation	Chloropromazine
	Acetanilide		Thioanisole
	Zoxazolamine		Thiophene
	Chlorzoxazone		Methylthiopurine
Aliphatic hydroxylation	Hexobarbital	S-Dealkylation	Phenacyl phenylsulfide
	Testosterone and other steroids	Dehalogenation	Methoxyflurane Halothene
	Octene		
	Cyclohexane		
Olefin epoxidation	Styrene	O-Dealkylation	*p*-Nitroanisole
	Cyclohexene		Acetophenetidin
Aromatization	Androstenedione	N-Dealkylation	3-Methyl-4-monomethylaminoazobenzene
N-Hydroxylation	*N*-Ethylaniline		Aminopyrine
	Aniline		Meperidine
	2-Acetylaminofluorene		*N*-Methylbarbital
			N,N-Dimethylaniline

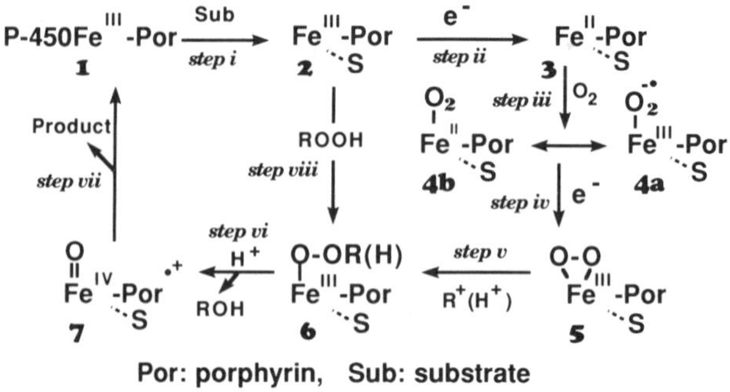

P-450Fe^{III}-Por $\xrightarrow[\text{step i}]{\text{Sub}}$ Fe^{III}-Por $\xrightarrow[\text{step ii}]{e^-}$ Fe^{II}-Por
　1　　　　　　　　　　　**2** ⋯S　　　　　　**3** ⋯S

Product ↑ | step vii

O_2 | step iii ↓ O_2

ROOH | step viii

Fe^{II}-Por ⟷ Fe^{III}-Por
4b ⋯S　　　$O_2^{\cdot-}$ **4a** ⋯S

step iv ↓ e^-

$\underset{\textbf{7}}{\overset{O}{\underset{\|}{Fe^{IV}\text{-Por}}}}$⋯S $\xleftarrow[\text{ROH}]{\text{step vi}}$ $\overset{O\text{-OR(H)}}{\underset{\textbf{6}}{Fe^{III}\text{-Por}}}$⋯S $\xleftarrow[R^+(H^+)]{\text{step v}}$ $\overset{O\text{-O}}{\underset{\textbf{5}}{Fe^{III}\text{-Por}}}$⋯S

Por: porphyrin, Sub: substrate

SCHEME II. Proposed catalytic cycle of *P*-450.

III. Mechanism of Oxygen Activation by *P*-450

The current understanding of oxygen activation by *P*-450 is summarized as follows (Scheme II) (*1c, 5a, b*): (*i*) incorporation of a substrate to the resting ferric state (**1**) of the active site of the enzyme to afford the ES complex (**2**); (*ii*) one-electron reduction of the heme from NAD(P)H via an associated reductase enzyme; (*iii*) reaction of the reduced heme (**3**) with O_2 to form an oxy complex (**4a, 4b**); (*iv*) one-electron reduction of the oxy complex to yield a peroxo complex (**5**); (*v*) protonation (or possibly acylation) of the peroxo oxygen; (*vi*) the formation of active species, the so-called oxenoid [FeO^{3+} (**7**)], by heterolytic O–O bond cleavage of (**6**); (*vii*) oxygen transfer to the substrate. Thus, the overall stoichiometry can be expressed as in Eq. (2), where R is H$^+$ or R'C(O)$^+$.

$$S + O_2 + 2e^- + R^+ + H^+ \rightarrow SO + ROH \tag{2}$$

A. SUBSTRATE BINDING AND ONE-ELECTRON REDUCTION

The *P*-450 reaction cycle begins with binding of the substrate at the active site. This process can be observed spectroscopically since the incorporation of the substrate eliminates the iron-coordinated water molecule from the active site, causing spin-state changes in some cases (*16, 17a–c*). The X-ray crystal structure of *P*-450$_{cam}$ (Fig. 2) shows the presence of the substrate (*d*-camphor) at the active site, whereas the crystal structure of substrate free *P*-450$_{cam}$ indicates the presence of several water molecules at the same place (*14*). On substrate binding to *P*-450, the reduction potential of the heme is increased (*18, 19*), i.e., the reduction of the heme becomes easier when substrates are present, as listed in Table II. Apparently, nonproductive consumption of NAD(P)H by *P*-450 can be prevented in the absence of the substrate.

In model systems, the effect of substrates on the redox potential of heme has

TABLE II

P-450 System	Midpoint (mV)
P-450$_{cam}$ − substrate	−300
P-450$_{cam}$ + substrate (*d*-camphor)	−170
P-450$_{scc}$ − substrate	−412
P-450$_{scc}$ + cholesterol	−305
P-450$_{scc}$ + dihydrocholesterol	−306
Hepatic microsomal *P*-450 − substrate	−300
Hepatic microsomal *P*-450 + hexobarbital	−237
Hepatic microsomal *P*-450 + benzphetamine	−225

not been examined; however, synthetic ferric heme complexes are readily reduced either chemically or electrochemically to give ferrous complexes. The reduction potentials of Fe(III)TPP with several axial ligands in organic solvents were determined in studies by Kadishi and co-workers (Table III) (*20a, b*). A Soret band at 451 nm characteristic of the RS−hemin−CO$_2$ complex was demonstrated both by Stern and Peisach and Chang and Dolphin (*21, 22*). Furthermore, Collman *et al.* (*23*) reported that EPR spectra of the FeTPP(SC$_6$H$_5$) complex are very similar to the substrate-bound high-spin *P*-450$_{cam}$. Since then, many studies have been carried out for ferrous and ferric forms of Fe−RS$^-$ porphyrin complexes (*24, 25*). More recently, Weiss and co-workers reported crystal structures of FeTPP(SEt) and FeTPP(SEt)(CO) in which Fe−S distances were reported to be 2.360(2) and 2.352(2) Å, respectively (*26*). The former bond length is somewhat longer than that present in Fe(PPIXDME)(SC$_6$H$_4$-*p*-NO$_2$) (PPIXDME: protoporphyrin IX dimethylester) [2.324(2) Å] (*24*).

B. Formation of Oxy Complexes

The reduced form of *P*-450 is able to coordinate O$_2$ reversibly to yield the oxy complex (**4**) [Eq. (3)]. Ishimura *et al.* (*27*) and Tyson and co-workers (*28*) found spectral changes when O$_2$ was introduced to the reduced form of *P*-450$_{cam}$ to form the oxy complex. The oxy complexes of other isozymes of *P*-450 are less stable and are observable only at low temperature or with stopped-flow techniques (*29*). Mössbauer studies of the oxy complex of *P*-450$_{cam}$ indicated that the complex is diamagnetic, very similar to oxymyoglobin (*17c*).

$$P\text{-}450\text{--}Fe^{III}(O_2^-) + e^- \rightarrow Fe^{III}(O_2^{2-}) \begin{array}{c} \nearrow \text{SO} \\ \\ \searrow \text{H}_2\text{O}_2 \end{array} \qquad (3)$$

$$\mathbf{4} \qquad\qquad\qquad \mathbf{5}$$

TABLE III

REDUCTION POTENTIALS OF Fe^{III} TPP BEARING SEVERAL LIGANDS[a]

Solvent	Donor number	Dielectric constant	$E_{1/2}$ (X)				
			ClO_4^-	Br^-	Cl^-	N_3^-	F^-
$EtCl_2$	0.0	10.7	0.24	-0.19	-0.31	-0.38	-0.47
CH_2Cl_2	0.0	8.9	0.22	-0.21	-0.29	-0.42	-0.50
CH_3NO_2	2.7	35.9	0.10	—[b]	—[b]	—[b]	—[b]
ϕ-CN	11.9	25.2	0.20	-0.18	-0.34	-0.39	-0.57
CH_3CN	14.1	37.5	0.11	—[b]	—[b]	—[b]	—[b]
PrCN	16.6	20.3	0.13	-0.15	-0.27	-0.33	-0.45
$(CH_3)_2CO$	17.0	20.7	0.09	-0.16	-0.28	-0.34	-0.43
THF[e]	20.0	7.6	0.17	-0.24	-0.34	-0.38	-0.47[a]
DMF[e]	26.6	38.3	-0.05	-0.05	-0.18	-0.25	-0.40
DMA[e]	27.8	37.8	-0.04	-0.05	-0.15	-0.24	-0.36
DMSO[e]	29.8	46.4	-0.09	-0.09	-0.10	—[c]	-0.09 (-0.40)[d]
Py[e]	33.1	12.0	0.15	0.17	0.16 (-0.25)[d]	0.15 (-0.28)[d]	0.16 (-0.46)[d]

[a] Half-wave potential (V versus SCE) for the Fe(III)/Fe(II) redox couple of TPPFeX in selected solvents. Data taken from L. A. Bottomley and K. M. Kadish, Ref. 20a.

[b] Complex was insoluble in solvent system.

[c] Reduction consisted of two overlapping processes.

[d] Second reduction process.

[e] THF, Tetrahydrofuran; DMF, dimethylformamide; DMSO, dimethylsulfoxide; DMA, dimethylacetamide; Py, pyridine.

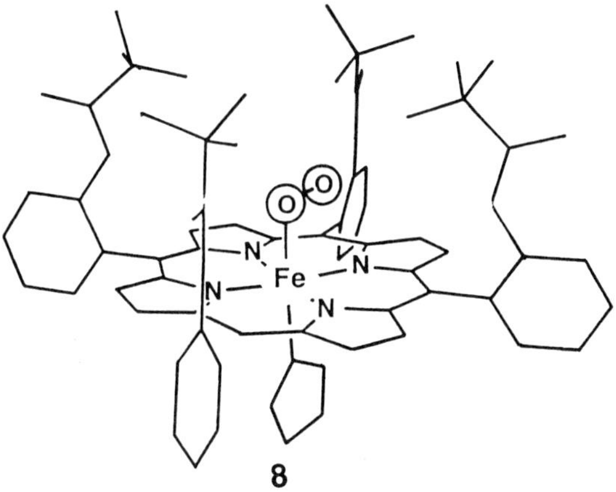

8

FIG. 3. Fe(O$_2$)(TpivPP) complex.

Collman and co-workers prepared the picket fence porphyrin iron complex, Fe(TpivPP) and showed that the reduced form of the porphyrin can reversibly pick up O$_2$ to afford the stable oxy complex (**8**) (*6a, b, 30*). The X-ray crystal structure of **8** is presented in Fig. 3. The Fe–O and O–O distances were determined to be 1.898(7) and 1.22(2) Å (*6b, 31*), respectively. The Fe–O–O angle was 129(1)°. Table IV shows infrared O–O stretching frequencies of some O$_2$–metal complexes (*30, 32*). Appearance of the O–O stretch of **8**-*N*-MeIm, **8**-*N*-TrIm, HbO$_2$, and MbO$_2$ in the range of 1160–1130 cm^{-1} is indicative of the formation of a superoxo ligand. The isomer shift and quadruple doublet observed in the Mössbauer spectra of **8**-*N*-MeIm and **8**-*N*-BuIm were consistent with this assignment and indicate that the central heme is low-spin Fe(III) (*33*). Thus, the spin coupling of Fe(III) ($S = \frac{1}{2}$) with O$_2^-$ ($S = \frac{1}{2}$) makes the complex diamagnetic (*34*).

TABLE IV

COMPARISON OF νO$_2$ FOR DIOXYGEN COMPLEXES

Compound[a]	νO$_2$ (cm^{-1})
Fe(TpivPP)(*N*-MeIm)(O$_2$)	1159
Fe(TpivPP)(*N*-TrIm)(O$_2$)	1163
Co(TpivPP)(*N*-MeIm)(O$_2$)	1150
Cr(TPP)(Py)(O$_2$)	1142
HbO$_2$	1107
MbO$_2$	1103
CoHbO$_2$	1106

[a] *N*-MeIm, *N*-methylimidazole; *N*-TrIm, *N*-tritylimidazole; Hb, hemoglobin; Mb, myoglobin.

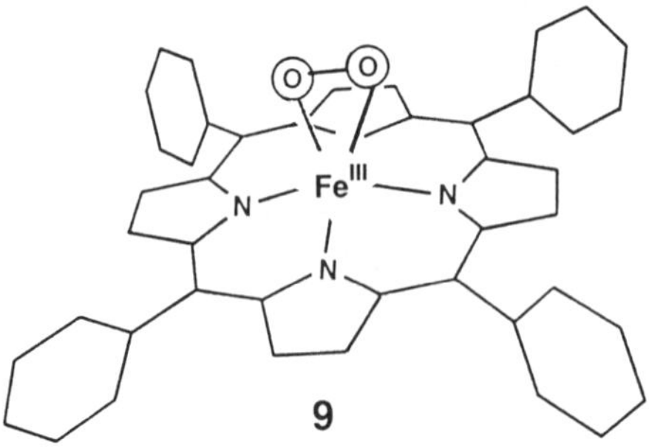

9

FIG. 4. Proposed structure for **9**.

C. ONE-ELECTRON REDUCTION OF OXY COMPLEXES

The rate-determining step in the oxygen activation cycle of P-450$_{cam}$ has been considered to be the reduction of the oxy complex (**4**) (28, 35). Therefore, steps (iv)–(vii) and the intermediates of **5–7** in Scheme II are not observable in the P-450$_{cam}$ system. Nordblom and Coon have reported that the active turnover of a fully reconstituted P-450$_{LM}$ system involves production of hydrogen peroxide; as much as 55% of the consumed oxygen appears as hydrogen peroxide in the presence of substrate (or essentially 100%, with no substrate) (36). These observations suggest the formation of hydrogen peroxide via protonation of the peroxo oxygen in the absence of substrate [Eq. (3)].

McCandlish *et al.* have isolated a peroxo–iron(III) complex (**9**) (Fig. 4) in the reaction of Fe(III)TPP(Cl) and KO$_2$ according to Eq. (4) (37). The Soret band of **9** appears at 437 nm with unusually red-shifted α- and -b-bands (565 and 609 nm in DMSO). The EPR spectrum of **9** at 77K showed a relatively narrow, sharp resonance at $g = 4.2$ and weak resonances at $g = 2$ and $g = 8$, typical of rhombic high-spin ferric complexes such as Fe^{III}EDTA (38). Such a spectrum is not typical of high-spin ferric porphyrin complexes, which usually show resonance at $g = 2$ and 6, indicative of axial symmetry. An IR band at 806 cm^{-1} was observed to shift to 759 cm^{-1} when K^{18}O$_2$ rather than K^{16}O$_2$ was used to prepare solutions of **9**; these observations suggested the side-on bonding formulation illustrated in Fig. 4. Extended X-ray absorption fine structure (EXAFS) studies of **9** also

$$\text{Fe}^{III}\text{TPP(Cl)} + \text{KO}_2 \xrightarrow[\substack{-\text{KCl} \\ -\text{O}_2}]{} \text{Fe}^{II}\text{TPP} \xrightarrow{\text{KO}_2} [\text{Fe}^{III}\text{TPP(O}_2{}^{2-})]^-\text{K}^+ \qquad (4)$$

supported the side-on structure (39). Complex **9** could also be prepared either in the reaction of O$_2$ with Fe(I) (40) or reduction of the oxy complex, Fe^{III}(O$_2{}^-$), at low temperature (41) [Eq. (5)]. The latter system is consistent with the proposed reductive activation of oxygen by *P*-450, although there is no sixth axial ligand to iron in these complexes.

$$
\left.
\begin{array}{l}
P\text{–}Fe^{I} + O_2 \\
P\text{–}Fe^{II} + O_2^{-} \\
P\text{–}Fe^{III}(O_2^{-}) + e^{-}
\end{array}
\right\}
\longrightarrow
\begin{array}{c}
O\text{–}O \\
\diagdown\diagup \\
P\text{–}Fe^{III}
\end{array}
\tag{5}
$$

Although no X-ray crystal structure determination of **9** is available, the structure of a related manganese species, [peroxo(tetraphenylporphyrinato)]manganese(III) [K(K222)][MnTPPP(O₂)], complex **10**, has recently been reported by Valentine and co-workers (*42*) by the two-step reaction of Mn^{III}TPP(Cl) with 2 equivalents of KO₂ [Eq. (6)].

$$
Mn^{III}TPP(Cl) + KO_2 \xrightarrow{-KCl,\ -O_2} Mn^{II}TPP \xrightarrow{+KO_2} \underset{\textbf{10}}{MnTPP(O_2)}
\tag{6}
$$

Two possible structures for **10** are given by **11** and **12**:

$$[Mn^{II}TPP(O_2^{-})]^{-}K^{+} \qquad [Mn^{III}TPP(O_2^{2-})]^{-}K^{+}$$
$$\textbf{11} \qquad\qquad\qquad \textbf{12}$$

When the reaction of Mn^{III}TPP with KO₂ was reported, the reaction product was believed to be either a manganese(II) superoxo (**11**) or manganese(III) peroxo complex (**12**) (*43*). Shirazi and Goff subsequently studied this oxygen complex by deuterium NMR and observed a downfield shift of the pyrrole hydrogens (*44*). Such shifts had been shown to be characteristic of an unpaired electron in the $d_{x^2-y^2}$ orbital (Table V) (*45*). Since this *d* orbital is usually highest in energy for metalloporphyrin complexes, these authors concluded that the central manganese must be d^5, i.e., Mn(II). On the other hand, the magnetic moment of **10** (5.0 μ_B) (*46*) could be explained either by a Mn(II) ($S = \frac{5}{2}$) structure, **11**, with spin coupling to coordinated superoxide ion ($S = \frac{1}{2}$) or a high-spin Mn(III) ($S = 2$) structure, **12**. The result of the X-ray crystal structure of **10** supports the latter electronic structure, **12**.

The manganese in **12** is bonded to four pyrrole nitrogens with the bidentate peroxo ligand (Fig. 5). The manganese atom lies 0.764 Å above the least-square plane based on the pyrrole nitrogens to the peroxo ligand, whreas the manganese atom of Mn^{II}TPP(Cl) is displaced only 0.641 Å from the plane. The O–O bond distance is 1.421(5) Å and the Mn–O distances are 1.901(4) and 1.888(4) Å. The peroxo ligand is bound "side-on" to the manganese, eclipsing two of the manganese–pyrrole nitrogen bonds. The large out-of-plane Mn displacement and strong d–$O_2\pi_g$ mixing result in an unusual *d*-orbital pattern for **12**: $d_{yz} - O_2\pi_{gz} < 40\%\ d_{xy} + O_2\pi_{gx} < 60\%\ d_{xy} - 40\%\ O_2\pi_{gz} < d_{xz}, d_{z^2}, d_{x^2-y^2}, d_{yz} + O_2\pi_{gz}$ (Fig. 6). Thus, the highest orbital is no longer $d_{x^2-y^2}$ but a $d_{yz} + O_2\pi_{gz}$. These considerations are fully consistent with the NMR results described above (*42*).

TABLE V

Contact Shifts and Electronic Structures

Metal	Pyrrole H	Pyrrole CH_2	meso- H	d_{xy}	d_{xz}	d_{yz}	d_{z^2}	$d_{x^2-y^2}$	Spin
Iron(II)									
$S = 2$	47	10	4	2	1	1	1	1	σ
$S = 1$	-25	17	32	2	1	1	2	0	π
Iron(III)									
$S = \frac{1}{2}$	61	32	-80	1	1	1	1	1	σ
$S = \frac{1}{2}$	-19	5	2	2	$\left\{\begin{matrix}1\\3\end{matrix}\right.$	1	0	0	π
Manganese(III)									
$S = 2$	-31	18	42	1	1	1	1	0	π
Cobalt(II)									
$S = \frac{1}{2}$	-2	-1	4	2	2	2	1	0	None

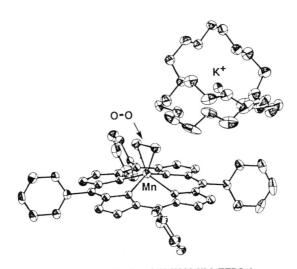

Fig. 5. ORTEP plot of [K(K222)][MnTTPO$_2$].

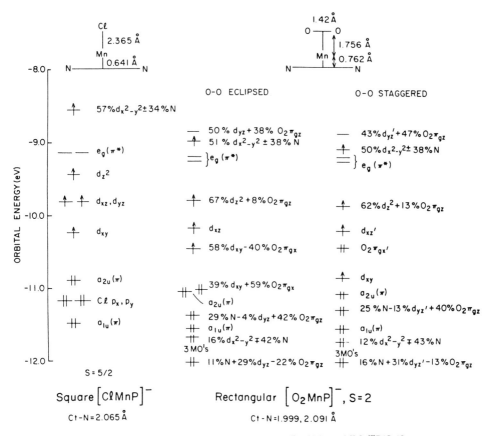

Fig. 6. IEH calculated orbital energies of [Mn^{II}P(Cl)]$^-$ and [Mn^{III}P(O$_2$)]$^-$.

417

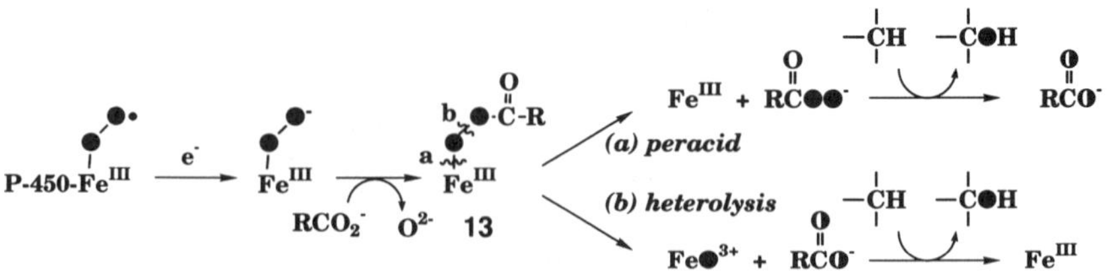

SCHEME III. Acylation of heme-bound dioxygen.

D. FORMATION OF ACTIVE SPECIES RESPONSIBLE FOR OXYGENATION

The peroxo complex of P-450 (**5**) (refer to Scheme II for species **1–7**) is generally believed to react with a proton to form the hydroperoxo–Fe^{III} intermediate [**6**, R = H], followed by second protonation and subsequent loss of water to produce the active species, $Fe^{IV}(=O)P^{+}$ (**7**) or more generally $(FeO)^{3+}$ (*1c, 5a, b, 46*). An alternative possibility is alkylation or acylation of **5** accompanied with protonation to release the corresponding alcohol or carboxylic acid. The latter mechanism was proposed by Sligar *et al.* (*47*) based on the following observation. When catalytic oxygenation of d-camphor by P-450$_{cam}$ took place under an $^{18}O_2$ atmosphere in the presence of dihydrolipoic acid, about the same amount of ^{18}O incorporation in both 5-*exo*-hydroxy-d-camphor and lipoic acid was indicated by the mass spectra of the products. This suggests that the oxygen activation mechanism for P-450$_{cam}$ could involve the transient formation of the acylperoxo intermediate (**13**) shown in Scheme III. However, no dependence on dihydrolipoic acid has been observed with other forms of P-450.

A very similar acylperoxo (or alkylperoxo) intermediate has also been postulated in the peroxidase reactions (*48*). For example, horseradish peroxidase (HRP) is oxidized by hydrogen peroxide to give Compound I, which has been characterized as an oxo porphyrin π-cation radical, $Fe^{IV}(=O)P^{+}$ [Eq. (7)] (*49*). The major difference between HRP and P-450 is the fifth ligand, the former having imidazole and the latter a thiol from cysteine.

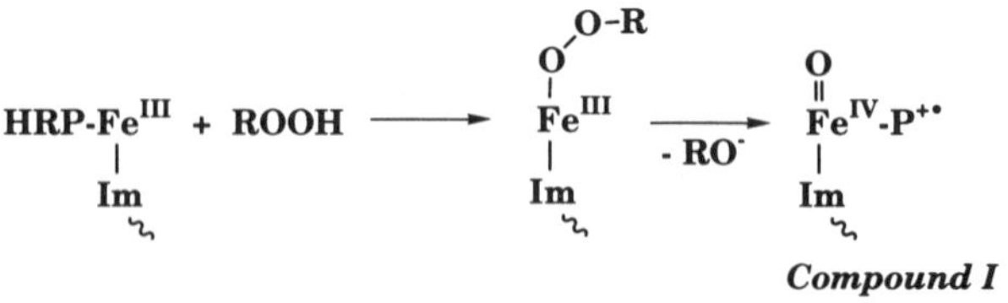

Compound I

Im: imidazole of histidine residue; P: porphyrin

In Eq. (7), Im is the imidazole of the histidine residue and P is porphyrin. Compound I in peroxidase systems is observable (*50*) even in the case of chloroperoxidase (CPO), which has a thiolate as the fifth ligand (*1d, 51*). These results for the peroxidase reactions suggest that the active species of *P*-450 may be similar to Compound I. Another result in support of the formation of a Compound I-type intermediate in *P*-450 is that *P*-450 is also able to catalyze the oxidation of organic substrates at the expense of a number of single-oxygen donors, such as hydrogen peroxide (*52*), peracids (*53*), sodium chlorite (*53*), iodosoarene (*54*), sodium periodate (*55*), and iodosobenzene diacetate (*56*).

McCarthy and White have noted the ability of a series of seven hemoproteins, $P\text{-}450_{LM2}$, $P\text{-}450_{LM4}$, $P\text{-}420_{LM2}$, HRP, CPO, catalase, and metmyoglobin, as well as hemin to catalyze a set of five oxidative reactions (*57, 58*). These reactions are typical of the peroxidase reaction [oxidation of pyrogallol to purpurogallin, Eq. (8)] and three characteristic *P*-450 reactions (aliphatic hydroxylation, aromatic hydroxylation, and olefin epoxidation).

$$2 \quad \text{(pyrogallol)} \quad + \ 3\,H_2O_2 \ \longrightarrow \ \text{(purpurogallin)} \quad + \ CO_2 + 5\,H_2O \qquad (8)$$

In addition, the ability to decarboxylate a peroxy acid was measured. All hemoproteins were able to carry out peroxidation, but three (HRP, CPO, and catalase) were much better catalysts than the others. Only *P*-450 enzymes were competent catalysts for hydroxylation and epoxidation. Furthermore, the decarboxylation reaction was strictly limited to the *P*-450 enzymes (Table VI).

Since the decarboxylation of peroxy acids is diagnostic of the formation of acylperoxyl radical intermediates, these results indicate a fundamentally different manner of processing of peroxides by *P*-450 than by the peroxidases. A

TABLE VI

DECARBOXYLATION OF PPAA BY HEMOPROTEINS

Heme protein	Heme (*nmol*)	Product Yield (*nmol*)		Benzyl alcohol/ purpurogallin (*mol/mol*)
		Benzyl alcohol	Purpurogallin	
$P\text{-}450_{LM2}$	1.4	$7.4 \pm 0.3\ (n = 2)$	$27 \pm 2\ (n = 2)$	0.27
$P\text{-}450_{LM4}$	2.2	$7.9 \pm 0.9\ (n = 3)$	$20 \pm 1\ (n = 3)$	0.40
$P\text{-}450_{LM2}$	1.4	$0.3 \pm 0.1\ (n = 2)$	$19 \pm 4\ (n = 2)$	0.02
Chloroperoxidase	0.9	0	$31 \pm 2\ (n = 3)$	0.00
Horseradish peroxidase	1.4	0	$30 \pm 3\ (n = 3)$	0.00
Catalase	1.0	0	0	
Metmyoglobin	2.4	0	$39 \pm 2\ (n = 2)$	0.00
Hemin	0.8	0	$74 \pm 1\ (n = 2)$	0.00

plausible explanation is as follows: (i) in the reaction of peroxidase with per-acids, an acylperoxo–iron(III) complex (13) [see Scheme IV, pathway (A)]

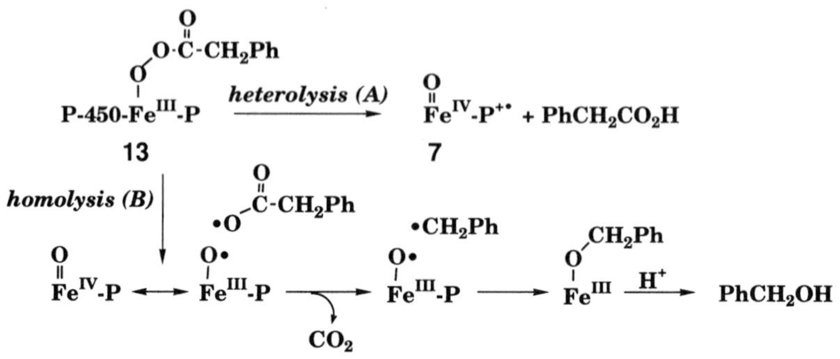

SCHEME IV. Homolysis versus heterolysis mechanism.

might be the initial intermediate followed by the ionic O–O bond cleavage (het-erolysis) to generate 7, similar to Compound I and the corresponding carboxylic acid; (ii) when P-450 was employed, two different modes of O–O bond cleavage in the acylperoxo–iron(III) complex competed; i.e., one is the heterolytic mecha-nism employed by the peroxidases and the other is a homolytic radical process to afford an acyloxy radical, which readily proceeds via the decarboxylation reaction [Scheme IV, pathway (B)]. For example, when phenylperoxyacetic acid was used, benzyl alcohol, a decarboxylation product, was observed (57, 58).

These results for the peroxidase reaction of P-450 support the proposed mecha-nism of the oxygen activation, namely, protonation (or acylation) of the peroxo oxygen in 5 (Scheme II) to give the hydroperoxo–iron(III) [or acylperoxo–iron(III)] complex (6), then O–O bond scission to afford the active iron–oxo species. However, a question remains as to where either the proton or the acylat-ing reagent comes from. The X-ray crystal structure of P-450$_{cam}$ with d-camphor lacks water molecules in the active site. It is also not yet clear what factor(s) control(s) the relative rate of homolysis and heterolysis of acylperoxo–iron(III) complex. The active site of peroxidase has a number of functional groups, in-cluding arginine, which could facilitate heterolysis by donating a proton. By contrast the *only* functional group near the O$_2$, the binding site in P-450$_{cam}$, is Thr-252.

E. ROLES OF THE HYDROXYL GROUP OF THREONINE-252 IN THE ACTIVE SITE

The recent X-ray crystal structure of P-450$_{cam}$ indicates that the local environ-ment of the active site is almost completely nonpolar. Thr-252 and Tyr-96 punc-tuate the largely aliphatic reactive cavity (13, 14). Peroxidases are known to catalyze the heterolytic O–O bond cleavage exclusively. A major difference between P-450$_{cam}$ and peroxidase active sites is that the latter have a number of polar functional groups. In the case of cytochrome c peroxidase (CCP), according to Poulos and Finzel (*59*), the proximal histidine binds strongly to heme iron, likewise HRP, as an electron donor to destabilize the O–O bond (push) (*60*). At the same time, the distal histidine provides a proton and an arginine, which makes the distal side substantially more polar, assisting in the charge separation of the O–O bond (pull). The role of Arg and His in assisting heterolytic O–O bond cleavage is outlined in Scheme V. Traylor and Popovitz-Biro have reported a very similar hydrogen bonding effect of a proximal imidazole on the peroxidase activity of iron(III) porphyrin complexes (*61*). In the case of P-450, the thiol group has been considered to be a strong internal electron donor to facilitate heterolytic O–O bond cleavage (*62*). Further, a recent site-directed mutation study of P-450$_{cam}$ has highlighted a possible proton source in P-450.

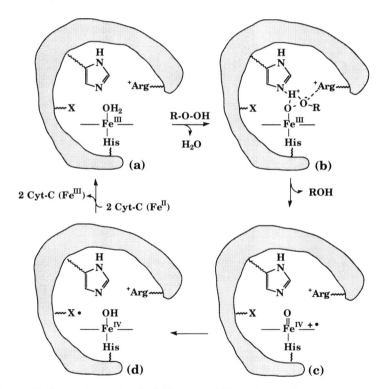

SCHEME V. Proposed mechanism for CCP compound I formation. (a) The native enzymes; (b) the activated complex with the distal histidine operating as an acid–base catalyst and the active site arginine stabilizing a developing negative charge on RO–OFe; (c) the hypothetical oxene intermediate; (d) Compound I after the intramolecular electron rearrangement of (c) to give Fe(IV) and a free radical, X.

TABLE VII

OXYGEN AND NADH CONSUMPTION RATES OF WILD-TYPE AND MUTANT P-450$_{cam}$
IN RECONSTITUTED SYSTEMS AND AMOUNT OF PRODUCTS FORMED[a]

Mutant	Rate of consumption (μM/min/μM heme)		Product (%) formed per O$_2$ consumed	
Site Residue	Oxygen	NADH	5-OH-CAM[b]	H$_2$O$_2$
252 Thr (wild)	1350	1380	97	3
252 Ala	1150	1180	5	89
252 Gly	1090	1090	3	88
252 Ser	830	830	85	15
252 Cys	690	700	7	86
252 Asn[c]	420	440	57	12
252 Val	260	250	24	51
252 Pro	100	100	10	77

[a] The reaction was carried out as described in Ref. 63. O$_2$ consumption and NADH oxidation were measured simultaneously. Other details of the determinations were also in Ref. 63.
[b] 5-OH-CAM: 5-exo-hydroxycamphor.
[c] The reaction was carried out in the presence of glycerol.

Using site-directed mutagenesis, Imai et al. replaced Thr-252 of P-450$_{cam}$ with alanine, glycine, serine, cysteine, asparagine, valine, and proline (Table VII). Whereas the optical absorption spectra of the mutant enzymes were almost identical, their catalytic properties were markedly altered. Rates of O$_2$ consumption and NADH oxidation were measured simultaneously in a reconstituted system composed of putidaredoxin and its reductase (63). As shown in Table VII, both O$_2$ and NADH were consumed at essentially the same rate, indicating that two electron equivalents of NADH were used for the reduction of O$_2$ both in the wild-type and in all the mutant enzyme reactions. As seen, the Ala enzyme consumed O$_2$ at a rate comparable to that of the wild-type enzyme. However, the product analysis showed that almost all of the oxygen consumed was reduced to H$_2$O$_2$ without forming the hydroxylated product. The loss of monooxygenase activity on substitution of alanine for the threonine was also observed by Martinis et al. (64).

Several lines of evidence have indicated that H$_2$O$_2$ is formed through the decomposition of the peroxo intermediate produced subsequent to the formation of the oxygenated form, and not through the autooxidation of the oxygenated enzyme (63). Production of H$_2$O$_2$ was also found in the Gly, Cys, Val, and Pro enzymes. The Ser and Asn enzymes, on the other hand, exhibited high monooxygenase activities. Thus, the presence of the HO and H$_2$N group in the 252-amino acid residue is indispensable for an efficient d-camphor monooxygenation. These findings, together with the essential role of Thr-301 in rabbit P-

$450_{\varpi-1}$ reported by Imai and Nakamura (65), indicate that the threonine in the active site is essential for oxygen activation in the P-450-catalyzed monooxygenation reaction; possibly the HO or H_2N side chain of the amino acid at position 252 serves as the proton donor for the peroxo oxygen to yield the hydroperoxo–iron intermediate. In addition, a careful inspection of the active site of P-450_{cam} reveals a series of water molecules underneath the porphyrin ring (66). These water molecules will also serve to supply protons in the oxygen activation processes by utilizing the HO group of Thr-252.

F. MODEL STUDIES RELATED TO MECHANISMS OF P-450 REACTIONS

The first successful application of synthetic iron porphyrins for the oxidation of alkanes and alkenes was reported by Groves et al. in 1979 (10). The key development was the use of iodosylbenzene as an oxygen donor to Fe(III)TPP(Cl). In 1981, Groves et al. prepared the active species responsible for the oxidations in a similar system and characterized it as an oxoferryl porphyrin cation radical, $Fe^{IV}(=O)P^{\cdot +}$ (**14**) [see Eq. (9)] (67). This reactive complex was generated by the reaction of Fe(III)TMP and mCPBA in dichloromethane at $-78°C$ (67). The ^1H NMR spectrum of **14** at $-78°C$ in dichloromethane–methanol showed absorbances at δ 68 (m-H), 26 and 24 (o-methyl), 11.1 (p-methyl), and -27 (β-pyrrole H). The magnetic susceptibility of **14** was determined by the Evans method to be 4.2 μ_B, slightly larger than expected for an $S = \frac{3}{2}$ system. The visible spectrum of **14** showed a broad Soret band at 405 nm and another broad band centered at 645 nm, characteristic of the porphyrin cation radical. The Mössbauer spectrum derived from ^{57}Fe-enriched Fe(III)TMP(Cl) showed a quadruple doublet centered at 0.05 mm/sec and an isomer shift of 1.49 mm/sec. Decoupling of the porphyrin spin from the two unpaired electrons on Fe(IV) with a small magnetic field supported this electron structure of **14** (68). The small isomer shifts in the Mössbauer spectrum of **14** are similar to those reported for HRP Compound I and are in the range expected for iron(IV) (69). The recent EXAFS study of **14** indicated that the Fe–O distance is 1.62–1.65 Å, as expected for the double bond formation between Fe and O (70). Kitagawa and co-workers have recently assigned the resonance Raman band of $Fe^{IV}=^{16}O$ stretching for the ferryl porphyrin π-cation radical to be 828 cm^{-1} (71).

Traylor et al. have described studies designed to trap the active species in the catalytic reaction of several iron porphyrins with peracid at room temperature using 2,4,6-tri-tert-butyl phenol (**15**) and observed an important role of acid on the rate of the oxidation of **15** (72). Lee and Bruice have used a similar method to examine the effect of peroxides on the rate of formation of the corresponding phenoxy radical catalyzed by metalloporphyrins (73). A good correlation between the pK_a of the leaving group, a substituted benzoic acid, and the oxidation rate of **15** ($\beta = 0.35$) was observed. The substituent effect for the reaction was considered to result from the heterolytic O–O bond cleavage of the postulated acylperoxo–iron porphyrin intermediate.

The reductive activation of molecular oxygen by a picket fence porphyrin–Fe complex/H$_2$–colloidal Pt system has been described by Tabushi *et al.* (*74*). Though small amounts of olefin epoxidation were observed in the system, the addition of acylating reagents to this system caused remarkable improvement in the epoxidation reaction (*74*). The effect of acylating reagents on the formation of epoxide could be due to the facilitated formation of the active species by acylation of peroxo–metal intermediates (*75*). Tabushi has also described a biomimetic system for the controlled two-electron reduction of dioxygen at a metalloporphyrin center (*76*). To accomplish this, the H$_2$ colloidal platinum was replaced with *N*-methyldihydronicotinamide/flavin mononucleotide. The new reduction system showed a 10-fold greater efficiency for olefin epoxidation catalyzed by MnTPP. Khenkin and Shteinman have also reported the addition of acylating reagents to a peroxo–iron(III) complex formed as the oxidation product in an electrolytic process (*77*). These results suggest that the role of acylating reagent in the activation of the peroxo–iron(III) complex is to form an acylperoxo–iron(III) complex.

However, several questions remain to be answered: (1) Is the postulated formation of an acylperoxo–iron(III) complex correct? (2) If the answer is yes, how can it be proved? (3) Does the peroxo–iron(III) porphyrin complex afford an oxoferryl porphyrin cation radical? Recently such an acylperoxo–iron(III) porphyrin complex (**16**) in solution at low temperature has been directly observed and characterized (*78*). It was found that **16** did indeed react to yield the oxoferryl porphyrin cation radical **14**.

G. Preparation, Characterization, and Reaction of Acylperoxo–Iron(III) Porphyrin Complexes

When the reaction of Fe(III)TMP(Cl) (**17**) and mCPBA in dichloromethane at low temperature was directly monitored by UV–Vis spectroscopy, it was found that the kinetic profile for the formation of **14** showed a pronounced induction period (*79*). Figure 7 shows the time-dependent visible spectral changes in the reaction of **17** and mCPBA. Absorbance changes monitored in the reaction are also illustrated, and the inset in Fig. 7 shows the effect of acid on the kinetic behavior. The addition of acid (mCBA in this case) clearly accelerated the reaction. Inspection of spectral changes, especially the absorbance changes around 390 nm, indicates the absence of isosbestic points in the spectral changes. These observations are indicative of the formation of an intermediate in the reaction of

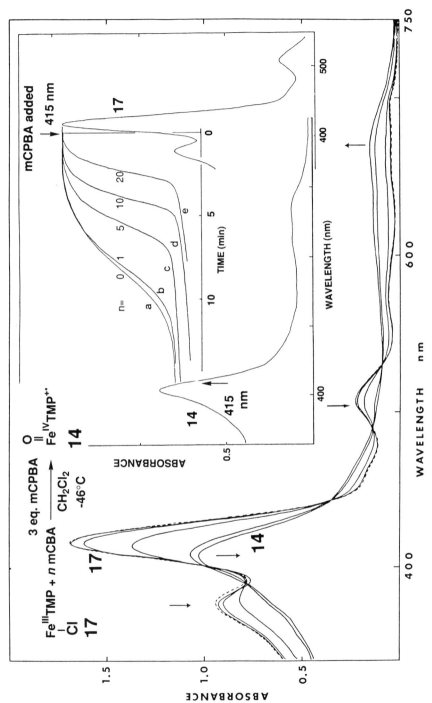

FIG. 7. Visible spectral changes in the reaction of Fe^{III}TMP(Cl) and mCPBA.

17 and mCPBA to give **14**. A likely intermediate is the acylperoxo–iron(III) complex, **16**, shown in Eq. (9).

$$
\underset{\textbf{17}}{\overset{\underset{\text{Cl}}{\mid}}{Fe^{III}TMP}} + \underset{}{\overset{\overset{\text{O}}{\overset{\|}{}}}{ArCOOH}} \rightleftharpoons \underset{\textbf{16}}{\overset{\overset{\overset{\text{O}}{\overset{\|}{\text{O-C-Ar}}}}{\overset{\text{O}}{\mid}}}{Fe^{III}TMP}} + HCl \xrightarrow{H^+} \underset{\textbf{14}}{\overset{\overset{\text{O}}{\overset{\|}{}}}{Fe^{IV}TMP^{+\bullet}}} + Ar\text{-}CO_2H \qquad (9)
$$

According to Eq. (9), HCl is expected to accumulate on the formation of **16**. Therefore, an acid-catalyzed reaction would be autocatalytic and the kinetic profile could be complicated. If Eq. (9) is correct, it would be very difficult to observe intermediate **16**, since the initial lignad exchange in Eq. (9) would be reversible, and high-spin Fe(III) porphyrin complexes usually give very similar visible spectra. However, Fe(III) porphyrin complexes with *hydroxo* ligands show very different spectroscopic behavior compared to other typical ligands. Furthermore, the ligand exchange reaction between Fe(III)TMP(OH), **18**, and peroxy acid to give **16** was found to be irreversible. Changes in the acidity of the medium are minimal during the formation of **16** since water is produced on ligand exchange. Furthermore, the acylperoxo ligand in **16** could not be displaced by a water molecule to produce **18**.

Visible spectral changes observed in the reaction of **18** with peracid at $-46°C$ in dichloromethane are shown in Fig. 8 (*78*). It is clear that rapid formation of an intermediate whose visible spectrum is characteristic of a high-spin Fe(III)

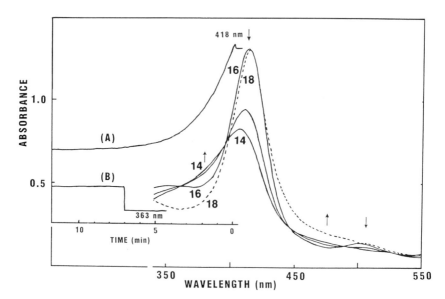

FIG. 8. The reaction of Fe^{III}TMP(OH) with mCPBA.

species is followed by a relatively slow O–O bond cleavage to afford **14**. Lines A and B in Fig. 8 illustrate the time course of absorbance changes observed at 418 and 363 nm on the addition of peroxy acid. The formation of the oxo-iron(IV) species, **14**, was found to be first order in the precursor (**16**). The reaction required 1.2 equivalents of peroxy acid when *p*-nitroperoxybenzoic acid was employed. These results offer strong evidence for the intermediate formation of an acylperoxo–Fe(III) complex.

The effect of acid on the formation of **14** was examined by varying the initial amount of peroxy acid [*n* in Eq. (10)]. The rate of the formation of **14** (k_{obs}) was found to correlate with [H$^+$]. Furthermore, for a series of peroxybenzoic acids, the substituent effect for the decomposition of **16** showed a good correlation between k_{obs} and Hammett σ ($\rho = 0.5$) (*79*). In summary, the reaction of Fe(III)TMP with peroxy acids has been shown to give an oxoferryl porphyrin cation radical, **14**, via O–O bond cleavage of an acylperoxo–iron(III) precursor, **16**. This O–O bond cleavage is heterolytic and is catalyzed by acid.

$$
\underset{\textbf{18}}{Fe^{III}TMP} + n\ Ar\overset{\overset{\textstyle O}{\|}}{C}OOH \xrightarrow[k_{obs}]{-\ H_2O} \underset{\textbf{16}}{Fe^{III}TMP} \xrightarrow{n\text{-}1\ ArCOOH} \underset{\textbf{14}}{Fe^{IV}TMP^+} \tag{10}
$$

Finally, the sterically hindered porphyrin, hydroxo[tetrakis(triphenyphenyl)-porphyrinatoiron(III)], Fe(III)TTPPP(OH) (**19**), was prepared and the reaction of **19** with mCPBA yielded the acylperoxo derivative (**20**), which was stable at low temperature (Scheme VI) (*80*). An infrared band corresponding to the carbonyl of **20** was observed at 1744 cm^{-1} whereas the C=O bands of free mCPBA and Fe(III)TTPPP(*m*-chlorobenzoate) appeared at 1735 and 1656 cm^{-1}, respectively. A very similar IR spectrum (1730 cm^{-1}) has been reported for (Ph$_3$P)$_2$PtOOC(O)Ph (*81*).

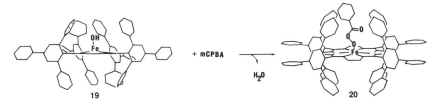

Scheme VI. Preparation of Fe^{III}TTPPP having the *m*-chloroperoxybenzoate ligand.

H. IRON PORPHYRIN *N*-OXIDE

The high reactivity of oxo-iron(IV) porphyrins such as **14** [Eq. (9)] toward hydrocarbons has given strong experimental support to the idea that such a species can be the active oxygen species in the *P*-450 cycle. IEHT (IEHT: iterative exchange-energy corrected version of EH theory) calculations by Loew *et al.* (*82*) have indicated an electronic structure bearing a net negative charge on the oxo-oxygen and positive charge on porphyrin ring. While the Fe–O distance assumed in this calculation was probably too long [1.75 Å; 1.62 Å from the EXAFS data on **14** (*70*)], considerable covalency was observed in the Fe–O π-bond. The electrophilicity of the FeO^{3+} moiety was attributed to the low-lying half-filled π^* iron–oxygen orbitals. Several more recent calculations by Tatsumi and Hoffmann, by Strich and Veillard, and by Jørgensen have suggested an iron porphyrin *N*-oxide as an alternative structure for **14** (*83–85*). Porphyrin *N*-oxides have been shown to coordinate copper and nickel (*86*) whereas iron porphyrins, with either nitrene (**21**) (*87*) or carbene (**22**) (*88*) bridging the iron to the pyrrole nitrogen (Fig. 9), have been reported as relatively stable species.

The oxidation of Fe(III)TMP(*m*-chlorobenzoate) with 2 equivalents of mCPBA in toluene has been reported by Groves and Watanabe to give iron(III) porphyrin *N*-oxide (**23**, (*89*) Fig. 9). The reaction proceeded quantitatively only at low concentrations of the heme. The presence of an acid such as benzoic acid drastically depressed the formation of **23**. The EPR spectrum, ^1H NMR chemical shifts, and solution magnetic moment (5.4 μ_B) indicated that **23** was a high-spin ferric complex. Inspection of the reaction mechanism indicated that the reaction proceeded via the formation of the acylperoxo–iron(III) precursor **16**, similar to the reaction carried out in dichloromethane.

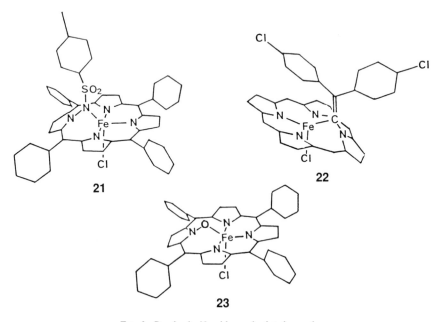

FIG. 9. Porphyrin *N*-oxides and related complexes.

Several crucial differences in the reaction of **16** in toluene and in dichloromethane have suggested a change in the mechanism of O–O bond cleavage (Scheme VII). Homolysis of the O–O bond of **16** to give **23** was indicated from the following observations. (1) The decomposition of **16** to afford **23** was accelerated by electron-donating groups on the aromatic ring of peroxybenzoic acid derivatives, whereas electron-withdrawing groups retarded the reaction ($\rho = -0.4$). Similar reactivity has been reported for the thermal decomposition of substituted dibenzoylperoxides in which homolytic O–O bond cleavage certainly occurs. (2) When phenylperoxyacetic acid was employed as the peroxy acid, $Fe^{IV}(=O)$ TMP (**29**) was obtained instead of **23** along with decarboxylation products of phenyacyloxyl radical, toluene, benzyl alcohol, and benzaldehyde. The reaction to form **23** required 2 equivalents of peroxy acid and led to the concomitant formation of dibenzoyl peroxide.

A mechanism to account for these observations is shown in Scheme VII. Homolysis of the O–O bond of **16** to produce an acyloxyl radical and **24** is considered to be a rate-determining step. Rapid decarboxylation affording $Fe^{IV}(=O)$TMP (**29**) would be expected when phenylperoxyacetic acid was used, since the rate of decarboxylation of the acyloxyl radical is known to be nearly diffusion controlled (*90*). Decarboxylation of benzoyloxyl radicals is much slower ($k_{dec} = \sim 10^4$ sec^{-1}) (*91*); accordingly the recoupling process to produce **25** could be faster. The *N*-oxyacyl group in **25** should be a reactive electrophile (*92*). Thus, the reaction of peroxy acid remaining in the system with **25** would yield both the *N*-oxide, **23,** and dibenzoyl peroxide. Coordination of *m*-chlorobenzoate in **23** was confirmed by the IR spectrum. Olefins were readily oxidized by **14,** but **23** was found to be much less reactive. The *N*-oxide **23** did not epoxidize olefins at all, but it did oxidize triphenylphosphine to the corresponding oxide (*79*). Accordingly, it is hard to believe that an iron(III) porphyrin *N*-oxide such as **23** could be the active species in the *P*-450 reaction.

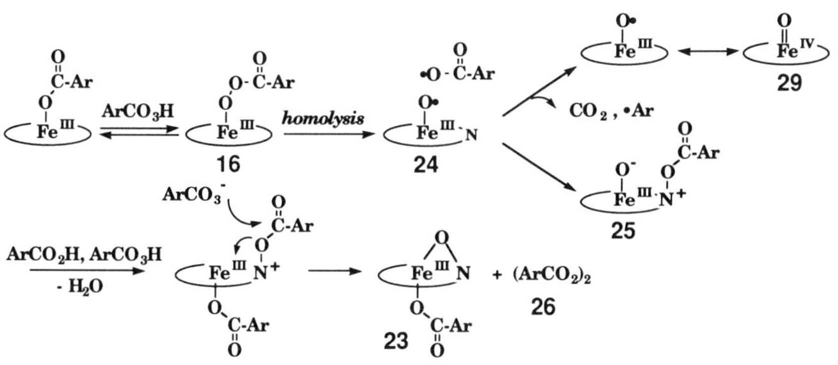

SCHEME VII. Proposed mechanism for the Fe^{III}TMP *N*-Oxide formation.

Thus, two distinct reactions are observed for acylperoxo–iron(III) complexes involving homolytic and heterolytic O–O bond cleavage processes. This implies that the enzymic system would have similar pathways. No iron porphyrin N-oxides have been observed with heme proteins although two competing pathways for the reaction of P-450 with peroxy acids have been described (57, 58, 93). When the reaction of mCPBA and Fe(III)TPP was carried out in toluene at − 40°C, no formation of the N-oxide was observed, but degradation of FeTPP was observed. These results suggested that the protection of the meso position of the porphyrin ring would be important to stabilize an N-oxyacyl intermediate. In fact, whereas the tetrakis(pentachlorophenyl)porphyrinato–iron(III) complex gave the N-oxide, the corresponding pentafluoro complex did yield the N-oxide, with a substantial amount of heme degradation. Acylation of octaethylporphyrin N-oxide was reported to afford 5-acetoxyoctaethylporphyrin (Scheme VIII) and treatment with a trace amount of HCl gave 5-chlorooctaethylporphyrin (94). If a similar process proceeds for FeTPP, subsequent oxidation could cause heme destruction.

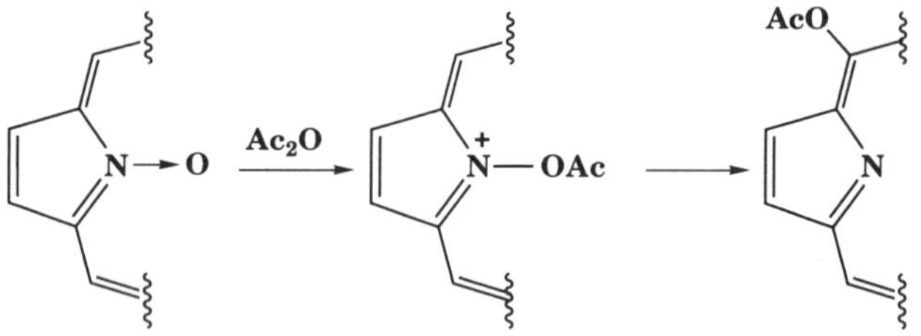

SCHEME VIII. Acylation of octaethylporphyrin N-oxide.

I. MANGANESE PORPHYRIN COMPLEXES

Manganese porphyrins have also been examined as model systems for P-450 (75, 95). In 1980, Hill and Schardt and Groves et al. reported that Mn(III)TPP(Cl) was able to catalyze hydroxylation and epoxidation with PhIO as the oxidant (96, 97). Several Mn(IV) porphyrin complexes were subsequently isolated and structurally characterized (Fig. 10) (98). Other oxidants, such as NaOCl, have also been shown to afford efficient hydrocarbon oxygenation with Mn(III) porphyrins (99). A reactive oxo-Mn(V) species (27) (see Scheme IX) has been postulated to be responsible for oxygen transfer; however, there has yet to be a definitive characterization of any oxo-manganese(V) complex. A brown complex isolated from the treatment of Mn^{III}TPP(Cl) with NaOCl has properties consistent with a manganese(IV) bishypoclorite (98–100). Manganese porphyrin complexes have been very useful in relating the chemistry of peroxo- and

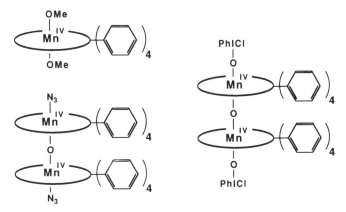

FIG. 10. Manganese(IV) porphyrin complexes.

oxo-metalloporphyrin complexes. The peroxo complex, $Mn^{III}TMP(O_2^{2-})$ (Scheme IX) (**12**), can be prepared according to the method of Valentine (*42, 43*). Acylation of the peroxo ligand of **12** was found to take place smoothly to yield the acylperoxo–Mn(III)TMP complex (**29**). The independent preparation of **29** was achieved by the reaction of $Mn^{III}TMP(OH)$ with 1 equivalent of peroxy acid in acetonitrile. When this reaction was carried out in dichloromethane at low temperature, **29** was observed to decompose spontaneously to give a higher valent manganese porphyrin complex, **27** (*75*). Complex **27** was shown to be EPR silent and was able to oxidize Mn(III) porphyrins to Mn(IV), suggesting that the complex was $Mn^V(=O)TMP$ (*95*).

Recently, the first authentic oxo-maganese(IV) porphyrin complex,

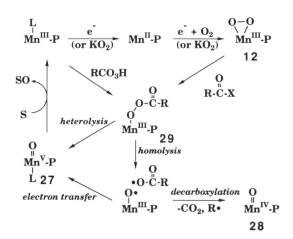

SCHEME IX. *P*-450-type oxygen activation by MnTMP. P, Porphyrins.

$Mn^{IV}(=O)TMP$ (**28**) was isolated and characterized (*101*). Infrared and Raman data have shown that the manganese–oxygen bond was unusually weak ($\nu_{Mn=O}$ 574). The weakening of the $Mn=O$ π-bond was ascribed to the high-spin d^3 electronic structure of **31** and the special nature of the resulting half-filled t_{2g} orbital subset. Interestingly, an isoelectronic nitrido-iron(V) porphyrin recently prepared by low-temperature matrix photolysis of an iron(III) azide shows an unusually weak iron–nitrogen bond (876 cm^{-1}) (*102*).

Studies of the stereochemistry of olefin epoxidation by manganese porphyrins have shown that there are two reactive species. The oxo-Mn(IV)TMP species, **28**, will epoxidize *cis-β*-methylstyrene with loss of the original double bond configuration but without appreciable ^{18}O incorporation from added $H_2^{18}O$. By contrast, the fleeting oxo-manganese(V) complex **27** reacted rapidly with olefins even at low temperature to give epoxides by a stereoselective pathway (Scheme X). Further, the cis epoxide produced in this way showed significant ^{18}O incorporation from added water (*103*). It was concluded from these results that **27**, formally a catalytic complex $[Mn^V(=O)TMP]^+$, exchanged its oxo-oxygen readily with water and reacted stereospecifically with olefins.

Taken together, the various reactions and interconversions of these manganese porphyrin complexes have allowed the examination of each step in the activation of molecular oxygen by the mechanism suggested for *P*-450. Detailed mechanistic studies of the O–O bond cleavage event in **29** by kinetics, substituent effects, and product analysis showed that the reaction proceeds via heterolysis to produce **27** when acid is present, whereas homolysis is predominant in the absence of acid but in the presence of hydroxide ion (*95*). Under basic conditions, homolytic cleavage of the O–O bond of **29** forms $Mn^{IV}(=O)TMP$ (**28**) and an acyloxyl radical. Thus, when an alkyl peroxy acid is employed, decarboxylation competes with electron transfer, as shown in Scheme IX, to afford a mixture of **27** and **28**. Yuan and Bruice have proposed a similar heterolysis mechanism based on the kinetic analysis for the reaction of mCPBA with catalytic amounts of $Mn^{III}TPP$ (*104*).

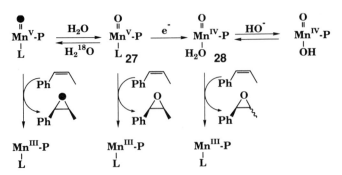

SCHEME X. Oxidation of *cis-β*-methylstyrene by Mn^{IV} and Mn^V porphyrin complexes.

J. OXIDATION OF SUBSTRATES

In this section, oxidation reaction mechanisms catalyzed by *P*-450 are reviewed.

1. *Aliphatic Hydroxylation*

The selective conversion of C–H bonds to alcohols by *P*-450 is certainly a remarkable transformation. With the proper choice of substrates, hydroxylation has been mechanistically very revealing. Some forms of *P*-450, particularly those of the adrenal cortex, which mediate steroidal metabolism (*105*), are highly regioselective and stereoselective (Scheme XI). Whereas reactions of bio-

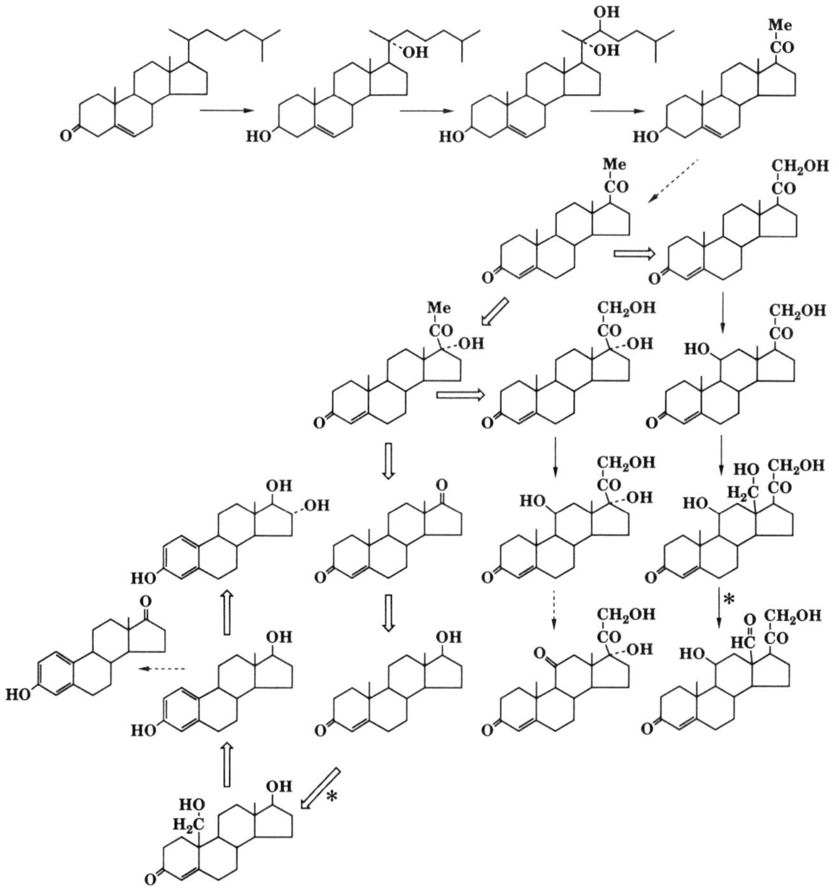

SCHEME XI. Biosynthesis of steroid hormones. →, *P*-450-catalyzed monooxygenase reactions in mitochondria; ⇒, *P*-450-catalyzed monooxygenase reactions in microsomes; -----→, reactions in microsomes; *, undecided.

synthetic P-450 enzymes are not described in this chapter, they have been well documented in a recent review article ($5a$).

Some years ago we showed that the hydroxylation of norbornane by P-450$_{LM2}$ gave a stereoisomeric mixture of alcohol ($106a$). Selective deuteration showed that the stereochemistry of hydrogen removal and that of alcohol formation were chemically distinct events and that the C–H bond cleavage proceeded with a very large isotope effect ($k_H/k_D \approx 11$). Accordingly, the reaction must be stepwise involving a carbon radical intermediate that lives long enough to epimerize partially at the active site. In related studies, the allylic hydroxylation of cyclohexene was shown to be accompanied by allylic rearrangement and the hydroxylation of phenylethane was shown to be stereochemically indiscriminate ($106b$, $106c$).

In contrast to these results, the ω-hydroxylation of octane by a rat liver microsomal system containing P-450 produced 1-octanol with predominant retention of configuration (107). These results can also be accommodated by a hydrogen abstraction–radical recombination mechanism if collapse of the primary carbon radical is fast due both to its relative instability and to its lack of steric configuration.

The X-ray crystal structure of P-450$_{cam}$ clearly shows the substrate, d-camphor, bound close to the heme with hydrogen bonding extending from Tyr-96 to the carbonyl oxygen of the substrate (14). Interestingly, this enzyme has been shown by Sligar and co-workers to remove either the 5-exo or 5-$endo$ hydrogen. However, 5-exo alcohol was the only product (108). Thus, P-450$_{cam}$ exerts its product-determining specificity on the intermediate carbon radical and not on the starting substrate (108). These observations can also be accommodated by the stepwise hydroxylation pathway outlined in Scheme XII.

Stearns and Ortiz de Montellano have suggested a single-electron transfer mechanism for the oxidation of quadricyclane (**30**) by P-450 (109). this substrate has a very low oxidation potential for an alkane (0.9 V), and the oxidation product (**31**) was characteristic of an intermediate cation radical (Scheme XIII) (109). The oxidation of norbornadiene also gave **31**. Similar electron transfer mechanisms have been proposed for several other oxidations (*vide infra*).

The hydroxylation of cycloheptane by Fe^{III}TPP(Cl)/PhIO in the presence of bromotrichloromethane gave 24% cycloheptanol and 18% bromocycloheptane (110). The hydrogen isotope effect for cyclohexane hydroxylation by this model system was found to be 12.9, similar to that of the enzyme system ($106a$, 111). A mechanism for hydroxylation proposed by Groves and Nemo on the basis of these results involves initial hydrogen abstraction from the alkane and rapid collapse of this radical to afford the product alcohol (Scheme XIV) (110). Lindsay-Smith *et al.* have also reported a large kinetic isotope effect in O-demethylation by the Fe^{III} porphyrin/PhIO system (112). When Mn^{III} porphyrin/PhIO was applied to hydroxylation in halogenated solvents, incorporation of the halogen in

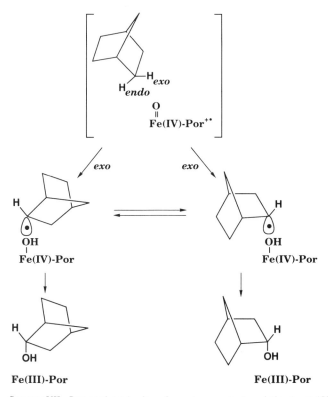

SCHEME XII. Proposed mechanisms for norbornane hydroxylation by *P*-450.

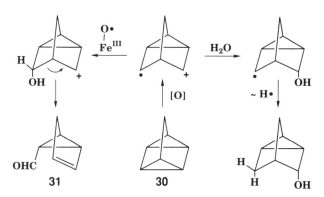

SCHEME XIII. A single-electron transfer mechanism proposed for the oxidation of quadricyclane by *P*-450.

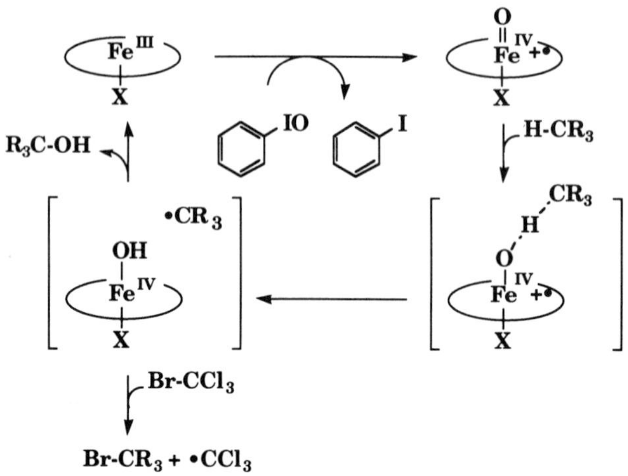

SCHEME XIV. Formation of bromoalkane in the hydroxylation reaction.

the products from the solvent was also observed (*96, 97, 113*). Moreover, the oxidation of norcarane (**32**) gave products derived from the ring opening of the exocyclic C–C bond in addition to cyclopropyl carbon products (Scheme XV) (*97*). Such a rearrangement is characteristic of a carbon radical intermediate and is counterindicative of a carbonium ion.

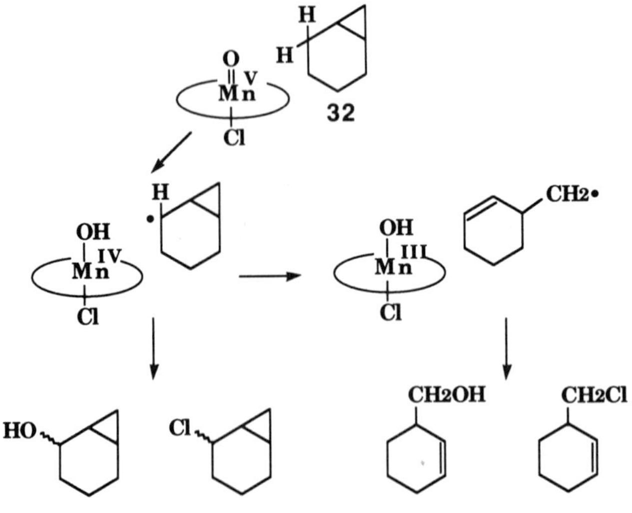

SCHEME XV. Oxidation of norcarane catalyzed by the MnTPP complex.

2. *Olefin Epoxidation Reaction*

P-450 has been shown to catalyze epoxidation with retention of the olefin configuration (*114*). Ortiz de Montellano and co-workers have shown that heme N-alkylation accompanies epoxidation when terminal olefins are oxidized by *P*-450 (*115*). Further, the oxidation of 1,1,2-trichloroethylene is known to give trichloroacetaldehyde along with epoxide (*116, 117*). A mechanism that explains simultaneous epoxidation, heme alkylation, and halogen migration is depicted in Scheme XVI (*117*). In this process, initial electron transfer affords a transient π-radical cation that can collapse with C–O bond formation to give either radical or cation intermediates.

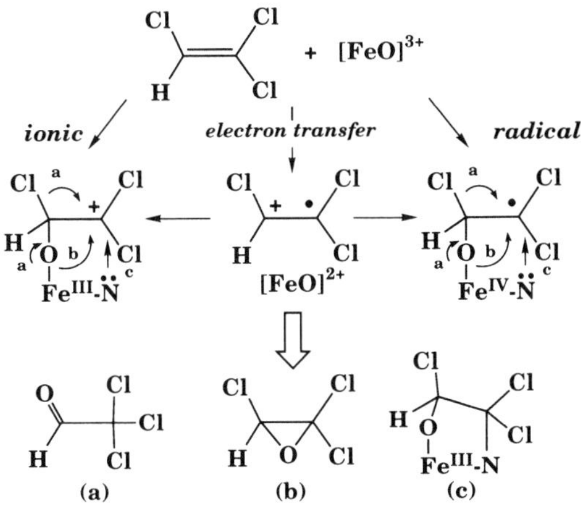

Scheme XVI. Proposed epoxidation mechanism by *P*-450.

Groves and co-workers have reported that the epoxidation of *trans*-1-deuterio-propylene by a reconstituted system with purified P-450$_{LM2}$ proceeded with significant loss of the deuterium labels (*118*). Further, the epoxidation of propylene by this enzyme system in D_2O afforded predominantly *trans*-1-deuteriopropy-lene oxide. When NADPH/O_2 was replaced with PhIO in the enzyme system, no incorporation of deuterium from D_2O in recovered propylene was observed.

In order to explain these observations, a metallacycle intermediate (**33**) (Scheme XVII), which is an alternative way of representing the σ-alkylporphyrin (**36**), was proposed. The metallacycle can be reversibly deprotonated to a carbene complex (**34**) (Scheme XVII [A]). On the other hand, Dolphin *et al.* have suggested that the first step in the deuterium exchange reactions could involve electron transfer from the olefin and N-alkylation of the subsequent cation to give β-hydroxy-N-alkylporphyrin (*119*). Once N-alkylporphyrins were formed,

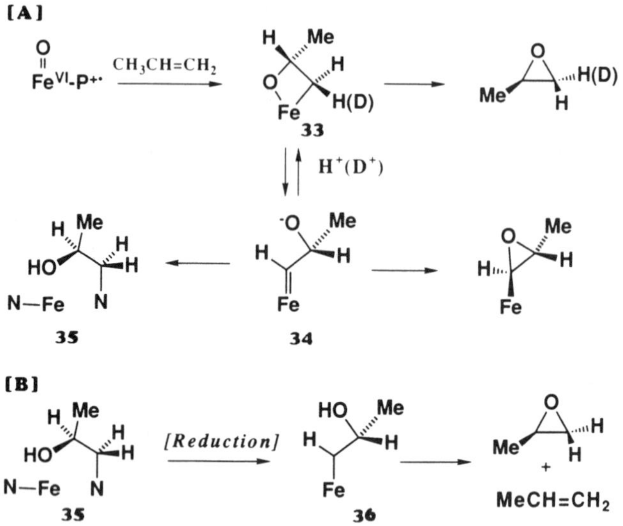

[A]

SCHEME XVII. Reactions of metallacycles.

the migration to form σ-alkylporphyrin (**36**) was observed under reducing conditions (**35**) (Scheme XVII [B]).

The olefin epoxidation reaction has also been extensively studied with model metalloporphyrin systems. The epoxidation of olefins by Fe^{III} porphyrin/PhIO systems generally proceeds by a stereo-retention pathway (Table VIII) (*10, 120a,c*). Asymmetric epoxidations of prochiral olefins have also been demon-

TABLE VIII

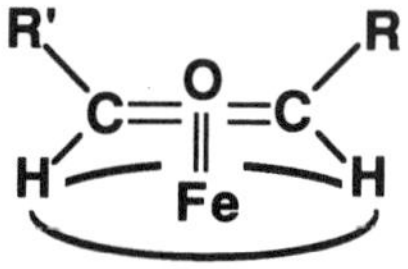

EPOXIDATION OF OLEFINS CATALYZED BY FeTPP(Cl)

Substrate	Product	Yield (%)
cis-Stilbene	*cis*-Oxide	79
trans-Stilbene	*trans*-Oxide	Trace
cis-4-Methyl-2-pentene	*cis*-Oxide	51
trans-4-Methyl-2-pentene	*trans*-Oxide	13
	2-methyl-3-penten-2-ol } 4-methyl-2-penten-1-ol }	2

TABLE IX

Epoxidation of *cis*- and *trans*-Cyclodecene
by Ferric Porphyrin Complexes

Ferric porphyrin	*cis/trans*-Epoxide	Yield (%)
FeTPP(Cl)	1.55	33
FeTNP(Cl)	2.76	98
FeTPPP(Cl)	1.52	55
FeTMP(Cl)	8.92	51
FePPIX-DME	1.27	16
mCPBA	0.59	No data

strated, with modified iron porphyrins having optically active functionalities at the meso positions (*120b, d–f*). Various substituted styrenes and aliphatic olefins were epoxidized, with enantiomeric excesses varying between 0% for 1-methyl-cyclohexene oxide and 51% for *p*-chlorostyrene oxide. Furthermore, *cis*-olefins were found to be more reactive than *trans*-olefins (Table VIII) (*120a*). Thus, whereas *cis*-stilbene was epoxidized by the FeTPP(Cl)/PhIO in good yield, the trans isomer was found to be unreactive under these conditions. The degree of *cis/trans* selectivity in the epoxidation by some Fe porphyrin complexes has been studied. The *cis* selectivity was found to be a sensitive function of the steric bulk of the porphyrin periphery (Table IX) (*120a*). These results suggest an approach of the double bond from the side of the iron-bound oxygen and parallel to the porphyrin plane, as shown in Table VIII. Accordingly, if the epoxidation is initiated by electron transfer from the olefin to the oxo-metallopor-phyrin, close approach of the double bond to the ferryl group is still required for reaction.

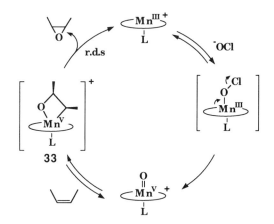

Scheme XVIII. Collman's mechanism; r.d.s., rate-determining step.

The apparent Michaelis–Menten behavior of olefin epoxidation by manganese porphyrins with hypochlorite in water/dichloromethane led Collman *et al.* to suggest that an intermediate, perhaps the oxo-metallocycle **33**, was formed during the reaction and that decomposition of the intermediate was rate limiting (Scheme XVIII) (*121*). Nolte *et al.* (*122*) have investigated this system and have found that an Mn(IV) oxo dimer accumulates. They have suggested that this process is an alternative explanation for the results of Collman *et al.* In a second study, a suspension of pentafluoroiodosylbenzene was stirred in dichloromethane with iron porphyrins and alkenes, and a similar kinetic pattern was observed (*123*). The kinetics, including the inhibition of cyclooctene epoxidation by norbornene and the much slower rate of norbornene epoxidation, were again interpreted as evidence for metallacycle formation.

Traylor *et al.* reached different conclusions in studies of alkene epoxidation. In homogeneous solutions in which iodosylbenzene was solubilized in dichloromethane/alcohol/water, very rapid epoxidation (300 turnovers/sec) was found to be independent of substrates or concentration of alkane (*124*). No dependence on alkene concentrations was reported by Dicken *et al.* when they employed *p*-cyanodimethylaniline *N*-oxide (*125*). Meanwhile, Traylor *et al.* found an accumulation of *N*-alkyliron porphyrin complexes (**35**) during the course of epoxidation of norbornene (*126*), and showed that it is the only intermediate accumulated in either the homogeneous system or the heterogeneous system; **35** was found to form with a bimolecular rate constant of 450 M^{-1} sec^{-1} and decomposes at a rate of 0.07 sec^{-1}. Epoxide formation occurs with a rate constant of at least 10^4 M^{-1} sec^{-1} in these conditions. Thus **35** is not an intermediate in the production of epoxide; however, it could account for at least a small percentage of the products, since **35** was found to catalyze alkene epoxidation by utilizing PhIO (Scheme XIX).

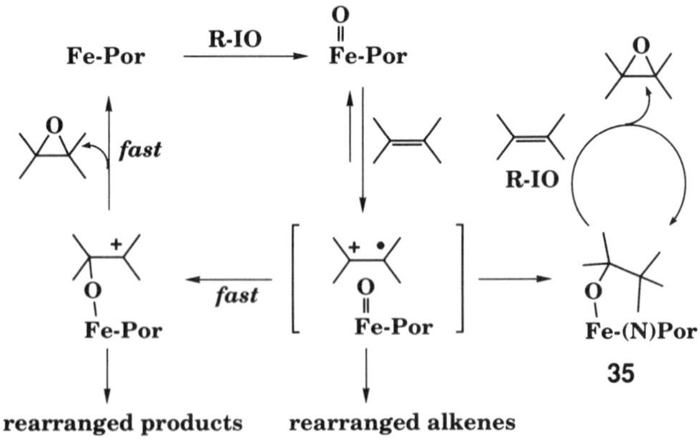

SCHEME XIX. Traylor's *N*-alkyl mechanism.

In order to examine the mechanism of oxygen transfer to olefins from oxo-metalloporphyrins, we examined the stoichiometric reaction of $Fe^{IV}(=O)TMP^{+}$ (**14**) [Eq. (10)] with a series of olefins (*127*). The reaction of **14** with cyclooctene was found to be biphasic, i.e., a rapid spectral change ($t_{1/2} \sim$ 50 sec) gave a dark green intermediate **36**; slow decomposition of **36** ($t_{1/2} \sim$ 1 hr) gave $Fe^{III}TMP$ (**37**) (Fig. 11). ¹H NMR, EPR, and UV–Vis spectra of **36** showed that it was very similar to **14**. The decomposition of **36** was found to afford epoxide; however, quenching **36** with iodide afforded no epoxide. A metallacycle complex such as suggested by Collman *et al.* and a charge transfer

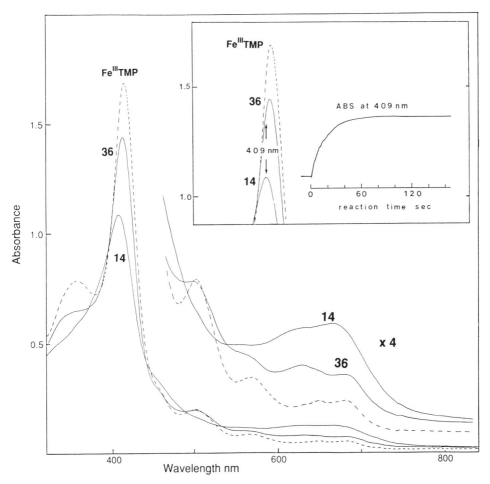

FIG. 11. Spectral changes in the reaction of cyclooctene and $Fe^{IV}(O)TMP^{+\cdot}$

complex are both plausible candidates for this intermediate adduct (*124*). Kinetic studies of the formation for **36** indicate that the reaction was first order in olefin and in porphyrin. No saturation kinetic behavior was observed with excess olefin. Substituent effects on the formation of **36** were examined by comparing the rates of substituted styrenes. A good correlation with the Brown–Okamoto σ^+ ($\rho = -1.9$) was observed.

When imidazole or methanol was introduced to the dichloromethane solution of **36,** instantaneous decomposition of **36** and the epoxide formation were observed. Thus, in the presence of either methanol or imidazole, the rate-determining step in the reaction of **14** and olefin was changed to the formation of **36.** Under these conditions, secondary deuterium kinetic isotope effects on epoxidation were examined by α- and β-deuterio-*p*-chlorostyrenes. For both the α- and the β-positions of styrene, $k_H/k_D = 1$ was observed. The isotope effect and substituent effect on the formation of **36** suggest that both the α- and β-carbons remain planar (sp^2 hybridized) at the transition state and that a positive charge forms on the α-carbon. Accordingly, the formation of an olefin cation radical by an electron transfer from the olefin to **14** is indicated in the formation of **36** (Scheme XX).

SCHEME XX. Reactions of intermediate **36.**

Indeed, oxidation of hexamethyl Dewar benzene (**37**) catalyzed by iron porphyrin yields hexamethylbenzene (**38**) as a product and a small amount of epoxide **39** (Scheme XXI) (*128*). In a similar manner the diene, 1,4,4a,5,8,8a-hexahydro-1,4,5,8-*endo,endo*-dimethanonaphthalene (**B**), was found close to the birdcage hydrocarbon (**C**) under these conditions. This diene also brought about some N-alkylation of the catalyst during the reaction. The formation of **38** also supported the involvement of a cation radical during the epoxidation (Scheme XXI). Now it becomes clear that the epoxidation reaction is not as simple as previously considered. However, deuterium isotope effects, substituent effects, rearrangement, and N-alkylation might be explained by the one-electron transfer mechanism followed by coupling of the oxo-oxygen on metals with the olefin cation radical to afford either a radical or a carbonium intermediate, as shown in Scheme XXII. Then, epoxide formation, rearrangement, or N-alkylation might

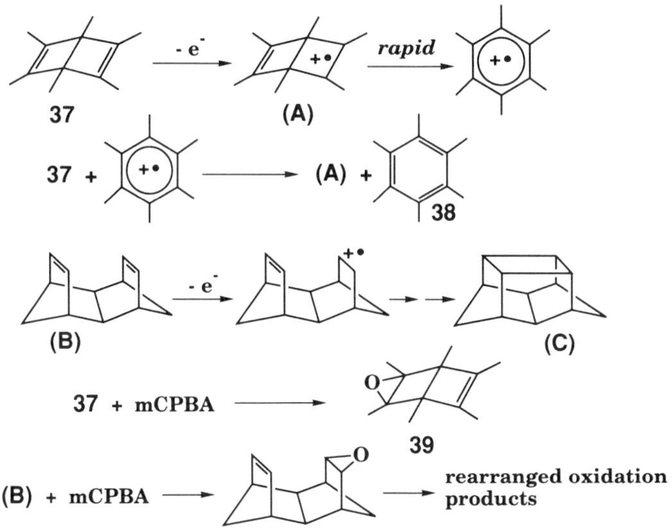

SCHEME XXI. Oxidation of hexamethyl Dewar benzene.

compete. It has been also shown in this section that the rate-determining steps of the epoxidation by the model systems are very dependent on the nature of the metalloporphyrins and solvents.

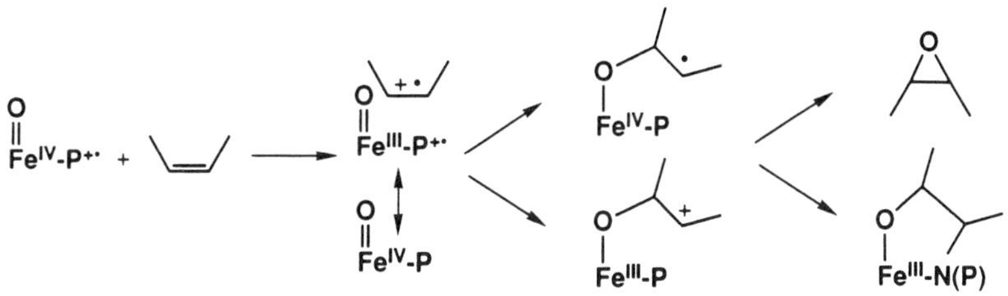

SCHEME XXII. Proposed oxidation mechanism of olefins.

3. Oxygenation of Organic Compounds Having Heteroatoms

The *P*-450 enzyme is known to mediate N- and O-dimethylation (dealkylation) and S-oxidation (Table I) in addition to hydroxylation and epoxidation. Watanabe *et al.* have suggested an electron transfer process for S-oxidation by $P\text{-}450_{LM2}$ based on the following observations (*129a, b*): (1) the oxidation product of *P*-450 is generally *S*-oxide; (2) S-dealkylation competes when sulfides

$$
\begin{array}{c}
& & & & & \overset{O}{\underset{\uparrow}{R\text{-}S\text{-}CH_2R'}} \\
& & & & \nearrow \\
R\text{-}S\text{-}CH_2R' + [FeO]^{3+} \xrightarrow[\text{transfer}]{\textit{electron}} R\text{-}\overset{+\bullet}{S}\text{-}CH_2R' + [FeO]^{2+} \\
& & \mathbf{40} \\
& & & \searrow_{-H^+} R\text{-}S\text{-}\overset{\bullet}{C}HR' + [FeO]^{2+}
\end{array}
$$

$$
\xrightarrow{} \underset{R\text{-}S\text{-}\overset{|}{C}HR'}{\overset{O\text{-}Fe}{}} \xrightarrow{H_2O} \left[\underset{R\text{-}S\text{-}\overset{|}{C}HR'}{\overset{OH}{}} \right] + Fe^{III} \longrightarrow R\text{-}SH + R'CHO \xrightarrow{[O]} 1/2\ RSSR
$$

SCHEME XXIII. Oxidation mechanism of sulfides catalyzed by *P*-450.

bearing active α-hydrogens, such as phenyl phenacyl sulfide, were employed;
(3) the extent of S-dealkylation increased with the acidity of the α-hydrogens of
the sulfide; (4) the dealkylaton activity was depressed if sulfides having substituents that destabilize α-radical intermediates were employed. Scheme XXIII
summarizes these results. When the sulfenium radical (**40**) was prepared by the
reaction of the hydroxy radical and sulfides, the same product distribution as that
observed in the *P*-450 system was obtained (*129a*). Furthermore, kinetic studies
of the enzymic S-oxygenation of substituted thioanisoles showed correlation of
$V_{\max}$ with the Brown–Okamoto σ^+ ($\rho = -0.16$) (*129b, 130*), whereas the oxidation by an electrophilic oxidant such as mCPBA gave only *S*-oxide, with relative rates that correlate with Hammett σ-values (*131*). The oxidation of sulfides
by a model porphyrin system, $Fe^{III}TPP(Cl)/H_2O_2/$imidazole, showed similar
reactivity (Table X) (*132, 133*).

TABLE X

RATIOS BETWEEN S-OXIDATION AND S-DEALKYLATION IN OXIDATION OF
PHENACYL PHENYL SULFIDE WITH VARIOUS OXIDANTS

$$
\underset{(A)}{\overset{O}{\underset{\|}{Ph\text{-}S\text{-}CH_2CPh}}} \longrightarrow \underset{(B)}{\overset{O}{\underset{\uparrow}{Ph\text{-}\overset{}{S}\text{-}CH_2\overset{\|}{C}Ph}}} + 1/2\ PhSSPh + \underset{(C)}{\overset{O}{\underset{\|}{PhCCHO}}}
$$

Oxidation system	Solvent	Ratio (B:C)
P-450/NADPH/O_2	Buffer (pH 7.4)	48 : 52
FeTPP(Cl)/H_2O_2/imidazole	CH_3CN	86 : 14
FeII(CO$_4$)$_2$/H_2O_2	Methanol	49 : 51
FAD-monooxygenase/NADPH/O_2	Buffer (pH 7.4)	100 : 0
4a-FlEt-OOH	Dioxane	100 : 0
mCPBA	CH_2Cl_2	100 : 0
H_2O_2	Acetic acid	100 : 0

Watanabe *et al.* (*134*) and Miwa *et al.* (*135*) have suggested a single-electron transfer mechanism for the enzymic N-demethylation reaction based on the small kinetic isotope effect, since the deprotonation of the α-hydrogen of the aminium radicals prepared by several means is known to proceed with smaller k_H/k_D values (*134, 135*). The oxidation of cyclopropylamine by *P*-450 has been shown to afford products believed to be derived from the aminium radical intermediate (**41**) (Scheme XXIV) (*136*).

By contrast, O-demethylation is considered to begin with direct hydrogen abstraction because of a relatively large kinetic isotope effect ($k_H/k_D \approx 5$–14) (*135–137*). The isotope effect for the deprotonation of an ether cation radical to

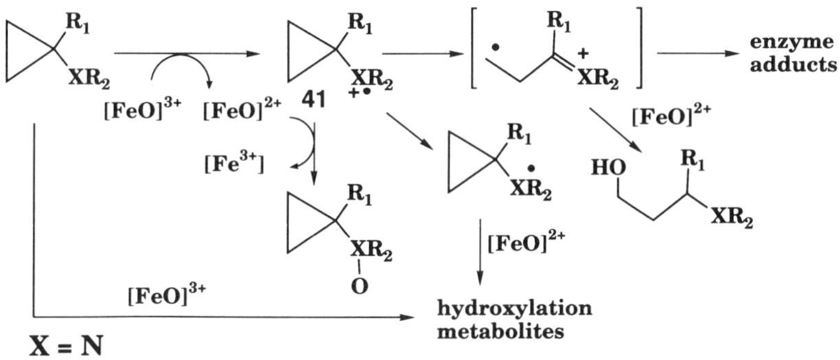

SCHEME XXIV. Proposed oxidation mechanism for cyclopropylamine by *P*-450.

form an ether α-radical is expected to be small (~ 2) (*138*). Thus, V_{max} values obtained for the enzymic O-demethylation of substituted anisoles were independent of their oxidation potentials (*123*). The O-demethylation of ethers by $Fe^{III}TPP/PhIO$ also gave a large kinetic isotope effect (*112*).

IV. Summary

In this review we have discussed the mechanistic features of *P*-450 reactions by comparing enzymic reactions with those of model porphyrin systems. Model porphyrin systems have allowed the complete dissection of the proposed mechanism of oxygen activation and transfer by *P*-450. Putative intermediates have been characterized, and each primitive transformation has now been observed. The similarities and differences in the oxygen activation between HRP and

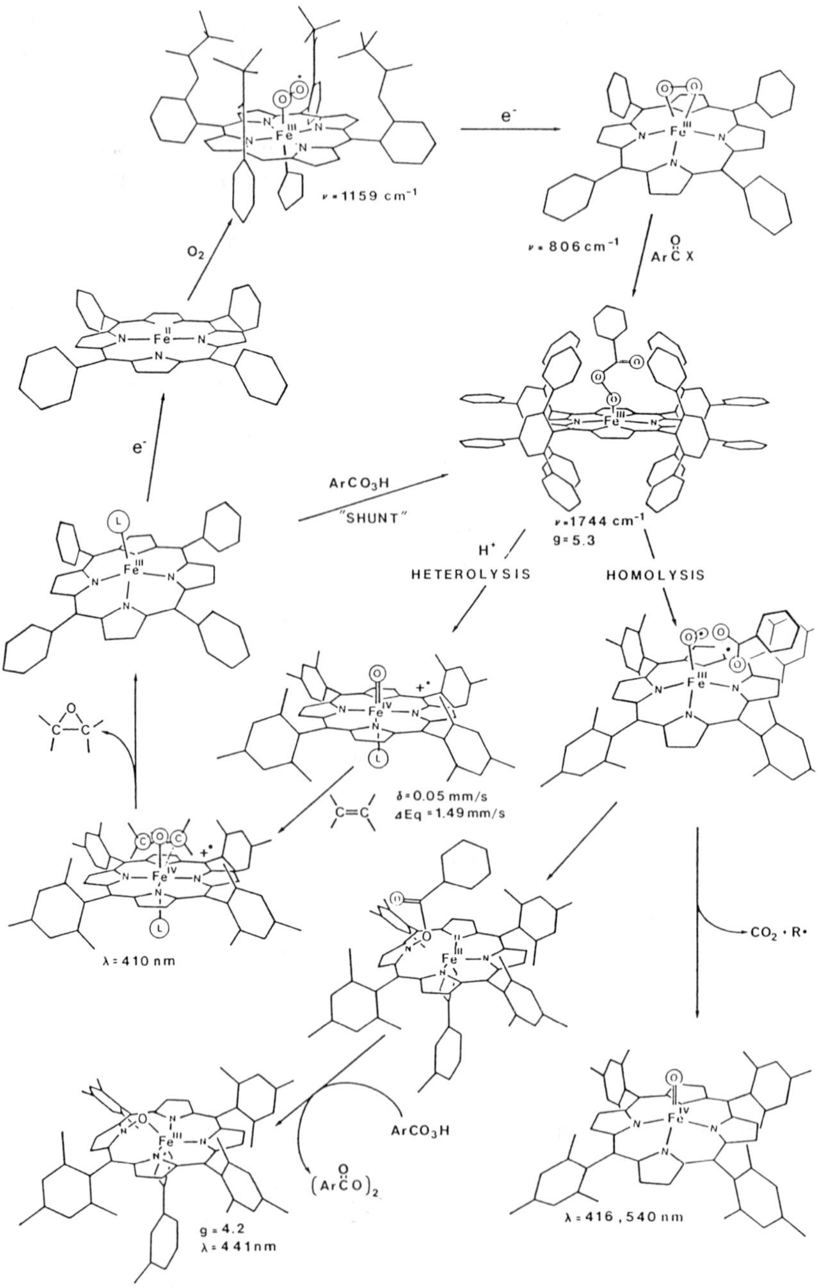

SCHEME XXV. Schematic view of *P*-450-type oxygen activation by Fe porphyrin complexes.

P-450 have also been discussed, especially homolysis and heterolysis of the acylperoxo–iron(III) complex. In both cases, we consider the active species to be oxoferryl porphyrin cation radicals; however, many different reactivities exist between peroxidases and *P*-450. Ortiz de Montellano *et al.* have proposed that the position of substrates in the active site might depend on the spatial characteristics of the individual enzymes and influence the detailed course of the reaction (*139*). These propositions should be carefully examined. Scheme XXV illustrates all of the intermediates that have been observed and/or proposed in the oxygen activation mechanism by *P*-450 and that have been prepared by the model systems.

ACKNOWLEDGMENTS

This article is dedicated to the memory of the late Professor Iwao Tabushi, whose work helped to pioneer this field of study.

Research described herein, which was carried out at the University of Michigan and Princeton University, was generously supported by the National Institutes of Health. The authors are grateful to Professors Yuzuru Ishimura, Ryu Makino (Keio University), and David Dolphin (University of British Columbia) for their helpful discussions.

REFERENCES

1a. Sato, R., and Omura, T. (eds.) (1978). "Cytochrome P-450." Kodansha, Tokyo.
1b. White, R. E., and Coon, M. J. (1980). *Annu. Rev. Biochem.* **50,** 315.
1c. Ortiz de Montellano, P. R. (ed.) (1986). "Cytochrome P-450." Plenum, New York.
1d. Dawson, J. H., and Sono, M. (1986). *Chem. Rev.* **87,** 1255.
2. Omura, T., and Sato, R. (1962). *J. Biol. Chem.* **237,** 1375; Omura, T., and Sato, R. *J. Biol. Chem.* **239,** 2370.
3. Gasgunitim Y., Yamano, T., and Mason, H. S. (1962). *J. Biol. Chem.* **237,** 3843.
4. Estabrook, R. W., Cooper, D. Y., and Rosenthal, Q. (1963). *Biochemistry* **338,** 741.
5a. Ortiz de Montellano, P. R. (1986). *In* "Cytochrome P-450" (P. R. Ortiz de Montellano, ed.), p. 217. Plenum, New York.
5b. Makino, R., and Ishimura, Y. (1982). *Taisha* **19,** 27.
6a. Collman, J. P., Gagne, R. R., Reed, C. A., Halbert, T. R., Lang, G., and Robinson, W. J. (1975). *J. Am. Chem. Soc.* **97,** 1427.
6b. Jameson, G. B., Molinaro, F. S., Ibers, J. A., Collman, J. P., Brauman, J. I., Rose, E., and Suslick, K. S. (1980). *J. Am. Chem. Soc.* **102,** 3224.
7. Udenfriend, S., Clark, C. T., Axelrod, J., and Brodie, B. (1954). *J. Biol. Chem.* **208,** 731.
8. Hamilton, G. A. (1964). *J. Am. Chem. Soc.* **86,** 3391; Hamilton, G. A., Workman, R. J., and Woo, L. (1964). *J. Am. Chem. Soc.* **86,** 3390.
9. For example: Matsuura, T. (1977). *Tetrahedron* **33,** 2869.
10. Groves, J. T., Nemo, T. E., and Myers, R. S. (1979). *J. Am. Chem. Soc.* **101,** 1032.

11. For example: Black, S. D., and Coon, M. J. (1986). *In* "Cytochrome P-450" (P. R. Ortiz de Montellano, ed.), p. 161. Plenum, New York.

12. Katagiri, M., Ganguni, B. N., and Gunsalus, I. C. (1968). *J. Biol. Chem.* **243**, 3543.

13. Poulos, T. L., Finzel, B. C., Gunsalus, I. C., Wagner, G. C., and Kraut, J. (1984). *J. Biol. Chem.* **260**, 16122.

14. Poulos, T. L., Finzel, B. C., and Howard, A. J. (1986). *Biochemistry* **25**, 5314.

15. Conney, A. H. (1976). *Pharmacol. Rev.* **19**, 317; Hanzlik, R. P., Shearer, G. O., Hamburg, A., and Gillesse, T. (1978). *Biochem. Pharmacol.* **27**, 1435; Hanzlik, R. P., and Shearer, G. O. (1978). *Biochem. Pharmacol.* **27**, 1441; Baggett, B., Engel, L. L., Savard, K., and Dorfman, R. (1956). *J. Biol. Chem.* **221**, 931; Ryan, K. J. (1959). *J. Biol. Chem.* **234**, 786.

16. Horie, S. (1978). *In* "Cytochrome P-450" (R. Sato and T. Omura, eds.), p. 73. Kodansha, Tokyo.

17a. Tsai, R., Yu, C.-A., Gunsalus, I. C., Peisack, J., Blumberg, W., Orme-Johnson, W. H., and Beinert, H. (1970). *Proc. Natl. Acad. Sci. U.S.A.* **66**, 1157.

17b. Peterson, J. A. (1971). *Arch. Biochem. Biophys.* **144**, 678.

17c. Sharrock, M., Munck, E., Debrunner, P., Lipscomb, J. P., Mahskall, V., and Gunsalus, I. C. (1973). *Biochemistry* **12**, 253.

18. Yu, C.-A., Gunsalus, I. C., Katagiri, M., Suhara, K., and Takemor, S. (1974). *J. Biol. Chem.* **249**, 94; Gunsalus, I. C., Mecks, J. R., Lipscomb, J. D., Debrunner, P., and Munck, E. (1974). *In* "Molecular Mechanisms of Oxygen Activation" (O. Hayaishi, ed.), p. 559. Academic Press, New York.

19. Sligar, S. G., Cinti, D. L., Gibson, G. G., and Shenkman, J. B. (1979). *Biochem. Biophys. Res. Commun.* **90**, 925; Light, D. R., and Orme-Johnson, N. R. (1981). *J. Biol. Chem.* **256**, 343; Guengerich, F. P., Ballou, D. P., and Coon, M. J. (1975). *J. Biol. Chem.* **250**, 7405.

20a. Bottomley, L. A., and Kadishi, K. M. (1981). *Inorg. Chem.* **20**, 1348.

20b. Kadishi, K. M. (1982). *In* "Iron Porphyrins (A. B. P. Lever and H. B. Gray, eds.), Vol. 1, p. 161. Addison-Wesley, Reading, Massachusetts.

21. Stern, J. O., and Peisach, J. (1974). *J. Biol. Chem.* **249**, 7495.

22. Chang, C. K., and Dolphin, D. (1975). *J. Am. Chem. Soc.* **97**, 5984.

23. Collman, J. P., Sorrell, T. N., and Hoffman, B. M. (1975). *J. Am. Chem. Soc.* **97**, 913; Collman, J. P., and Sorrell, T. N. (1975). *J. Am. Chem. Soc.* **97**, 4133; Collman, J. P., and Groh, S. E. (1982). *J. Am. Chem. Soc.* **104**, 1391.

24. Tang, S. D., Koch, S., Papaefthymious, G. C., Foner, S., Frankel, R. B., Ibers, J. A., and Holm, R. H. (1976). *J. Am. Chem. Soc.* **98**, 2414.

25. Cramer, S. P., Dawson, J. H., Hodgson, K. O., and Hager, L. P. (1978). *J. Am. Chem. Soc.* **100**, 7282.

26. Caron, C., Mitschler, A., Ricard, G. R. L., Schappacher, M., and Weiss, R. (1979). *J. Am. Chem. Soc.* **101**, 7401.

27. Ishimura, Y., Ullrich, V., and Peterson, J. A. (1971). *Biochem. Biophys. Res. Commun.* **42**, 140.

28. Tyson, C., Lipscomb, J. D., and Gunsalus, I. C. (1972). *J. Biol. Chem.* **247**, 5777.

29. Bonfils, C., Debey, P., and Maurel, P. (1979). *Biochem. Biophys. Res. Commun.* **88**, 1301; Oprian, D. D. and Coon, M. J. (1982). *In* "Microsomes, Drug Oxidations, and Drug Toxicity" (R. Sato and R. Kato, eds.), p. 139. Japan Scientific Press, Tokyo; Tuckey, R. C., and Kamin, H. (1982). *Fed. Proc.* **41**, 1405.

30. Collman, J. P., Brauman, J. E., Rose, E., and Suslick, K. S. (1978). *Proc. Natl. Acad. Sci. U.S.A.* **75**, 1052.

31. Collman, J. P. (1977). *Acc. Chem. Res.* **10**, 265.

32. Cheung, S. K., Grimes, C. J., Wong, J., and Reed, C. A. (1986). *J. Am. Chem. Soc.* **98**, 5028; Barlow, C. H., Maxwell, J. C., Wallace, W. J., and Caughey, W. S. (1973). *Bio-*

chem. Biophys. Res. Commun. **55,** 91; Maxwell, J. C., Volpe, J. A., Barlow, C. H., and Caughey, W. S. (1974). *Biochem. Biophys. Res. Commun.* **58,** 166.

33. Spaetalian, K., Lang, G., Collman, J. P., Gange, R. R., and Reed, C. A. (1975). *J. Chem. Phy.* **63,** 5375.

34. Reed, C. A., and Cheung, S. K. (1977). *Proc. Natl. Acad. Sci. U.S.A.* **74,** 1780.

35. Pederson, T. C., Austine, R. H., and Gunsalus, I. C. (1977). *In* "Microsomes and Drug Oxidation" (V. Ullich, I. Roots, A. G. Hildebrant, R. W. Estabrook, and A. H. Conney, eds.), p. 275. Pergamon, Oxford; Hui Bon Hoa, G., Begard, E., Debey, P., and Gunsalus, I. C. (1978). *Biochemistry* **17,** 2835.

36. Nordblom, G. D., and Coon, M. J. (1977). *Arch. Biochem. Biophys.* **180,** 343.

37. McCandlish, E., Miksztal, A. R., Nappa, M., Sprenger, A. R., Valentine, J. S. Strong, J. D., and Spiro, T. G. (1980). *J. Am. Chem. Soc.* **102,** 4268.

38. Asaka, R. (1970). *J. Chem. Phys.* **52,** 3919; Asaka, R., Malmstrom, B. G., Saltman, P., and Vanngard, T. (1963). *Biochim. Biophys. Acta* **75,** 203.

39. Friant, P., Goulon, J., Fischor, J., Ricard, L., Schappacher, M., Weiss, R., and Momenteau, M. (1985). *Nouv. J. Chim.* **9,** 33.

40. Reed, C. A. (1982). *In* "Electrochemical and Spectrochemical Studies of Biological Redox Complex: Advances in Chemistry 201" (K. M. Kadishi, ed.), p. 333. American Chemical Society, Washington, D.C.

41. Welborn, C. H., Dolphin, D., and James, B. R. (1981). *J. Am. Chem. Soc.* **103,** 2869.

42. Van Atta, R. B., Strouse, C. E., Hanson, L. K., and Valentine, J. S. (1987). *J. Am. Chem. Soc.* **109,** 1425.

43. Valentine, J. S., and Quinn, A. E. (1976). *Inorg. Chem.* **15,** 1997.

44. Shirazi, A., and Goff, H. M. (1982). *J. Am. Chem. Soc.* **104,** 6318.

45. Goff, H. M. (1982). *In* "Iron Porphyrins" (A. B. P. Lever and H. B. Gray, eds.), Vol. 1, p. 237. Addison-Wesley, Reading, Massachusetts.

46. McMurry, T. J., and Groves, J. T. (1986). *In* "Cytochrome P-450" (P. R. Ortiz de Montellano, ed.), p. 1. Plenum, New York.

47. Sligar, S. G., Kennedy, K. A., and Person, D. C. (1980). *Proc. Natl. Acad. Sci. U.S.A.* **77,** 1240.

48. Poulos, T. L., and Kraut, J. (1980). *J. Biol. Chem.* **255,** 8199.

49. Walsh, C. (1979). "Enzymatic Reaction Mechanisms," p. 448. Freeman, San Francisco, California; Hewson, W. P., and Hager, L. P. (1979). *In* "The Porphyrins" (D. Dolphin, ed.), Vol. 7, p. 295. Academic Press, New York.

50. Chance, B. (1949). *Arch. Biochem. Biophys.* **21,** 416; Asada, R., Vanngard, T., and Dunford, H. B. (1975). *Biochim. Biophys. Acta* **39,** 259; Robert, J. E., Hoffman, B. N., Rutter, R., and Hager, L. P. (1981). *J. Biol. Chem.* **256,** 2118; Dolphin, D., Forman, A., Borg, D. C., Fajar, J., and Felton, R. H. (1971). *Proc. Natl. Acad. Sci. U.S.A.* **68,** 641.

51. Hollenberg, P. F., and Hager, L. P. (1974). *J. Biol. Chem.* **248,** 2630; Palic, M. M., Rutter, R., Araiso, T., Hager, L. P., and Dunford, H. B. (1980). *Biochem. Biophys. Res. Commun.* **94,** 1123.

52. Kadlibar, F. F., Morton, K. C., and Ziegler, D. N. (1973). *Biochem. Biophys. Res. Commun.* **54,** 1225.

53. Nordblom, G. D., White, R. E., and Coon, M. J. (1976). *Arch. Biochem. Biophys.* **175,** 524.

54. Lichtenberger, F., Nastainczyk, W., and Ullrich, V. (1976). *Biochem. Biophys. Res. Commun.* **70,** 939.

55. Hrycay, E. G., Gustafsson, J.-A., Ingelman-Sundburg, M., and Ernster, L. (1975). *Biochem. Biophys. Res. Commun.* **66,** 209.

56. Gustafsson, J.-A., Rondahl, L., and Bergman, J. (1979). *Biochemistry* **18,** 865; Berg, A., Ingelman-Sundberg, M., and Gustafsson, J.-A. (1979). *J. Biol. Chem.* **254,** 5264.

57. McCarthy, M. B., and White, R. E. (1983). *J. Biol. Chem.* **258**, 9153.
58. McCarthy, M. B., and White, R. E. (1983). *J. Biol. Chem.* **258**, 11610.
59. Poulos, T. L., and Finzel, B. C. (1984). *In* "Peptide and Protein Reviews" (M. W. Hearn, ed.), p. 115. Dekker, New York.
60. Chin, D.-H., Balch, A. L., and La Mer, G. N. (1980). *J. Am. Chem. Soc.* **102**, 1446.
61. Traylor, T. G., and Popovitz-Biro, R. (1988). *J. Am. Chem. Soc.* **110**, 239.
62. Dawson, J. R., Holm, R. H., Trudell, J. R., Barth, G., Linder, R. E., Bunnenberg, E., Djerassi, C., and Tang, S. C. (1976). *J. Am. Chem. Soc.* **98**, 3707.
63. Imai, M., Shimada, H., Watanabe, Y., Matsushima-Hibiya, Y., Makino, R., Koga, H., Horiuchi, T., and Ishimura,Y. (1989). *Proc. Natl. Acad. Sci. U.S.A.* **86**, 7823; Shimada, H., Makino, R., Imai, M., Horiuchi, T., and Ishimura, Y. (1990). "Yamada Conference XXVII," p. 107. International Symposium on Oxygenases and Oxygen Activation, Kyoto, Japan.
64. Martinis, S. A., Atkins, W. M., Stayton, P. S., and Sligar, S. G. (1989). *J. Am. Chem. Soc.* **111**, 9252.
65. Imai, Y., and Nakamura, M. (1989). *Biochem. Biophys. Res. Commun.* **158**, 717.
66. Poulas, T. L., and Raag, R. (1990). "Yamada Conference XXVII," p. 103. International Symposium on Oxygenases and Oxygen Activation, Kyoto, Japan.
67. Groves, J. T., Haushalter, R. C., Nakamura, M., Nemo, T. E., and Evans, B. J. (1981). *J. Am. Chem. Soc.* **103**, 2884.
68. Boso, B., Lang, G., McMurry, T. J., and Groves, J. T. (1983). *J. Chem. Phys.* **79**, 1122.
69. Schultz, C. E., Devaney, P. W., Winkler, H., Debrunner, P. G., Doan, N., Chiang, R., Rutter, R., and Hager, L. P. (1979). *FEBS Lett.* **103**, 102; Moss, T. H., Ehrenberg, A., and Beraden, A. J. (1968). *Biochemistry* **8**, 4159.
70. Penner-Hahn, J. E., Elbe, K. S., McMurry, T. J., Renner, M., Balch, A. L., Groves, J. T., Dawson, J. H., and Hodgson, K. O. (1986). *J. Am. Chem. Soc.* **108**, 7819.
71. Hashimoto, S., Tstsuno, Y., and Kitagawa, T. (1987). *J. Am. Chem. Soc.* **109**, 8096.
72. Traylor, T. G., Lee, W. A., and Stynes, D. V. (1984). *J. Am. Chem. Soc.* **106**, 755.
73. Lee, W. A., and Bruice, T. C. (1985). *J. Am. Chem. Soc.* **107**, 153.
74. Tabushi, I., Kodera, M., and Yokoyama, M. (1985). *J. Am. Chem. Soc.* **107**, 4466.
75. Groves, J. T., Watanabe, Y., and McMurry, T. J. (1983). *J. Am. Chem. Soc.* **105**, 4489.
76. Tabushi, I., and Kodera, M. (1986). *J. Am. Chem. Soc.* **108**, 1101.
77. Khenkin, A. M., and Shteinman, A. A. (1984). *J. Chem., Soc. Chem. Commun.* **20**, 1219.
78. Groves, J. T., and Watanabe, Y. (1986). *J. Am. Chem. Soc.* **108**, 7834.
79. Groves, J. T., and Watanabe, Y. (1988). *J. Am. Chem. Soc.* **110**, 8443.
80. Groves, J. T., and Watanabe, Y. (1987). *Inorg. Chem.* **26**, 785.
81. Chen, M. J. Y., and Kochi, J. K. (1977). *J. Chem. Soc., Chem. Commun.* p. 204.
82. Loew, G. H., Kent, C. J., Hjelmeland, L. M., and Kirehner, R. F. (1977). *J. Am. Chem. Soc.* **99**, 3534.
83. Tatsumi, K., and Hoffmann, R. (1981). *Inorg. Chem.* **20**, 3771.
84. Strich, A., and Veillard, A. (1983). *Nouv. J. Chim.* **7**, 347.
85. Jørgensen, K. A. (1987). *J. Am. Chem. Soc.* **109**, 698.
86. Balch, A. L., Chan, Y. W., Olmstead, M. M., and Renner, M. W. K. (1985). *J. Am. Chem. Soc.* **107**, 2393; Balch, A. L., Chan, Y. W., Olmstead, M. M., and Renner, M. W. K. (1985). *J. Am. Chem. Soc.* **107**, 6510.
87. Mahy, J.-P., Battioni, P., and Mansuy, D. (1986). *J. Am. Chem. Soc.* **108**, 1079.
88. Chavrier, B., Weiss, R., Lange, M., Chottard, J.-C., and Mansuy, D. (1981). *J. Am. Chem. Soc.* **103**, 2899; Mansuy, D., Morgenstern-Badarau, I., Lange, M., and Ganz, P. (1982). *Inorg. Chem.* **21**, 1427; Olmstead, M. M., Change, R.-J., and Balch, A. L. (1982). *Inorg. Chem.* **21**, 4243.
89. Groves, J. T., and Watanabe, Y. (1986). *J. Am. Chem. Soc.* **108**, 7836.

90. Braun, W., Rajbenbach, L., and Eiricj, F. R. (1962). *J. Phys. Chem.* **66,** 1591.
91. Bevington, J. C., and Toole, J. (1958). *J. Polym. Sci.* **28,** 413; Detar, D. F. (1976). *J. Am. Chem. Soc.* **89,** 4058.
92. Katritzky, A. R., and Lagowski, J. M. (1971). *In* "Chemistry of the Heterocyclic N-Oxides," p. 147. Academic Press, London.
93. White, R. E., Sligar, S. G., and Coon, M. J. (1980). *J. Biol. Chem.* **255,** 11108.
94. Andrews, L. E., Bonnett, R., and Ridge, R. J. (1983). *J. Chem. Soc., Perkin Trans. 1* p. 103.
95. Groves, J. T., and Watanabe, Y. (1986). *Inorg. Chem.* **25,** 4808.
96. Hill, C. L., and Schardt, B. C. (1980). *J. Am. Chem. Soc.* **102,** 6374.
97. Groves, J. T., Kruper, W. J., and Haushalter, R. C. (1980). *J. Am. Chem. Soc.* **102,** 6375.
98. Smegel, J. A., Schardt, B. C., and Hill, C. L. (1983). *J. Chem. Soc.* **105,** 3510; Smegel, J. A., and Hill, C. L. (1983). *J. Am. Chem. Soc.* **105,** 2920; Camenzid, M. J., Hollander, F. J., and Hill, C. K. (1982). *Inorg. Chem.* **21,** 4301; Camenzid, M. J., Hollander, F. J., and Hill, C. K. (1983). *Inorg. Chem.* **22,** 3776.
99. Guilmet, E., and Meunier, B. (1980). *Tetrahedron Lett.* **21,** 4449; Meunier, B., Guilmet, E., De Carvalho, M.-E., and Poilblanc, R. (1984). *J. Am. Chem. Soc.* **106,** 6668.
100. Bortolini, O., Ricci, M., and Meunier, B. (1986). *Nouv. J. Chim.* **10,** 39.
101. Groves, J. T., and Stern, M. K. (1987). *J. Am. Chem. Soc.* **109,** 3812; Czernuszewicz, R. S., Su, Y. O., Stern, M. K., Macor, K. A., Kim, D., Groves, J. T., and Spir, T. G. (1988). *J. Am. Chem. Soc.* **110,** 4158; Groves, J. T., and Stern, M. K. (1988). *J. Am. Chem. Soc.* **110,** 8628.
102. Wagner, W.-D., and Nakamoto, K. (1989). *J. Am. Chem. Soc.* **111,** 1590.
103. Groves, J. T., and Stern, M. K. (1987). *J. Am. Chem. Soc.* **109,** 3812.
104. Yuan, L.-C., and Bruice, T. C. (1985). *Inorg. Chem.* **24,** 987.
105. For example: Takemore, S., and Kominami, S. (1984). *Trend Biochem. Sci.* **9,** 1; Light, D. R., White-Stevens, R. W., and Orme-Johnson, W. H. (1979). *J. Biol. Chem.* **254,** 2103; Nakajima, S., Shinoda, M., Haniu, M., Shively, J. E., and Hall, P. F. (1984). *J. Biol. Chem.* **259,** 3971.
106a. Groves, J. T., McClusky, G. A., White, R. E., and Coon, M. J. (1978). *Biochem. Biophys. Res. Commun.* **81,** 154.
106b. Groves, J. T., and Subramanjan, D. V. (1984). *J. Am. Chem. Soc.* **106,** 2177.
106c. White, R. E., Bhattacharyya, A., Miller, J. P., and Favreau, L. V. (1985). *Fed. Proc.* **44,** 474.
107. McMahon, R. E., Sullivan, H. R., Craig, J. C., and Pereira, W. E. (1969). *Arch. Biochem. Biophys.* **132,** 575; Hamberg, M., and Bjorkhem, I. (1971). *J. Biol. Chem.* **246,** 7411; Shapiro, S., Piper, J. U., and Caspi, E. (1982). *J. Am. Chem. Soc.* **104,** 2301.
108. Gelb, M. J., Heimbrook, D. C., Malkonen, P., and Sligar, S. G. (1982). *Biochemistry* **21,** 370.
109. Stearns, R. A., and Ortiz de Montellano, P. R. (1985). *J. Am. Chem. Soc.* **107,** 4081.
110. Groves, J. T., and Nemo, T. E. (1983). *J. Am. Chem. Soc.* **105,** 6243.
111. Hjelmeland, L. M., Aronow, L., and Trudell, J. R. (1977). *Biochem. Biophys. Res. Commun.* **76,** 541.
112. Lindsay-Smith, J. R., and Sleath, P. R. (1983). *J. Chem. Soc., Perkin Trans. 2* p. 621; Lindsay-Smith, J. R., and Mortimer, D. N. (1985). *J. Chem. Soc., Chem. Commun.* **20,** 64.
113. Hill, C. L., Smegal, J. A., and Henly, T. J. (1983). *J. Org. Chem.* **48,** 3277.
114. Watabe, T., and Akamatsu, K. (1972). *Biochem. Pharmacol.* **23,** 1079; Watabe, T., Ueno, Y., and Imazumi, J. (1971). *Biochem. Pharmacol.* **20,** 912.
115. Ortiz de Montellano, P. R., and Kunze, K. L. (1980). *J. Biol. Chem.* **255,** 5578; Ortiz de Montellano, P. R., and Kunze, K. L. (1981). *Biochemistry* **20,** 7266; Ortiz de Montellano, P. R., Beilan, H. S., Kunze, K. L., and Mico, B. A. (1981). *J. Biol. Chem.* **265,** 4395; Ortiz de Montellano, P. R., Kunze, K. L., Beilan, H. S., and Wheeler, C. (1982). *Biochemistry* **21,**

1331; Ortiz de Montellano, P. R., Mangold, B. L. K., Wheeler, C., Kunze, K. L., and Reich, N. O. (1983). *J. Biol. Chem.* **258**, 4208.

116. Miller, R. E., and Guengerich, F. P. (1982). *Biochemistry* **21**, 1090.

117. Guengerich, F. P., and MacDonald, T. L. (1984). *Acc. Chem. Res.* **17**, 9.

118. Groves, J. T., Avaria-Neisser, G. E., Fish, K. M., Imachi, M., and Kuczkowshi, R. L. (1986). *J. Am. Chem. Soc.* **108**, 3837.

119. Dolphin, D., Matsumoto, A., and Shortman, C. (1989). *J. Am. Chem. Soc.* **111**, 411.

120a. Groves, J. T., and Nemo, T. E. (1983). *J. Am. Chem. Soc.* **105**, 5786.

120b. Groves, J. T., and Myers, R. S. (1983). *J. Am. Chem. Soc.* **105**, 5791.

120c. Lindsay-Smith, J. R., and Sleath, P. R. (1982). *J. Chem. Soc., Perkin Trans 2* p. 1009.

120d. Naruta, Y., Tani, F., Maruyama, K. (1989). *Chem. Lett.* **XX**, 1269.

120e. O'Malley, S., and Kodadek, T. (1989). *J. Am. Chem. Soc.* **111**, 9116.

120f. Groves, J. T., and Viski, P. (1990). *J. Org. Chem.* **55**, 3628.

121. Collman, J. P., Kodadeck, T., Raybuck, S. A., and Meunier, B. (1983). *Proc. Natl. Acad. Sci. U.S.A.* **80**, 7039; Collman, J. P., Brauman, J. I., Meunier, B., Hayashi, T., Kodadeck, T., and Raybuck, S. A. (1985). *J. Am. Chem. Soc.* **107**, 2000.

122. Nolte, R. J. M., Razenberg, J. A. S. J., and Schuraman, R. (1986). *J. Am. Chem. Soc.* **108**, 2751.

123. Collman, J. P., Kodadeck, T., Raybuck, S. A., Brauman, J. I., and Papazian, I. M. (1985). *J. Am. Chem. Soc.* **107**, 4343.

124. Traylor, T. G., Marster, J. C., Jr., Nakano, T., and Dunlap, B. E. (1985). *J. Am. Chem. Soc.* **107**, 5537.

125. Dicken, C. M., Woon, T. C., and Bruice, T. C. (1986). *J. Am. Chem. Soc.* **108**, 1636.

126. Mashiko, T., Dolphin, D., Nakano, T., and Traylor, T. G. (1985). *J. Am. Chem. Soc.* **107**, 3735; Traylor, T. G., Nakano, T., Miksztal, A. T., and Dunlop, B. E. (1987). *J. Am. Chem. Soc.* **109**, 3625.

127. Groves, J. T., and Watanabe, Y. (1986). *J. Am. Chem. Soc.* **108**, 507.

128. Traylor, T. G., and Miksztal, A. R. (1987). *J. Am. Chem. Soc.* **109**, 2770.

129a. Watanabe, Y., Numata, T., Iyanagi, T., and Oae, S. (1981). *Bull. Chem. Soc. Jpn.* **54**, 1163.

129b. Watanabe, Y., Iyanagi, T., and Oae, S. (1980). *Tetrahedron Lett.* **21**, 3685.

130. Watanabe, Y., Iyanagi, T., and Oae, S. (1982). *Tetrahedron Lett.* **23**, 533.

131. Modena, G., and Mariola, L. (1957). *Gazz. Chim. Ital.* **87**, 1306.

132. Oae, S., Watanabe, Y., and Fujimori, K. (1982). *Tetrahedron Lett.* **23**, 1189.

133. Asada, K. (1983). MS Thesis. The University of Tsukuba Japan; Oae, S., and Fujimori, K. (1983). *J. Synth. Org. Chem. Jpn.* **41**, 848.

134. Watanabe, Y., Oae, S., and Iyanagi, T. (1982). *Bull. Chem. Soc. Jpn.* **55**, 188; Chow, Y. L., Danen, W. C., Nelsen, S. F., and Rosenblatt, D. H. (1978). *Chem. Rev.* **78**, 243.

135. Miwa, G. T., Walsh, J. S., and Lu, A. Y. H. (1984). *J. Biol. Chem.* **259**, 3000.

136. Hanzlik, R. P., and Tullman, R. H. (1982). *J. Am. Chem. Soc.* **104**, 2048; MacDonald, T. L., Zirvi, K., Burka, L. T., Peyman, P., and Guengerich, F. P. (1982). *J. Am. Chem. Soc.* **104**, 2050; Guengerich, F. P., Willard, F. J., Shea, J. P., Richards, L. E., and MacDonald, T. L. (1984). *J. Am. Chem. Soc.* **106**, 6446.

137. Foster, A. B., Jarman, M., Stevens, J. D., Thomas, P., and Westwood, J. H. (1974). *Chem.–Biol. Interact.* **9**, 327.

138. Boyd, J. W., Schumulzl, P. W., and Miller, L. L. (1980). *J. Am. Chem. Soc.* **105**, 3856.

139. Ortiz de Montellano, P. R., Choe, Y. S., DePillis, G., and Catalano, C. E. (1987). *J. Biol. Chem.* **262**, 1542 and 11641.

10

Mechanism of NAD-Dependent Enzymes

NORMAN J. OPPENHEIMER · ANTHONY L. HANDLON
Department of Pharmaceutical Chemistry
University of California, San Francisco
San Francisco, California 94143

In this chapter we will discuss the mechanisms by which NAD-dependent enzymes catalyze chemistry at two sites in the coenzyme molecule. First we discuss the dehydrogenase-catalyzed redox chemistry that occurs at the 4-position of the nicotinamide ring, and then cleavage of the C-1′–N ribosyl linkage cata-

THE ENZYMES, Vol. XX

lyzed by a variety of NAD$^+$ glycohydrolases and ADP-ribosyltransferases is assessed. Discussions will concentrate on the underlying chemistry and the various means by which these enzymes promote catalysis.

I. Mechanism of NAD-Dependent Dehydrogenases

The mechanism of dehydrogenases is intimately related to the chemistry of the pyridine coenzymes (for recent reviews of the chemistry and enzymology see Refs. *1–5*). Emphasis will be on dehydrogenases that catalyze oxidation of alcohol functionalities,* with specific emphasis on horse liver and yeast alcohol dehydrogenases and α-hydroxy-acid dehydrogenases such as lactate dehydrogenase. This group of dehydrogenases is the most extensively studied, and what is said about their mechanism should be applicable to the machinery of catalysis for other enzymes. The impact of the chemical and mechanistic constraints of the reactions of models for the evolution of dehydrogenase function will also be discussed.

A. ACTIVATION OF COENZYME FOR HYDRIDE TRANSFER

Activation of the coenzyme for hydride transfer involves a complex interplay among a number of factors. These include distortion of the nicotinamide/dihydronicotinamide ring, solvation, direct electrostatic interactions, conformation of the side chain at C-3, conformation around the glycosyl bond, and inductive effects directed at the sugar moiety. The following sections will provide a discussion of these interactions and their potential mechanistic impact.

1. *Nonplanarity of Dihydronicotinamide Ring (Puckering)*

The potential mechanistic involvement of nonplanar, puckered conformations of the dihydronicotinamide ring has been the subject of continuing speculation ever since it was suggested by Vennesland and Levy (*6*). The first experimental evidence for nonplanarity of the dihydronicotinamide ring came from analysis of the vicinal coupling constants between the N-5 and N-4 protons in the ^{1}H NMR spectrum of NADH (*7*) and other reduced coenzyme analogs (*8*). Importantly, the degree of puckering correlates with the extent and selectivity of intramolecular association and with previous evidence for stereoselective chemical reactivity (*8*). As shown in Fig. 1, dehydrogenases can potentially activate the desired C-4

---

* The term "alcohol" dehydrogenase with quotes will be used for a general designation of enzymes that oxidize —CHOH— functionalities to carbonyls and will include enzymes that oxidize substrates such as lactate to pyruvate. The term alcohol dehydrogenase without quotes will refer to enzymes that oxidize primary alcohols, secondary alcohols, or polyols.

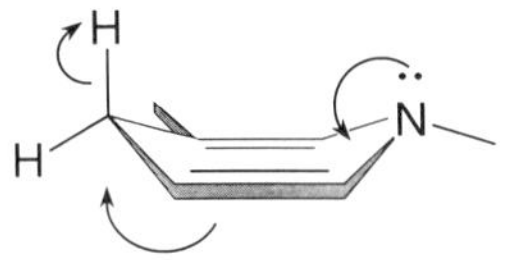

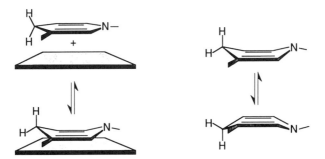

FIG. 1. Puckering of the dihydronicotinamide ring can lead to activation of the axial proton at the C-4 position.

proton by forcing the ring to pucker in a direction that specifically weakens the C–H bond of the hydride to be transferred. Such a model has both experimental (9) and computational (10) support. As shown in Fig. 2, the nature of the distortion remains unknown. Is the dihydronicotinamide ring intrinsically planar, but easily distorted, or is it intrinsically puckered, but with only a small energy barrier between two rapidly interconverting forms (a single versus double potential energy minimum model)? Calculations indicate that the dihydronicotinamide requires only 1.4 kcal/mol to distort from planarity to 160° (11). With such a small difference in energy there is little hope to distinguish the two models experimentally; even the unambiguous observation of a puckered form in an active site could be accounted for by either model.

No significant departure of the dihydronicotinamide ring from planarity has been observed from X-ray crystallography studies of model N-alkyldihydronicotinamides (12). These negative results are inconclusive. The molecules studied all are symmetrical to internal reflection through the plane of the dihydronicotinamide ring; thus in the absence of highly stable puckered forms, both faces will have identical interactions in the achiral crystal lattice, which decreases the likelihood of observable puckering. In contrast, both puckering of the ring and stereoselectivity of hydrogen abstraction are observed for model compounds where a dihydropyridine ring is rigidly held against a planar group (9). The

FIG. 2. Two models for puckering of the dihydronicotinamide ring are shown. (Left) Puckering is induced by interactions with other molecules or surfaces. (Right) The dihydronicotinamide ring flips rapidly between two puckered forms. Interactions with other molecules would alter the populations of an intrinsically puckered ring, but would not induce any net puckering of the ring.

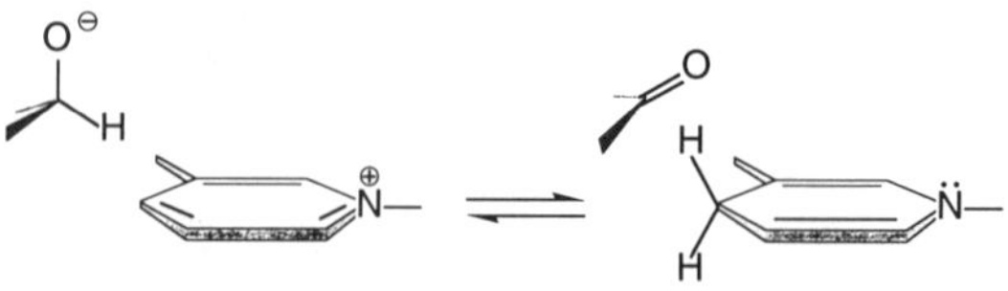

FIG. 3. Hydride transfer occurs with neutralization of the charges on the anionic alkoxide and the cationic pyridinium; thus the redox reaction will be sensitive to the dielectric constant of the active site.

current resolution of X-ray crystallographic structures of reduced coenzyme bound to dehydrogenases is insufficient to provide direct evidence either for or against puckering. Furthermore, the methods of refinement used in previous analyses mathematically suppress puckering even if it were to occur (D. Matthews, personal communication, 1989).

2. Solvation and Dielectric Environment of Active Site

The redox reaction of the pyridine coenzyme with an alcohol involves both a change from a cationic to a neutral species and the reversible loss of a proton, as shown in Fig. 3. Therefore, the external equilibrium constant of these reactions depends on solvent polarity, pH, and the structure of the substrate

$$NAD^+ + CN \rightleftharpoons NAD–CN$$

Reaction (1)

The addition of cyanide or other nucleophiles to the C-4 position of NAD^+ illustrates this point, as shown in reaction (1). Adduct formation becomes increasingly favorable as solvent polarity decreases. Burgner and Ray reported that lactate dehydrogenase activates NAD^+ toward cyanide adduct formation by 100 times relative to NAD^+ in solution (13). This difference is attributed to stabilization of the neutral adduct in the relatively nonpolar active site.

Isolating a cationic pyridinium in a nonpolar environment increases the electrophilicity of the ring, especially at the C-4 position. As electron density is increased on the substrate, such as by formation of an alkoxide, the substrate becomes, in effect, the nearest counterion. Hydride transfer can then be viewed as occurring with either collapse or formation of a charge pair in a hydrophobic pocket, as shown in Fig. 3. The transfer need not be in-line. Calculations on model systems suggest that hydride transfer can occur at C–H–C angles down to 158° (14).

Note that the nicotinamide group cannot be directly stabilized by H bonding except through the substituent at C-3 because none of the atoms in the ring can form conventional hydrogen bonds. In dihydrofolate reductase (DHFR) a series of polar groups that line the active site may be closer to the dihydronicotinamide ring than expected for van der Waals contact, consistent with the possible involvement of C–H ⋯ O hydrogen bonds (15). The highly polar environment in the nicotinamide binding site of DHFR is distinct from the hydrophobic pocket found in "alcohol" dehydrogenases. DHFR, however, promotes the reduction of the substrate and formation of $NADP^+$; hence these interactions may be present to facilitate hydride transfer from the less polar reduced coenzyme by stabilizing the cationic pyridinium of $NADP^+$.

3. Syn versus Anti Rotamers

There is limited direct experimental evidence for redox-dependent alterations in the populations of rotational isomers around the nicotinamide–ribosyl bond. Measurements of the nuclear Overhauser effect (NOE) in nicotinamide mononucleotide (NMN^+) indicate that the populations of syn and anti rotamers are nearly equal (16). No comparable (NOE) studies have been conducted on the reduced nicotinamide mononucleotide (NMNH) or NADH.* Investigations of the concentration dependence of the ^1H NMR chemical shifts of NADH resonances, addition of nucleophiles to the C-4 position of NAD^+, temperature dependence of the methylene C-4 proton resonances of NADH, and pH-dependent alterations in the chemical shifts in the ^1H NMR spectrum of NADPH are consistent with both the oxidized and reduced coenzymes both favoring an anti rotamer in the folded geometry (8). These results must be treated as only suggestive and not definitive, but they do, however, indicate that redox-dependent differences in the syn/anti ratio are significantly less than a factor of 10, i.e., ≤1 kcal. Any discussion of the possible thermodynamic significance of syn versus anti conformations on the functioning of dehydrogenases (18) must be based on the properties of the dinucleotides. It is their thermodynamic properties in solution that would provide any putative energy differences as far as functional models are concerned.

4. Stereochemistry and Bound Coenzyme Conformation

The X-ray crystallographic structures of lobster glyceraldehyde-3-phosphate dehydrogenase (G3PDH) (19) and dog fish lactate dehydrogenase (LDH) (20) show that the stereospecificity of hydride transfer is a function of the rotational

---

* Indirect evidence for small redox-dependent differences has been presented based on titration studies of the 5′-phosphate for NMN^+ and NMNH (17). Such determinations, however, are highly suspect because the effects of titration of the 5′-phosphate are not limited to through-space interactions. Chemical shifts also depend on orbital overlap of the intervening bonds and localized bond polarization, neither of which is a direct function of the interatomic distances (4).

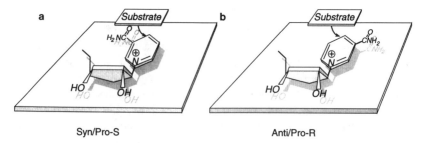

FIG. 4. The relative orientations of the substrate and nicotinamide lead to a correlation between rotamers and stereospecificity. When the substrate binds on top of the nicotinamide ring, then the syn rotamer will lead to *pro-S* specificity of hydride transfer (a); however, when the anti rotamer of the nicotinamide ring is bound, then the enzyme will transfer the *pro-R* proton (b).

isomer around the glycosyl bond. G3PDH uses the *pro-S* proton of NADH and binds the coenzyme in a syn conformation, whereas LDH uses the *pro-R* proton and binds the coenzyme in an anti conformation (see Figs. 4 and 5). The syn/ *pro-S* : anti/*pro-R* correlation has been found for all "alcohol" dehydrogenases studied to date. The correlation has a simple structural origin. It will apply to all enzymes in which the substrate binding domain is on top of the coenzyme. Largely overlooked in the literature is a second relationship, that the opposite correlation will apply to all enzymes that bind the *coenzyme on top of the functionality that is to undergo redox chemistry.* Glutathione dehydrogenase is an example of this latter behavior. The coenzyme binds on top of the flavin ring; thus the enzyme is *pro-S* specific but with an anti conformation of the nicotinamide ring (*21*). The fact that the anti/*pro-R* relationship holds for "alcohol" and "aldehyde" dehydrogenases has more to do with the constraints imposed on the location of substrate binding sites, because they utilize freely dissociating substrates, than on any functional requirements or stereochemical imperatives.

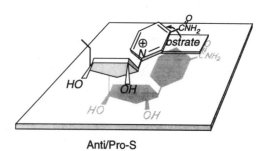

Anti/Pro-S

FIG. 5. For enzymes that bind the coenzyme on top of the redox-active functionality, then the opposite correlation to that shown in Fig. 4 will hold.

B. STEREOSPECIFICITY OF DEHYDROGENASES

1. *Stereospecificity of Hydride Transfer by Dehydrogenases*

The demonstration of the stereospecific hydride transfer by alcohol dehydrogenase was a turning point in the appreciation of the stereochemical properties
of enzymes (*22*). The initial methods employed, however, were of limited accuracy ($\pm 2\%$ at best). Recently Anderson and LaReau (*23*) demonstrated that
in pig heart lactate dehydrogenase the incidence of hydride transfer with the
opposite specificity occurs in less than 2×10^8 turnovers. This result establishes
that the energies of the transition states corresponding to *pro-R* and *pro-S* transfers must differ by more than 10 kcal/mol. The origins of this remarkable degree
of specificity are not obvious. Steric exclusion of the wrong conformer is questionable. LDH can function with substituents attached to the 5-position of the
nicotinamide ring that serve as substrates for an enzyme-catalyzed internal redox
reaction (*24*). LDH can also function with α-NADH as a coenzyme (*25*) and,
like other dehydrogenases (*26*), it is not sensitive to alterations in the stereochemistry of the sugar moiety or the configuration of furanosyl linkage to the
nicotinamide ring. Finally, specific interactions with the amino group of the
3-carboxamide are questionable because LDH can function with coenzyme analogs such as the 3-acetylpyridine or the 3-pyridinecarboxaldehyde analogs.
Clearly the interactions governing the high degree of stereospecificity are both
substantial and remain to be defined.

2. *Epimerases*

Epimerases represent an opposite extreme of stereospecificity (for a recent
review see Ref. *27*). They are chiral catalysts designed to function achirally. One
class of epimerases contains a tightly bound molecule of NAD^+. The proposed
mechanism involves the following reactions. First, an alcohol functionality on
the sugar is oxidized to a ketone (without regard to the stereoconfiguration at
that site), then the hydride is reintroduced to either face of the carbonyl. The
most extensively studied example is uridine diphosphogalactose 4-epimerase.
Frey and co-workers have obtained clear spectrophotometric and ^{31}P NMR spectroscopic evidence for the catalytically competent involvement of a transient formation of NADH (*28, 29*). No evidence has been found for either exchange of
the bound keto sugar intermediate with exogenous compounds or an ability of
the enzyme to show net dehydrogenase activity. The suggested mechanism is
shown in Fig. 6.

3. *Orientation and Electronic and Steric Importance of 3-Substituents*

The substituent at C-3 of the pyridine ring can serve both as a handle for
enzyme binding and as a mediator of the microscopic redox potential. The importance of these interactions for coenzyme function has been probed by inves-

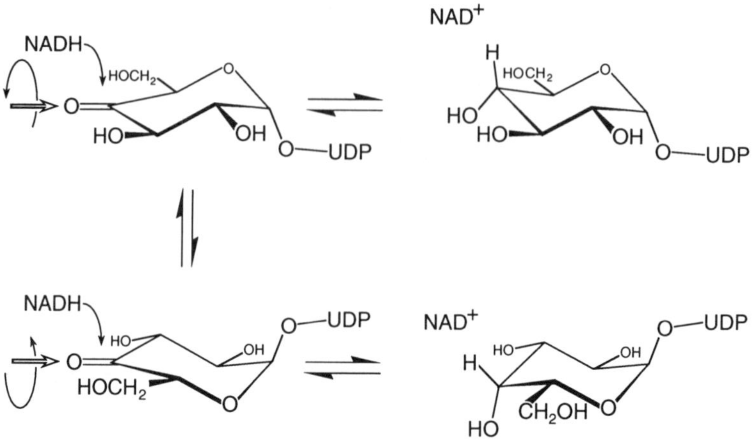

FIG. 6. The epimerase can reversibly reduce the tightly bound NAD$^+$, with the transient genera-tion of a 4-keto sugar. Viewing the keto sugar along an axis through the carbonyl suggests that rotation around this axis would allow either face to be presented to the bound NADH, thus randomly regenerating the sugar hydroxyl.

tigations of the properties of a wide range of analogs synthesized through the use of NAD$^+$ glycohydrolases (for recent reviews on the properties of NAD analogs see Refs. 30 and 31).

The carboxamide is skewed from planarity by 23° in nicotinamide and the nitrogen is oriented toward C-2. In N-substituted pyridinium compounds and 1,4-dihydronicotinamides (12, 32), the planar geometry is favored with the nitrogen oriented toward the C-4. The stability of these geometries is borne out by theoretical calculations (33). Determining the geometry of the carboxamide when bound to dehydrogenases is more difficult. Current X-ray crystallographic data for the carboxamide are unable to distinguish between the oxygen and nitro-gen, which precludes direct observation of the carboxamide orientation. Orien-tations can be inferred from the groups that interact with the carboxamide. In the refined X-ray structures of HL-ADH, the C-2–C-3–C-7–O-7 bond angle θ is −150° (34) and for *Bacillus stearothermophilus* G-3-PDH θ is 149° (35). Thus in both cases the carboxamide is skewed from planarity by equal but opposite amounts.

A new approach to investigating interactions directed at the carboxamide is the use of resonance Raman spectroscopy. Callender and co-workers have found significant alterations in the carbonyl stretching frequencies attributed to the carboxamide of NADH on binding (36). Thus this technique holds promise for analyzing in detail the interactions directed toward the side chain and pos-

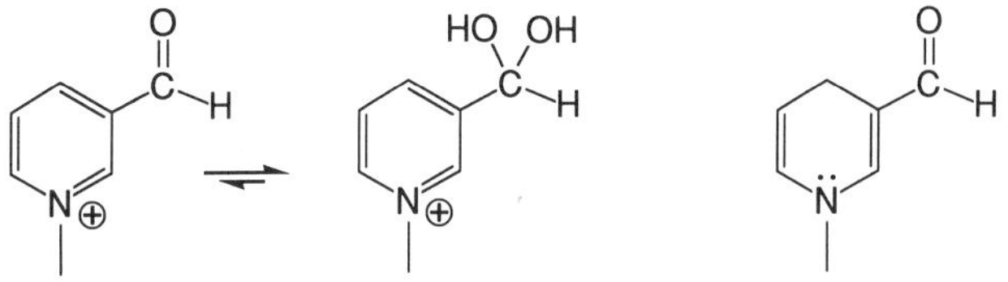

FIG. 7. The oxidized pyridine aldehyde analog is predominantly the *gem*-diol form, whereas for the reduced adduct only the carbonyl is observed. These results illustrate the redox-dependent differences in the chemical properties of the C-3 side chain.

sibly observing perturbations in the electronic structure of the dihydronicotinamide ring.

The side chain at C-3 undergoes profound redox-dependent alterations in its chemical properties. The electron orbitals of the carboxamide in NAD^+ are not delocalized into the pyridine ring. On reduction, however, the side chain becomes effectively a vinylogous urea. It shows strong resonance delocalization into the double bonds, with the carbonyl coplanar [although skewed conformations at $\pm 150°$ also represent local minima in the torsional conformation space (33)]. The impact of these differences is most dramatically illustrated in the pyridine 3-carboxaldehyde coenzyme analog (see Fig. 7). In its oxidized form the side chain behaves as an aldehyde. It is primarily hydrated* and it reacts rapidly with common aldehyde reagents (37). In contrast, the reduced form behaves as a vinylogous formamide. There is no detectable hydrate (<0.3%) and no evidence for reaction with aldehyde reagents (L. A. Wainschel and N. J. Oppenheimer, unpublished results). The potential influence on enzyme activity of the very different attributes of the side chain of oxidized versus reduced coenzymes remains to be explored.

The 3-acetylpyridine and 3-pyridinecarboxaldehyde analogs are both active with a wide range of enzymes, including lactate dehydrogenase. Their ability, as well as that of other of 3-substituted analogs, to function effectively with dehydrogenases raises questions as to the nature of the interactions responsible for the high degree of stereospecificity observed in hydride transfer (23) (see Section I,B,1).

C. COENZYME ANALOGS CONTAINING MODIFIED SUGARS

A growing number of coenzyme analogs have been prepared containing modifications in the nicotinamide sugar moiety. These modifications focus on three primary interactions: inductive effects, tolerance of sugar substituents (anomeric configuration), and nucleoside flexibility (the importance of an intact furanose

---

* The ratio of hydrate to aldehyde is 11:1 at 20°C (L. A. Wainschel and N. J. Oppenheimer, unpublished data, 1990).

ring). To date no systematic investigation of analogs with target dehydrogenases has been conducted, and a definitive analysis of these factors must await further study. Nonetheless, the limited results summarized in this section provide intriguing insights into the mechanism of dehydrogenases.

1. Inductive Effects from Sugar Moieties

The 2'-substituent of the nicotinamide nucleoside group has been implicated as part of a proton relay system in horse liver alcohol dehydrogenase (38). Such interactions can also serve to influence the redox potential of the coenzyme. As shown in Fig. 8, the log K_{eq} for cyanide adduct formation is linearly dependent on the inductive Taft σ_i value* (39). These results suggest that the 2'-substituent of the sugar moiety can transmit inductive effects into the nicotinamide ring from specific interactions with active site residues. The magnitude of the effect, however, should be modest. Importantly, the coenzyme redox potential could be influenced by the protonation state of the interacting residues, which in turn would depend on the timing of the transport of the proton from the substrate (see Fig. 9). Clearly such interactions are not mechanistically "essential" because HL-ADH can function effectively with analogs that lack a comparably located hydroxyl (see Section I,C).

2. Configurational Discrimination of Dehydrogenases

Reduced pyridine nucleotides are configurationally unstable and anomerize at significant rates under physiological conditions (25) to yield at equilibrium a 7:1 mixture of β/α-anomers of NADH.† Given the generally high stereoselectivity of dehydrogenases (see Section I,B,1), their ability to use α-NADH as a cofactor with the same stereochemistry as β-NADH is unexpected (25, 26). Indeed, the α-anomer of the arabinose analog of NADH has a V_{max} 1.7 times that of β-ribo-NADH with HL-ADH (40). The ability to function with α-NADH is also a property of both *pro-R* and *pro-S*-specific dehydrogenases (26). Therefore, maintenance of specific interactions with the nicotinamide sugar moiety is not essential for functioning of the dehydrogenases studied thus far. As shown in Fig. 10, the active site must accommodate the sterically more demanding binding geometry of the α-anomers without major interference in the proper seating of the nicotinamide ring.

---

* The correlation is with the value of σ_i for the arabinosides. Note that there is no dependence on the bulk of the substituents, i.e., the association constants for two smallest substituents, H and F, are fully consistent with the linear relation. The absence of steric effects for the cis substituents raises questions regarding recent suggestions that configurational differences in redox potential in α- and β-NAD$^+$ arise solely from steric interactions with the *cis*-hydroxyl in the α-anomer (12).

† The λ_{max} for β-NADH is 338 nm, for α-NADH it is 346 nm, and for the equilibrated mixture it is 340 nm; these are the usually cited maximum absorptions of NADH solutions.

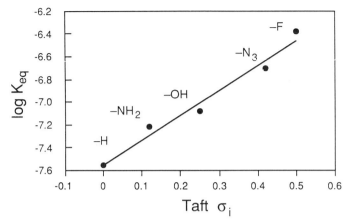

FIG. 8. Plot of log K_{eq} for cyanide adduct formation versus Taft σ_i.

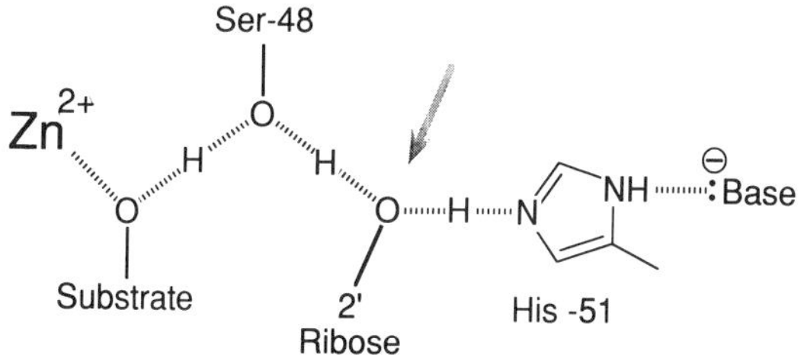

FIG. 9. The nicotinamide 2'-hydroxyl is implicated in a proton relay system for horse liver alcohol dehydrogenase (38). Thus electron density at the 2'-position induced during proton transfer can influence the microscopic redox potential by through-bond inductive effects.

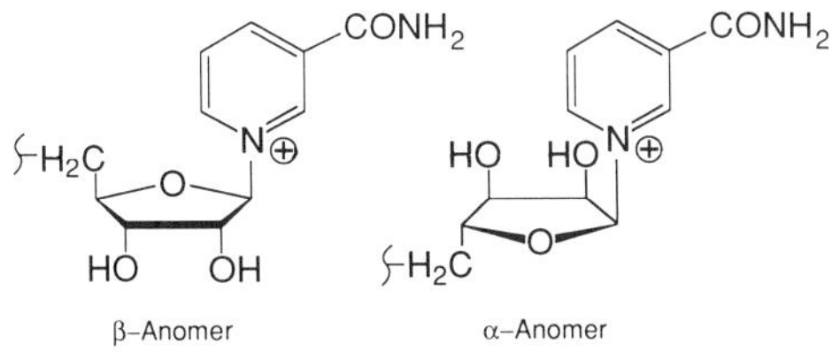

FIG. 10. Comparison of α- and β-anomers of NAD$^+$ illustrating the differences in the orientation of the furanosyl ring and 5'-substituents when viewed from the perspective of the nicotinamide ring in the active site.

3. Acyclic Analogs of NAD$^+$

The ability of dehydrogenases to function with α-anomers focuses attention on the requirement of the furanosyl ring for coenzyme activity. In order to study the ability of dehydrogenases to accommodate torsional flexibility, three categories of acyclic analogs have been synthesized: alkyl acyclic analogs, the products of periodate oxidation of NADP$^+$, and N-oxymethylnicotinamide analogs. All have been found to support to some extent redox activity with dehydrogenases. Acyclic alkylnicotinamide analogs (1) show poor activity with dehydrogenases, presumably because of their low redox potential of -388 mV (41). The corresponding 3-acetylpyridine analogs (2) have redox potential of approximately -316 mV* and show enzymic activity (42) (for a comprehensive listing of redox potentials for NAD analogs see Ref. 31). Activity depends on chain length. The n-butyl analog is active with a wide range of common dehydrogenases whereas the n-pentyl analog was only active with G3PDH and aldehyde dehydrogenase. In contrast, the reduced analogs containing n-propyl, n-butyl, or n-pentyl chains were reportedly active with a wide range of dehydrogenases (42).

(1) Acyclo-NAD$^+$ (2) (Tetramethylene-acetylpyridine)AD$^+$

NADP$^+$ containing an acyclic 2′,3′-"dialdehyde" nicotinamide sugar moiety (3) can be synthesized by periodate oxidation (43) and exists primarily as the hydrated gem-diol. This analog has been used as an affinity label [it inactivates glucose-6-phosphate dehydrogenase (44)], and is active with some dehydrogenases (45).

(3) 2′,3′-"dialdehyde"-NADP$^+$

---

* Interestingly, the 72-mV difference in redox potentials for these two N-alkylpyridiniums is identical to the difference for the corresponding nucleosides (-320 mV for β-NAD$^+$ and -248 mV for 3-acetylpyridine-AD$^+$).

TABLE I

V_{max} for Acyclo-NAD$^+$ and Horse Liver Alcohol Dehydrogenase[a]

Alcohol	$\dfrac{V_{max}\ \beta\text{-NAD}^+}{V_{max}\ \text{acyclo-NAD}^+}$	$\dfrac{V_{max}\ \text{alcohol}\ +\ \beta\text{-NAD}^+}{V_{max}\ \text{ethanol}\ +\ \beta\text{-NAD}^+}$
Ethanol	0.018	1
1-Propanol	0.26	0.72
1-Butanol	0.49	0.54
1-Hexanol	0.18	0.19
2-Propanol	<0.005	0.25
2-Butanol	<0.005	0.14
2-Pentanol	<0.005	0.18
Cyclohexanol	<0.005	0.94

[a] Unpublished data (O. Malver, M. Sebastian, and N. J. Oppenheimer) 1990.

Recently the N-oxymethylnicotinamide analog acyclo-NAD$^+$ has been synthesized (46). This analog is functional with horse liver ADH, as detailed in Table I, and its V_{max} is 50% that of NAD$^+$ using n-butanol as a substrate. More importantly, this analog alters the substrate specificity of HL-ADH. HL-ADH can catalyze oxidations of either primary or secondary alcohols with comparable values of V_{max} using β-NAD$^+$. It can also readily function with acyclo-NAD$^+$ using primary alcohols, but the rate of oxidation of secondary alcohols is undetectable ($<<10^{-3}$ the corresponding rate with β-NAD$^+$). *Therefore, a modification remote from the site of catalysis, which decreases the steric bulk and increases the torsional flexibility of the sugar moiety, alters the substrate specificity of the enzyme!*

A possible explanation for these results is shown in Fig. 11. Normally, the 2'-hydroxyl is hydrogen bonded to both His-51 and Ser-48. Note that these residues occur in the middle of a sequence that is intimately involved with defining the active site. When acyclo-NAD$^+$ is bound, these residues can relocate to the position formerly occupied by the 2'-hydroxyl. The attendant movement of the protein backbone would be transmitted into the active site via Cys-46 and His-67, which are directly coordinated to the catalytic zinc, and Asp-49, which is hydrogen bonded to His-67.* Arg-47, the counterion for the adenosine phosphate, and Val-57, part of the substrate binding pocket, could also participate in the structural reorganization. The sterically demanding secondary alcohols would then be more affected by conformational changes in the active site than the primary alcohols.†

* The interactions involving Asp-49 are important to enzyme catalysis. Replacement of the Asp-49 by Asn in the yeast ADH causes a nearly 1000-fold decrease in enzymic activity (47).

† The conformational changes associated with coenzyme binding and formation of ternary complexes are discussed in Ref. 48.

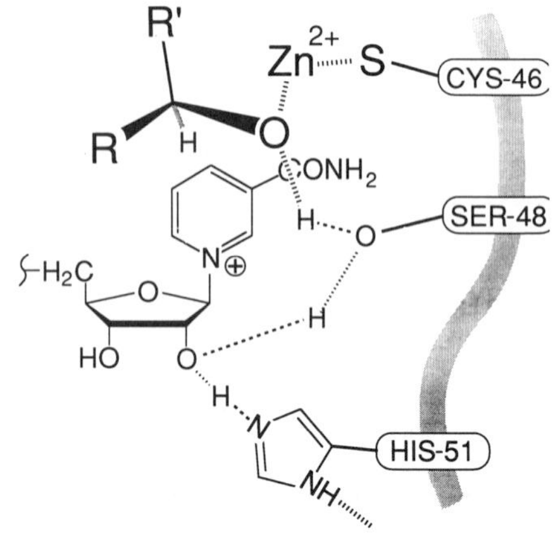

FIG. 11. Schematic representation of the interactions between the substrate, coenzyme, and the active site residues in horse liver alcohol dehydrogenase. Not shown are the interactions between Arg-47 and the pyrophosphate backbone, and Asp-49, which forms a salt bridge with His-57, another ligand of the zinc atom. Because of the close proximity to residues having obvious catalytically important functions, alterations in the interactions between the coenzyme and Ser-48 and His-51 that are anticipated from the binding of acyclo-NAD$^+$ could readily cause the observed changes in substrate specificity. Based on Ref. 38.

4. Miscellaneous Analogs Containing Modified Sugars

Two early analogs containing modified sugars have been reported, nicotinamide 6-phosphoglucoside (49) and nicotinamide 2′,3′-dideoxyriboside (50) (4 and 5). The glucosyl analog is inactive with common dehydrogenases but the dideoxyribosyl analog reportedly has weak but observable activity with liver ADH.

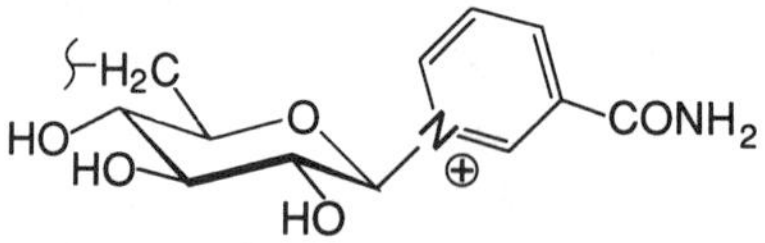

(4) Glucose 6-phosphate-NAD$^+$

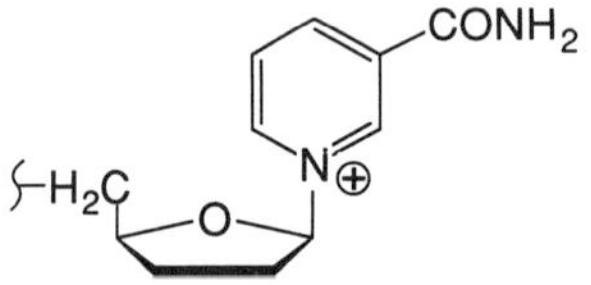

(5) 2′,3′-Dideoxy-NAD$^+$

Two new analogs have been reported, the nicotinamide *arabino* analog ara-NAD$^+$ (40) and the carbocyclio ribose analog carba-NAD$^+$ (51) (6 and 7). The *arabino* moiety has little impact on coenzyme function. With HL-ADH, the V_{max} for β-*ara*-NADH oxidation is six times faster than for β-NADH, and for α-*ara*-NADH oxidation the V_{max} is 1.7 times faster,* both anomers are also active with

---

* The value of V/K for β-*ara*-NADH is 1.2 times that for β-NADH, whereas the poorer K_m for α-ara-NADH leads to a V/K one-tenth that for β-NADH.

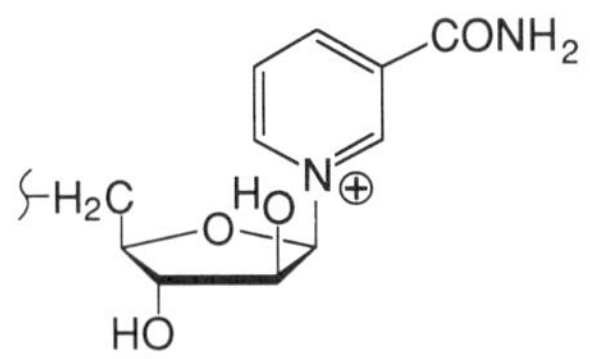

(6) *arabino*-NAD⁺ 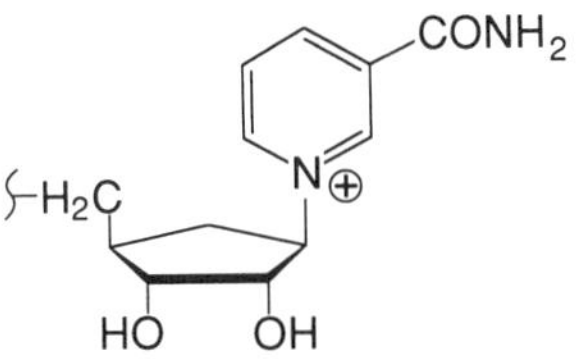   (7) Carba-NAD⁺

yeast ADH. The stereospecificity of hydride transfer is identical with both anomers; thus the nicotinamide ring must be seated congruently in the active site
regardless of the configuration at C-1'. In the recently synthesized carbocyclic
ribose analog carba-NAD⁺, the furanose ring oxygen is replaced with a methylene. The D configuration of the analog is reportedly active with both yeast and
horse liver alcohol dehydrogenase and has a redox potential of -340 mV,*
whereas the L configuration ψ-carba-NAD⁺ has no detectable activity (*51*).
Taken altogether, these results show that neither the furanose ring oxygen nor
configurations at C-1' or C-2' are essential for the redox functions of the
coenzyme.

D. ACTIVATION OF SUBSTRATE

1. *Chemical Mechanisms*

The analogy has been drawn between the chemistry catalyzed by dehydrogenases and the base-catalyzed hydride transfer chemistry of Meerwein–
Ponndon–Verley (MPV) reductions or the Cannizzaro reaction (for recent reviews of the chemistry of pyridine nucleotides see Refs. *1*, *3*, and *5*). In MPV
reductions [reaction (2)] aluminum or lithium salts of the alkoxide activate the
C–H bond for transfer of hydrogen to suitable acceptors. Both direct hydride
transfer (*52*) and radical intermediates (*53*) have been suggested for this reac-

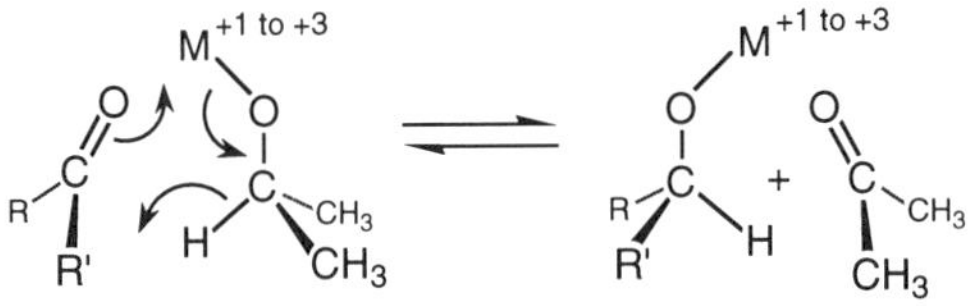

Reaction (2)

---

* This value of -340 mV for carba-NAD⁺ is considerably higher than the previously reported
redox potential of -388 mV for the chemically related acyclic pentamethylene-*N*-nicotinamide analogs (*41*), thus a general reassessment of these values is in order.

Reaction (3)

tion. The Cannizzaro reaction [reaction (3)] involves disproportionation of aldehydes to carboxylic acids and alcohols, and is thought to proceed via either direct hydride transfer from the *gem*-diol anion (or dianion form) (54) or by single-electron transfers (55). The essential feature for both MPV reductions and the Cannizzaro reaction is the activation of the C–H bond for transfer of the hydrogen atom (along with one or both of its electrons) by formation of an oxygen-based anion. There is no evidence for solvent exchange of the transferred proton. Work in aqueous methanol (52) clearly shows that the development of electron density on the carbon alone is sufficient to promote hydride transfer. Direct intervention of a specific metal ion in the transfer of hydrogen is not mechanistically essential. Unfortunately, quantitative substituent effects for these reactions are not available.

Most of the rate acceleration afforded by dehydrogenases can be conceptually accounted for by two key steps. In the direction of alcohol oxidation, dehydrogenases first promote formation of the equivalent of an alkoxide. That alkoxide is then juxtaposed to a nicotinamide ring (a group at the functional level of a protonated carbonyl). The magnitude of the anticipated rate acceleration for this reaction can then be imagined if it were possible in the Cannizzaro reaction to arrange for the *gem*-diol anion to encounter a *protonated* aldehyde (a situation that is kinetically unobservable in solution because of the extreme differences in the pK_a value of the reacting groups). Transfer of a hydride equivalent can then be viewed as occurring with neutralization of the separated charges.*

The controversy as to whether dehydrogenase-catalyzed reduction occurs by a one-electron or two-electron transfer will probably never be definitively settled to the satisfaction of all. Limits on the participation by radical intermediates in dehydrogenation have been made using cyclopropylmethanols as "radical clocks." Horse liver ADH oxidizes cyclopropylmethanol to the corresponding aldehyde with no detectable opening of the cyclopropyl ring, as would be anticipated if a radical were to be generated [reaction (4)]. This result establishes an upper limit of less than 10^{-8} sec for the lifetime of any radical intermediates formed in the enzyme-catalyzed reaction (56).† Although limited to HL-ADH,

---

* Direct modeling of hydride transfer from alkoxides to pyridinium compounds has not been explored because of the general instability of pyridinium compounds toward alkaline conditions.

† It can be argued that binding interactions in the active site could prevent the ring from opening. This argument is weakened by the broad substrate specificity of the liver ADH, which suggests that there are few constraints on the alkyl side chain.

$$\triangleright\!-\!\overset{\overset{\displaystyle OH}{|}}{CH_2} \;+\; \begin{matrix} NAD^+ \\ HL\text{-}ADH \end{matrix} \quad \longrightarrow \quad \triangleright\!-\!\overset{\overset{\displaystyle O}{\|}}{C}\!-\!H$$

$$\xrightarrow{\;\;\not{}\;\;}\;\; \overset{O}{\underset{}{\diagup\!\diagup\!-\!C\!-\!H}}$$

Reaction (4)

there is no compelling reason why their conclusion is not generally applicable to other dehydrogenases, subject to experimental verification.

2. Stereospecificity of Substrate Binding

The ability of alcohol dehydrogenases to distinguish the methylene protons of ethanol is a paradigm of enzymic stereospecificity. Many alcohol dehydrogenases can function with both primary and secondary alcohols; thus their ability to show selectivity toward the chirality of the secondary alcohols is remarkable. An example of the difficulties associated with stereospecificity of secondary alcohol dehydrogenases is illustrated by the thermophilic alcohol dehydrogenase from *Thermoanaerobacter ethanolicus* (57). At low temperatures the enzyme preferentially uses the S isomer of 2-pentanol or 2-butanol. The enantiomeric discrimination, however, is temperature dependent. Above a temperature T_r, at which there is no enantiomeric discrimination, the enzyme shows the opposite preference and uses the R isomer. The value of T_r is a function of the substrate; for 2-butanol T_r is 26°C and for 2-pentanol the extrapolated value is 77°C. The results can be rationalized if the binding of the alkyl groups is enthalpically driven in one site and entropically driven in the other. No experiments have been conducted to measure any temperature dependence in the stereospecificity of hydride transfer from ethanol.

3. Active Site Substituents

Two mechanistic motifs have been recognized for enzyme catalysis of hydride transfer as shown in Fig. 12. The first class of dehydrogenases utilizes imidazole as an acid/base to promote proton transfer to or from the oxygen functionality. These enzymes include lactate dehydrogenase and glyceraldehyde-3-phosphate dehydrogenase. The second class of enzymes consists of metallo dehydrogenases that utilize a water ligand to a tightly coordinated zinc as an acid/base. These enzymes include the alcohol dehydrogenases for yeast and liver and related enzymes (for recent reviews see Refs. 1 and 2).

In carbonyl reduction, the conjugate acid of the imidazole increases the electron deficiency of the carbonyl through hydrogen bonding to the oxygen, which promotes hydride transfer from the dihydronicotinamide ring. Likewise, coor-

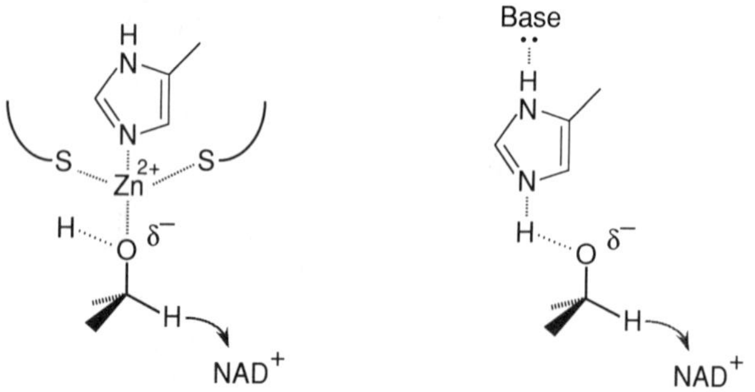

FIG. 12. A generalized representation of the active sites of the two mechanistically distinct forms of dehydrogenases. The coordination of an alcohol substrate in a metallo dehydrogenase, such as liver alcohol dehydrogenase, is shown on the left. The coordination of an alcohol substrate with an imidazole-base dehydrogenase, such as lactate dehydrogenase, is shown on the right.

dination of a carbonyl to the zinc atom Lewis acid greatly increases the electrophilicity of the carbon. The details of hydride transfer are still strongly debated and the discussion in Section I,G on the implications of the alternative chemistry will only add to the controversy. The parallels with either MPV or Cannizzaro chemistry, however, are readily apparent.

4. Active Site Thiols in Aldehyde Dehydrogenases

A major structural distinction between noncyclic "aldehyde" dehydrogenases and "alcohol" dehydrogenases is the presence of a thiol in the active site. "Aldehyde" dehydrogenases catalyze redox chemistry at the oxidation level of aldehyde/carboxylate as exemplified by glyceraldehyde-3-phosphate dehydrogenase. The active site thiol is traditionally viewed as serving two catalytic functions, as shown in Fig. 13. First, it forms a thiohemiacetal that anchors the aldehyde substrate into the active site. Second, the thiohemiacetal is viewed as activating the C-1 carbon to promote hydride transfer to NAD$^+$ with generation of a thioester. Note that a thiol is not essential for oxidation of aldehydic func-

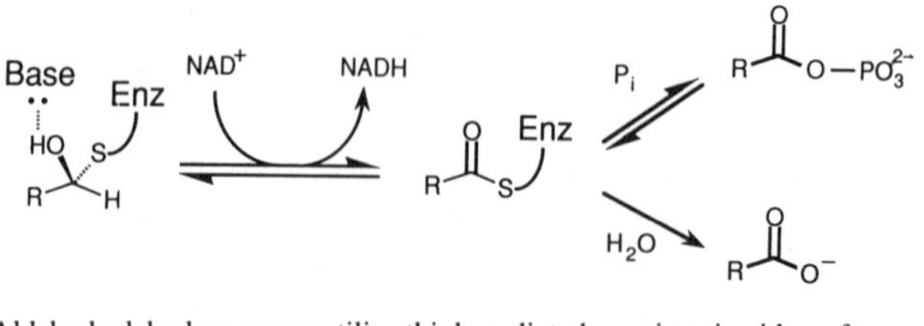

FIG. 13. Aldehyde dehydrogenases utilize thiol-mediated reactions in either of two ways. The top pathway demonstrates the capture of the energy of aldehyde oxidation through formation of a carboxylate–phosphoric acid anhydride, which can subsequently generate ATP. The bottom pathway shows the hydrolysis of the active site thioester to the thermodynamically favored carboxylate.

tionalities. Glucose-6-phosphate dehydrogenase catalyzes the oxidation of a pyranose to a glucolactone without apparent intervention of a covalent intermediate with any active site thio groups.

Thermodynamically the oxidation of aldehydes is strongly favored. The redox potential of acetaldehyde/acetate is -589 mV compared to only -320 mV for the coenzyme. Because of the large $-\Delta G$ of hydrolysis of thioesters, their formation sequesters the energy of oxidation in a form that can be used in subsequent reactions. For example, G3PDH-catalyzed production of 1,3-diphosphoglycerate is used to synthesize ATP. Alternatively, the thioester can be hydrolyzed, rendering a reaction irreversible, e.g., histidinol dehydrogenase formation of the carboxylate. Mechanistically, the operation of "aldehyde" dehydrogenases is conceptually no different than that for "alcohol" dehydrogenases except for the active site thiol. As will be discussed in Section I,E, the mechanistic distinction between these two classes of dehydrogenases is far less than is generally believed.

5. Dehydrogenases that Catalyze Sequential Oxidations

The distinction between "alcohol" and "aldehyde" dehydrogenases is clearly blurred with a major group of enzymes that can conduct the sequential oxidation of alcohols to carboxylates *in the same active site*. Among these dehydrogenases are L-histidinol:NAD^+ oxidoreductase (EC 1.1.1.23; histidinol dehydrogenase), hydroxylmethylglutaryl-CoA reductase (EC 1.1.1.34; HMG-CoA reductase), and uridine diphosphate D-glucose:NAD^+ 6-oxidoreductase (EC 1.1.1.22; UDPglucose dehydrogenase). Histidinol dehydrogenase [reaction (5)] and UDPglucose dehydrogenase [reaction (6)] share a common mechanistic scheme. These enzymes differ from most dehydrogenases by binding the substrate before binding the coenzyme. The first reaction, alcohol to aldehyde, is freely reversible, but

Reaction (5) Histidinol dehydrogenase

Reaction (6) UDP-Glucose dehydrogenase

the subsequent aldehyde oxidation is irreversible. Both oxidations occur at comparable rates. Although there is no evidence for release of the intermediate aldehyde, the enzyme is fully capable of oxidizing exogenous aldehydes at rates that are kinetically competent. Participation of an active site thiol is implicated in both reactions, presumably through formation of an enzyme-bound thiohemiacetal intermediate.

The primary function of hydroxylmethylglutaryl-CoA reductase is the reduction of a thio ester (HMG-CoA) to an alcohol (mevalonate) with release of HS-CoA via a putative thiohemiacetal intermediate, as shown in reaction (7). This reaction is fully reversible and in the presence of high concentrations of mevalonate and CoASH it can be run in the nonphysiological direction to generate the thioester from an alcohol. The reaction is mechanistically similar to that catalyzed by G3PDH, with the exception that for HMG-CoA reductase the thio group belongs to a soluble cofactor rather than to a cysteine side chain. Once the HS-CoA has dissociated, then aldehyde reduction can be viewed as comparable to that catalyzed by normal "alcohol" dehydrogenases lacking active site thiols.

$$\text{HOOC-CH}_2\underset{\overset{|}{\text{CH}_3}}{\overset{\text{OH}}{|}}\text{CH}_2\text{-CH}_2\text{OH} \quad \underset{}{\overset{\text{HS-CoA}}{\underset{+}{}}} \quad \xleftarrow{\overset{2\text{NAD}^+ \quad 2\text{NADH}}{}} \quad \text{HOOC-CH}_2\underset{\overset{|}{\text{CH}_3}}{\overset{\text{OH}}{|}}\text{CH}_2\text{-}\overset{\overset{\text{O}}{\|}}{\text{C}}\text{-S-CoA}$$

Reaction (7) HMG-CoA Reductase Reaction

E. SUBSTITUENT EFFECTS, ALTERNATIVE CHEMISTRY, AND MECHANISM OF DEHYDROGENASES

1. Substrate Substituent Effects

Physical organic chemical approaches to understanding the mechanism of dehydrogenases have raised important mechanistic questions. Klinman and co-workers observed that the rates of reaction for a series of para-substituted benzyl alcohols with Y-ADH were insensitive to the electronic effects of the substituent (58, 59). Their results indicate that there is no change in the charge of C-1 in the activated complex relative to the alcohol. Therefore, either hydride transfer involves an early transition state or the reaction is highly concerted, i.e., removal of the proton from the alcohol is concomitant with transfer of hydride. The corresponding reductions of para-substituted benzaldehydes by NADH show strong substituent effects indicative of a late transition state. A concerted mechanism is unlikely, however, because of the strong evidence of alkoxide formation prior to hydride transfer in HL-ADH (for recent reviews see Refs. 1 and 2). Further undermining any concerted mechanism is the ability of the enzyme to function with alternative substrates such as *gem*-diols and hemiacetals (see Section I,D). The maintenance of the coordination in transfer of charge without alteration in charge density at C-1 over the extreme range of substrates is highly unlikely. Therefore, the picture of the activated complex that emerges has the substrate as a zinc-ligated alkoxide and the coenzyme pyridinium-like, as shown in Fig. 14. Lactate dehydrogenases show a similar lack of sensitivity toward substituent effects and thus presumably have a similar transition state.

The observed primary and secondary deuterium kinetic isotope effects for alcohol dehydrogenases do not provide an internally self-consistent correlation. The inconsistencies have been partially reconciled, with the realization that secondary kinetic isotope effects have both vibrational and translational components in a coupled motion along the reaction pathway (*60, 62*). Recently Klinman and co-workers have proposed that hydride transfer occurs with substantial contributions from quantum mechanical tunneling of the hydride through the potential energy barrier (*63, 64*). Mechanistic participation by tunneling raises intriguing prospects for enzymic catalysis. Not only can catalysis by dehydrogenases involve lowering the activation barrier to facilitate hydride transfer by the chemical means discussed in previous sections, but catalysis can also involve enforcing a proximity that decreases the separation along the reaction coordinates, thereby increasing the probability that the hydride penetrates the barrier (*63*).

2. Reactions with Alternative Substrates

The ability of dehydrogenases to function with substrates that cover a much broader range of substituents than that implied by their name has been largely overlooked. Horse liver ADH, a metallo "long dehydrogenase" (*65–69*); *Drosophila* ADH (D-ADH), a nonmetallo "short dehydrogenase" (*70*); and pig heart lactate dehydrogenase (*71*) are all fully competent at conducting hydride

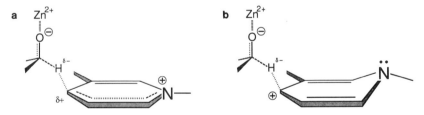

FIG. 14. Based on the insensitivity toward substituent effects, the transition state for horse liver alcohol dehydrogenase is expected to be reactant-like, and with little charge buildup on the hydrogen to be transferred. The electron distribution in the nicotinamide ring is not known. The ring could retain pyridinium character, with only modest positive charge density at C-4, as shown in (a). Alternatively the charge could even reside entirely at C-4, as shown in (b) following rehybridization at N-1. Secondary ^{15}N kinetic isotope effects have been reported to support the latter picture (*60*); however, further studies have failed to corroborate these results (*61*) (W. W. Cleland, personal communication), leaving the question open.

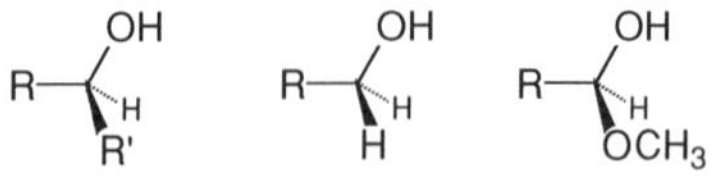

Fig. 15. Range of substrates for the horse liver and *Drosophila* alcohol dehydrogenases (from left to right): secondary alcohols, primary alcohols, hemiacetals, and *gem*-diols.

transfer for the range of substrates shown in Fig. 15.* Although no direct measurement has been made of the rate of hydride transfer for these substrates, the overall enzymic kinetics differ little (see Table II). Indeed, based on V_{max}, pig heart lactate dehydrogenase would be better designated as a glyoxalate dismutase! The catalytic efficiency of *Drosophila* ADH toward aldehyde oxidation is essentially the same as that for alcohol oxidation, raising questions as to the true metabolic function of this enzyme (70).† Thus representative members of three distinct dehydrogenase families catalyze hydride transfer with little regard to the substituent at C-1. Translation of binding energy into activation of the scissile bond cannot account for the insensitivity because, for HL-ADH, alcohols and aldehydes (as their *gem*-diol forms) have similar binding constants.

3. Dismutation Reactions

The most common reaction catalyzed by "alcohol" dehydrogenases is dismutation of aldehydes to equal molar mixtures of a carboxylic acid and the corresponding alcohol, as shown in reaction (8). Dismutation in its simplest form

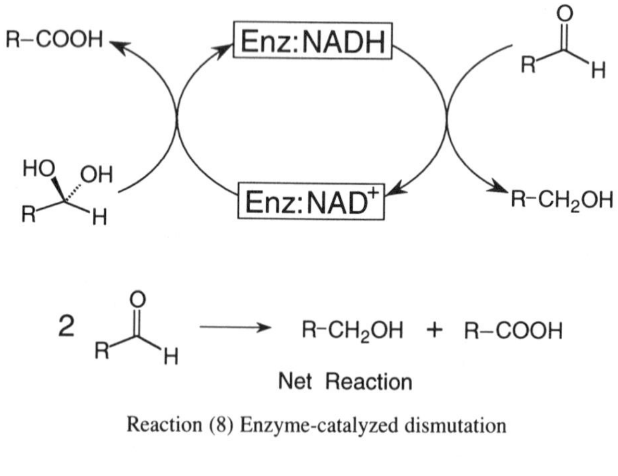

Net Reaction

Reaction (8) Enzyme-catalyzed dismutation

* Care must be exercised in evaluating reports of no activity for a substrate, e.g., trifluoroethanol with HL-ADH. The experiments that have been conducted (72) cannot differentiate between the intrinsic inability of a substrate analog to function and its binding in a nonproductive orientation that would preclude hydride transfer.

† D-ADH oxidation of aldehydes to carboxylates is so facile that the author has suggested that this reaction is the "natural" reaction and that secondary alcohols are the "alternative" substrate.

TABLE II

KINETIC AND THERMODYNAMIC PARAMETERS FOR ALTERNATIVE REACTIONS CATALYZED BY DEHYDROGENASES

Source	Enzyme	$\log K_{eq}$	Activity	K_m	mol sec^{-1}/ mol enzyme	Ref.
Pig heart	Lactate DH	-17.5	Glyoxalate reduction		430	71
		-11.6	Pyruvate reduction/lactate oxidation		192/85	
		$\sim -0.9^a$	Glyoxalate oxidation		115	
					Rel V_{max}	
Horse liver	ADH	-11.2	Octanol oxidation	$14 \mu M$	1.0	66
		-7.6	Secondary alcohol oxidation	n.r.^{c}	n.r.	
		$+2.1^b$	Octanal oxidation	$29 mM$	0.07	
				K_m	V/K_{max}	
Drosophila	ADH	-11.2	Ethanol DH	$3.2 mM$	31	70
		-7.6	Secondary alcohol DH	n.r.	n.r.	
		$+2.1^b$	Acetaldehyde oxidation	$1.0(0.65)^d$	$18(29)^d$	

^{a} Value calculated from redox potential of glyoxalate/oxalate (-0.50 mV).
^{b} Value calculated from redox potential of acetaldehyde/acetate (-0.589 mV).
^{c} n.r., Not reported.
^{d} Values in parentheses are corrected for the amount of the *gem*-diol form.

involves no net formation of NADH. Only transient amounts of NADH are needed to shuttle hydride equivalents between the substrates. In the absence of net NADH synthesis, dismutation is also spectrophotometrically silent and unless formation of the acid is monitored, the reaction can be easily overlooked.

The oxidation of aldehydes to carboxylic acids has been most extensively investigated with horse liver alcohol dehydrogenases (65–67, 69, 73). There are two distinct reactions: the direct oxidation of aldehydes as their hydrated gem-diol form [reaction (9)] and the oxidation of hemiacetals to esters [oxidative esterification, reaction (10)].

$$\text{HO, OH} \quad \xrightarrow[\text{NAD}^+ \quad \text{NADH}]{} \quad R-C(O^-)O$$

Reaction (9) Enzyme-catalyzed oxidation of gem-diols to carboxylic acids

$$\text{H}_3\text{CO, OH} \quad \xrightarrow[\text{NAD}^+ \quad \text{NADH}]{} \quad R-C(O)OCH_3$$

Reaction (10) Enzyme-catalyzed oxidation of hemiacetals to esters

Kendal and Ramanathan (73) first reported that mammalian liver alcohol dehydrogenases can catalyze the dismutation of formaldehyde to formic acid and methanol. Subsequently Abeles and Lee (69) showed that dismutation was a true function of HL-ADH and not the result of any contaminating aldehyde oxidase activities. Indeed, with n-octanol as a substrate, HL-ADH generates octanoic acid as the primary product after catalyzing a transient formation of octanal (66). Trapping with semicarbazide is only partially effective in inhibiting octanoic acid formation, which suggests that octanal remains bound while the coenzyme exchanges, a result consistent with the comparable K_m values for NADH and octanal of about 10^{-5} M. The V_{max} for octanal oxidation is 7% of that for oxidation of octanol. This value must, however, be viewed as a conservative estimate of the ability of HL-ADH to oxidize octanal because the results are not corrected for the concentration of the putative gem-diol substrate. If the octanal generated by octanol oxidation does not dissociate, then this scheme also predicts that hydration must occur in the active site, presumably catalyzed by the same proton donation that activates the system to accept hydride. The overall rate of octanol to octanoic acid then would have to be further corrected for the rate of this hydration reaction (see Fig. 16).

Conditions that lead to dismutation of formaldehyde also result in the isolation of methyl formate, presumably by oxidation of formaldehyde–methyl hemiacetal (73). Subsequently Hinson et al. (66) found that the net reduction of NAD$^+$

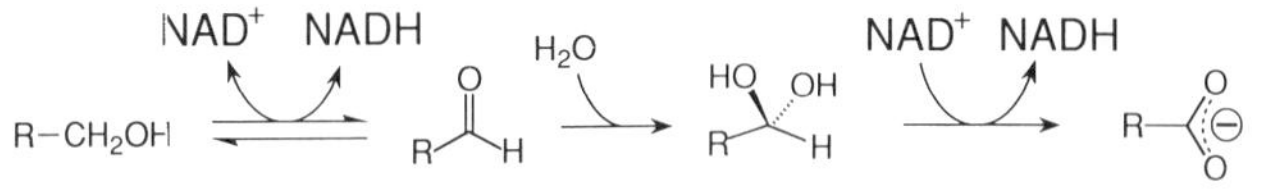

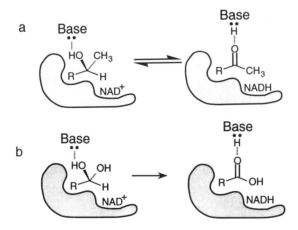

FIG. 16. The three steps in the sequential enzyme-catalyzed oxidation of a primary alcohol to a carboxylic acid. The first step is the reversible oxidation of the primary alcohol. On dissociation of NADH, the enzyme must catalyze hydration of the aldehyde. Finally, binding of NAD⁺ leads to oxidation of the *gem*-diol to the carboxylate.

to NADH was enhanced by addition of primary alcohols and that the product was the corresponding ester. Low-molecular-weight aldehydes are especially susceptible to this form of oxidative esterification, which suggests that binding of the hemiacetal and not catalytic proficiency is the main determinant for the reaction. Binding specificity may also determine which enzymes will catalyze these reactions. As seen in Fig. 15, hemiacetals are structural mimics of secondary alcohols. HL-ADH, which functions nearly equally well with primary and secondary alcohols, catalyzes both dismutation and oxidative esterification of aldehydes. There are no corresponding reports for yeast alcohol dehydrogenase, which shows a strong preference for primary alcohols.

The dismutation reaction has important practical implications. The coupling of the thermodynamically favorable oxidation of a secondary alcohol to drive the reduction of an aldehyde is a common strategy for dehydrogenase-mediated specific reductions in "bioreactors." These conditions, however, are ideal for dismutation of the aldehyde. As shown in Fig. 17, the Enz–NAD⁺ complex is

FIG. 17. Comparison of the functioning of an alcohol dehydrogenase with a secondary alcohol (a) to its functioning with a *gem*-diol (b). Note that although they differ in their respective inductive effects, the hydroxyl and methyl groups are isosteric. No stereochemistry for the hydride transfer should be inferred.

equally capable of binding either the secondary alcohol or the aldehyde. In many cases the generally poorer binding constants for secondary alcohols mean that they are not present in saturating concentrations, thus the extent of dismutation becomes a function of the ratio of the respective values of k_{cat}/K_m for the corresponding secondary alcohol and the *gem*-diol of the aldehyde. For example, attempts to produce chirally labeled [1-^{2}H]butanol by *in situ* reduction of butanal with 2-[2-^{2}H]propanol using HL-ADH led to recovery of fully protio *n*-butanol (^{2}H incorporation of <1% as determined by mass spectrometry) and an equal molar amount of butyric acid; only dismutation was observed (O. Malver and N. J. Oppenheimer, unpublished results).

4. α-Ketoacid Dehydrogenases

Lactate dehydrogenases from a number of vertebrate sources have been shown to catalyze the dismutation of glyoxalate (*71, 74*). For pig heart LDH isozyme 1, the steady-state inhibition patterns shown by oxamate and oxalate toward the glyoxalate reaction are similar to those observed for pyruvate/lactate, which suggests that the active site is similar, if not identical, for the two enzymes, as shown in Fig. 18. At their respective pH maxima, the specific activities for glyoxalate are comparable to those for the "natural" substrates pyruvate/lactate (the V_{max} for glyoxalate/NADH was 2.2 times greater than for pyruvate/NADH, and for glyoxalate/NAD$^+$ it was 1.3 times greater than lactate/NAD$^+$) (*71*). Rabbit muscle LDH had similar properties with a V_{max} for glyoxalate reduction only three times greater than for glyoxalate oxidation (*74*).

5. Thiols and Aldehyde Oxidation

The catalysis of the oxidation of aldehydes to carboxylates by "alcohol" dehydrogenases raises questions regarding the function of the active site thiols found in most "aldehyde" dehydrogenases. Clearly a free thiol is not mechanistically essential for aldehyde oxidation. For example, pig heart lactate dehydrogenase catalyzes the facile oxidation of glyoxalate to oxalate (*71*), glucose-6-

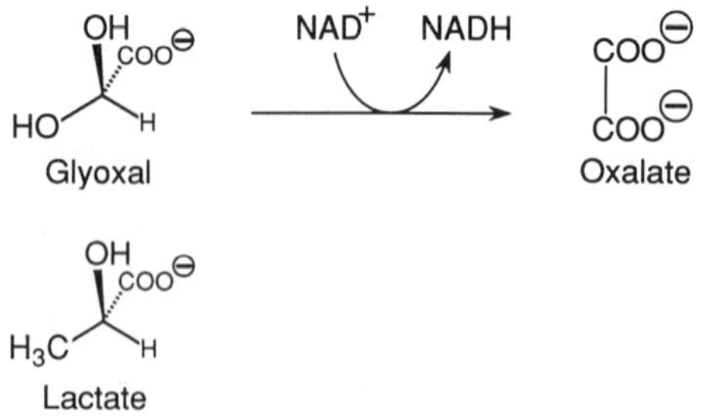

FIG. 18. Lactate dehydrogenases can catalyze the oxidation of glyoxalate to oxalate, as shown in the top reaction, and also the reduction of glyoxalate to glycolic acid (not shown). Note that glyoxalate differs from lactate by the isosteric replacement of the methyl group by a hydroxyl.

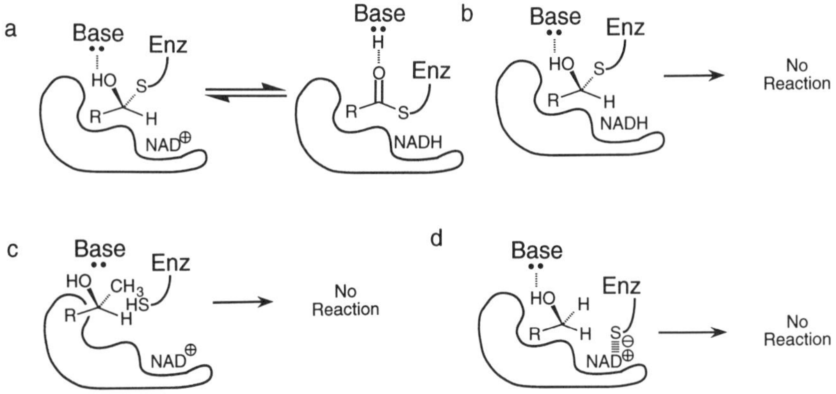

FIG. 19. The traditional function for a thiol in an active site is shown in (a). The following alternative functions are illustrated. (b) Prevention of aldehyde reduction in the presence of bound NADH by thiohemiacetal formation. (c) Steric exclusion of secondary alcohols to prevent activity as a secondary alcohol dehydrogenase. (d) Formation of an intimate interaction between the thiol and NAD$^+$ such as occurs in G3PDH to prevent functioning as an alcohol dehydrogenase.

phosphate dehydrogenase catalyzes the formation of a lactone, and HL-ADH conducts the sequential oxidation of histidinol to histidine (75).*

The function of thiols in active sites of "aldehyde" dehydrogenases needs to be reconsidered in view of the ability of the catalytic machinery of many "alcohol" dehydrogenases to conduct comparable oxidations without invoking an obligatory participation of thiols. As shown in Fig. 19, for "aldehyde" dehydrogenases, rather than just anchoring the substrate, the thiol can serve to focus the chemistry on aldehyde oxidation. Thiohemiacetal intermediates are resistant to reduction by NADH; thus the reaction of an aldehyde with an active site thiol precludes reduction by NADH and thus prevents the enzyme from serving as an aldehyde reductase. Thiols in an active site can also sterically block binding of secondary alcohols, thus preventing the enzyme from serving as a secondary alcohol dehydrogenase. In this latter case, however, it may still be capable of functioning as a primary alcohol dehydrogenase.†

* Cys-174, one of the ligands to zinc, has been proposed to form a thiohemiacetal intermediate with the histidinal; however, catalysis would require functioning of the enzyme with an altered metal coordination. It is difficult to imagine how activity would be maintained if such a major reorganization were to occur.

† These ideas give a new and different perspective on the function of the interaction between NAD$^+$ and Cys-149 in G3PDH that is responsible for the "Racker band" (76) (for a review of optical spectroscopy of pyridine nucleotide-dependent enzymes see Ref. 77). By forming a charge transfer complex with the nicotinamide ring, the thiol can be viewed as sequestering the NAD$^+$ in a nonreactive form, which renders it unavailable to participate in dehydrogenation reactions with primary alcohols. Formation of a thiohemiacetal on binding of aldehydes releases the NAD$^+$, which allows it to serve as a redox coenzyme.

F. Evolution, Stereochemistry, and Mechanism of Dehydrogenases

The provocative proposals by Benner and co-workers (*18, 78, 79*) regarding the mechanistic and functional significance of the stereospecificity of hydride transfer catalyzed by dehydrogenases raise intriguing questions about fundamental aspects of the mechanisms of these enzymes. The functional model attempts to treat various dehydrogenase attributes, but primarily stereospecificity, as manifestations of underlying chemical imperatives. The following sections will address the points raised by their hypotheses and analyze possible relations among dehydrogenase function, mechanism, and stereospecificity.

1. *Stereospecificity and Stereochemistry*

The stereospecificity of hydride transfer is a highly conserved property among related dehydrogenases (*80–82*). Even the example of *Drosophila* ADH versus yeast ADH proposed by Benner *et al.* as evidence for heterogeneity among alcohol dehydrogenases (*79*) illustrates the extreme conservation of specificity. Based on sequence analyses, Jörnval and co-workers (*83*) have designated two major categories of alcohol dehydrogenases: the "short dehydrogenase," characterized by a molecular weight of ~25,000, *pro-S* stereospecificity, and no requirement for divalent metal ions (*Drosophila* ADH), and "long dehydrogenase," with a molecular weight of ~35,000–40,000, *pro-R* stereospecificity, and an absolute requirement for zinc (yeast and horse liver ADH). Within each group the stereospecificity has been conserved over numerous phyla and among enzymes with a wide range of substrates (see Table III). For example, short alcohol dehydrogenase function as general alcohol dehydrogenases in *Drosophila*, whereas enzymes derived from long dehydrogenases have analogous functions in mammalian livers. This appears to be a case of convergent utilization of enzymes. Any relatedness of these two classes of dehydrogenases exists at a level

TABLE III

PROPERTIES OF LONG AND SHORT DEHYDROGENASES[a]

Property	Long dehydrogenases	Short dehydrogenases
M_r	35,000–40,000	25,000–27,000
Metal requirement	Mechanistically essential zinc	No mechanistically essential metal requirement
Stereospecificity	*pro-R* specific	*pro-S* specific
Sources and examples	Yeast alcohol dehydrogenase, liver alcohol dehydrogenases, sorbitol dehydrogenase (sheep)	*Drosophila* alcohol dehydrogenase, glucose dehydrogenase (*Bacillus megaterium*), ribitol dehydrogenase (*Klebsiella aerogenes*)

[a] Based on Refs. *83, 85,* and *86.*

of fundamental properties shared by all dehydrogenases and also with other nucleotide binding proteins (*83–88*).

2. Functional Model for Dehydrogenases

Many of the observed attributes of enzymes arise by natural selection in order to help the host organism survive and reproduce. Benner *et al.* have proposed that one such attribute, the stereospecificities of dehydrogenases, has functional significance based on stereochemical arguments (*18, 79*). The central features of their functional model can be summarized as follows. The stereospecificities of dehydrogenases acting on alcohols are correlated with the equilibrium constant for the alcohol–carbonyl redox reaction as listed in Table IV (*18*). Enzymes catalyzing reactions where the K_{eq} is $<10^{-11}$ M transfer the *pro-S* proton from NADH; when K_{eq} is $>10^{-11}$ M, the *pro-R* proton is transferred. Thus the more readily reduced carbonyl compounds use the *pro-R* proton, but the more difficult to reduce carbonyl compounds use the *pro-S* proton. The proposed correlation is restricted to "simple" aldehydes and ketones (i.e., without additional chemistry that would influence the equilibrium constant, such as cyclizations of polyols or formation of lactones). The "natural" substrate of the enzyme must be well

TABLE IV

STEREOSPECIFICITY OF DEHYDROGENASES ORGANIZED BY SUBSTRATE CLASS[a]

log K_{eq}	Enzyme	Substrate class	Stereospecificity
−17.5	Glyoxalate reductase[b]	α-HAld	*pro-R*
−13.5	Tartronate semialdehyde reductase	α-HAld	*pro-R*
−12.8	Glycerate DH	α-HAc	*pro-R*
−12.1	Malate DH	α-HAc	*pro-R*
−11.6	L-Lactate DH	α-HAc	*pro-R*
−11.6	D-Lactate DH	α-HAc	*pro-R*
−11.2	Ethanol DH	Primary alcohol	*pro-R*
−11.1	Glycerol DH	Primary alcohol	*pro-R*
−10.9	Homoserine DH	Primary alcohol	*pro-S*
−10.9	Carnitine DH	β-HAc	*pro-S*
−10.5	3-Hydroxyacyl-CoA DH	β-HAc	*pro-S*
−8.9	3-Hydroxybutyrate DH	β-HAc	*pro-S*
−7.6	3-Oxoacyl-ACP Reductase	β-HAc	*pro-S*
−7.7	Estradiol 17-DH	Secondary alcohol	*pro-S*
−7.6	3-Hydroxysteroid DH	Secondary alcohol	*pro-S*

[a] α-HAld, α-Hydroxy aldehyde; α-HAc, α-hydroxy acid; β-HAc, β-hydroxy acid; DH, dehydrogenase. Values for K_{eq} are from Ref. *78*. Malic enzyme (*pro-R*) was included in the original correlation; however, it catalyzes both redox chemistry and decarboxylation, and it is not subject to a simple analysis of its external equilibrium constant [this point has been invoked for polyol dehydrogenases (*78*)].

[b] "Glyoxalate reductase" for which the stereospecificity was measured is an activity of glycerate dehydrogenase and not a separate enzyme (*89*).

defined because the correlation will only be relevant for substrate–dehydro-genase pairs that have been "optimized" by evolutionary selective pressure.* Finally, the external equilibrium constant for the substrate must be a factor of 10 from the "breakpoint" at 10^{-11} M.

3. Correlation of Substrate Properties and Dehydrogenase Stereochemistry

Two points must be kept in mind when any correlation is proposed. First, correlations do not necessarily imply causality. Second, the factors being corre-lated may be ancillary to the primary factors responsible for the correlation. The correlation shown in Table IV of the stereochemistry versus external equilibrium constants of substrates at first glance is impressive. Consider, however, that the external equilibrium constants for the alcohol functionalities in question reflect the thermodynamic driving force of those reactions. The equilibrium constants for alcohol oxidation are governed, however, by the nature of the flanking sub-stituents. Alcohols with electron-donating substituents, such as hydrogen or methyl (primary or secondary alcohols), are easier to oxidize than alcohol func-tionalities having more electron-withdrawing substituents, such as carboxylates (α-hydroxy acids). *Therefore, the external equilibrium constant for a given al-cohol functionality depends on the nature of the neighboring substituents, that is, on its structure.* The following corollaries can be made. (1) Chemically related compounds will share similar external equilibrium constants. (2). Struc-turally related compounds will be more likely to share evolutionarily related enzymes. The latter corollary is a restatement of "Bentley's first rule" that de-hydrogenases utilizing similar substrates will have the same stereospecificity (*80*). Corollary 2, however, puts the cart before the horse: The more closely dehydrogenases are evolutionarily related, the more likely they are to function on similar substrates. This follows from the simple axiom of protein chemistry that sequence defines structure defines function.

If the dehydrogenases listed in Table IV (*89*) are viewed not by the K_{eq} of their substrates, but rather by the chemical nature of their substrates, then the correlation takes on a very different meaning. As shown in Table IV, all the

---

* Arguments based on the kinetic parameters to define the "natural " substrate for a given enzyme can be difficult. For example, consider that the k_{cat} for broad-spectrum alcohol dehydrogenases, such as the liver ADH, is about 100 times slower than for highly specific and metabolically unidirectional enzymes, such as yeast ADH. In the Theorell–Chance model for HL-ADH, coenzyme dissociation is rate limiting. Because of this, the turnover number for HL-ADH appears deceptively low when assayed in one direction with a single substrate such as ethanol. When HL-ADH is assayed with a coupled oxidation–reduction reaction, e.g., ethanol + lactaldehyde $\leftrightharpoons$ acetaldehyde + 1,2-dihydroxypropane, much higher catalytic activities are found. This is rationalized by having the coenzyme remain bound while substrates and products exchange rapidly (*68*). Therefore the argu-ments for the inclusion of testosterone dehydrogenase and the exclusion of HL-ADH based on their relative rates (*79*) are misleading.

enzymes listed above the breakpoint are either α-hydroxy-acid or α-hydroxyaldehyde dehydrogenases. Below the breakpoint they are either secondary alcohol or β-hydroxy-acid dehydrogenases. The middle ground is covered by primary and secondary alcohol dehydrogenases. Thus the dehydrogenases listed in Table IV fall into a limited number of categories based on substrate structure. Clearly the enzymes processing the α-hydroxy acids, malate and lactate, are structurally related (90, 91). Underscoring that relationship, a functional MDH can be derived from *B. stearothermophilus* LDH with only three amino acid replacements (92). A much more meaningful correlation would be to group enzymes into families based on relations among their primary sequences such as discussed in Section I,F,2 for long and short dehydrogenases.

4. Natural Substrates of Dehydrogenases

In the functional model the observed stereospecificity derives from selective pressures on the "natural" substrate; thus the decision to include or exclude a given dehydrogenase in the functional model correlation degenerates into arguments over the "natural" substrate of the enzyme and ignores obvious relationships among the sequences of the enzymes. For example, the presence of the *pro-R*-specific yeast *ethanol* dehydrogenase (a long dehydrogenase) and the *Drosophila* "ethanol" dehydrogenase (a short dehydrogenase) was cited as a manifestation of the ambiguity for enzymes catalyzing reactions with K_{eq} near the "breakpoint" of the correlation (the K_{eq} for ethanol is at the breakpoint) (79). The two major families of alcohol dehydrogenases involved, however, have members that are specific for primary alcohols, secondary alcohols, both primary and secondary alcohols, and polyols (83). All of these enzymes maintain the identical stereospecificity with respect to hydride transfer that is characteristic of their respective dehydrogenase family.* Also, both families contain enzymes that catalyze redox chemistry on substrates having a much wider range of external equilibrium constants.

---

* Note that the two dehydrogenases extensively studied by Benner and co-workers to support the functional model, lactaldehyde dehydrogenase and *Drosophila* ADH, do not meet their own criteria for inclusion because of the lack of a "defined" substrate. *Drosophila* ADH is argued to be an "ethanol" dehydrogenase on the grounds that ethanol is the selective agent. (79, 93). However, *in vitro*, D-ADH has an extremely broad substrate specificity. The catalytic efficiency for secondary alcohols exceeds that for ethanol and other primary alcohols (88), and D-ADH can conduct the facile oxidation of aldehydes to acids. Indeed, as discussed in Section I,E,2, this latter reaction has been suggested to be the true metabolic function of D-ADH (70). Prediction of the stereospecificity of lactaldehyde dehydrogenase was given as a test of the functional model (78); however, as discussed by You (82), lactaldehyde dehydrogenase (EC 1.1.1.55) is identical to both glucuronate dehydrogenase (EC 1.1.1.19) and mevaldate dehydrogenase (EC 1.1.1.33). All these enzymes are now designated as a nonspecific, high-K_m NADP$^+$-dependent alcohol dehydrogenase (EC 1.1.1.2).

5. *Conformation, Analogs, and Stereoelectronic Effects*

The functional model is based on stereoelectronic effects involving the putative dependence of the redox potential on the torsional geometries around the nicotinamide ribosyl bond (*12, 78*). Specific χ angles are invoked to promote $n-\sigma\star$ overlap between the carbon–oxygen bond of the sugar ring, with the lone pair on the ring nitrogen of the dihydronicotinamide ring as a means of influencing the coenzyme redox potential in the active site. This model has been largely unsuccessful in accounting for experimental results.

As discussed in Section I,C,1, the association constant for formation of the cyanide adduct correlates with the inductive effects of cis 2'-substituents and not with their steric bulk. That latter correlation would have been anticipated if, as suggested by Benner, the torsional angle around the glycosidic bond influences the redox potential (*12*). Given the 20-mV difference in redox potential between the α- and β-anomers of NADH (*94*) or *arabino*-NADH (*40*), any significant correlation of redox potentials with changes in χ related to the 2'-substituent would have been readily observable.

The functioning of dehydrogenases with α-anomers, acyclic analogs, or carbocyclic analogs is inconsistent with the stereoelectronic models.* For example, as shown in Fig. 20, the orientation of the carbon–oxygen bond in α-anomers is, at best, orthogonal to the lone pair on nitrogen and possibly even aligned in the opposite sense as the stereoelectronic effect invoked for the β-anomer.† Finally, the positioning of the transition state along the reaction coordinates will determine the importance of such stereoelectronic effects. For enzymes such as Y-ADH, where the transition state for hydride transfer from the dihydronicotinamide is apparently late and productlike (see Section I,E,1), the proposed orbital effects would not be germane.

Although any functional origins of the *pro-R* versus *pro-S* specificities in early dehydrogenases are subject to debate, once established, this property is not easily transmuted and no examples of altered specificity within a major family have been demonstrated. This is easily explained. Alterations in interactions re-

---

* The significance of these activities has been questioned because they involve coenzyme analogs whereas the active sites have been optimized for the "natural" coenzyme. This argument is less than compelling. The fact remains that dehydrogenases can use coenzymes containing modifications that make their functioning inconsistent with the stereoelectronic model. These results do not rule out stereoelectronic effects; rather they indicate that such effects are not of transcending importance and should be treated as just one of the many ways the microscopic redox potential can be influenced in the active site.

† The interpretation that the anomeric configuration is irrelevant to stereoelectronic considerations (*12*) appears to be based on a mistaken assumption that the orientation of the lone pair of electrons on nitrogen with respect to the carbon–oxygen bond is the same for either α- or β-anomers in the active site of a dehydrogenase.

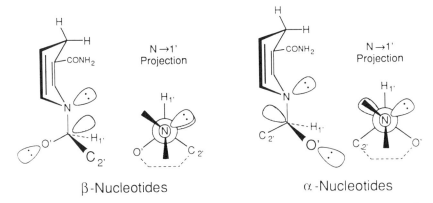

β-Nucleotides   α-Nucleotides

FIG. 20. Comparison of the alignment of the orbitals in the α- and β-anomers of NADH when bound in an active site. In the β-anomer, this conformation allows for overlap between the lone pair on nitrogen with the $n–\sigma^*$ antibonding orbital of the carbon–oxygen bond, as shown on the left. Maintenance of the same geometry for the α-anomer, as required by its functioning with the identical stereochemistry as the β-anomer, leads to the lone pair on nitrogen being, at best, orthogonal to the $n–\sigma^*$ orbital. (It is impossible for the sugar ring of the α-anomer to maintain the same interactions in the active site as the β-anomer while also having the dihydronicotinamide ring bind productively.)

sponsible for controlling the stereospecificity would, by necessity, involve the same residues also responsible for activating the coenzyme for catalysis. Any intermediate enzyme forms would therefore lose catalytic efficiency without any selectable advantage. Compare this to the observed alterations involved with *in vivo* selection for an altered substrate specificity. Selective pressure on the ribitol dehydrogenase of *Klebsiella aerogenes* grown on xylitol yields intermediate enzymes with relaxed substrate specificity (*95*). What is lost in enzyme efficiency with the original substrate is gained by increased activity with the new substrate. Therefore, clear pathways can be envisioned for altering substrate specificity even where multiple mutations are required.*

* The argument has been made that if the stereospecificity of hydride transfer of dehydrogenases is a neutral feature subject to no obvious selective pressure, then it should be randomly distributed (*96*). This is a thermodynamic argument for what is a kinetically controlled process. In the course of evolution many choices have been made among alternatives that are isoenergetic, including the specific genetic code, D versus L amino acids, D versus L sugars, etc. Such decisions are necessitated by the inherent diastereomeric catastrophe that would follow if organisms tried to operate with mixtures of enantiomers as molecular building blocks. Although theories have been proposed to explain (rationalize?) such choices (for a recent example see Ref. *97*), clearly once such a choice has been made it becomes increasingly irreversible. The barrier today to the generation of "mirror image" life is kinetic and not thermodynamic. Any attempt at the gradual creation of such life by natural selection is impossible, not because mirror life itself is impossible, but rather because intermediate forms would be nonfunctional (the "you can't get there from here" syndrome).

G. INTERNAL EQUILIBRIUM CONSTANTS AND MECHANISM

There have been extensive discussions in the literature regarding maximization of the catalytic efficiency of enzymes and the value of their "internal" equilibrium constants (K_{int} is the equilibrium constant between substrates and products of the enzyme when all are bound productively) (98–102). For example, the value of K_{int} is near unity for both liver alcohol dehydrogenase (78) and lactate dehydrogenase (103) when measured with their natural substrates. The ability of these same enzymes to function with alternative substrates with widely differing external equilibrium constants raises important questions regarding the relationships of the internal thermodynamics of such reactions.

Three scenarios will be considered. In the first, K_{int} for the reaction with the alternative substrates could also be near unity. This is unlikely, however, because the large difference in external equilibrium constants between the natural and alternative substrates would require unrealistic binding constants ($<<10^{-15}\ M$) that are not supported by their experimentally observed K_m values or the overall turnover of the enzyme-catalyzed reaction. The observed independence of the rate of hydride transfer on the external equilibrium constant raises questions regarding the importance of maintaining K_{int} near unity. In the second scenario the similar rates for the substrates could arise from a compensating shift in the transition state as predicted by the Hammond postulate (104). As the energy difference between reactants and products increases, the transition state occurs at an earlier point in time; thus the activation barrier could remain nearly constant over the wide range of substrate external equilibrium constants. How this would operate is not clear. The transition state is already suggested to be so early for alcohol dehydrogenases oxidizing primary alcohols that no substituent effects are observed. How much earlier can the reaction get? Such an early transition state also requires an extremely electrophilic nicotinamide moiety—far more electrophilic than is currently envisioned. Can all the mechanisms discussed in this chapter generate such a reactive species without making the nicotinamide extremely sensitive to the addition of external nucleophiles? Also, are such early transition states consistent with the primary isotope effects observed for the hydride transfer reaction? The final scenario is that quantum mechanical tunneling is far more important to the mechanism of dehydrogenases than is currently accepted. Can dehydrogenases move hydride *through* the energy barrier in such a way as to be insensitive to the height of that barrier? What is the precedent for this mechanism?

H. MECHANISTIC OVERVIEW OF DEHYDROGENASES

The ability of dehydrogenases to conduct reactions on alternative substrates, especially involving "unnatural" reactions, has been explored with only a few dehydrogenases. Nonetheless, based on the extensive sequence identity among the long alcohol dehydrogenases represented by HL-ADH, short alcohol dehydrogenases represented by D-ADH, or the α-hydroxy-acid dehydrogenases represented by LDH, it is anticipated that many related enzymes will possess, to a greater or lesser degree, the same abilities. The high degree of sequence identity also suggests that these reactions do not represent aberrant behavior on the part of the specific dehydrogenases investigated. Rather the reactions represent a fundamental ability of the catalytic machinery of these enzymes to conduct a wide range of dehydrogenation reactions subject only to limitations imposed by the ability of substrates to bind. The conclusion is that the central catalytic site of each of these enzymes can conduct with near equal facility the transfer of hydride equivalents to or from oxygen-functionalized carbons, whether alcohol $\rightleftharpoons$ aldehyde, aldehyde (as the *gem*-diol) $\rightarrow$ acid, or hemiacetal $\rightarrow$ ester! These results impose important constraints on any proposed mechanisms and raise serious questions regarding the current models to explain the functioning of dehydrogenases. Clearly the ability of these enzymes to operate with such a wide range of substrates requires new thinking on the mechanism of dehydrogenases.

II. Mechanism of Enzymes That Cleave Nicotinamide–Glycosyl Bonds of NAD $^+$

The ADP-ribosyltransferases and NAD$^+$ glycohydrolases represent a diverse group of enzymes that catalyze the cleavage of the nicotinamide–glycosyl bond of NAD$^+$, as shown in Fig. 21. These enzymes are found throughout prokaryotic and eukaryotic life. Their function derives from the large free energy of hydrolysis of the nicotinamide–glycosyl bond that is comparable to that for the hydrolysis of the γ-phosphate of ATP. Thus NAD$^+$ can be viewed as an "activated" ADP-ribose. This portion of the chapter will first provide a brief description of the enzymes involved and then discuss the chemistry and enzymology of ribosyl bond cleavage, including stereochemical and mechanistic aspects (for recent reviews on biological and biochemical aspects of these enzymes see Refs. *105* and *106*).

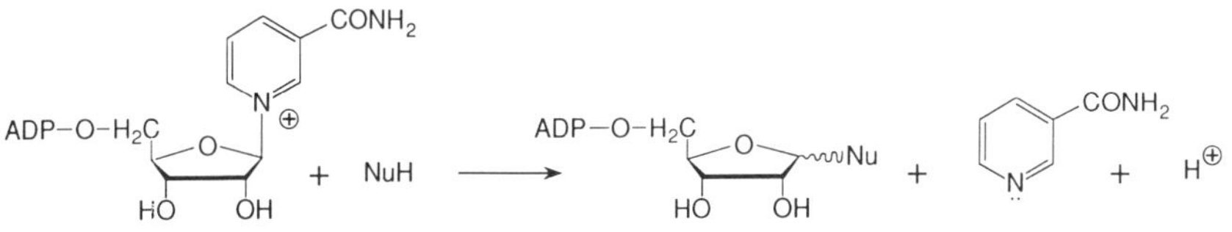

FIG. 21. Generalized cleavage of the nicotinamide–ribosyl bond with formation of a linkage to a new nucleophile.

The enzymes that catalyze the cleavage of the nicotinamide–ribosyl bond are divided into two major classes; ADP-ribosyltransferases and NAD$^+$ glycohydrolases. Their distinction is based on the chemistry that occurs after cleavage of the glycosyl bond.

ADP-ribosyltransferases catalyze the formation of an ADP-ribosyl bond to a specific amino acid residue on a target protein or other acceptor molecule. The covalent modification then alters the activity or function of the target protein. This class of proteins includes bacterial toxins, bacterial activating proteins, endogenous ADP-ribosyltransferases, and the eukaryotic poly(ADP-ribose) synthase.

The NAD$^+$ glycohydrolases constitute the second major class of proteins involved in nicotinamide–glycosyl bond cleavage. This activity is widely distributed in mammalian tissue as well as in other prokaryotic and eukaryotic organisms, although its biological function is not understood. There are two mechanistically distinct categories of NAD$^+$ glycohydrolases, strict hydrolases (nicotinamide insensitive) and those that conduct transglycosidation (nicotinamide sensitive). All NAD$^+$ glycohydrolases catalyze the hydrolytic cleavage of the nicotinamide–glycosyl bond yielding ADP-ribose and nicotinamide. Those that catalyze transglycosidation also exchange the nicotinamide moiety of the ADP-ribose moiety with an exogenous nucleophilic acceptor or another pyridine base. These enzymes have been invaluable in the preparation of NAD$^+$ analogs (*30, 31, 107*).

A.   ADP-RIBOSYLTRANSFERASES

1.   *Mono-ADP-ribosylation*

Mono-ADP-ribosyltransferases are found in both eukaryotes and prokaryotes. The bacterial toxins are the best understood in terms of structure and function. These toxins include cholera, diphtheria, pertussis, botulinum D, and functionally related toxins. In all cases studied thus far, the covalent attachment of the ADP-ribose group to the target proceeds with inversion at the (nicotinamide)-ribose C-1 to generate α-linkages. In the absence of appropriate acceptors the mono-ADP-ribosyltransferases also catalyze a much slower hydrolysis of NAD$^+$. This latter reaction occurs without detectable methanolysis, which precludes stereochemical analysis of the hydrolytic reaction. The enzymes and acceptor preferences for these enzymes are summarized in Table V (*108–120*).

2.   *Cholera Toxin*

Cholera toxin catalyzes the attachment of ADP-ribose to an arginine side chain of the $G_{s\alpha}$ subunit of the guanine nucleotide binding protein in the adenylate

TABLE V

ACCEPTOR SPECIFICITIES FOR ADP-RIBOSYLTRANSFERASES

Transferase	Acceptor	Ref.
Bacterial		
Cholera toxin	Arginine	*108, 109*
Pertussis toxin	Cysteine	*110, 111*
Diphtheria toxin	Diphthamide	*112, 113*
Pseudomonas aeruginosa		
exotoxin A	Diphthamide	*114, 115*
Botulinum C1 toxin	Asparagine	*116*
C2 toxin	Arginine	*117*
Eukaryotic		
NAD:arginine ADPRTase	Arginine	*108, 118*
NAD:cysteine ADPRTase	Cysteine	*119*
Poly(ADP-ribose) synthase	Glutamic acid γ-COOH (histone H1), adenonsine ribose 2′-OH, and (nicotinamide) ribose 2′-OH	*120*

cyclase regulatory system (*108, 121*) (**8**). This covalent modification disrupts the GTPase activity of the subunit, leading to sustained activation of adenylate cyclase. The toxin consists of two subunits; the A subunit possesses the enzymic activity and the B subunit is responsible for binding of the toxin to the cell surface, but is otherwise nontoxic (*120*). *In vitro*, arginine and other guanidino compounds act as acceptors of ADP-ribose (*121*), and these compounds can inhibit the ADP-ribosylation of proteins by competing for the active site of the toxin (*108*). Cholera toxin also ADP-ribosylates arginine in transducin, a protein of retinal rod outer segments, which is homologous to G_s in both structure and function (*109*). Cholera toxin initially generates α-ADP-ribosylarginine, i.e., the reaction proceeds with inversion of configuration at the ribose C-1 (*122*). Under optimal conditions for ADP-ribosyltransferase activity of cholera toxin, the rate of ADP-ribosylation is 20 times that of NAD^+ hydrolysis (*123*).

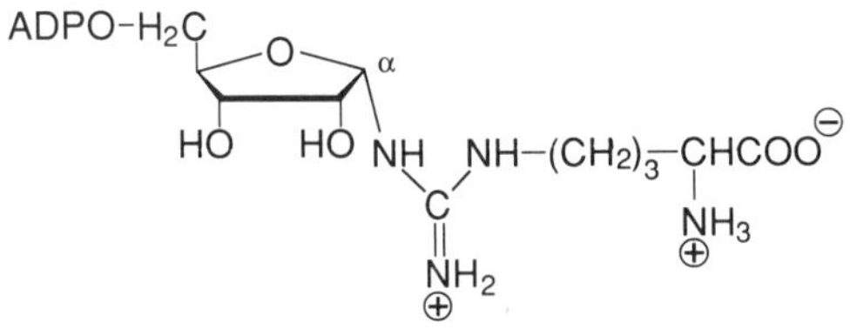

(**8**) α-ADP-ribosylarginine

3. Pertussis Toxin

Pertussin toxin was first investigated as an activator of pancreatic islet cells (*108*). The toxin ADP-ribosylates G_i, the inhibitory guanine nucleotide binding protein of the adenylate cyclase regulatory system, and leads to cAMP accumulation inside the cell (*108*). Pertussis toxin reduces or abolishes G_i-mediated signal transduction from a wide range of inhibitory receptors, including opiate, prostaglandin E_1, α_2-adrenergic, muscarinic–cholinergic, dopamine, and adenosine in a variety of mammalian cells (*120*).

Pertussis toxin consists of five subunits, S1 to S5. The S1 subunit (or A protomer) is responsible for the ADP-ribosyltransferase activity and also possesses NAD^+ glycohydrolase activity (*111*). Cysteine has been identified as the amino acid residue in transducin that is ADP-ribosylated by the toxin (*110*); recently thiols such as cysteine and dithiothreitol have been reported to be ADP-ribosylated by pertussis toxin (*111*). No data on the stereochemistry of this linkage are available. Several groups are attempting to characterize the structure of the S1 subunit (*124, 125*) and to identify the amino acid residues essential for enzymic activity (*126, 127*). Interestingly, there is a sequence identity in a portion of the active subunits of cholera and pertussis toxins, and antibodies raised to residues 6–17 of the pertussis subunit will also bind to the A subunit of cholera toxin (*124*). The two toxins are clearly distinct with respect to their target specificity, yet both exhibit NAD^+ glycohydrolase activity, which suggests that the homologous regions may be responsible for binding and labilizing NAD^+.

4. Diphtheria Toxin

Diphtheria toxin disrupts cell function by ADP-ribosylating a unique, posttranslationally modified histidine acceptor diphthamide (**9**) found only in eukaryotic elongation factor-2 (EF-2) (Fig. 18) (*112*). Nuclear Overhauser enhancement experiments established that the imidazole N-1 of diphthamide is covalently linked to ADP-ribose via an α-glycosidic linkage (*113*).

(**9**) Diphthamide

Exotoxin A produced by the aerobic bacterium *Pseudomonas aeruginosa* (*128*) possesses the same catalytic activity as diphtheria toxin (*120, 129*), including ADP-ribosylation of EF-2 on the same site (*115*). The reaction is reversible because ADP-ribosylated EF-2 produced by exotoxin A can be deribosylated (functional activity restored) by incubating with excess nicotinamide and either excess exotoxin A or subunit A of diphtheria toxin. NAD$^+$ was identified as the sole product of this reverse reaction (*115*). These results also establish that the configuration and site of ADP-ribosylation catalyzed by exotoxin A are identical to those of diphtheria toxin.

Irradiation of the NAD$^+$–diphtheria toxin binary complex at 254 nm results in the photoinactivation of the toxin and fragmentation of the NAD$^+$ via attachment at the C-6 position of the nicotinamide ring to the γ-methylene of Glu-148 (*130*). Glu-148 was subsequently found to be essential for catalysis (without affecting the binding of NAD$^+$) by substituting Glu-148 with Asp using site-directed mutagenesis (*131, 132*). *Pseudomonas aeruginosa* exotoxin A showed an analogous photoaffinity labeling with NAD$^+$; however, the nicotinamide was attached to the decarboxylated side chain of Glu-553 (*133*). Site-specific substitution of Glu-553 for aspartic acid results in a 1800-fold decrease in enzymic activity and confirms that the exact positioning of the carboxyl group is crucial to catalysis (*134*). Little sequence identity has been noted between the two proteins using conventional comparisons, but when the active site glutamic acid residues are aligned, regions of identity in the primary structure of the active sites become apparent (*133*).

5. *Botulinum Toxins*

Strains of *Clostridium botulinum* produce toxins with ADP-ribosylation activity. Botulinum C1 toxin ADP-ribosylates a guanine nucleotide binding protein (G$_b$) that is widely distributed in various mammalian tissues (*135*). Recent results suggest that the site of ADP-ribosylation is an asparagine residue (*116*). Botulinum C2 toxin ADP-ribosylates cytoplasmic actin at Arg-177, but the stereochemistry of addition has not been established (*117*). This result is significant because it represents the first time that an NAD-dependent toxin has been directed at a target other than a GTP-binding protein and thus may signal the presence of a new range of toxin targets. Botulinum toxin type C3 has recently been purified and characterized (*136*). This ADP-ribosyltransferase targets a 21- to 24-kDa protein in human platelets, fibroblasts, and several other tissues. The substrate of ADP-ribosylation has been purified from pig brain and has been identified as a GTP-binding protein (*137*). The specific amino acid residue that is ADP-ribosylated has not been determined. Botulinum toxin type D ADP-ribosylates a 21-kDa protein from beef adrenal gland, which has a molecular mass similar to the masses of some recently discovered G proteins (*138*). ADP-ribosyltransferase activity of the type D toxin is inhibited by agmatine and L-arginine methyl ester, which suggests that the ADP-ribosyl acceptor is also the guanidino group of arginine. Unlike cholera toxin, however, type D toxin does not ADP-ribosylate arginine or agmatine when incubated with NAD$^+$ (138).

6. Other ADP-ribosyltransferase Activities

Endogenous mono-ADP-ribosyltransferase activities have also been found in prokaryotes and eukaryotes. In eukaryotes, these proteins represent only "activities"; their physiological purpose is not understood. Arginine-dependent ADP-ribosyltransferases have been found in a wide range of animal tissues, including turkey erythrocytes, hen liver, rat liver, and pig skeletal muscle (108). Similar to cholera toxin, these enzymes activate the nicotinamide–glycosyl bond of NAD$^+$ and transfer the ADP-ribose group to arginine and other guanidines in vitro. Also like cholera toxin, the NAD:arginine ADP-ribosyltransferase from turkey erythrocytes catalyzes formation of an α-glycosidic linkage between arginine and ADP-ribose (118). A separate enzyme has been isolated from turkey erythrocytes that cleaves the arginine–ribose bond of α-ADP-ribosylarginine (139), which provides the basis for a possible regulatory cycle although no target metabolic pathway has been identified in eukaryotes. A hen liver nuclear enzyme that ADP-ribosylates agmatine, arginine, and histones has been purified to homogeneity (120). Its specific target protein(s) and biological function are unknown. The only such regulatory system that has all three components (a well-defined target protein, a specific ADP-ribosyltransferase, and an ADP-ribosyl hydrolase) has been identified for the prokaryotic regulation of bacterial nitrogenase activity based on reversible ADP-ribosylation at arginine (140).

ADP-ribosyltransferases specific for EF-2 have been isolated from rat liver (141) and beef liver (120). The fact that intoxication with only one molecule of diphtheria toxin A fragment is sufficient to kill a cell suggests that ADP-ribosylation of diphthamide is irreversible, with no endogenous enzymic activity that cleaves the α-ribosyl–diphthamide bond. Therefore, the regulatory function of an endogenous protein that inactivates EF-2 is obscure. Recently, a cysteine-specific ADP-ribosyltransferase was purified from human erythrocytes that also catalyzes the ADP-ribosylation of G_i (the target of pertussis toxin). Pretreatment of G_i with pertussis toxin and NAD$^+$ inhibits the human enzyme, which suggests that both enzymes are specific for the same acceptor residue (119). Whether the ADP-ribosylation of G_i is the primary physiological role remains to be determined.

B. POLY(ADP-RIBOSYLATION)

Poly(ADP-ribose) synthase (120, 142) is a eukaryotic nuclear enzyme that catalyzes three reactions: the transfer of the ADP-ribose group of NAD$^+$ to other nuclear proteins (including itself), the polymerization of ADP-ribose units, and branching of the ADP-ribose polymers. Poly(ADP-ribose) is composed of 2 to 200 repeating adenosine diphosphate ribose units (*120*) in which the adenosine ribose 2'-OH is linked to the next ADP-ribose group through an α-1'',2'-glycosidic bond (Figs. 22–24) (*143*). Pulse-chase experiments with radiolabeled NAD$^+$ confirm that the polymerization occurs by a mechanism in which the first ADP-ribose unit is covalently attached to the protein followed by successive additions of ADP-ribose to the adenosine 2'-OH of the previous unit (*144*). Typically 2–3% of the ADP-ribose moieties in a polymer are subject to branch-

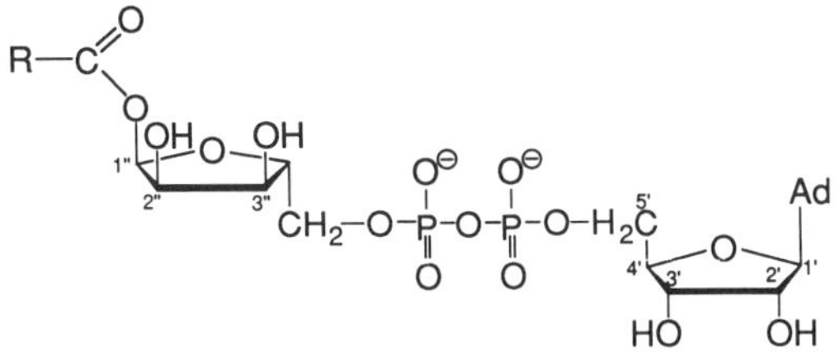

FIG. 22. Structure of the ADP-ribosyl linkage to the γ-carboxylate of glutamate that attaches the poly(ADP-ribose) chain to a protein.

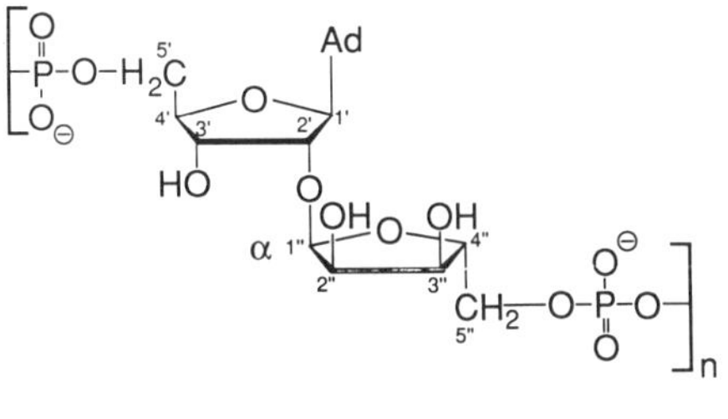

FIG. 23. Structure of AMP-RP (adenosine monophosphate-α-$O^{2'},O^{1'}$-ribose 5-phosphate), the monomeric unit of poly(ADP-ribose) containing an α-glycosyl linkage to the adenine 2'-hydroxyl

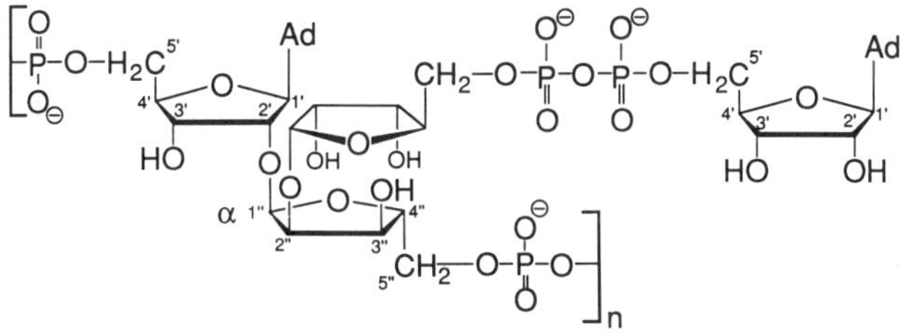

FIG. 24. Structure of a branch point in poly(ADP-ribose) formed by attachment at the 2''-hydroxyl (formerly the nicotinamide ribose) of an ADP-ribose monomeric unit.

ing in which the ADP-ribose unit is linked to the 2'-OH of the (nicotinamide) ribose group instead of the adenosine ribose group (*120*).

Stereochemically and functionally the synthase is a remarkable enzyme. As outlined in Figs. 22–24, all evidence points to a single protein conducting the three distinct reactions, presumably in the same active site (although the size of the protein does not rule out multiple catalytic domains). The cleavage of the nicotinamide–glycosyl bond (Fig. 22) generates an acylal linkage to a glutamate or aspartate carboxylate (stereochemistry unknown). Note that the attacking nucleophile is a carboxylate at the end of a flexible side chain. Elongation (Fig. 23) involves cleavage of the nicotinamide–glycosyl bond with formation of an α-linkage to the adenosine 2'-hydroxyl. The nucleophile is a 2'-hydroxyl trans to the adenine ring. Finally, branching (Fig. 24) occurs through formation of an α-glycosyl linkage with the ribose 2'-hydroxyl of the polymer (the former nicotinamide ribose) cis to an adenosine moiety. The great disparity in the steric and electronic attributes of the nucleophiles engaged in the three reactions makes this either a truly unique single active site, or makes this enzyme a prime candidate for the involvement of multiple catalytic domains.

C.   NAD⁺ GLYCOHYDROLASES

NAD⁺ glycohydrolases (NADases) have been isolated from a diverse range of animal tissues, including calf spleen (*145*), pig brain (*146*), bull semen (*147*), and snake venom (*148*). In contrast to the proteins already considered, very little is understood about the biological roles of the NAD⁺ glycohydrolases (for reviews see Refs. *31* and *107*). The primary reaction catalyzed by these enzymes is hydrolysis of the nicotinamide–ribose bond of β-NAD⁺. The NADases have been divided into two mechanistically distinct categories, those that are nicotinamide sensitive and those that are not. The nicotinamide-sensitive NADases also catalyze alcoholysis and transglycosidation, such as exchange of the nicotinamide group for exogenous pyridine bases or other nucleophilic heterocycles (*30, 31*). The NAD⁺ glycohydrolase from calf spleen proceeds by a Uni Bi mechanism, with nicotinamide being released first, followed by ADP-ribose release (*149*). Catalysis of transglycosidation implies the presence of a relatively stable "high-energy" ADP-ribosyl·enzyme intermediate of unknown structure.

The stereochemical outcome of the NAD⁺ glycohydrolase reaction is opposite to that of the ADP-ribosyltransferase reaction. Hydrolysis of β-NAD⁺ by calf spleen NAD⁺ glycohydrolase in aqueous methanol solutions occurs with ≥99% retention of configuration for the resulting 1'-*O*-methyl-ADP-ribose, although a small, and possibly significant, amount of inverted product is also recovered (*150*). Methanolysis catalyzed by NAD⁺ glycohydrolase from pig brain also proceeds with retention of configuration (*151*).

D.   Coenzyme Analogs and Inhibitors

Attempts have been made to design mechanism-based inhibitors of NAD^+ glycohydrolases based on the intermediacy of an oxocarbocation. Potent inhibitors of NAD^+ glycohydrolase would be valuable for investigation of the biological role of these enzymes and for further elucidation of their enzymic mechanism. The first such compound was ADP-ribonolactone (**10**). The K_i for ADP-ribonolactone is 115 μM, only nine times better than ADP-ribose (K_i = 1.0 mM) despite the planar atom at C-1' (*152*).* The carbocyclic analog of NAD^+ containing a 1-N-nicotinamide-2,3-dihydroxycyclopentane adenine dinucleotide (carba-NAD^+) is a structural analog of the substrate containing a C–N bond to the nicotinamide ring that should be resistant to hydrolysis (**7**) (*51*). As anticipated, it is not a substrate for NAD^+ glycohydrolase, but it is also not a particularly good inhibitor; its inhibition constant is comparable to that for ADP-ribonolactone (**10**) (*154*). Interestingly, the L-carbocyclic ribose analog ψ-carba-NAD^+ (**11**) is a better inhibitor, with a K_i of 6.7 μM for the cow brain NADase.

(**10**) Ribonolactone

(**11**) ψ-Carba-NAD^+

Unlike dehydrogenases, enzymes cleaving the glycosyl bond are extremely sensitive to modifications of the sugar moiety. The arabino analogs, *arabino*-NAD^+ and 2'-fluoro-*arabino*-NAD^+ (**12**) are both potent, slow-binding competitive inhibitors of calf spleen NADase with K_i values of 170 nM and 2 μM,

* A lactone has been successfully used to inhibit lysozyme, another enzyme that cleaves glycosyl bonds and is thought to generate an oxocarbocation intermediate (*153*).

(**12**) 2'-Fluoro-2'-deoxynicotinamide arabinoside adenine dinucleotide

respectively (*155*). The rate of enzyme-catalyzed hydrolysis of 2'-fluoro-*ara-bino*-NAD$^+$ is less than 10^{-6} that for NAD$^+$ hydrolysis and the upper limit for the *arabino*-NAD$^+$ analog is less than 10^{-5} that of NAD$^+$ (*155*). The results are summarized in Table VI (*156*).

The lack of activity of the 2'-fluoro-*arabino*-NAD$^+$ analog can be rationalized in part by the inductive destabilization of the putative oxocarbocationic inter-mediate by the neighboring fluorine atom. The substituent at the 2'-position of the sugar ring exerts a powerful inductive effect on the stability of the glycosyl bond. The nonenzymic hydrolysis of 2'-fluoro-*arabino*-NAD$^+$ is 3% the rate of NAD$^+$ (*157*). On the other hand, no such explanation can be offered for the *arabino*-NAD$^+$ analog, which hydrolyzes nonenzymically at 33% the rate of NAD$^+$. Thus the intrinsic stability of the nicotinamide–glycosyl bond in the *arabino*-NAD$^+$ analogs alone does not explain why they fail to be substrates for the calf spleen NADase. Clearly alterations at the 2'-position enhance binding, but render the coenzyme dysfunctional with respect to cleavage of the glycosyl bond. This suggests that the 2'-hydroxyl of the ribose has a key function in either catalysis or productive binding in the active site.

E. MECHANISTIC OVERVIEW OF GLYCOHYDROLASES

Mechanistically, the stereochemical outcome of the reactions catalyzed by the ADP-ribosyltransferases and NAD$^+$ glycohydrolases defines the relative orien-

TABLE VI

RATES OF HYDROLYSIS AND INHIBITORY CONSTANTS FOR NAD ANALOGS

Inhibitor (or substrate)	K_i or (K_m) (μM)	Dissociation rate		Nonenzymic hydrolysis: $k_h \times 10^8$ at 37°C (sec^{-1})
		$k_{off} \times 10^3$ (sec^{-1})	$t_{1/2}$ (min)	
2'-fluoro-*arabino*-NAD$^+$	0.3	3.45	3.35	1.6
arabino-NAD$^+$	11.9	3.88	2.97	15.8
(NAD$^+$)	$(60.0)^a$	—	—	48.3

[a] Calf spleen NAD$^+$ glycohydrolase (*156*).

tations of the reactants in the active site. Thus far, all the ADP-ribosyltransfer-ases catalyze the formation of α-ribosidic linkages. These results establish that the acceptor must be bound distal to the departing nicotinamide group. Although such an arrangement would be stereochemically conducive to an in-line associative displacement reaction, the mechanisms of cleavage of the nicotinamide–ribose bond and formation of the linkage to the acceptor remain to be determined. The observation of hydrolysis in the absence of acceptors suggests that ADP-ribosyltransferases labilize the nicotinamide–ribosyl bond without the need or participation of a specific acceptor nucleophile. How toxins can conduct hydrolysis but not methanolysis remains an open question, especially in view of the latter's greater nucleophilicity (158, 159). This discrimination is consistent neither with water attacking from the site formerly occupied by the departed nicotinamide (retention), nor with water attacking in place of the acceptor moiety (inversion). Because both nicotinamide and the acceptors are considerably larger than methanol, steric exclusion of methanol seems unlikely.

The hydrolysis of β-NAD$^+$ catalyzed by pig brain NADase (151) and calf spleen NADase (150) occurs with retention of configuration. These results establish that the incoming nucleophile attacks from the same side as that formerly occupied by the nicotinamide ring. Such an arrangement is fully consistent with the Ping-Pong kinetics and the enzyme-bound intermediate observed for the reaction (149).

Substantial secondary deuterium kinetic isotope effects have been observed for the hydrolysis of [1'-^{2}H]-NAD$^+$ and related compounds by both the NADases that catalyze transglycosidation and those that are strict hydrolases. These results are consistent with appreciable sp^3 to sp^2 rehybridization during the cleavage of the glycosyl bond (146) and point to an oxocarbocationic intermediate. Linear free energy relationships for the calf spleen NADase-catalyzed hydrolysis of NAD$^+$ analogs with various substituents at either the C-3 or C-4 of the pyridine ring show that the rate of hydrolysis is inversely correlated with the pK_a of the departing pyridine (156). A Brønsted plot of log(V_{rel}) versus pK_a gives a slope of $\beta_{lg} = -0.9$ over 4 pK_a units. The sign and magnitude of this slope indicate the development of electron deficiency with a large degree of bond breaking in the transition state. Interestingly, only NAD$^+$ and NMN$^+$ deviated from the plot and were hydrolyzed 14 and 2.8 times faster, respectively, than anticipated from their pK_a values. These results suggest that interactions at the carboxamide, such as increased electron withdrawal by strong hydrogen bonding to a proton donor, might be used to labilize further the glycosidic bond. Likewise, the rate of hydrolysis is strongly influenced by alterations in the substituent at the 2'-position on the sugar. As shown in Fig. 25, a plot of the log of the rate of hydrolysis versus the Taft inductive σ_i for 2'-substituted nicotinamide ribo- and arabino-furanosides gives a slope of -6.7. Such a strong dependence of the rate of hydrolysis is also consistent with an oxocarbocation intermediate.

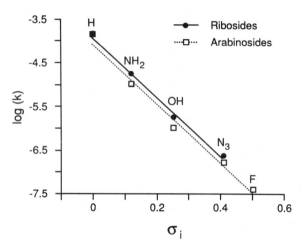

FIG. 25. The plot of the rate of hydrolysis versus the Taft inductive σ_i for 2'-substituted nicotin-amide *ribo-* and *arabino*-furanosides gives slope -6.7, which is consistent with an oxocarbocation intermediate.

Cleavage of the glycosyl bond by calf spleen NADase differs fundamentally from the nonenzymic reaction with respect to reactivity with exogenous nucleophiles. Nonenzymic hydrolysis of β-NAD$^+$ shows no significant discrimination on the basis of nucleophilicity of the attacking group (*159, 160*). In contrast, the calf spleen NADase discriminates strongly between acceptor nucleophiles and reacts 50 times faster with methanol than with water. This suggests that the ADP-ribosyl·enzyme intermediate generated in the reaction is equilibrated with its potential nucleophiles (*159*). Given the vanishingly short lifetime expected for oxocarbocations (*161, 162*), the nature of stabilization in the active site is of considerable interest.

There is a mechanistic dilemma for the enzyme-catalyzed hydrolysis of NAD$^+$. As discussed by Cordes and co-workers, an enzyme conducting cleavage of a nicotinamide–glycosyl bond via a dissociative mechanism has few possible means to promote catalysis (*146*). As they state, general acid catalysis is ruled out because the leaving group is already cationic. General base or nucleophilic catalysis is excluded by the large secondary kinetic isotope effects. Finally, because both the transition state and substrate possess positive charges, the importance of their differential electrostatic stabilization is greatly diminished. In the absence of viable mechanistic alternatives, catalysis by distortion was invoked in order to explain how the enzyme accelerates the rate of cleavage of the glycosyl bond.

The inability of these enzymes to cleave the furanosyl bond in the *arabino* analogs argues against catalysis by strain. There is no obvious reason why an *arabino* configuration should so effectively preclude the requisite distortions, especially in view of their high affinity. The extreme sensitivity of the reaction to the sugar configuration suggests mechanistic involvement of the 2′-hydroxyl. Recent studies of the nonenzymic hydrolysis of NAD⁺ at high pH provide an alternative mechanism for catalysis (160). The pH profile for the aqueous hydrolysis of NAD⁺ shows that the alkaline lability of NAD⁺ does not stem from direct attack of hydroxide. Rather, ionization of the ribose diol ($pK_a = 11.9$ at 37°C) accelerates hydrolysis by over four orders of magnitude relative to hydrolysis in the neutral sugar. The rate of hydrolysis for the diol anion was attributed to electrostatic stabilization of an oxocarbocation intermediate. Direct anchimeric attack by the anion cannot be ruled out, although the intermediacy of 1,2-anhydro sugars provides a less satisfactory explanation for most of the experimental evidence, as discussed in Ref. 160.

The thermodynamic activation parameters for the chemical hydrolysis of the neutral diol and diol anion are compared in Table VII. Ionization of the diol results in a nearly complete rupture of the nicotinamide–ribosyl bond ($\Delta H^{\ddagger}$ decreases to 11.8 kcal/mol) that is largely compensated for by the decrease in $\Delta S^{\ddagger}$ to -30.8 eu; thus the overall $\Delta\Delta G^{\ddagger}$ is only 5.2 kcal/mol, which corresponds to a rate increase of approximately 10^4. These results provide the basis for a plausible mechanism of enzyme catalysis by labilizing the glycosyl linkage through electron donation directed at the 2′-hydroxyl. A simple model would involve an interaction with a carboxylate side chain as shown in Fig. 26. Regardless of how it is accomplished, however, if the enzyme were to cause an increase in electron density on the diol equivalent to that of ionization, then a corresponding decrease in $\Delta H^{\ddagger}$ would be anticipated. If the enzyme were also to provide an entropically structured environment for the activated complex, the result would be to bring the large $-\Delta S^{\ddagger}$ toward zero. The overall consequence would be a rate acceleration far larger than the four orders of magnitude increase in the rate of hydrolysis as the result of diol ionization observed for the solution reaction.

Such a mechanism based on entropic control would circumvent the Cordes dilemma and would account for the mechanistic importance of the 2′-ribo hydroxyl. In the absence of the properly configured 2′-hydroxyl, electrostatic interactions with basic groups (carboxylates?) in the active site would be redirected to the nicotinamide ring. Such interactions would then serve both to promote

TABLE VII

ACTIVATION PARAMETERS FOR NAD HYDROLYSIS[a]

pH	$\Delta H^{\ddagger}$ (kcal/mol)	$\Delta S^{\ddagger}$ (eu)	$\Delta G^{\ddagger}$ (kcal/mol)
6.2	25.2 ± 1.7	-3.3 ± 2.5	26.2 ± 2.5
13.4	11.8 ± 0.5	-30.8 ± 1.5	21.0 ± 0.9

[a] At 25°C. From Ref. 160.

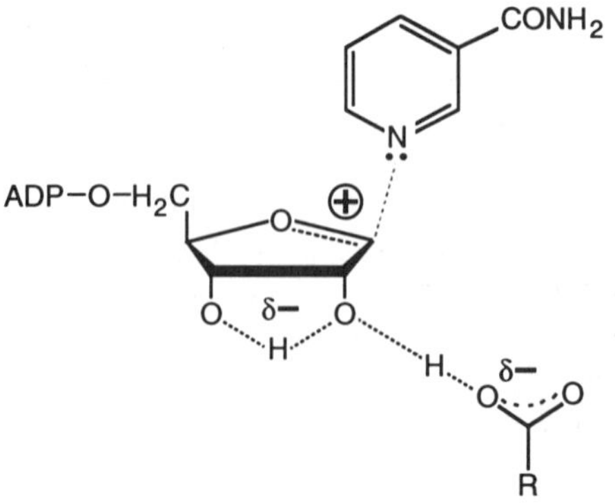

FIG. 26. Proposed mechanism for enzyme-catalyzed hydrolysis involving electron withdrawal by an active site carboxylate promoting cleavage of the glycosyl bond through inductively stabilizing an oxocarbocationic intermediate.

binding and to stabilize the glycosyl bond. Photoaffinity reactions of NAD^+ combined with site-directed mutagenesis have shown the presence of a carboxylate in the active sites of diphtheria toxin and *Pseudomonas* exotoxin A (*133*). The principles being elucidated for the NADases should be generally applicable to all glycohydrolases that operate by a dissociative mechanism and have similarly acidic functional groups at the 2'-position. A classic example of an enzyme meeting these criteria is lysozyme. It has two carboxylates in the active site and a substrate with a 2'-substituent having a pK_a of approximately 12. Present theories for catalysis primarily involve "catalysis by strain or distortion" arguments (for a recent review see Ref. *163*) and do not satisfactorily address the chemical nature of catalysis. Therefore, lysozyme and related glycosidases are likely candidates for a mechanism similar to that outlined here.*

REFERENCES

1. Ohno, A., and Ushio, K. (1987). *In* "Redox Reactions of Pyridine Nucleotides" (D. Dolphin, R. Poulson, O. Avramovic, eds.), pp. 276–331. Wiley, New York.
2. Pettersson, G. (1987). *Crit. Rev. Biochem.* **21**, 349.
3. Oppenheimer, N. J. (1987). *In* "Chemical Stability and Reactivity of Pyridine Nucleotides" (D. Dolphin, R. Poulson and O. Avramovic, eds.), pp. 323–365. Wiley, New York.

* Interestingly, the parameters for the β-galactosidase-catalyzed hydrolysis to the 3-chloro-1-(β-D-galactopyranosyl)pyridinium are $\Delta H^{\ddagger} = 11.6 \pm 0.8$ kcal mol^{-1} and $\Delta s^{\ddagger} = -11.8 \pm 2.7$ cal mol^{-1} °K^{-1} (*164*). However, corresponding values for the fully base-catalyzed reaction (with possible involvement of a sugar anion) are not available. Note that the activation parameters for the base-catalyzed portion of the pH profile are ambiguous. If base catalysis is due to a sugar anion, then the large temperature dependence of the pK_a of the ionizing group has to be taken into account. For NAD^+ this means measuring these values above the pK_a of the diol in a pH region where the reaction is pH independent (*160*).

4. Oppenheimer, N. J. (1987). *In* "Nuclear Magnetic Resonance Spectroscopy of Pyridine Nucleotides" (D. Dolphin, R. Poulson, and O. Avramovic, eds.), pp. 186–230. Wiley, New York.

5. Westheimer, F. H. (1987). *In* "Mechanism of Action of the Pyridine Nucleotides" (D. Dolphin, R. Poulson, and O. Avramovic, eds.), pp. 253–322. Wiley, New York.

6. Vennesland, B., and Levy, H. R. (1957). *J. Biol. Chem.* **228,** 85.

7. Oppenheimer, N. J., Arnold, L. J., Jr., and Kaplan, N. O. (1971). *Proc. Natl. Acad. Sci. U.S.A.* **68,** 3200.

8. Oppenheimer, N. J., Arnold, L. J., Jr., and Kaplan, N. O. (1978). *Biochemistry* **17,** 2613.

9. Rob, F., Ramesdonk, H. J., van Gerresheim, W., van Bosma, P., Scheele, J. J., and Verhoeven, J. W. (1984). *J. Am. Chem. Soc.* **106,** 3826.

10. Ohno, A., Ohara, M., and Oka, S. (1986). *J. Am. Chem. Soc.* **108,** 6438.

11. Raber, D. J., and Rodriquez, W. (1985). *J. Am. Chem. Soc.* **107,** 4146.

12. Glasfeld, A., Zbinden, P., Dobler, M., Benner, S. A., and Dunitz, J. D. (1988). *J. Am. Chem. Soc.* **110,** 5152.

13. Burgner, II, J. W., and Ray, W. J., Jr. (1984). *Biochemistry* **23,** 3636.

14. Wu, Y.-D., and Houk, K. N. (1987). *J. Am. Chem. Soc.* **109,** 2226.

15. Filman, D. J., Bolin, J. T., Matthews, D. A., and Kraut, J. (1982). *J. Biol. Chem.* **257,** 13663.

16. Egan, W., Forsen, S., and Jacobus, J. (1975). *Biochemistry* **14,** 735.

17. Sarma, R. H., and Mynott, R. J. (1973). *J. Am. Chem. Soc.* **95,** 1641.

18. Benner, S. A. (1982). *Experientia* **38,** 633.

19. Moras, D., Olsen, K. W., Sabesan, M. N., Buehner, M., Ford, G. C., and Rossmann, M. G. (1975). *J. Biol. Chem.* **250,** 9137.

20. Adams, M. J., Buehner, M., Chandrasekhar, K., Ford, G. C., Hackert, M. L., Anders, L., Rossman, M. G., Smiley, I. E., Allison, W. S., Everse, J., Kaplan, N. O., and Taylor, S. S. (1973). *Proc. Natl. Acad. Sci. U.S.A.* **70,** 1968.

21. Pai, E. F., Karplus, P. A., and Schulz, G. E. (1988). *Biochemistry* **27,** 4465.

22. Fisher, H. F., Ofner, P., Conn, E. E., Vennesland, B., and Westheimer, F. H. (1953). *J. Biol. Chem.* **202,** 687.

23. Anderson, V. E., and LaReau, R. D. (1988). *J. Am. Chem. Soc.* **110,** 3695.

24. Kapmeyer, H., Pfleiderer, G., and Trommer, W. (1976). *Biochemistry* **15,** 5024.

25. Oppenheimer, N. J., and Kaplan, N. O. (1975). *Arch. Biochem. Biophys.* **166,** 526.

26. Oppenheimer, N. J. (1986). *J. Biol. Chem.* **261,** 12209.

27. Frey, P. A. (1987). *In* "Complex Pyridine Nucleotide-Dependent Transformations" (D. Dolphin, R. Poulson, and O. Avramovic, eds.), pp. 461–511. Wiley, New York.

28. Arabshahi, A., Flentke, G. R., and Frey, P. A. (1988). *J. Biol. Chem.* **263,** 2638.

29. Konopka, J. M., Halkides, C. J., Vanhooke, J. L., Gorenstein, D. G., and Frey, P. A. (1989). *Biochemistry* **28,** 2645.

30. Anderson, B. M., and Kaplan, N. O. (1987). *In* "Model Studies and Biological Activity of Analogs" (D. Dolphin, R. Poulson, and O. Avramovic, eds.), pp. 569–611. New York.

31. Woenckhaus, C., and Jeck, R. (1987). *In* "Preparation and Properties of NAD and NADP Analogs" (D. Dolphin, R. Poulson, and O. Avramovic, eds.), pp. 449–568. Wiley, New York.

32. Karle, I. L. (1961). *Acta Crystallogr.* **14,** 497.

33. Perahia, D., Pullman, B., Saran, A., Sundaralingam, M., and Rao, S. T. (1975). *In* "Structure and Conformation of Nucleic Acids and Protein-Nucleic Acid Interactions" (M. Sundaralingam, and S. T. Rao, eds.) pp. 685–708. Univ. Park Press, Baltimore, Maryland.

34. Eklund, H., Samama, J.-P., Wallén, L., Brändén, C.-I., Åkeson, Å., and Jones, T. A. (1981). *J. Mol. Biol.* **146,** 561.

35. Biesecker, G., Harris, J. I., Thierry, J. C., Walker, J. E., and Wonacott, A. J. (1977). *Nature (London)* **250,** 9137.

36. Deng, H., Zheng, J., Sloan, D., Burgner, J., and Callender, R. (1989). *Biochemistry* **28,** 1525.

37. Kaplan, N. O., Ciotti, M. M., and Stolzenbach, F. E. (1956). *J. Biol. Chem.* **221**, 833.
38. Eklund, H., Plapp, B. V., Samama, J.-P., and Bränden, C.-I. (1982). *J. Biol. Chem.* **257**, 14349.
39. Hine, J. (1975). "Structural Effects on Equilibria in Organic Chemistry," p. 98. Wiley, New York.
40. Kam, B. L., Malver, O., Marschner, T. M., and Oppenheimer, N. J. (1987). *Biochemistry* **26**, 3453.
41. Göbbeler, K. H., and Woenckhaus, C. (1966). *Liebigs Ann. Chem.* **700**, 180.
42. Jeck, R. (1977). *Z. Naturforsch.* **32C**, 550.
43. Lamed, R., Levin, Y., and Wilchek, M. (1973). *Biochim. Biophys. Acta* **304**, 231.
44. White, B. J., and Levy, H. R. (1987). *J. Biol. Chem.* **262**, 1223.
45. Chang, G.-G., and Huang, T.-M. (1979). *Biochem. Biophys. Res. Commun.* **86**, 829.
46. Malver, O., and Oppenheimer, N. J. (1992). *Nucleosides Nucleotides* (in press).
47. Ganzhorn, A. J., and Plapp, B. V. (1988). *J. Biol. Chem.* **263**, 5446.
48. Cedergren-Zeppezauer, E. (1983). *Biochemistry* **22**, 5761.
49. Woenckhaus, C., Volz, M., and Pfleiderer, G. (1964). *Z. Naturforsch. B: Anorg. Chem. Org. Chem. Biochem. Biophys. Biol.* **19B**, 467.
50. Woenckhaus, C., and Jeck, R. (1970). *Liebigs Ann. Chem.* **700**, 180.
51. Slama, J. T., and Simmons, A. M. (1988). *Biochemistry* **27**, 183.
52. Swain, C. G., Powell, A. L., Lynch, T. J., Alpha, S. R., and Dunlap, R. P. (1979). *J. Am. Chem. Soc.* **101**, 3584.
53. Ashby, E. C., Goel, A. B., and Argyropoulos, J. N. (1982). *Tetrahedron Lett.* **23**, 2273.
54. Swain, C. G., Powell, A. L., Sheppard, W. A., and Morgan, C. R. (1979). *J. Am. Chem. Soc.* **101**, 3576.
55. Ashby, E. C., Coleman III, D. T., and Gamasa, M. P. (1983). *Tetrahedron Lett.* **24**, 851.
56. MacInnes, I., Nonhebel, D. C., Orszulik, S. T., and Suckling, C. J. (1983). *J. Chem. Soc., Perkin Trans 1* p. 2777.
57. Pham, V. T., Phillips, R. S., and Ljungdahl, L. G. (1989). *J. Am. Chem. Soc.* **111**, 1935.
58. Klinman, J. P. (1976). *Biochemistry* **15**, 2018.
59. Klinman, J. P. (1981). *Crit. Rev. Biochem.* **10**, 39.
60. Cook, P. F., Oppenheimer, N. J., and Cleland, W. W. (1981). *Biochemistry* **20**, 1817.
61. Burgner II, J. W., Oppenheimer, N. J., and Ray, W. J., Jr. (1987). *Biochemistry* **26**, 91.
62. Kurtz, L. C., and Freiden, C. (1980). *J. Am. Chem. Soc.* **102**, 4198.
63. Cha, Y., Murray, C. J., and Klinman, J. P. (1989). *Science* **243**, 1325.
64. Klinman, J. P. (1989). *Trends Biochem. Sci.* **14**, 368.
65. Hinson, J. A., and Neal, R. A. (1972). *J. Biol. Chem.* **247**, 7106.
66. Hinson, J. A., and Neal, R. A. (1975). *Biochim. Biophys. Acta* **384**, 1.
67. Gupta, N. K. (1970). *Arch. Biochem. Biophy.* **141**, 632.
68. Gupta, N. K., and Robinson, W. G. (1966). *Biochim. Biophys. Acta* **118**, 431.
69. Abeles, R. H., and Lee, H. A. (1960). *J. Biol. Chem.* **235**, 1499.
70. Eisses, K. T. (1989). *Bioorg. Chem.* **17**, 268.
71. Warren, W. A. (1970). *J. Biol. Chem.* **245**, 1675.
72. Anderson, D. C., and Dahlquist, F. W. (1980). *Biochemistry* **19**, 5486.
73. Kendal, L. P., and Ramanthan, A. N. (1952). *Biochem. J.* **52**, 430.
74. Duncan, R. J. S., and Tipton, K. F. (1969). *Eur. J. Biochem.* **11**, 58.
75. Dutler, H., and Ambar, A. (1983). *Nato Adv. Study Inst. Ser. Ser. O* **100**, 135.
76. Racker, E., and Krimsky, I. (1952). *J. Biol. Chem.* **198**, 731.
77. Rizzo, V., Pande, A., and Luisi, P. L. (1987). *In* "Optical Spectroscopy of Pyridine Nucleotides" (D. Dolphin, R. Poulson, and O. Avramovic, eds.), pp. 99–161. Wiley, New York.
78. Nambiar, K. P., Stauffer, D. M., Kolodziej, P. A., and Benner, S. A. (1983). *J. Am. Chem. Soc.* **105**, 5886.

79. Benner, S. A., Nambiar, K. P., and Chambers, G. K. (1985). *J. Am. Chem. Soc.* **107**, 5513.
80. Bentley, R. (1970). *In* "Molecular Asymmetry in Biology," pp. 1–89. Academic Press, New York.
81. You, K.-S., Arnold, L. J., Jr., Allison, W. S., and Kaplan, N. O. (1978). *Trends Biochem. Sci.* **3**, 265.
82. You, K.-S. (1985). *Crit. Rev. Biochem.* **17**, 313.
83. Jörnvall, H., Persson, M., and Jeffery, J. (1981). *Proc. Natl. Acad. Sci. U.S.A.* **78**, 4226.
84. Jeffery, J., and Jörnvall, H. (1988). *Adv. Enzymol.* **61**, 47.
85. Jörnvall, H., Persson, B., and Jeffery, J. (1987). *Eur. J. Biochem.* **167**, 195.
86. Jörnvall, H., von Bahr-Lindström, H., Jany, K.-D., Ulmer, W., and Fröschle, M. (1984). *FEBS Lett.* **165**, 190.
87. Jörnvall, H., Höög, J.-O., von Bahr-Lindström, H., and Vallee, B. L. (1987). *Proc. Natl. Acad. Sci. U.S.A.* **84**, 2580.
88. Schneider-Bernlöhr, H., Adolph, H.-W., and Zeppezauer, M. (1986). *J. Am. Chem. Soc.* **108**, 5573.
89. Kohn, L. D., Warren, W. A., and Carroll, W. R. (1970). *J. Biol. Chem.* **245**, 3821.
90. Rossman, M. G., Liljas, A., Braenden, C.-I., and Banaszak, L. J. (1975). *In* "Evolutionary and Structural Relationships Among Dehydrogenases" (P. D. Boyer, ed.), pp. 61–102. Academic Press, New York.
91. Birktoft, J. J., and Banaszak, L. J. (1983). *J. Biol. Chem.* **258**, 472.
92. Wilks, H. M., Hart, K. W., Feeney, R., Dunn, C. R., Muirhead, H., Chia, W. N., Barstow, D. A., Atkinson, T., Clarke, A. R., and Holbrook, J. J. (1988). *Science* **242**, 1541.
93. Allemann, R. K., Hung, R., and Benner, S. A. (1988). *J. Am. Chem. Soc.* **110**, 5555.
94. Kaplan, N. O. (1960). "Enzymes" (P. D. Boyer, H. Lardy, and K. Myrback, eds.) Vol. 3B, pp. 105–169. Acad. Press, New York.
95. Dothie, J. M., Giglio, J. R., Moore, C. B., Taylor, S. S., and Hartley, B. S. (1985). *Biochem. J.* **230**, 569.
96. Benner, S., and Ellington, A. D. (1988). *Crit. Rev. Biochem.* **23**, 369.
97. Hegstrom, R. A., and Kondepudi, D. K. (1990). *Sci. Am.* **262**, 108.
98. Albery, W. J., and Knowles, J. R. (1976). *Biochemistry* **15**, 5631.
99. Chin, J. (1983). *J. Am. Chem. Soc.* **105**, 6502.
100. Ellington, A., and Benner, S. A. (1987). *J. Theor. Biol.* **127**, 491.
101. Burbaum, J. J., and Knowles, J. R. (1989). *Biochemistry* **28**, 9306.
102. Burbaum, J. J., and Knowles, J. R. (1989). *Biochemistry* **28**, 9293.
103. Gutfreund, H. (1975). *Prog. Biophys. Mol. Biol.* **29**, 161.
104. Hammond, G. S. (1955). *J. Am. Chem. Soc.* **77**, 334.
105. Jacobson, M., and Jacobson, E. (eds.) (1989). "ADP-Ribose Transfer Reactions," pp. 1–531. Springer-Verlag, New York.
106. Moss, J., and Vaughn, M. (eds.) (1990). "ADP-Ribosylation and G Proteins," pp. 1–568. American Society for Microbiology, Washington, D. C.
107. Anderson, B. M. (1982). *In* "Analogs of Pyridine Nucleotide Coenzymes" (J. Everse, B. M. Anderson, and You, K.-S., eds.), pp. 91–133. Academic Press, New York.
108. Moss, J., and Vaughan, M. (1987). *Adv. Enzymol.* **61**, 303.
109. Van Dop, C., Tsubokawa, M., Bourne, H. R., and Ramachandran, J. (1984). *J. Biol. Chem.* **259**, 696.
110. West, R. E., Moss, J., Vaughan, M., Liu, T., and Liu, T. Y. (1985). *J. Biol. Chem.* **260**, 14428.
111. Lobban, M. D., and von Heyningen, S. (1988). *FEBS Lett.* **233**, 229.
112. Van Ness, B. G., Howard, J. B., and Bodley, J. W. (1980). *J. Biol. Chem.* **255**, 10717.
113. Oppenheimer, N. J., and Bodley, J. W. (1981). *J. Biol. Chem.* **256**, 8579.
114. Pastan, I., and Fitzgerald, D. (1989). *J. Biol. Chem.* **264**, 15157.

115. Iglewski, B. H., Liu, P. V., and Kabat, D. (1977). *Infect. Immun.* **15**, 138.
116. Sekine, A., Fujiwara, M., and Narumiya, S. (1989). *J. Biol. Chem.* **264**, 8602.
117. Vandekerckhove, J., Schering, B., Barmann, M., and Aktories, K. (1988). *J. Biol. Chem.* **263**, 696.
118. Moss, J., Stanley, S. J., and Oppenheimer, N. J. (1979). *J. Biol. Chem.* **254**, 8891.
119. Tanuma, S., Kawashima, K., and Endo, H. (1988). *J. Biol. Chem.* **263**, 5485.
120. Althaus, F. R., and Richter, C. (1987). "ADP-Ribosylation of Proteins," pp. 20–53. Springer-Verlag, Berlin.
121. Moss, J., and Vaughan, M. (1979). *Annu. Rev. Biochem.* **48**, 581.
122. Oppenheimer, N. J. (1978). *J. Biol. Chem.* **253**, 4907.
123. Osborne, J. C., Stanley, S. J., and Moss, J. (1985). *Biochemistry* **24**, 5235.
124. Burns, D. L., Hausman, S. Z., Lindner, W., Robey, F. A., and Manclark, C. R. (1987). *J. Biol. Chem.* **262**, 17677.
125. Pizza, M., Bartoloni, A., Prugnola, A., Silvestri, S., and Rappuoli, R. (1988). *Proc. Natl. Acad. Sci. U.S.A.* **85**, 7521.
126. Locht, C., Capiau, C., and Feron, C. (1989). *Proc. Natl. Acad. Sci. U.S.A.* **86**, 3075.
127. Kaslow, H. R., and Lesikar, D. D. (1987). *Biochemistry* **26**, 4397.
128. Palleroni, N. J. (1975). In "General Properties and Taxonomy of the Genus *Pseudomonads*" (P. H. Clarke and M. H. Richmond, eds.), pp. 1–36. Wiley, New York.
129. Thompson, M. R., and Iglewski, B. H. (1982). In "*Pseudomonas aeruginosa* Toxin A and Exoenzyme S" (O. Hayaishi and K. Ueda, eds.), p. 698. Academic Press, New York.
130. Carroll, S. F., McCloskey, J. A., Crain, P. F., Oppenheimer, N. J., Marschner, T. M., and Collier, R. J. (1985). *Proc. Natl. Acad. Sci. U.S.A.* **82**, 7237.
131. Carroll, S. F., and Collier, R. J. (1984). *Proc. Natl. Acad. Sci. U.S.A.* **81**, 3307.
132. Tweten, R. K., Barbieri, J. T., and Collier, R. J. (1985). *J. Biol. Chem.* **260**, 10392.
133. Carroll, S. F., and Collier, R. J. (1987). *J. Biol. Chem.* **262**, 8707.
134. Douglas, C. M., and Collier, R. J. (1987). *J. Bacteriol.* **169**, 4967.
135. Morii, N., Sekine, A., Ohashi, Y., Nakao, K., Imura, H., Fujiwara, M., and Narumiya, S. (1988). *J. Biol. Chem.* **263**, 12420.
136. Aktories, K., Rosener, S., Blaschke, U., and Chhatwal, G. S. (1988). *Eur. J. Biochem.* **172**, 445.
137. Braun, U., Habermann, B., Just, I., Aktories, K., and Vandekerckhove, J. (1989). *FEBS Lett.* **243**, 70.
138. Ohashi, Y., and Narumiya, S. (1987). *J. Biol. Chem.* **262**, 1430.
139. Moss, J., Oppenheimer, N. J., West, R. E., Jr., and Stanley, S. J. (1986). *Biochemistry* **25**, 5408.
140. Lowery, R. G., and Luddon, P. W. (1988). *J. Biol. Chem.* **263**, 16714.
141. Sayhan, O., Ozdemirli, M., and Bermek, E. (1986). *Biochem. Biophys. Res. Commun.* **139**, 1210.
142. Ueda, K., and Hayaishi, O. (1985). *Annu. Rev. Biochem.* **54**, 73.
143. Ferro, A. M., and Oppenheimer, N. J. (1978). *Proc. Natl. Acad. Sci. U.S.A.* **75**, 809.
144. Taniguchi, T. (1987). *Biochem. Biophys. Res. Commun.* **147**, 1008.
145. Schuber, F., and Travo, P. (1976). *Eur. J. Biochem.* **65**, 247.
146. Bull, H. G., Ferraz, J. P., Cordes, E. H., Ribbi, A., and Apitz-Castro, R. (1978). *J. Biol. Chem.* **253**, 5186.
147. Yuan, J. H., and Anderson, B. M. (1973). *J. Biol. Chem.* **248**, 417.
148. Yost, D. A., and Anderson, B. M. (1983). *J. Biol. Chem.* **258**, 3075.
149. Schuber, F., Travo, P., and Pascal, M. (1976). *Eur. J. Biochem.* **69**, 593.
150. Pascal, M., and Schuber, F. (1976). *FEBS Lett.* **66**, 107.
151. Oppenheimer, N. J. (1978). *FEBS Lett.* **94**, 368.
152. Schuber, F., and Pascal, M. (1977). *FEBS Lett.* **73**, 92.

153. Secemski, I. I., Lehrer, S. S., and Lienhard, G. E. (1972). *J. Biol. Chem.* **247,** 4740.
154. Slama, J. T., and Simmons, A. M. (1989). *Biochemistry* **28,** 7688.
155. Schuber, F., Muller, H. M., Handlon, A. L., and Oppenheimer, N. J. (1989). Unpublished results.
156. Tarnus, C., and Schuber, F. (1987). *Bioorg. Chem.* **15,** 31.
157. Handlon, A. L., and Oppenheimer, N. J. (1990). Unpublished results.
158. Sinnot, M. L., and Viratelle, O. M. (1973). *Biochem. J.* **133,** 81.
159. Tarnus, C., Muller, H. M., and Schuber, F. (1988). *Bioorg. Chem.* **16,** 38.
160. Johnson, R. W., Marschner, T. M., and Oppenheimer, N. J. (1988). *J. Am. Chem. Soc.* **110,** 2257.
161. Jencks, W. P. (1980). *Acc. Chem. Res.* **13,** 161.
162. Knier, B. L., and Jencks, W. P. (1980). *J. Am. Chem. Soc.* **102,** 6789.
163. Sinnott, M. L. (1987). *In* "Glycosyl Group Transfer" (M. I. Page and A. Williams, eds.), pp. 259–297. Burlington House, London.
164. Jones, C. C., Sinnott, M. L., and Souchard, I. J. L. (1977). *J. Chem. Soc., Perkin Trans. 2* p. 1191.

Author Index

Numbers in parentheses are reference numbers and indicate that an author's work is referred to although the name is not cited in the text. Numbers in italics refer to the page numbers on which the complete reference appears.

A

Subject Index:

A

O

P